Handling and Management of Hazardous Materials and Waste

Handling and Management of Hazardous Materials and Waste

Theodore H. Allegri, Sr.

Chapman and Hall
New York • London

First published 1986
by Chapman and Hall
29 West 35 Street, New York, N.Y. 10001

Published in Great Britain by
Chapman and Hall Ltd
11 New Fetter Lane, London EC4P 4EE

© 1986 Chapman and Hall

Printed in the United States of America

Library of Congress Cataloging-in-Publication Data

Allegri, Theodore H. (Theodore Henry), 1920–
    Handling and management of hazardous materials and
wastes.

    Bibliography: p.
    Includes index.
    1. Hazardous substances—United States. 2. Hazardous
wastes—United States. 3. Hazardous substances—Law and
legislation—United States. 4. Hazardous wastes—Law
and legislation—United States. I. Title.
T55.3.H3A45 1986   604.7   86-17187
ISBN 0-412-00751-7

Table of Contents

Preface

This book deals with the safe and legal handling of hazardous materials and waste from the manufacturer's plant through the storage, transportation and distribution channels to the user, and, ultimately, to the disposal of the product or waste materials. There is increasing pressure today from the public, academia, government at all levels, and industry to improve the handling and management of hazardous materials. A knowledge of the methods required to safely handle and manage those materials in all of their various aspects, together with an understanding of the many governmental regulations that apply to those materials in the various stages of the distribution chain, is absolutely essential to their proper handling and disposal.

Efficient handling and the safe management of hazardous materials requires an expertise in the skills and techniques of the latest innovations, which in turn are often based upon the firm foundation of data and experience in those areas. Personal and public safety require that the information concerning hazardous materials be disseminated as widely as possible.

This document should *not* be used to determine compliance with the U.S. DOT hazardous material regulations, or with any other regulations imposed by local, State, or Federal regulatory bodies.

T.H. Allegri, Sr.
April, 1986.

1

Introduction and Background History

A. Introduction

In Houston, Texas, a tank truck of anhydrous ammonia fell from an expressway onto a populated freeway, dumping the contents of the tanker. Six people were killed, and scores of people were injured by the fumes engulfing the area.

Waverly, Tennessee, was the scene of another catastrophe when a tank of liquid propane (LP-gas) burst open a few days after a railroad derailment. The resulting BLEVE (boiling liquid-expanding vapor explosion) killed 15 people, and several score were injured.

The Love Canal in the Niagara Falls area demonstrated, quite vividly, the results of indiscriminate dumping of toxic wastes that eventually poisoned ground waters, streams, and rivers. More recently, in Bhopal, India, methyl isoscyanate, a toxic ingredient of certain pesticides, was accidentally released into the air, killing several thousand persons and injuring hundreds of thousands more in one of the worst manufacturing disasters in modern times.

The problems of what to do with toxic wastes and the methods and means for handling dangerous chemicals are not confined to the United States alone. Ample evidence points to the worldwide concern surrounding the use of hazardous materials. Let it suffice to say that our country itself generates 90 billion pounds of hazardous waste each year, and many more billion pounds are used, transported, stored, and spilled.

B. Background History

Public awareness of the dangers inherent in hazardous materials has increased with the passage of the Williams-Steiger Occupational Safety and Health Act of 1970, or OSHA, as it is commonly designated. It was concerned with the safety of employees in the workplace. It requires that employers furnish their employees places and conditions of employment free from recognized hazards that are causing or are likely to cause death or serious physical harm.

On February 26, 1980, President Jimmy Carter signed Executive Order 12196 reinforcing OSHA Section 19, which covers all Federal employees. In 1-602(c) of E.O. 12196, the General Services Administration was authorized to "procure and provide safe supplies, devices, and equipment, and establish and maintain a product safety program for those supplies, devices, equipment and services furnished to agencies, including the issuance of Material Safety Data Sheets when hazardous materials are furnished to them."

1

What are hazardous materials? They have been defined in Title 49, Code of Federal Regulations (49 CFR), Parts 100-177 and 178-199, as substances or materials that have been determined to be capable of posing an unreasonable risk to health, safety, and property when transported in commerce. Only a few people realize how many industries depend on hazardous materials in their manufacturing processes. Hazardous materials are pervasive and necessary to our high standard of living. During the Colonial period of our country's development there were very few hazardous materials. A few come to mind, such as lye used in soap making, various tars or pitch with which to caulk the seams of wood boats, and sulfur with its many uses.

Today we find many uses for hazardous materials in our everyday life in hospitals, laboratories, and workshops. Oxygen and flammable gases for anesthesia are used in hospitals as well as radioactive substances for treatment and diagnosis. In sewage treatment plants and swimming pools, chlorine, a poisonous gas, is used in large quantities. Calcium hypochlorite, another hazardous material, is also used in swimming pools.

Farmers make extensive use of fertilizers that are composed of nitrates and ammonia. Moreover, chlorinated hydrocarbons and organic phosphates are widely used in agriculture nowadays. Petroleum products are used everywhere in our society. They provide fuel for motor vehicles; they provide heat for homes, office buildings, and factories. They are used extensively in dry-cleaning operations and in plastics manufacturing.

In automobile air-conditioning units and in our home refrigerators we use fluorinated hydrocarbons. In large refrigeration plants, office buildings, and ice-skating rinks we also use fluorinated hydrocarbons, ammonia, and methyl chloride.

If we examine industrial uses of hazardous materials, we find that acids and alkalies are used for cleaning metals in solvent baths and for plating metals, such as galvanizing processes to prevent the rusting of metal components wherever they are used. In fabricating metals, industrial uses require the use of acetylene and methyl-acetylene propadiene (MAPP) gas for cutting and welding. Sometimes MAPP gas is used instead of acetylene because it is more chemically stable.

If we were to look at the entire picture of industrial and agricultural uses of hazardous materials, we would find that more than 2 billion tons of hazardous materials are moved each year in the United States alone in all modes of transportation. Since the volume of hazardous materials transported and used in this country and the world increases annually, we must once and for all think about the potential harm that may arise for people, property, and the environment. It is only with the most dedicated effort on the part of everyone concerned in the processing, manufacturing, and distribution chain that the majority of the materials movements made arrive at their destinations safely or are processed without catastrophe or disaster.

While hazardous materials regulations have existed for many years, the Federal Government in 1974 passed the Transportation Safety Act. In that measure, the Department of Transportation (DOT) has been made responsible for regulating the transportation of hazardous materials by rail, air, water, highway, and pipelines. The DOT regulations are described in Title 49 of the Code of Federal Regulations (49 CFR). The laws cover every phase of a shipment even to the point of determining which materials may or may not be transported.

Certain forms of specific chemicals cannot be conveyed, such as dry fulminate of mercury. Other chemicals, such as nitroglycerin, can be moved only by certain types of carriers. And other chemicals, such as acids or ethers, are not permitted to be carried in aircraft. The reasons for those determinations rest with the highly hazardous and unstable chemical nature of the chemicals. Also, there is the constant danger the chemicals present to humans if the substances are released, that is, spilled in confined quarters, such as the cargo hold of an airplane, or interspersed with general cargo or handled by inexperienced personnel.

Title 49 of the CFR also contains requirements for the training of personnel involved in the handling and shipping of hazardous materials. The regulations of 49 CFR are quite stringent and require that the manufacturer, shipper, and the carrier, usually a common carrier, train personnel in the proper preparation of materials for shipment and for strict adherence to the regulations. In the context of those regulations, the manufacturer, shipper, or carrier may be fined for failure to train their personnel in the requirements for the transportation of hazardous materials.

Specific regulations govern the packaging of hazardous materials for shipment. The DOT also has delineated strict container specifications for the transportation of hazardous materials. Some of the materials may be shipped only in hermetically sealed drums that are manufactured in accordance with DOT specifications. They govern material strength, thickness, method of fabrication, and the capacities of the container.

After the hazardous materials have been packaged, each container must be labeled. Also, bulk freight containers, or vehicles carrying containers of hazardous materials, must carry warning signs or placards that are visible on the exterior surfaces of the vehicle. The DOT has standardized labeling and placarding in order to develop a system that warns of the presence of hazardous materials.

Over the course of years, the DOT has added to its regulations and codified them in such a manner that safety considerations are promoted and accidents and disasters have been minimized. While some may think that the regulations are too stringent, the vast majority of the people working with hazardous materials have a very high regard for the standards and are alert and attentive to the requirements, however strict they may be.

During the course of this text we shall explore in greater detail the regulations and requirements that have been briefly described in this introduction. We shall be quite explicit concerning the regulations so that the reader will have a working knowledge with which to approach the subject of hazardous materials with their many requirements. We shall also discuss another important function of the DOT, which is to provide information for emergency response, whenever a spill or other disaster involving hazardous materials arises.

2

The Code of Federal Regulations

A. The Federal Register

The Federal Register is a document that is published daily, Monday through Friday, by the Office of the Federal Register, an agency of the General Services Administration. It contains not only Federal regulations but also executive orders (EO) of the President of the United States, proposed rules or regulations being considered in the U.S. Congress, legal and informational notices, and other Federal agency documents of public importance. it also serves notice to all interested parties—civilian, military, and governmental—of items that may affect them in one way or another. In this manner, they may influence Federal legislation in some degree by their comments or appearances before the appropriate bodies. Those notices are usually given with ample time to prepare responses to the particular or specific legislation.

The Federal Register is generally sold to the public on a semiannual and annual mail subscription basis. One may also obtain copies of individual issues from the U.S. Government Printing Office by writing the Superintendent of Documents, Washington, D.C. 20402, or by calling (202) 783-3238.

Your local or main library often carries copies of the Federal Register, or you might ask your local Congressional office for assistance in obtaining copies or information.

B. Explanation of the Code

The Code of Federal Regulations is a codification of the general and permanent rules published in the Federal Register by the Federal Government. The code is divided into 50 "titles," which represent the complete spectrum of areas subject to Federal regulation. Every title is divided into chapters, each of which usually bears the name of the agency that has the responsibility for administering the subject field. Every chapter is further subdivided into "parts" covering specific regulatory areas.

The Federal agency administering the subject field originally obtains its mandate, or responsibility, for the area through congressional action or legislation. One of the primary functions, therefore, of the Federal Register is to publish or publicize the legislative interpretations of the Federal bureaucracy so that they do not overstep their bounds. Therefore, as indicated in A, above, ample opportunity is given to interested parties to respond to pending legislative or regulatory actions promulgated by these agencies of the Federal Government.

C. Code Issue Dates

Each volume of the code is revised at least once each calendar year and released on a quarterly basis approximately as follows:

Title 1 through Title 16 . . . . as of January 1
Title 17 through Title 27 . . . as of April 1
Title 28 through Title 41 . . . as of July 1
Title 42 through Title 50 . . . as of October 1

D. Some Further Facts About the Code

1. Legal Status

The contents of the Federal Register are required to be judicially noticed (44 U.S.C. 1507). The Code of Federal Regulations is prima facie evidence of the text of the original documents (44 U.S.C. 1510).

2. Code Continuity

The Code of Federal Regulations is kept up to date by the individual issues of the Federal Register. *It is important to note that these 2 publications must be used together to determine the latest version of any given rule.*

To determine whether a code volume has been amended since its revision date, consult the "List of CFR Sections Affected (LSA)," which is issued monthly, and the "Cumulative List of Parts Affected," which appears in the Reader Aids section of the daily Federal Register. Those 2 lists will identify the Federal Register page number of the latest amendment of any given rule.

Each volume of the Code contains amendments published in the Federal Register since the last revision of that volume of the Code. Source citations for the regulations are referred to by date of publication, volume number, and page number of the Federal Register. Publication dates and effective dates are usually not the same, and the reader must exercise care in determining the actual effective date. In instances where it is beyond the cutoff date for the code, a note will be inserted to reflect the future effective date.

3. Obsolete Provisions

Provisions that become obsolete before the revision date stated on the cover of each volume are not carried. Code users may find the text of provisions in effect on a given date in the past by using the appropriate numerical list of sections affected. For the period before January 1, 1964, the user should consult the "List of Sections Affected, 1949–1963" published in a separate volume. For the period beginning January 1, 1964, a "List of CFR Sections Affected" is published at the end of each CFR volume.

4. Incorporation by Reference

Incorporation by reference was established by statute and allows Federal agencies to meet the requirement to publish regulations in the Federal Register by referring to materials already published elsewhere. For an incorporation to be valid, the director of the Federal Register must approve it. The legal effect of incorporation by reference is that the material is treated as if it were published in full in the Federal Register (5 U.S.C. 552(a)). That material, like any other properly issued regulation, has the force of law.

The director of the Federal Register will approve an incorporation by reference only when the requirements of 1 CFR Part 51 are met. Here are some of the elements on which approval is based:

1. The incorporation will substantially reduce the volume of material published in the Federal Register.
2. The matter incorporated is in fact available to the extent necessary to afford fairness and uniformity in the administrative process.
3. The incorporating document is drafted and submitted for publication in accordance with 1 CFR 51.4.

Properly approved incorporations by reference are listed in the Finding Aids at the end of each volume.

If you have any problem in locating or obtaining a copy of material listed in the Finding Aids of any volume as an approved incorporation by reference, please contact the agency that issued the regulation containing that incorporation. If, after contacting the agency, you find the material is not available, please notify: Director of the Federal Register, National Archives and Records Service, Washington, D.C. 20408 or call (202) 523-4534.

(See Appendix B for a list of the organizations that distribute material which has been incorporated into the regulations by reference.)

5. Indexes and Tabular Guides to the CFR

A subject index to the Code of Federal Regulations is contained in a separate volume, revised semiannually as of January 1 and July 1, entitled CFR Index and Finding Aids. This volume contains the Parallel Table of Statutory Authorities and Agency Rules (Table I) and Acts Requiring Publication in the Federal Register (Table III). A list of CFR Titles, Chapters, and parts and an alphabetical list of agencies publishing in the CFR are also included in this volume.

The Federal Register Index is issued monthly in cumulative form. This index is based on a consolidation of the "Contents" entries in the daily Federal Register.

In addition, a list of CFR Sections Affected (LSA) is published monthly, keyed to the revision dates of the 50 CFR titles.

6. Interpretations of the Code

In general, when the reader wishes to obtain a summary, a legal interpretation, or an explanation of any regulation that appears in the code, he or she should inquire at the Federal agency that has issued the regulation.

7. To Obtain Copies of the Code

The reader who wishes to obtain copies of the Code may purchase them directly from: the Superintendent of Documents, Government Printing Office, Washington, D.C. 20402. Also, it is possible to obtain copies of the code through GPO bookstores across the country from Alabama to Washington. (See Appendix A.)

8. Title 49—Transportation (49 CFR)

The 49 CFR is composed of 8 volumes. The "parts" in the volumes are arranged in the following order:

Parts 1–99, Subtitle A—Office of the Secretary of Transportation

Parts 100–177 and Parts 178–199, Chapter I—Research and Special Programs Administration

Parts 200–399, Chapter II—Federal Railroad Administration (DOT), and Chapter III—Federal Highway Administration (DOT)

Parts 400–999, Chapter IV—Coast Guard (DOT), Chapter V—National Highway Traffic Safety Administration (DOT), Chapter VI—Urban Mass Transportation Administration (DOT), Chapter VIII—National Transportation Safety Board, and Chapter IX—United States Railway Association

Parts 1000–1199, Parts 1200–1299, and Part 1300, Chapter X—Interstate Commerce Commission

3

THE REGULATION OF HAZARDOUS MATERIALS

A. Basic Guidelines

According to the laws of the land, all industrial, Federal, state, and local government employers and employees engaged in the handling, shipping, storage, and disposal of hazardous materials and waste are legally obligated to be familiar with, and to comply with, applicable DOT regulations.

Those regulations pertain to the labeling, marking, packaging, and preparation of shipping papers for hazardous materials that are transported in interstate commerce.

It is necessary, therefore, that anyone engaged in any of the activities related to hazardous materials be aware of the need to review the latest provisions of 49 CFR, as printed in the Federal Register periodically. Severe monetary penalties may result from noncompliance with the regulations.

B. General Instructions

In order to determine the correct marking, labeling, packaging, and shipping document preparation for hazardous materials, a logical, programmed sequence must be followed. If the technician omits a phase of the program, or if he or she takes a step out of sequence, then it is highly probable that the requirements obtained will be in error.

The following program will illustrate the correct methodology to be employed in arriving at or conforming to the regulations.

The technician is directed to 49 CFR, "Subpart B—Tables of Hazardous Material, Their Description, Proper Shipping Name, Class, Label, Packaging and Other Requirements, Section 172.100." The purpose and use of the hazardous materials tables are described in detail.

We shall lead the reader through several applications:

1. Selecting the Proper Shipping Name

After having located 172.101 of CFR 49, locate the proper DOT shipping name. In the example, below, there are 3 columns. The first column gives the common name, the second contains the proper DOT shipping name, and the third gives the DOT hazard classification.

Sometimes a hazardous material may not be listed under its common name.

However, since it is still hazardous, it could be listed in the table by a different one. Therefore, the proper shipping name may be found by using the generic name based upon the hazardous material's chemical family, its end use, or its hazard class.

As an example:

Column (1)\nCommon Name	Column (2)\nProper DOT\nShipping Name	Column (3)\nDOT Hazard\nClassification
Pitch Oil	Creosote, Coal Tar	Combustible Liquid

Note: If the material meets more than one DOT class definition, then the highest one on the list in Section 173.2(a) is the one to be assigned unless the exceptions in Section 173.2(b) apply; however, even when a material is classified using the highest classification definition, the other classifications cannot be ignored. It must comply with the regulations as specified in Section 172.402, "Additional Labeling Requirements."

The location of classification definition paragraphs are in List I, below:

LIST I

Classification	Definition
Combustible Liquid	173.115(b)
Compressed Gas	173.300(a)
Corrosive Material	173.240
Etiologic Agent*	173.386
Explosives	173.58, 88, 100
Flammable Compressed Gas	173.300(b)
Irritating Materials	173.381
Nonflammable Gas	173.300
Organic Peroxide	173.151(a)
ORM-A, B, C, D**	173.500, 1200
Oxidizer	173.151
Poison A	173.389
Flammable Liquid	173.115(a)
Poison B	173.343
Radioactive Material	173.389

*Etiology is the study of the cause of diseases. An "etiologic agent" means a viable microorganism, or its toxin, which causes or may cause human disease and is limited to those agents listed in 42 CFR 72.25(c) of the regulations of the Department of Human Resources.

**ORM = Other Regulated Materials that do not meet the Department of Transportation definitions for the other hazardous materials classes but do possess enough hazard characteristics in transport to require some regulation. ORMs will be discussed in greater detail in the next chapter.

2. Labeling

Labeling is covered in Subpart E of Section 172.400 of 49 CFR. When the required label is formed under column (4) of the Hazardous Materials Table in Section 172.101 of 49 CFR, it is very explicit.

When a material that is not listed in the Hazardous Materials Table is corrosive and meets the definition of Poison B, then the material must be labeled "Corrosive Material" and "Poison B," as explained in Section 172.402 of 49 CFR.

a. Exceptions.   As can be seen in column 5(a) of the Hazardous Materials Table, there are directions to specific sections of the CFR for "exceptions." They usually apply to limited quantities of the substance.

Also, examples of general types of exceptions are indicated in List II, below. They are based upon the volume of product to be shipped and the type of packaging that is used. They are, essentially, DOT-approved packages and differ from DOT specification packages or containers because they are not fully specified in Part 178 of 49 CFR. Material packaged in accordance with those exceptions are given relief from labeling (*except for air shipments*) and from using a DOT specification package.

LIST II

Hazard Classification	Exceptions
Flammable Liquids	173.118
Flammable Solids, oxidizing materials and organic peroxide	173.153
Corrosive Materials	173.244
Compressed gases: flammable gas and nonflammable gas	173.306
Poison B Liquids	173.345
Poison B Solids	173.364

Included in many limited-quantity exceptions is a reference to an ORM-D (consumer commodity) class that allows further exceptions if the commodity meets the criteria for reclassification in Sections 173.500 and 173.1200. Those commodity reclassification paragraphs are illustrated in List III, shown below.

LIST III

Hazard Classification	ORM-D Reference
Flammable Liquids	173.118(d)
Corrosive Liquids	173.244(b)
Corrosive Solids	173.244(b)
Flammable Solids	173.153(c)
Oxidizers	173.153(c)
Organic Peroxides	173.153(c)
Poison B Liquids	173.345(b)
Poison B Solids	173.364(b)
Compressed Gases	173.306(a)

Note: Unless one of the specific ORM-D exceptions above apply, then no further exceptions are allowed for reclassification. The ORM-D category has a gross weight limitation of 65 pounds for each package. Specific commodity exceptions are often found in the packaging requirements column (5)(a) of the Hazardous Materials Table 172.101 of 49 CFR. Those exceptions only give relief from using a DOT specification package, such as is required by column (5)(b) of the same table.

b. Packaging Requirements. If the material quantities or type of packaging do not qualify for the exceptions, then it is necessary to follow the instructions for the specific type of packaging required. The references for the packaging directions are found in column (5)(b) of the Hazardous Materials Table (49 CFR, 172.101). The sections listed in column (5)(b) specify the approved methods for packaging and packing materials. Those sections give no relief from marking, from labeling, or from providing the correct shipping papers unless there are provisions for a specific exception.

(i) Packing Container Weight Limitations

The specified DOT containers shall be embossed on the head with raised marks or by embossing or die stamping on the footing of drums; in addition, printed symbols shall be placed in rectangular borders for fiberboard boxes, as follows:

DOT-37B*** or, DOT-12B***; asterisks are to be replaced by the authorized gross weight at which the container was type tested, for example: DOT-37B450, or DOT-12B40, etc. A DOT container should not be filled with regulated material to a weight greater than the design load limit.

4

Shipping and Marking Hazardous Materials and Hazardous Waste

A. Requirements

The DOT hazardous materials regulations apply primarily to international and interstate, i.e., transportation across state lines. They apply only to intrastate carriers when those carriers also haul materials in interstate commerce, i.e., from one state into other states.

While the above is true, most states have adopted the Federal regulations, such as 49 CFR in whole or in part for their intrastate carriers. To determine what regulations apply to the transportation of hazardous materials within your state, it is a simple matter to check with state or local authorities. A list of those agencies is as follows:

1. The state department of transportation,
2. The police department,
3. The fire department,
4. The state fire marshall's office, and
5. The state department of commerce.

1. Shipping Papers

The Department of Transportation requires that shipping documents indicate the quantity being shipped, the proper DOT shipping name for the material, the identification number, and the hazard class.

Let us emphasize the 4 most important elements of the shipping document:

1. Quantity
2. DOT shipping name
3. I.D. number
4. Hazard class

"Subpart C—Shipping Papers" of Section 172.200 indicates the following regulations for a hazardous substance or a hazardous waste:

i. The hazardous materials must be listed first or be in a contrasting color or be in an HM*-headed column bordered with a contrasting color.

*Henceforth in this book the abbreviation HM shall mean "hazardous material."

ii. The name of the HM should be spelled out; i.e., codes and unauthorized abbreviations cannot be used.

iii. The HM description is not required for ORM-A, B, or C, unless it is offered for transportation by air or water. In that event, it is subject to regulations pertaining to transportation by air or water, as indicated in column 1 of the HM Table of Section 172.101 of 49 CFR. (See Table 4-1 for a sample page of the HM Table of Sec. 172.101 of 49 CFR.)

In addition, if the word *waste* is not included in the hazardous material description in the table, the proper shipping name for a hazardous waste must include the word *Waste* preceding the shipping name of the material; for example, "Waste acetone."

iv. The HM description is not required for ORM-D unless it is shipped by air when it must be described, as follows:

"Consumer Commodity" and "ORM-D."

v. Include the shipping name (column 2 of the HM Table), the hazard class (column 3 of the HM Table), and "limited quantity" if applicable (from column 6 of the HM Table).

2. Marking

The general marking requirements for HM and HMW (hazardous material waste) are to be found in Subpart D—Marking, of Section 172.300 of 49 CFR.

Each person who offers an HM for transportation must mark each package, freight container, and transport vehicle containing the HM according to this subpart. The same is true for each carrier-transporting HM.

Table 4-1. A Sample Page of the Hazardous Materials Table of Sec. 172.101 of 49 CFR

§ 172.101 Hazardous Materials Table

(1) +/ E/ A/ W	(2) Hazardous materials descriptions and proper shipping names	(3) Hazard class	(3A) Identi- fication number	(4) Label(s) required (if not excepted)	(5) Packaging (a) Exceptions	(b) Specific require- ments	(6) Maximum net quantity in one package (a) Passenger carrying aircraft or railcar	(b) Cargo only aircraft	(7) Water shipments (a) Cargo vessel	(b) Pas- senger vessel	(c) Other requirements
	Accumulator, pressurized *(pneumatic or hydraulic), containing nonflammable gas*	Nonflammable gas	NA1956	Nonflammable gas	173.306		No limit	No limit	1,2	1,2	
	Acetal	Flammable liquid	UN1088	Flammable liquid	173.118	173.119	1 quart	10 gallons	1,3	4	
E	Acetaldehyde *(ethyl aldehyde) (RQ-1000/454)*	Flammable liquid	UN1089	Flammable liquid	None	173.119	Forbidden	10 gallons	1,3	5	
EA	Acetaldehyde ammonia *(RQ-1000/454)*	ORM-A	UN1841	None	173.505	173.510	No limit	No limit	1,2	1,2	
E	Acetic acid *(aqueous solution) (RQ-1000/454)*	Corrosive material	UN2790	Corrosive	173.244	173.245	1 quart	10 gallons	1,2	1,2	Stow separate from nitric acid or oxidizing materials
E	Acetic acid, glacial *(RQ-1000/454)*	Corrosive material	UN2789	Corrosive	173.244	173.245	1 quart	10 gallons	1,2	1,2	Stow separate from nitric acid or oxidizing materials. Segregation same as for flamma- ble liquids
E	Acetic anhydride *(RQ-1000/454)*	Corrosive material	UN1715	Corrosive	173.244	173.245	1 quart	10 gallons	1,2	1,2	
	Acetone	Flammable liquid	UN1090	Flammable liquid	173.118	173.119	1 quart	10 gallons	1,3	4	
E	Acetone cyanohydrin *(RQ-10/4.54)*	Poison B	UN1541	Poison	None	173.346	Forbidden	55 gallons	1	5	Shade from radiant heat. Stow away from cor- rosive materials
	Acetone oil	Flammable liquid	UN1091	Flammable liquid	173.118	173.119	1 quart	10 gallons	1,2	1	
	Acetonitrile	Flammable liquid	NA1648	Flammable liquid	173.118	173.119	1 quart	10 gallons	1	4	Shade from radiant heat
	Acetyl acetone peroxide, *in solution with not more than 9% by weight active oxygen.* See Organic peroxide, liquid or solution, n.o.s.		UN2080								
	Acetyl acetone peroxide *with more than 9% by weight active oxygen*	Forbidden									
	Acetyl benzoyl peroxide, *not more than 40% in solution.* See Acetyl benzoyl peroxide solution, not over 40% peroxide		UN2081								
	Acetyl benzoyl peroxide, *solid, or more than 40% in solution*	Forbidden									
	Acetyl benzoyl peroxide solution, *not over 40% peroxide*	Organic peroxide	UN2081	Organic peroxide	None	173.222	Forbidden	1 quart	1,2	1	
E	Acetyl bromide *(RQ-5000/2270)*	Corrosive material	UN1716	Corrosive	173.244	173.247	1 quart	1 gallon	1	1	Keep dry. Glass carboys not permitted on pas- senger vessels
E	Acetyl chloride *(RQ-5000/2270)*	Flammable liquid	UN1717	Flammable liquid	173.244	173.247	1 quart	1 gallon	1	1	Stow away from alcohols. Keep cool and dry. Separate longitudinally by an intervening complete compartment or hold from explo- sives

a. Proper Shipping Name. Every HM must be marked with the Proper Shipping Name, including ORM-D materials for which the Shipping Name is "Consumer Commodity," as described in Section 172.101 of 49 CFR.

b. Name and Address of Consignee or Consignor. As indicated in 172.306 of 49 CFR, each package containing an HM offered for transportation must be marked with the name and address of the consignee or consignor. There are 3 exceptions to that rule, as follows:

(1) Except when the package is transported by highway and will not be transferred from 1 motor carrier to another;

(2) Except when the package is part of a carload lot, truckload lot, or freight container load and the entire contents of the rail car, truck, or freight container are tendered from 1 consignor to 1 consignee; or

(3) Except when a portable tank, cargo tank, or tank car is used to transport the material.

c. Liquid HM. Unless exempt by the regulations, liquid HM, including ORM-D (Consumer Commodities), must observe the following regulations:

(1) They must be packed with the closures upward; and

(2) They must be legibly marked "THIS SIDE UP" or "THIS END UP" as appropriate to indicate the upward position of the inside packaging.

An arrow symbol indicating "This Way Up" as specified in American National Standards Institute (ANSI) publication MH6.11968 entitled "Pictorial Marking for Handling of Goods" should be used in addition to the marking of Section 173.312 and Section 173.25 of 49 CFR.

(3) Exceptions to (1) and (2), above, are that when offered for transportation by air, limited quantities of flammable liquids packed in inside packagings of 1 quart or less are excepted and, when offered for transportation by air, are excepted from the above regulations when packed with inside packagings of 1 quart, or less, with sufficient absorption material between the inner and outer packaging to completely absorb the liquid contents.

d. ORM Packagings. As indicated in an earlier chapter of this text, an explanation of ORMs is necessary for understanding the significance of ORM packagings.

ORM is the abbreviation for Other Regulated Materials. ORMs are materials that do not meet the DOT definitions for other HM classes but possess sufficient hazardous characteristics in transportation to require regulation. That applies to ORM-A, B, C, and E only. An ORM-D is a limited quantity of a hazardous material.

(1) ORM-A:

An ORM-A is a material that has an anesthetic, irritating, noxious, toxic, or other similar property and can cause extreme annoyance or discomfort to passengers or crew in the event of leakage.

(Examples of ORM-As are *carbon tetrachloride*, used, generally, as a solvent; *chloroform*, used as a solvent or in the manufacture of fumigants; *naphthalene*, used as a moth repellent or in the manufacture of dye materials.)

(2) ORM-B:

An ORM-B is a material capable of causing significant damage to a transport vehicle or vessel from leakage during transportation.

(Examples of ORM-B's are *quicklime*, used in iron and steel production and in agriculture, and *gallium metal*, used in the fabrication of optical mirror instruments and in chemical processing.)

(3) ORM-C:

An ORM-C is a material that is unsuitable for shipment unless properly identified and prepared for transportation.

(Examples of ORM-Cs are *excelsior*, a packaging material, and *magnetized material*, used in the manufacture of magnetic devices.)

(4) ORM-D:

ORM-Ds are any consumer commodities that meet the definitions of HM but present limited hazard during transportation because of form, quantity, and, most important of all, packaging.

(Examples of ORM-D's are *household cleaning supplies*, which are used in homes and which are ordinarily purchased in drugstores, supermarkets, groceries, and hardware stores.)

(5) ORM-E:

An ORM-E is a material that is not included in any other hazard class but is subject nevertheless to regulation. Materials in this class include hazardous waste and hazardous substances as defined by DOT and EPA regulations.

(Examples of ORM-Es are *waste acetone*, a waste produced by chemical processing, and *endrin mixture in liquid form*, used in the production and manufacture of pesticides.)

Each packaging having a rated capacity of 110 gallons or less and containing any of the above materials classed as ORM-A, B, C, D, or E must be plainly, durably, and legibly marked on at least 1 side or end with the appropriate ORM designation immediately following or below the Proper Shipping Name of the material.

The appropriate ORM designation must be placed within a rectangle that is approximately ¼-inch (6.3 mm) larger on each side than the designation.

The following denotes the accepted designations:

1. ORM-A;
2. ORM-B—KEEP DRY for an ORM-B that is a solid and is corrosive only to aluminum when wet;
3. ORM-B for any other ORM-B;
4. ORM-C;
5. ORM-D-AIR for an ORM-D that is prepared for air shipment and packaged according to Section 173.6 of 49 CFR. (This section is contained in Appendix C.);
6. ORM-D for any other ORM-D; note that when the ORM-D marking including the Proper Shipping Name cannot be affixed to the package surface, it may be attached on a tag;
7. ORM-E

The marking of ORM-A, B, C, D, or E is the certification by the person offering the package for transportation that the material is *properly described, classed, packaged, marked,* and *labeled.*

This form of certification does not preclude the requirement, however, for a certificate on a shipping paper when required by Subpart C.

3. Shipper's Certification

Each person who offers an HM for transportation must certify that the material offered for transportation is in accordance with the regulations (see Section 172.204 of 49 CFR) by printing by hand or mechanically the following statement on the shipping paper that contains the required shipping description:

> "This is to certify that the above-named materials are properly classified, marked, and labeled and are in proper condition for transportation according to the applicable regulations of the Department of Transportation."

> Note: In line 1 of the certification the words "herein-named" may be substituted for "above-named."

There are several exceptions to the above rule, and we shall discuss them in the following paragraphs. One applies to hazardous waste and another to air transportation. Unfortunately or fortunately, as the case may be, the DOT has seen fit to publish other exceptions.

Except for hazardous waste (HW), no certification is required for an HM offered for transportation by motor vehicle and transported, as follows:

(1) In a cargo tank supplied by the carrier; and
(2) By the shipper as a private carrier except for an HM that is reshipped or transferred from 1 carrier to another.

No certification is required for the return of an empty tank car that previously contained an HM and that has not been cleaned or purged.

4. Transportation by Air

A Shipper's Certification containing the following language may be used in place of the above certification, as follows (Sec. 172.204 (2)(c) of 49 CFR):

> "I hereby certify that the contents of this consignment are fully and accurately described above by proper shipping name and are classified, packed, marked, and labeled and in proper condition for carriage by air according to applicable national government regulations."

It is also required that 2 copies of this certification be supplied to the carrier. In addition, when either passenger or cargo aircraft will be the carrier, the shipper has to add an additional statement to the certification that indicates the regulations have been observed. That statement is as follows:

1. "This shipment is within the limitations prescribed for passenger aircraft," or
2. "This shipment is within the limitations prescribed for cargo-only aircraft."

5. Radioactive Material

When a shipment of radioactive material is offered for transportation aboard a passenger-carrying aircraft, the person requesting the transportation must sign a printed certificate stating that the shipment contains radioactive material intended for use in, or incidental to, research or medical diagnosis or treatment. Thus, the Shipper's Certification statement should read as follows:

"This is to certify that the above-named materials are properly classified, described, packaged, marked, and labeled and are in proper condition for transportation according to the applicable regulations of the Department of Transportation."

and, also:

"This shipment is within limitations prescribed for passenger/cargo aircraft only."
(delete nonapplicable)

Also, if the shipment is permitted on passenger aircraft (by special requirements for air), the following statement must be added:

"This shipment contains radioactive material intended for use in, or incident to, research or medical diagnosis."

Each certification must be legibly signed by an employee of the shipper. Since an air shipment is involved, the shipper must provide at least 2 signed copies to the air carrier.

Regardless of whether or not air shipment is the mode of transportation, each of the Shipper's Certifications must be:

(1) Legibly signed by a principal, officer, partner, or employee of the shipper or his agent; and
(2) May be legibly signed manually, by typewriter, or by other mechanical means.

6. Hazardous Waste Manifest

When a shipper is to move hazardous waste, in whatever manner authorized by law, a manifest must be prepared, signed, carried, and given to the recipient. We shall discuss this subject, in depth, in the chapter covering 40 CFR. This chapter contains a description of regulations governing hazardous wastes.

5

Hazardous Materials in Transit

A. Coping with HM and Community Awareness

Rules and regulations that are formulated by the Federal, state, and local governments have attempted to protect the public and employees working in commerce and industry from the dangers inherent in the handling, storage, and transportation of hazardous substances and hazardous wastes. Despite all of the regulations, however, the transportation of hazardous materials and waste presents risks that cannot be totally eliminated by those or any other means.

Although most of the accidents that occur are minor ones and do not require the massive use of emergency efforts, the potential for property loss, for injury, and for death is present in every accident occurrence, no matter how small or insignificant it may appear at the time.

One of the first steps to be learned in coping with a potential hazardous materials accident is to determine the presence of HM in or near your community.

In general, interstate highways and arterial roadways used by trucks are prime locations for the transportation of HM, as are railroads, airports, port areas, and inland waterways that serve as the shipping networks for hazardous substances.

Air, trucking, rail, and marine port terminals, industrial sites, and manufacturing facilities are all part of the industrial complex that serves as a vehicle for the potential accident. Certainly, the volume of business activity of a community serves to increase the possibility of transportation hazards that may lead to an increase in the frequency of HM accidents. Also, one must not be lulled into complacency just because one does not reside in a largely industrial neighborhood. Of course, the perils are greater in an industrial area. However, hospitals, gasoline service stations, storage warehouses, auto body shops, dry-cleaning plants, and paint stores nearby or in residential areas must share also in the risk for potential hazards.

To be aware of all the areas of a community where potential accidents involving hazardous substances may occur requires a complete listing on the part of a survey team. That evaluation is a major part of a program that should become the viable portion of the emergency response awareness of a community. Therefore, placing "emergency response" in its proper context an enlightened community should first survey the potential hazards that exist and then take the proper steps to minimize the risk of accident.

B. Emergency Response

After it is determined that sufficient high-risk hazards exist that may require immediate action upon notification of an occurrence, the basic community program of emergency response should include, among other necessities, an evacuation plan. This text does not attempt to resolve community problems concerning hazardous substances. However, it would be irresponsible not to point out the underlying considerations of which well-intended citizens of any community should be aware. In this way, future catastrophes may be avoided or their consequences minimized. Also, it should be determined that emergency response personnel in the community have the equipment, the knowledge, the training, and the essential planning and experience to have an impact on potential problems involving hazardous substances.

C. Knowledge, Training, and Planning

As an example of placing a community Emergency Response plan into action effectively, it is necessary that the ER team know the locations of hazardous materials. It should also understand the dangerous properties of hazardous materials and be able to recognize a hazardous material in storage and in transit. In addition, it should be able to properly identify hazardous substances.

Each community should endeavor to assure itself that its emergency response forces are adequately trained and that such training is an ongoing program. The training should consist of practicing emergency response drills of the techniques used to prevent death and injury during a hazardous material accident. It would follow that cleaning up after a hazardous materials spill is important. Also, a constant refinement of skills and techniques used and equipment used for handling hazardous materials should be developed.

The importance of planning before an accident or incident involving hazardous materials can hardly be overemphasized. The planning function is often overlooked in many programs, yet it should be a fundamental concept of every good program, since the difference between an incident and a disastrous accident may hang in the balance. Such events as formulating an evacuation plan, deciding on an emergency response team plan, and obtaining technical assistance during an emergency should be considered. Adopting a plan and organizing will encourage teamwork in all of the emergency services that may be available or could be called upon during an emergency.

The National Transportation Safety Board is an independent Federal agency that investigates accidents involving hazardous materials in other areas of transportation safety.

Among the many accidents that the National Transportation Safety Board has investigated, 2 in particular stand out. The reason is that they demonstrated so forcefully what can happen to increase the severity of an incident especially when the trinity of knowledge, training, and planning is lacking.

A passenger automobile carrying 8 persons made a left turn at an intersection and was immediately struck by a large, tractor-van semitrailer combination that was attempting to overtake and pass the automobile.

The force of the collision propelled each of the vehicles into a nearby roadside ditch.

Several loose, large steel cylinders, containing a mixture of methyl bromide and chloropicrin were knocked out of the trailer. The containers were damaged in the collision, and the chemical started to leak.

The driver of the truck and 4 of the automobile's occupants managed to escape, unaided, from the contaminated area. The other 4 occupants remained in or near the automobile and were exposed for 30 minutes to high concentrations of the toxic vapors of methyl bromide and chloropicrin. They perished in the ordeal.

As a summary of this accident, in addition to the 4 fatalities, 14 of the on-site rescue personnel and bystanders were contaminated through skin contact with the chemical and through inhalation of the vapors.

In another serious accident, a Volkswagen sedan traveling west crossed the centerline of a 2-lane highway and collided head-on with an east-bound tractor, semitrailer carrying 25,000 pounds of explosives.

Fire broke out immediately along the left and front sides of the tractor. Fire fighters were called and quickly responded to the emergency. When they arrived at the scene of the accident, they started to attempt to put out the fire. In the meantime, the truck driver tried to persuade onlookers to move away from the burning wreckage. The cargo detonated approximately 10 to 15 minutes after the collision.

The driver of the Volkswagen was, apparently, fatally injured in the collision. Fortunately, the truck driver was unharmed. Nevertheless, 2 of the fire fighters, a wrecker driver, and 2 bystanders were killed by the explosion, and 33 people were injured. Damage to property was estimated to have exceeded $1 million.

The National Transportation Safety Board, upon investigation, determined that the lack of training by local emergency and rescue personnel was an underlying cause of the severity of the accident.

It was also found that the failure to notify the police promptly and accurately about the nature of the hazards involved was due to poor planning involving emergency procedures. With better planning authoritative crowd control measures could have been taken.

The local community authorities, it is assumed, had not been taught how to identify alternative courses of actions in dealing with explosives under emergency conditions, nor did the emergency service personnel fully understand the hazards of the explosives contained in the trailer.

In contrast to the above stories, the following case history demonstrates how knowledge, training, and planning can favorably affect the outcome of a hazardous materials accident.

On a clear, hot, sunny day in Texas, a passing motorist noticed a vapor cloud poised over a butane (liquid fuel) company plant. He immediately called the dispatcher at the police and fire headquarters. An alarm was sounded to alert the 50 volunteer fire fighters of the nearby community fire house.

Fire fighters arriving at the scene heard several small explosions and found 3 delivery trucks in flames. The butane cylinders mounted on the trucks were rupturing, and flames were threatening an 18,000-gallon butane tank, two 4,000-gallon butane tanks, and a 6,000-gallon tank. Just imagine the conflagration 28,000 gallons of butane could have caused!

Fortunately, an alert police officer had already proceeded to evacuate a nearby apartment building. Significantly enough, if the emergency response team had not been trained to fight a butane-type fire, they might have decided to attack the blaze immediately. Since they had seen and studied a film slide presentation of an LP-gas (liquid propane) explosion in Kingman, Arizona, where 13 persons had died, they realized that they were faced with a similar potential for disaster.

Based upon what they had learned in the training session, however, the fire fighters decided to evacuate everyone within 2,000 feet of the butane plant fire. Policemen from the city, officers from the Sheriff's Department, and highway patrolmen directed the orderly procedure. The fire fighters concentrated on protecting exposures rather than on trying to save the butane tanks.

Within 20 minutes of the alarm, the 18,000-gallon tank and the two 4,000-gallon tanks exploded, releasing a fireball 150-to-200-feet in diameter. Pieces of steel from the tanks were hurled as far as 1,500 feet. The concomitant blast of the explosion damaged several buildings within the 2,000-foot-perimeter area. The only fire damage outside the butane plant occurred when a tent-type skating rink structure ignited.

Several smaller fires were quickly extinguished. The property damage that occurred was unavoidable; however, it was kept to a minimum by the fire fighters' decision to protect exposures. Above all, 600 people had been safely evacuated from the danger zone within minutes, and there were no injuries among the civilians or the emergency response team personnel.

The conclusions to be derived from the above scenario reinforce the thesis that successful emergency response requires adequate preparation. In the successful reaction of an emergency response team that was depicted in the above account, this preparation included the knowledge of the characteristics and behavior of hazardous materials and their containers under various emergency conditions. It also illustrated how training can modify emergency response capabilities, since the ER team did not rush headlong into the fire-fighting phase of the approach. There was a deliberate, decision-making factor that was important to the lifesaving characteristics of the exercise, which is of considerable importance in dealing with hazardous materials. That one factor is of major importance for either paid or volunteer fire fighters when dealing with emergencies.

And, lastly, the interaction of such emergency services as the city police, the fire department, the Sheriff's Department, and the highway patrol represents the finest tribute possible to planning for emergency response.

6

Hazardous Classes and Properties

This chapter has been written to give the reader specific illustrations of what truly constitutes hazardous materials and their properties. The many varieties of hazardous materials and the general uncertainty of their behavior under varying conditions present a considerable problem.

The U.S. Department of Transportation has developed an extremely useful and logical classification system. It provides a description of each hazard class and the major or significant hazardous properties of materials in each class.

The information in this chapter also is presented in general terms. For more information, the reader is advised to refer to Title 49, Code of Federal Regulations (49 CFR), Parts 100–177 and 178–179 and other authorities on hazardous materials. In addition, this chapter discusses the multiple-hazard factor of many materials and the effect it may have on the handling of emergency response.

A. Hazard Classes

As discussed in an earlier chapter, according to 49 CFR, a hazardous material is a substance that has been determined to be capable of posing an unreasonable risk to health and safety or property when it is transported in commerce.

The health and safety of people, the safety of the environment with its precarious ecological balance, and the safety of property are all risks that can be endangered by hazardous materials. It is important to note that the *quantity* and the *form* of a hazardous material that is being transported have a large impact on the degree of risk involved.

As an example, all whiskeys contain flammable liquids in addition to other ingredients. The DOT, however, does not consider that 1 quart of whiskey packed in a traveler's suitcase is hazardous, because the whiskey is not in a hazardous quantity or form.

There are several thousands of hazardous materials differing in chemical name and chemical formula. No one person can be expected to know and understand all of their names and chemical compositions. Some materials, however, share common identifiable physical and chemical properties that impart to them their hazardous characteristics. The DOT uses those properties to group materials into understandable hazard classes:

The elements are as follows:

1. Physical properties,
2. Chemical properties, and
3. Hazardous characteristics.

1. Physical properties

The main physical property of a hazardous material is the state in which it exists. That property is common to all substances, since all matter exists as a solid, a liquid, or a gas.

Other physical properties further describe the solid, liquid, or gas. They are as follows:

1. *Vapor density:* The substance is either lighter or heavier than air;
2. *Water solubility:* It may be dissolved in water;
3. *Specific gravity:* It is either lighter or heavier than water; and
4. *Boiling point:* The temperature at which a liquid turns into gas.

The state in which a substance exists, i.e., a solid, liquid, or gas can be considered its main physical property. Therefore, all physical properties are important considerations in a transportation emergency and are required information when technical assistance is desired. The properties affect the physical behavior of a substance, both when it is inside a container or when it is released outside the container after an accident. For example, if a gas, such as LP-gas, has a vapor density greater than air, it will hug the ground and accumulate in low pockets when it is released from its container. Physical properties of a material, therefore, are related to its physical state and form.

2. Chemical Properties

The principal chemical properties of a hazardous material are those that are related to toxicity and may affect health, flammability, or chemical reactivity, in which the substance may react to whatever it comes in contact with. Chemical properties, therefore, include such elemental factors as toxicity levels, corrosiveness, flash point, ignition temperature, oxidizing ability, instability, or reactivity with air or water. (See the Glossary in the Appendix section for definitions of those terms.)

As indicated in Chapter 5, chemical properties are extremely important in terms of emergency response. Those characteristics or properties of a chemical reflect a hazardous material's capability of producing toxic substances harmful to man and the environment and its capability of burning or exploding.

A checklist of chemical properties is as follows:

1. *Ignition temperature:* the temperature at which combustion occurs,
2. *Oxidizing ability:* the ability of a substance to combine with oxygen,
3. *Toxicity:* the degree to which the substance may be harmful to health and the environment,
4. *Instability:* the tendency of a substance to decompose readily,
5. *Corrosiveness:* The capacity of a substance to destroy human skin or metals,
6. *Flash point:* The point at which the substance releases enough gas to ignite, and
7. *Reactivity:* The reaction of a substance to combine chemically with other materials or decompose.

B. DOT System of Hazard Classes

As indicated, above, physical and chemical properties can suggest many important characteristics of a particular substance. From a practical standpoint, however, it is not strictly functional or logical to categorize hazardous materials into classes based solely on physical or chemical properties.

Therefore, in the regulation of hazardous materials in transit, the DOT uses a Hazard Class System. It incorporates aspects of the physical and chemical properties of a material, but, more significant, it classifies materials according to their *major hazardous characteristics.*

Nevertheless, it is important to remember that quite a few hazardous materials possess more than 1 hazardous characteristic. It is extremely difficult to obtain in this complex field a single classification system that is easily understood and that will be able to indicate all the hazards of all materials. Thus, in dealing with a hazardous material emergency response situation, it is necessary to remember to use extreme caution, since many substances have a multiple hazardous characteristic, simply named, a *multiple hazard factor.*

The multiple hazard factor is not reflected in the hazard class system because materials are classified according to their major hazardous characteristic. As an example, a liquid may be both flammable and toxic, but the danger of flammability clearly outweighs the danger of toxicity. Therefore, according to the DOT system, this hazardous material would be classified as a flammable liquid, and the DOT hazard class name would be "flammable liquid." Flammability would be its major hazardous characteristic.

Acrylonitrile, which is a chemical used in plastics processing and fumigation, is classed as a "flammable liquid." It is also a very poisonous substance, and, in addition, it is thermally unstable. In other words, the container and its contents may explode under certain conditions. Those *multiple hazard factors,* however, are not reflected in the DOT hazard class system, because materials are classed according to a major hazardous characteristic. For those reasons, then, it is always necessary to consider the possibility of a multiple hazard before commencing any major control action involving an emergency response.

An emergency response team arriving at the scene of a transportation emergency may find that a tank truck bearing a sign identifying the contents as "flammable" has begun to leak on the pavement or roadway. Before the crew initiates any major control action, it would be well advised to determine the exact nature and contents of the tank truck.

The multiple hazard factor is extremely important to remember when one is dealing with any of the DOT hazard classes. A number of hazardous characteristics are included in each class of materials so that when an emergency response team, or a shipper, or other involved persons has the necessity of handling 1 of these materials, it is necessary to remember to consider all of the material's potential hazards.

According to the DOT system materials have been grouped, in general, into 11 major hazard classes. For the purposes of our discussion, these are as follows:

1. Explosives
2. Compressed gases
3. Flammable and combustible liquids
4. Flammable solids
5. Oxidizers
6. Organic peroxides
7. Poisonous and irritating materials
8. Etiologic agents
9. Corrosive materials

10. Radioactive materials
11. Other Regulated Materials (ORMS)

1. Explosives

The first major group of materials is explosives. The DOT defines an explosive as any chemical compound mixture or device whose primary purpose is to function by explosion.

A few of these materials can be given as an example:

1. Dynamite,
2. TNT (trinitrotoluene),
3. Black powder
4. Nitroglycerine, and
5. Ammunition.

Explosives are further subdivided in the DOT system into several hazard classes, although the differences between classes are not always very distinctive or clear-cut. These hazard classes are, as follows:

1. Class A explosives,
2. Class B explosives,
3. Class C explosives, and
4. Blasting agents.

a. Class A Explosives. Class A explosives are sensitive to heat and shock and thus present a maximum hazard. The Class A explosives include dynamite, which is commonly used in demolition work, mining, and quarrying. TNT is also a Class A explosive, and it is used for making military explosives, and industrial dyes and in fabricating photographic supplies. Another Class A material is black powder used in making home-made ammunition and safety-fuses. Military ammunition also may be made from Class A explosives.

b. Class B Explosives. Class B explosives are sensitive to heat and present a flammable hazard. They are used in making display fireworks, rocket motors, and some types of military ammunition.

c. Class C Explosives. Class C explosives contain some quantities of either Class A or Class B explosives or both but present a minimum hazard. Class C explosives include detonating fuses, common fireworks, and small arms ammunition.

d. Blasting Agents. As the name implies, blasting agents are designed for blasting. They are so insensitive that there is very little probability of an accidental explosion. The materials are nitro-carbonitrate and water gels and are used primarily in demolition, in mining, and in quarrying rock.

An explosive presents multiple hazards when significant quantities of it are considered. There need not be a large quantity of an explosive for it to be considered hazardous. Even a few ounces of some substances should be considered lethal. If an emergency arises, as in the event of an accident of some kind, some of the factors to be considered are (a) the possibility of an explosion, (b) the creation of blast overpressure, or a shock wave, (c) the hazard of scattering ruptured fragments over a wide area, and (d) the starting of additional fires.

The safest way to handle explosives in order to avoid an explosion is to protect the materials from exposure to fire, heat, static electrical charge, shock, or contamination from other materials. If any of those catastrophic conditions are present, the best and safest response is to evacuate all personnel as far away as possible from the center of reaction and as fast as possible since the materials could explode at any time.

2. Compressed Gases

The second major hazard class according to the DOT system is compressed gases. A compressed gas is a gas within a container under many times the normal atmospheric pressure and at normal temperature. The DOT regulations separate compressed gases into "Flammable" and "Nonflammable" gases. Those divisions are based upon the chemical properties of the gases.

A "Flammable Gas" is a "Compressed Gas" that will burn. On the other hand, "Nonflammable Gases" are "Compressed Gases" that are not classed as "Flammable." The nonflammable gases normally will not burn, but since they may support combustion, there is a danger that they may cause an explosion.

Compressed gases are widely used in industry, in medicine (as in hospitals), and as fuels. Table 6-1, which follows, lists some common compressed gases and classifies them as either "Flammable" or "Nonflammable". (See Table 6-1, Compressed Gases.)

The hazards of flammable gases such as propane and butane are obvious. Several of the nonflammable gases, however, present other serious hazards. Chlorine and anhydrous ammonia are severely corrosive, and oxygen speeds up some chemical reactions.

Table 6-1.   COMPRESSED GASES

Flammable	Nonflammable
Hydrogen (used in welding, hydrogenation of vegetable oils, and chemical processing)	Carbon dioxide (used for carbonating beverages and as an extinguishing agent)
Acetylene (used for welding and the production of synthetic chemicals and plastics)	Oxygen (used for the production of steel and in life-support systems)
Propane and Butane (used as fuels)	Nitrogen (used in fertilizer production, in refrigeration, and in making things inert)
Vinyl Chloride (used in processing plastics)	Anhydrous Ammonia (used for refrigeration and in making fertilizers.
	Chlorine (used for chemical processing, bleaching, and water purification)

Compressed gases are usually compressed to the point of liquification; that is, they are converted under pressure from a gas to a liquid. The reason for that conversion is to make their transportation both efficient and economical. That is why propane, butane, vinyl chloride, anhydrous ammonia, chlorine, and carbon dioxide are usually transported and stored as liquified gases. Nitrogen and oxygen are among the gases called cryogens, which means mixtures at low temperatures. Those gases become liquified at very low temperatures and thus can be economically transported.

a. Multiple Hazards of Compressed Gases. The physical and chemical properties of a compressed gas are reflected in the several possible multiple hazards that may be involved when an accident that might involve an emergency response occurs. The multiple hazards may include the following destructive and unpleasant results:

1. BLEVE—Boiling Liquid-Expanding Vapor Explosion,
2. An open-air combustion explosion,
3. Asphyxiation,
4. Toxicity or corrosiveness,
5. Frostbite (from the cryogens),
6. The ability to move along the ground for considerable distances, and
7. Reactivity

(1) BLEVE
The threat of potential energy is especially severe with liquified gas containers. As an example, the volume of propane as a gas is 270 times greater than its volume in the liquified state.

Liquified gases are characterized by rapid vaporization of the liquid should the container rupture. That hazard involving a container failure is best described by the term "Boiling Liquid-Expanding Vapor Explosion," or "BLEVE." A BLEVE, therefore, is a major container failure, into 2 or more pieces, at the instant when the contained liquid is at a temperature well above its boiling point at normal atmospheric pressure. At the time of the BLEVE, the ensuing explosion hurls pieces of the fragmented container like shrapnel, a vapor cloud rapidly forms, and a shock wave follows.

Since a BLEVE is the result of an accident involving a boiling liquid, compressed gases in the gaseous state are not subject to this hazard. Cryogens, which are compressed gases in a liquid state at very low temperatures are generally transported in insulated or refrigerated containers, thus substantially reducing the chances for a BLEVE. Should the insulation fail or if the refrigeration should cease functioning or become damaged, then the possibility of a BLEVE would increase dramatically.

(2) Open-Air Combustion Explosion
A BLEVE may result from impact or exposure of the container to fire. Either of those events might weaken the metal of the container in one area severely. Thus, if the gas in the container is flammable and the container failure has resulted from exposure to fire, then the flammable chemical will ignite immediately, producing a ground flash and rising fireball. If the container failure should result from impact, then delayed ignition could produce an open-air combustion explosion severe enough to demolish structures and knock personnel about.

(3) Asphyxiation

The potential for a BLEVE or a BLEVE-type explosion is perhaps the greatest danger in a transportation emergency involving compressed gases. Nevertheless, other hazards may be present. None of the gases, with the exception of oxygen, can support human or animal life; therefore, those gases present the hazard of asphyxiation by replacing breathing air.

(4) Toxicity or Corrosiveness

A large number of gases, such as chlorine, anhydrous ammonia, and hydrogen cyanide, are toxic or corrosive to the human respiratory system. Furthermore, lethal concentrations of those gases in the air may be invisible. As part of an emergency response team responding to a possible leak in a container of chlorine, it would be advisable to wear protective clothing equipped with self-contained breathing equipment.

(5) Frostbite

All liquified gases can cause severe frostbite when they escape from their containers. In addition, such objects as automotive truck tires exposed to the released, liquified gas may become brittle and shatter upon contact. It should be remembered that frostbite and low-temperature embrittlement hazards increase when certain gases are transported in the cryogenic state.

(6) Heavy Gases Move near the Ground

The vapor density of air is a reference point for determining whether a gas is lighter or heavier than air. Thus, the weight of the same volume of the pure vapor or gas is compared with the weight of dry air at the same temperature and pressure. Most gases are heavier than air at equal temperatures; therefore, when those gases are released from their containers, the heavier gases can move along the ground for great distances and can accumulate in low spots or pockets or depressions in the ground. Sometimes, gases that are lighter than air can also roll along the ground if they are colder than the ambient air. In that manner, the gases may reach an ignition source far from the original container. If the gases happen to be flammable, they will ignite with frequently disastrous results.

(7) Reactivity

Certain gases, such as vinyl chloride, react chemically with other substances or with themselves. The gases that are self-reactive produce excessive heat and other gaseous reaction products inside the container, presenting a potential BLEVE-like situation. In an effort to control that hazardous problem, stabilizing or inhibiting materials are usually added to the gas.

3. Flammable Liquids and Combustible Liquids

Flammable and combustible liquids are common to our everyday working and living days. Gasoline, paints, solvents, fuels, and liquors are everywhere around us, and it is with those widely used chemicals that many serious accidents involving emergency response occur.

The DOT divides those commonly used liquids into 2 separate hazard classes: flammable liquids and combustible liquids. A flammable liquid is any liquid with a "flash point" below 100° Fahrenheit. The *flash point* of a liquid is the lowest temperature at which a liquid gives off enough vapor to form an ignitable mixture with the air near the surface of the liquid. The lower the flash point, the more readily a liquid will ignite.

Gasoline used as a fuel and ethylene oxide used as a fumigant are flammable liquids. Ethyl alcohol used in liquor, in paint manufacture, and as a pharmaceutical solvent is also a flammable liquid. Another flammable liquid is toluene, which is used as a solvent and in the manufacturing of organic compounds. Gasoline, ethylene oxide, ethyl alcohol, and toluene are all hazardous materials with flash points below 100°F.

The DOT defines a combustible liquid as any liquid that has a flash point at or above 100°F and below 200°F. Despite the flash point definitions quoted above, one should not complacently believe the misconception that liquids with flash points above 200°F will not ignite. That could be fatal, because they most certainly will burn.

Domestic heating of fuel oil, diesel engine oil, and stoddard solvent, an industrial solvent, are widely used and readily available combustible liquids. Those liquids all have flash points at or above 100°F and below 200°F. Since the flash points of combustible liquids are above the normal temperature of the surrounding atmosphere, they can be ignited less easily than the flammable liquids. Also, once ignited, they can usually be extinguished with a lot less difficulty than flammable liquids because of their higher flash points.

The flammable and combustible liquids both require an external ignition source. One group of liquids, however, is capable of autoignition, or self-ignition, if heated sufficiently. Such liquids are called pyrophoric (meaning self-igniting). The DOT also classifies them as flammable liquids.

A pyrophoric liquid is any liquid that will ignite or burn spontaneously in dry or moist air at or below 130°F. Examples are certain aluminum alkyls and alkyl boranes, which are used as rocket propellants. Since a pyrophoric liquid exhibits the tendency to self-ignite, it presents a greater flammability hazard than do the liquids in the flammable or combustible classes. In order to reduce the hazard in transportation, a pyrophoric liquid is shipped in an airtight container.

Five of the 7 possible multiple hazards present in an emergency response situation involving flammable or combustible liquids are similar to those for compressed gases. They are as follows:

1. A BLEVE, or BLEVE-like explosion.
2. Combustion explosion,
3. Toxicity or corrosiveness,
4. Reactivity, or,
5. The ability to move considerable distances along the ground to a source of ignition.

With the type of hazardous liquids we have been discussing, the most obvious additional hazard is fire. That danger is always present in the event of a gasoline tank truck accident. It is a particular problem with the transportation of gasoline, since so many tankers are on the roads day and night and there are an increasing number of other cars and trucks to contend with; that, combined with the low flash point of gasoline and the carelessness attendant on its use because of increased familiarity with the product, is a factor to be taken into consideration when one studies the problem.

Another hazard in connection with the transportation of flammable or combustible liquids or involving an accident or emergency response is the contamination of drinking water supplies. Runoffs from the hazardous substance, either the material itself or the water used to control the emergency, may enter storm sewers or contaminate ground waters.

Several of the less commonly used flammable and combustible liquids are nevertheless dangerous and present serious multiple hazards because they are poisonous, corrosive, and thermally unstable. Those chemicals include acrolein, ethylene oxide, and carbon disulfide, which are often used in chemical processing and in fumigation treatments.

In the light of the many multiple hazard factors that are present in handling the hazardous liquids discussed above, one should take additional precautions when faced with an accident involving those substances or when an emergency response is required.

The following steps would provide the ER team with greater safety:

1. Identify the substance before taking action,
2. Avoid walking through the substance,
3. Use self-contained breathing apparatus, and
4. Eliminate as many sources of ignition as possible, if the material has not already ignited.

As a general precaution in dealing with whatever emergency response situation that may arise, the single, most important step toward increasing the safety of the operation is to properly identify the material before taking action. When you are certain of the material that is involved in the incident and knowledgeable about it, then and then only will you be able to identify the additional hazards that you may have to face.

4. Flammable Solids

In dealing with accidents or emergency response situations, water is the most abundant, readily available, and commonly used extinguishing agent. Despite those factors, it is often not advisable to use water on the fourth hazard class, flammable solids, because of sometimes violent chemical reactions.

As the name suggests, flammable solids are materials that are solid, other than explosives that ignite quite easily and burn vigorously. Some flammable solids are air-reactive or pyrophoric (self-igniting). Other flammable solids are water-reactive; some ignite by spontaneous combustion; and some burn with extreme rapidity.

An example of a flammable solid that is air-reactive is white phosphorous, which is used in the manufacture of common matches and rat poisons. It is packed in airtight containers, sometimes with a water seal, to prevent accidental ignition. A water-reactive flammable solid, on the other hand, will react with water when it gets wet or when in contact with humid or moist air. A water-reactive flammable such as this is metallic sodium, which is used in making motor-fuel antiknock compounds, indigo dyes, and nonglare streetlights.

Spontaneously combustible flammable solids can decompose and ignite either in the presence or absence of air. As an example, cotton waste, a textile manufacturing waste product, is a spontaneously combustible (flammable solid) solid. Although spontaneously combustible solids can ignite in the absence of air, air-reactive flammable solids can ignite only when exposed to air.

In handling emergency responses involving flammable solids, the ER team must consider the following 5 primary multiple hazards:

Flammable Solids—

1. Ignite easily and burn with explosive violence,
2. React with air,

3. React with water,
4. Ignite spontaneously, and
5. Produce toxic or corrosive compounds when they are in contact with water or air, or they are initially toxic or corrosive.

Flammable solids are generally characterized by their ease of ignition and their rapid burning. In fact, their burning speed is much greater than that of ordinary combustible substances. In confined areas the burning of a flammable solid can proceed with explosive violence. The explosion hazard of a flammable solid, while almost always present, is more than likely to be somewhat less than a Class A or Class B explosive.

As mentioned above, the application of water is one of the main techniques used in combating and extinguishing fires. That makes it extremely important to know, in advance of any emergency response action, the nature and characteristics of the material involved in an accident. For example, if the material is a flammable solid, it is urgently required that the substance be quickly identified, especially if it is water-reactive. Water reactivity is a multiple hazard that, like flammability, can be dealt with once the substance has been properly identified and its multiple hazard characteristics are determined.

Many flammable solids, besides being innately toxic or corrosive, may produce compounds that are toxic or corrosive when they come in contact with water or air. For example, calcium carbides generate acetylene upon contact with water. Metallic or nonmetallic chlorides form hydrochloric acid upon contact with water. Those examples, it is believed, are sufficient to illustrate the necessity for identifying the materials involved in an emergency response before using water to control the incident.

As sometimes happens, a derailment can be the beginning of a major railroad disaster. The DOT regulations, therefore, are extremely necessary to develop the proper emergency response to a situation, for example, which involves a derailment of tank cars. Imagine, if you will, a train of freight cars passing through a small community. At the outskirts a derailment occurs, and the railroad cars are strewn helter-skelter in a jumble along the tracks. A railroad car starts burning. When the members of an emergency response team cautiously approach the wreck, they see that one of the tank cars is placarded with a large sign reading "FLAMMABLE SOLID." Now the ER team can identify the material contained in the tanker and can take proper precautionary action. If it were not for the DOT system of hazard class identification, almost all ER actions would be even more dangerous and disastrous than they are at present.

5. Oxidizers

A few years ago an explosion killed four people and injured several others. It seems that broken and leaky boxes of sodium chlorate were being removed on wheeled dollies from the basement of a university building. From the friction of the dolly wheels rolling on the floor through the spilled sodium chlorate and dust, sufficient heat had built up to ignite the mixture. The resultant heat buildup ignited the sodium chlorate and caused the explosion. Sodium chlorate belongs in the fifth hazard class of substances called oxidizers.

Oxidizers usually contain large amounts of chemically bonded oxygen. The oxygen is readily released from the compound, especially when heated, and will encourage the burning of combustible materials. In the instance of the sodium chlorate explosion, noted above, the combination of the sodium chlorate, the dust on the floor, and the friction generated by the dolly wheels was enough to cause the fatal explosion.

The oxidizer, in this instance sodium chlorate, provided the oxygen to stimulate the burning of the dust.

An oxidizer may be defined as a substance that yields oxygen readily to stimulate the combustion of both organic and inorganic substances. An obvious oxidizer is oxygen itself. The DOT classifies oxygen as a nonflammable gas or compressed gas. Oxidizers are transported all over the country. Included in this hazard class are some concentrations of nitric acid used in making fertilizers, explosives, nitrated organic compounds, and etching metal, including printing plates. Other oxidizers are potassium and sodium nitrate, which are used in metal heat treating and in making fertilizers. Also in this hazard class as an oxidizer is calcium hypochlorite, used as a swimming pool chemical.

In addition to supplying oxygen and increasing the intensity of a fire, oxidizers have other known hazards. Some oxidizers are themselves explosively sensitive to heat, shock, and friction. Others will react with combustible organic materials rapidly enough to cause a spontaneous combustion. Most will form a readily ignitable or explosive mixture when combined with finely divided particles of organic materials. For that reason, especially, it is advisable not to store even small amounts of certain oxidizers in a warehouse building. Some of those oxidizers would be sodium nitrate, nitric acid, and calcium hypochlorite.

There are several excellent reasons why oxidizers should not be stored together with combustible materials. The multiple hazard factors that make them poor storage companions for combustibles are as follows:

1. They react with organic materials;
2. They form easily ignited mixtures;
3. They supply oxygen to increase the intensity of fire; and
4. They are explosively sensitive to heat, shock, and friction.

6. Organic Peroxides

The sixth hazard class in the DOT hazard class system is the organic peroxides. They have been given a separate hazard class because of their great destructive potential.

The consumption of organic peroxides has kept pace with the large-scale growth in the plastics industry. Nearly all of the organic peroxides are used in making plastic products, such as adhesives and resin-bonded fiberglass. Some examples are benzoyl and lauroyl peroxide.

Among other organic peroxides we find peracetic acid used as a fungicide and as a sterilizing agent. Flour and textiles are bleached using a number of bleaching peroxides; other peroxides are used in the pharmaceutical industry. By and large, however, the greatest volume of organic peroxides is used in the manufacture and processing of plastic products.

A growing number of small plastics-manufacturing plants use organic peroxides. They are useful in plastics manufacture because they are unstable and decompose readily. One of the hazards associated with them, however, is their instability and tendency to become explosive in the presence of heat, shock, and friction. They have other hazards:

1. Many are highly flammable;
2. During the decomposition process they release large amounts of heat and gaseous by-products, which are often toxic; and

3. Their exposure to fire may evaporate reactivity inhibitors that have been added to the chemical to lessen reactivity.

Since the organic peroxides benzoyl and lauroyl peroxide are often used in small-plastics-plant-manufacturing processes, an emergency response situation at one of those plants might involve a tractor trailer. If yellow smoke billows out from the burning tractor, it might be assumed to be one of the organic peroxides, since yellow smoke is one of their characteristics when they burn. Several multiple hazards would, of course, be involved in such a situation. Many organic peroxides are highly flammable; when exposed to heat, shock, or friction, some are unstable or explosive; gaseous products can be released in discomposition and can be toxic; and reactivity inhibitors may evaporate when exposed to fire.

7. Poisons

Hazardous, poisonous materials are classified in 3 groups of hazard classes according to the DOT hazard classification system, depending on the degree of health hazard concerned. The classes are as follows:

a. Class A Poisons. The greatest degree of hazard comes from Class A poisons, usually gases. Hydrogen cyanide, which is used as a fumigant and also as a death cell execution gas, is one example. Nitrogen tetroxide, also a Class A poison, is used as an oxidizer in rocket fuels. Phosgene is a substance used in chemical processing and also as a war gas. Those examples of Class A poisons have an extremely high degree of health hazard and can cause death by inhalation, absorption, or ingestion in relatively small quantities.

b. Class B Poisons. Class B poisons are only somewhat less hazardous than Class A poisons but still represent a significant health hazard, often lethal, if they should be released from their containers. These poisons are the chemical aniline used in making dyes and inks; tetraethyl lead used as an additive to make gasoline motor fuel antiknock compounds; parathion, used in pesticides; and sodium cyanide, which is used in metal heat-treating plants.

c. Irritating Materials. The third grouping of poisonous materials has been classified by the DOT as irritating materials. Those substances release dangerous or intensely irritating fumes. A good example is the tear gas grenade, which is probably familiar to us all. As irritating as tear gas is, it is only somewhat less hazardous than sodium cyanide pellets, which are categorized as a Class A poison.

d. Multiple Hazards Involving Poisons. There are significant multiple hazards to be considered during an emergency response situation in dealing with poisons. Among the multiple hazards are the following:

1. The poisonous chemical may enter the body by inhalation, absorption, or ingestion;
2. Fire may rupture the containers, thus releasing the contents;
3. Runoffs may contaminate the water supply; and
4. The poison may also be flammable.

There are 3 major ways by which poisons may enter the human system. Through the nose and mouth by inhalation, through the skin (epidermis) by absorption, and through the mouth by ingestion.

The inhalation of toxic fumes is an obvious hazard, particularly with poison gases. The hazard becomes more insidious and deadly, particularly if the substance is odorless. Some poisons, such as hydrogen cyanide and hydrogen sulfide, rapidly deaden the sense of smell so that a fatal dose may be inhaled before a person has time to react to the consequences of the fumes.

Absorption is the process by which the poisonous chemical enters the body through contact with the skin. Ingestion of a poisonous chemical occurs in a relatively innocent manner when the chemical substance is usually carried into the mouth by contaminated hands or other contaminated objects placed near or into the mouth.

The manner in which emergency response team personnel may be affected by responding to an accident situation serves to illustrate some obvious and not so obvious examples of the way in which serious results sometimes ride on the aftermath of a successful cleanup operation.

Death or injury from *inhalation* may occur when the worker smells something strange near a ruptured or leaking container or when he inhales an unusual odor from smoke.

Death or injury from *absorption* may occur when someone does not remove all his clothes when washing up after an accident response or does not throw away a good pair of work boots.

Death or injury from *ingestion* may result when someone smokes a cigarette after an operation is concluded and carries contaminated substances into the mouth or drinks a hot cup of coffee and touches the rim occasionally with hands or gloves that have been contaminated with the poisonous substance.

The multiple hazard factor of some poisonous chemicals can become life threatening and destructive to property. For example, if a poisonous substance that is also a compressed gas is transported in cylinders that have no pressure relief devices, then a serious accident may turn into a disaster. If a cylinder should rupture because of fire, then in addition to the exploding cylinder the poison gases would be released.

Hydrogen cyanide is transported and stored as a compressed gas. In the event of an accident an ensuing fire could cause cylinder rupture and the resultant release of the poisonous gas. Since the hydrogen cyanide is under pressure, a BLEVE-like explosion would more than likely occur.

Owing to the complexity of certain chemicals there are some flammable liquids that are also toxic, and vice versa. Thus, the further necessity of examining the multiple hazard factor when one deals with emergency responses that have to do with poison gases and solids.

An additional hazard of poisonous materials is the possibility of poisoning water supplies. Water runoff, which may occur during an emergency response, could be corrosive as well as toxic. When the ER team uses water, it could become highly contaminated and enter storm sewers or nearby lakes, streams, or rivers. The ER team should get in touch with the local water authorities as soon as possible and should make every attempt to dike the liquid runoff.

8. Etiologic Agents

Etiologic agents are the eighth hazard class of the DOT hazard class system. They have properties similar to those of poisonous materials. As indicated in a previous section, an etiologic agent is a living microorganism that may cause human disease. The principal etiologic agents that are transported are usually biological specimens and virus specimens used in testing and research. Measles and rabies virus, for example, are commonly transported to and between hospitals, research laboratories, disease control centers, and pharmaceutical companies.

The main hazards of an emergency response involving etiologic agents are as follows:

1. The disease-producing agent may enter the human body through inhalation, ingestion, or absorption; or
2. The disease-producing agent may contaminate the environment.

9. Corrosive Materials

The ninth hazard class of the DOT hazard class system consists of corrosive materials. They are either liquids or solids that can destroy human tissue or severely corrode metals.

Corrosive materials include sulfuric acid, which is widely used in manufacturing, for many chemical processes, and in automotive and industrial truck batteries. Hydrochloric acid, another corrosive substance, is used for soldering, in galvanizing steel, in the production of household cleaning fluids, and in many chemical manufacturing processes. Sodium hydroxide, which is also called lye or caustic soda, is a corrosive material used in the purification of petroleum products in the manufacture of soap, pulp, and paper. Potassium hydroxide is another widely used corrosive material. It is sometimes called caustic potash and is used for the manufacture of soft and liquid soaps and as an alkaline battery fluid.

Since the transportation of corrosive materials constitutes a large portion of the shipping volume in the chemical industry, an emergency response team might have to face the following hazards:

1. Destruction of human tissue may be caused by skin contact or by inhalation of the fumes and mists generated from ruptured containers;
2. Ignition of combustible materials may occur because some corrosive materials are oxidizers and may cause combustibles to ignite;
3. When diluted with water, corrosive materials release heat rapidly and are predisposed to splatter;
4. Some corrosive materials are toxic; and
5. Some corrosives are unstable and tend to decompose when heated.

10. Radioactive Materials (RAMs)

The tenth hazard class in the DOT system is Radioactive Materials (RAMs). They are widely used in medicine, industry, and research and in the fueling of nuclear power plants for the generation of electric power.

Radioactive iodine is used in medical treatment; radioactive cobalt is used in the practice of medicine and in industrial radiography. Enriched uranium is used as a fresh fuel element in nuclear power generation. Plutonium is a waste hazardous material that also comes from nuclear power plants. RAMs are being generated daily as waste products from hospitals and research laboratories.

The largest number of radioactive shipments involve small, carton-type packages used for medical purposes. Industrial shipments usually consist of devices used in radiography as a part of the manufacturing process and for quality control or inspection functions.

There are 3 multiple hazard factors that an emergency response team should consider:

1. The presence of 2 types of radioactive exposure, one of which is external where the x-rays pass through the body, and the other internal wherein particles are inhaled or ingested.
2. Contamination of personnel, property and the environment by smoke, steam, or water run-off during fire fighting.
3. Exposure of the radioactive substances if a fire should melt or destroy the protective lead shielding of the packages containing the hazardous materials.

Radioactive materials have some hazards in common with etiologic agents and poisons. The fire-fighting methodology using water can contaminate the environment extensively for all 3 hazard classes. In addition, internal exposure to humans may occur by means of inhalation, ingestion, or absorption.

The safest transportation of RAMs is accomplished by using approved packaging methods. The proper packaging securely encloses the hazardous substance and reduces the radiation emitted from the package to acceptable levels.

In the event of accidents involving RAMs, a qualified person should determine whether there is a radiation or contamination problem. The State Department of Health should be called during a radioactive materials emergency response in order to assess the extent of radiological contamination. The carrier of the shipment should notify the DOT Response Center (phone: 800-424-8802) if the incident involves the type, quantity, and form of RAM that is reportable.

11. Other Regulated Materials (ORMs)

There are 5 additional hazard classes called ORMs, Other Regulated Materials, ORM-A, -B, -C, -D, and -E, which have already been described in Chapter 4. ORMs, except for ORM-D, do not meet the definitions of any of the DOT hazard classes but possess enough hazardous characteristics in transport that they require some regulation.

The establishment of the ORM classification indicates a certain flexibility with which the DOT views the subject of hazardous materials. After all, many of the regulations were created to alert involved personnel to the presence of hazardous materials should an accident or other emergency occur.

To summarize the concepts developed in this chapter, we should reaffirm that the DOT hazard class system attempts to group materials with similar hazardous characteristics so that regulations that will provide safety in transportation can be established for those groups or classes. Any attempt to group hazardous materials, however, is bound to admit to limitations and exceptions that should be observed for the protection of all personnel involved in an emergency response.

It should be noted, that the DOT hazard class system does not recognize that the degree of hazard for materials within the same class may vary from slight to severe under emergency conditions.

In addition, some materials that are transported today are known to be hazardous but are not currently classified as such by the DOT. Certain cryogenic materials, such as liquified natural gas and liquified oxygen, although classified as hazardous, are not classified as cryogens. Nevertheless, cryogens require special handling by emergency response personnel. Despite the fact that several materials are known carcinogens, or cancer-causing agents, that hazard is not recognized now as a hazard class. Therefore, some of those materials are not regulated in transportation.

Perhaps the greatest limitation of the DOT hazard class system is that there is no relatively simple way to deal with the fact that many materials have more than one hazardous characteristic. Materials with multiple hazards may be highly flammable, thermally unstable, and poisonous at the same time, thus presenting additional difficulties when they are handled in manufacturing, in transportation, in storage, and in an emergency response situation. In dealing with hazardous materials, one must always be alert to the multiple hazard factor.

7

Recognizing and Identifying Hazardous Materials

A. Emergency Response Requirements

There are, generally speaking, a number of precautions that must be observed when an accident involving, or presumed to involve, hazardous materials takes place. If it were not for the complexity and danger from multiple hazards, we would not need to emphasize that in any major transportation accident the ER team usually can find sufficient information on the scene to take logical action.

The members of the ER team will need to know the following:

1. Are hazardous materials present?
2. If hazardous materials are involved, with what chemicals are they dealing?

1. On-The-Scene Information

As we have indicated, there are usually enough indications present at the accident scene to supply information that can be used before any major action is attempted.

Some of those indicators are the shipping containers themselves or the package markings. Also, there should be labels, placards on the vehicle or tank, identification numbers, and shipping papers. If for some reason, the above information is not readily available, then no time should be lost in calling the shipper, the carrier, or the consignee for the necessary information. With the availability of shipping papers, it is possible to identify the chemical being transported, the shipper and shipping point, and the consignee, together with the destination of the cargo.

B. Preliminary Actions to be Taken—Identification

Since every second counts during an emergency, we must quickly determine whether the materials are hazardous. Proper identification of the cargo, especially when combined with a knowledge of the physical and chemical properties of the hazardous materials involved, may save lives. It is obvious that this information is needed to select the methods used to control the hazardous materials incident and to take adequate measures to protect civilians and emergency response team personnel.

Although there are several ways to recognize and identify hazardous materials, a major device is the shipping container. Before any hazardous solid, liquid, or gas can be transported, it must be confined or protected by a container or both. DOT regulations, therefore, specify the type of container that must be used and in what manner the container should be labeled and marked.

Since there is a wide divergence in the physical and chemical properties of the many hazardous chemicals used in our modern society, there is also a wide difference in their shipping containers. For the purpose of clarification, we shall divide those containers into 2 categories: (a) large, *bulk containers*, and (b) *smaller containers.*

A "bulk container" is a cargo container used for transporting materials in bulk or large quantities. As an example, a railroad tank car or an over-the-road, tank truck.

A "smaller container" is a container, such as a box, or drum, a cylinder, etc., used to transport materials in small quantities.

The DOT has issued regulations governing the construction of both smaller and bulk containers used in the transportation of hazardous materials. There are several reasons why that was necessary. The first purpose was to improve the safety of transporting the materials. The second motive was to standardize the quality of the containers. And the third reason was to prevent the leakage of materials from these containers.

1. Smaller Containers

Examples of smaller containers are illustrated in Fig. 7-1. The figures show smaller containers constructed along the lines of the DOT regulation as follows:

   (a) wooden boxes,
   (b) metal drums,
   (c) fiberboard drums,
   (d) plastic pails,
   (e) glass carboys in a protective shell,
   (f) cylinders,
   (g) ton cylinders,
   (h) mailing tubes, and,
   (i) multi-wall paper sacks.

If a shipment includes a quantity of smaller containers, they may be shipped together in a van-type trailer or truck, a flat-bed trailer truck, a railroad box car, a cargo vessel, or in aircraft. The DOT regulations also require that those containers be braced and blocked so that shifting is precluded and any other such accidents in transport are eliminated.

Except for flat-bed trucks and trailers, many of the larger transport vehicles are completely enclosed. In order to determine whether they contain hazardous materials in individual containers, it would be necessary to open or to enter the carrier vehicle. At the time of an emergency situation, that would not be the most logical, practical, or safe approach. We shall discuss the methods to be used in recognizing and identifying the contents of a vehicle later in this chapter.

Let us, for the moment, proceed to the larger containers, or the so-called bulk containers.

Fig. 7-1   Illustration of Smaller Containers Constructed According to DOT Regulations

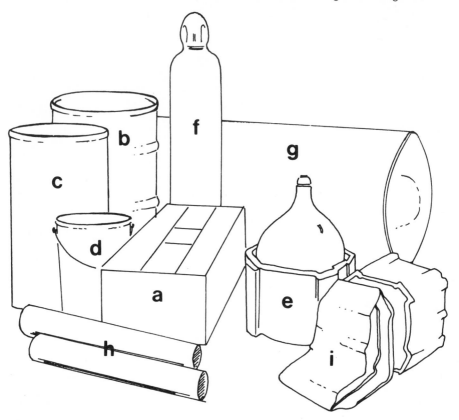

2. Bulk Containers—Tank Trucks

The majority of bulk containers used to transport hazardous materials are either tank trucks, tank railroad cars, tank barges, or large, portable tanks. The size and type of a bulk transport container is a real clue to the possible presence of hazardous materials. And because of the size of the container it is usually possible to make that identification at a safe distance, although there are more reliable methods for recognizing the presence of hazardous materials than just the size and shape of the bulk container or carrier.

However, if a tanker is involved in an accident, it is a wise precaution to proceed as if a hazardous material is present, until positive identification has determined the contents of the carrier. While perishable foodstuffs, such as milk, molasses, fruit beverages, and corn syrup, are often transported in tankers, it is advisable to assume the presence of hazardous materials, since many tank-type vehicles are used for transporting such hazardous materials as liquids, gases, or liquified gases, and a wide variety of petroleum-based products.

(See Figs. 7-2, 7-3, 7-4, 7-5, 7-6, and 7-7 for illustrations of various types of tank trucks.)

The shape of a tank truck may give a clue to the hazard class of the material being carried, as illustrated below:

Fig. 7-2   Gasoline and home heating oil are usually carried in a tank shaped liked this; however, flammable and combustible liquids may also be transported, in a tank shaped like that shown in Fig. 7-3.

Fig. 7-3   Flammable and combustible liquids are often carried in tanks shaped like this.

Fig. 7-4   A tank truck carrying corrosives will have a cylindrical cargo tank, most often with external rings. These tankers are smaller in cross-section than those carrying other liquids, because acids are usually heavier than other liquids.

Fig. 7-5   Another view of a tank truck carrying a flammable liquid. It is larger than a tanker carrying corrosives.

Fig. 7-6   Compressed Liquified Gasses are carried in pressurized cargo tanks that are usually shaped like the tanker shown above.

Fig. 7-7   Dry, bulk hazardous materials, such as some oxidizers, are carried in cargo tank trucks that look like a series of funnels. The tanker shown above is waiting to be hitched to a tractor.

Tankers or tank trucks are built with either 1 main compartment, or they may be divided into several. If a tank truck has a number of compartments, it may very well be that each is carrying a different hazardous material. Also, tankers that carry corrosives, such as acids, liquified gases, or other substances requiring pressurization, have to be specially fabricated with rigorous quality control efforts in order to provide safe transportation for hazardous materials.

In summary, all tank trucks are not of the same shape and size, and the shape and size of a tanker may sometimes give a clue about the presence of a hazardous material and its hazard class.

3. Bulk Containers—Rail

Hazardous materials are transported in bulk by rail in tank cars that are very similar to tank trucks. As with tank trucks, the shape and size of tank cars can provide effective clues that will enable one to determine the presence of hazardous materials from a safe vantage point. Therefore, whenever tank cars or trucks are involved in a transportation accident, it is advisable to assume that the cargo is hazardous unless it can be proved otherwise beyond any doubt.

The differences in configuration among the various classes of railroad tank cars are not so pronounced as they are among tank trucks. Fortunately, however, there are a number of other recognizable differences. Tank cars generally fall under several distinctions. Either they are pressurized or nonpressurized types. They may be insulated or noninsulated depending upon the type of materials to be transported. In an emergency response situation, such classification can be very important.

For example, in a pressurized tank car one might find compressed gases, unstable flammable liquids, or liquified gases. Since pressure-type tank cars transport their contents under pressure, one might expect to find as cargo some volatile flammable liquids and compressed or liquified gases. They would include fluorocarbons, butane, propane, and anhydrous ammonia. Pressure-type tank cars, therefore, carry hazardous materials any of which could BLEVE under certain conditions. (A BLEVE, as you may recall from an earlier chapter, is a

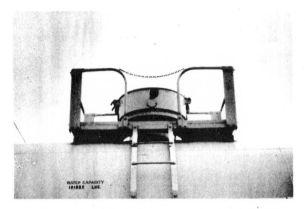

Fig. 7-8 Protective Dome Housing on a Pressurized Car

boiling liquid-expanding vapor explosion.) A BLEVE or BLEVE-like explosion is one of the main hazards that must be contended with in the hazard classes consisting of compressed gases, flammable gases, nonflammable gases, and flammable liquids.

4. Safety Relief Valves and Vents

Pressure tank trucks and cars usually have only 1 compartment or single unit. On those single-unit tankers there is a topside arrangement of safety valves and discharge pipes. On pressurized cars (a topside view is shown below in Fig. 7-8) all of those valves are contained within a protective dome housing.

A protective dome indicates a pressurized car. In addition, there are no bottom outlet valves on the underside of most pressurized cars.

On nonpressurized cars the safety vents or discharge connectors may be outside of and even a short distance away from the dome housing, as illustrated in a topside view, below, in Fig. 7-9.

Fig. 7-9 Safety vents on a nonpressurized car shown apart from the dome housing

Nonpressurized cars often have a bottom outlet valve. Therefore, if a bottom outlet valve is visible and there is a vent on a tank car that does not have a dome cover or housing, then more than likely it is a nonpressurized car.

Safety relief valves are located inside the dome housing on tank cars and other pressurized bulk cargo tanks carrying hazardous materials. They are set to operate whenever there is a buildup of pressure inside the tank. (See Fig. 7-10, below.)

When a safety relief valve operates under normal conditions, it usually releases small amounts of vapor into the atmosphere. A sufficient volume is released in order to prevent an overpressure condition. If an emergency situation or an accident should occur, the amount of material that is discharged could cause a problem. Also, if a fire should occur in the immediate vicinity of the tank, it might be sufficiently close to heat the vapor space in the tank; therefore, the operation of the safety valve may indicate that a BLEVE may happen at any minute.

5. Insulated Tank Cars

Pressurized and nonpressurized tank cars can either be insulated or noninsulated. When the tank car is insulated, it will usually be transporting materials that require protection from heat, for example, cryogenic or thermally unstable chemicals. Or the tank cars are insulated in order to keep the materials warm, as with such nonhazardous materials as foodstuffs or asphalt. On the other hand, the tank cars may need extra protection because of additional hazards that may arise, as with the transportation of chlorine.

When an insulated tank car transporting a hazardous material is involved in an emergency situation, the fact that the car is insulated may give the ER team some additional time to take major action. Of course, if the insulation has been damaged or destroyed in the accident, then the factor of added protection becomes largely academic.

Fig. 7-10 Safety relief valves are shown located inside the dome housing on a pressurized tank car. The cover has been raised to show the valves.

Fig. 7-11   A noninsulated railroad tank car showing the smooth, cylindrical body and rounded ends.

6. The Shape, Size, and Color of Tank Cars

a. Shape: As shown in Fig. 7-11, above, noninsulated tank cars have smooth, cylindrical bodies and rounded—that is, curved—ends.

Insulated tank cars often have recessed heads or horizontal seams that are clearly visible. (See Fig. 7-12, below.) A jacketed midsection may also be in evidence.

Tank cars usually have the same basic cylindrical cross section, but they may vary somewhat in the shape of the bottom part, the size of the car, and the color. For example, some tank cars have sloping bottom sections in order to increase the volume of material being carried. Such commodities as corn syrups and heavy fuel oils are usually transported in cars with sloping bottoms. Another advantage of the sloping bottom is that it permits an easier discharge of the material. Fig. 7-13 illustrates a tank car with a sloping bottom.

Fig. 7-12   An insulated railroad tank car.

Fig. 7-13   Tank car with sloping bottom.

b. Size.   Just as tank trucks are smaller than the more common tankers, railroad tank cars that transport corrosive materials are usually smaller in size in order to meet railroad load limits. Acids generally weigh much more than other liquids.

c. Color.   Noninsulated pressurized cars that carry liquified compressed gases and some volatile flammable liquids are required by DOT regulation to have at least the upper two-thirds of the tanker painted white. That does not mean, necessarily, that any white-colored tank car is carrying 1 of those hazardous products; however, if a white-painted tank car is also identified as a pressurized car, then it will, no doubt, be carrying those types of hazardous substances.

To summarize some of the above facts about truck and rail tank cars, we may say:

1. The specific hazardous material being carried cannot usually be recognized from the size, shape, or color of the tanker,
2. Tank cars may be nonpressurized and insulated,
3. Some noninsulated pressurized tank cars must have a portion of their bodies painted white,
4. Tank cars vary in size, shape, and color,
5. The shape of the tank car does not, necessarily, indicate its specific hazardous contents,
6. The colors of tank cars may vary,
7. It is usually possible to recognize a pressurized tank car from a safe distance.

7. Marine Bulk Containers

The bulk shipping container of marine vessels does not usually provide the recognition characteristics for identifying hazardous materials in the manner of tank cars and trucks. Therefore, it is more difficult to determine whether or not bulk quantities of hazardous materials are being carried. That is because, by and large, marine vessels transport a much larger variety of materials.

Water transportation on inland waterways is primarily accomplished by barge; however, hazardous materials are not carried in open hopper barges but in

Fig. 7-14   Several bulk tank containers in a barge.

covered dry cargo barges or in tank barges. An illustration of a tank barge may be seen in Fig. 7-14.

Several barges may be lashed together in a tow and are pushed or pulled, usually by 1 tugboat. Tank barges may carry all kinds of hazardous liquids, including flammable liquids and corrosives. A cylindrical tank-type barge could, also, carry flammable liquids and liquified gases under pressure. A good indication that a barge may be transporting a load of hazardous liquids is the presence of cargo piping on the vessel, as shown in Fig. 7-14. In Fig. 7-15, below, a string of tank barges is shown lashed together.

Although barges are the principal cargo vessels on inland waterways, a number of different types of ships and barges are used in oceanic transportation. Deep-water vessels that are capable of carrying hazardous materials are as follows:

1. Container ships that may easily carry hundreds of bulk shipping containers,
2. Liquified Natural Gas (LNG) carriers,

Fig. 7-15   Lashed barges in a tugboat tow.

3. Ocean-going barges,
4. Tankers that are used to transport petroleum products, caustic soda, liquified compressed gases, and other chemicals.

During the past several decades container ships, railroads, and trucks have combined to form an intermodal transportation system. For example, a bulk container is loaded in some foreign port and is then transported across the ocean to land at 1 or more port facilities. At the seaport the container is loaded onto a railroad train to travel to an inland freight terminal. After the container arrives at the terminal, it is off-loaded from the train to a waiting truck; thereupon begins its last haulage to some factory or plant that is not served by a railroad siding. Sometimes the bulk container may be placed on a set of bogie wheels at a seaport off-loading site and travel piggyback by train in a somewhat different variation of intermodality, but the end result is the same. The loaded container is handled several times by various modes of transportation equipment.

Since there are many variations in the types of transportation and constant innovation is taking place in industry, it is necessary that the ER team personnel should be prepared for hazardous incidents in all of the various modes. The innovative as well as the established modes of transportation may vary from 1 part of the country to another. For that reason, ER personnel should strive to increase their knowledge of the various kinds of transportation within their communities and the mutual aid districts where they live.

Both industrial and transportation personnel are willing to cooperate with any community safety effort. They should be able to provide ER team personnel with information and training about safety and emergency responses to apply when incidents arise involving their hazardous materials.

Some of the information that could be provided comes under the following headings:

1. Emergency operations of transportation modes,
2. Hazards of certain materials,
3. Innovations in methods of transportation,
4. Safety features of cargo containers,
5. Types of cargo containers,
6. Types of hazardous materials being transported.

C. Labels, Placards, and Identification Numbers

In addition to the characteristics of the shipping containers, a more positive means for identifying the presence of hazardous materials in an emergency response situation is the DOT identifying label, placard, and identification number.

1. Labels

Labels are 4-inch (10 cm) squares set on their points, or diamond-shaped. (see Chart 7-1.) They are required to be placed on individual containers of hazardous

Chart 7-1.

DOT Hazardous Materials Warning Labels

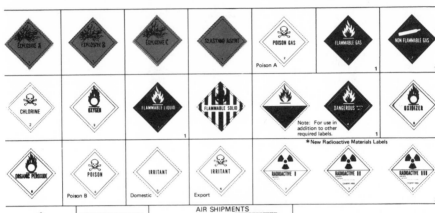

AIR SHIPMENTS

Etiological Agents

Magnetized Material

Cargo Aircraft Only

Notes: 1. Symbol and inscription *MUST be black or white.*

2. *Required as of July 1, 1983.*

General Guidelines on Use of Labels

1. The Hazardous Materials Tables, Sec. 171.101 and 172.102, identify the proper label(s) for the hazardous materials listed.

2. Any person who offers a hazardous material for shipment *must label* the package, if required. [Sec. 172.400(a)]

3. Labels *may* be affixed to packages (even though not required by the regulations) provided *each* label represents a hazard of the material in the package. [Sec. 172.401]

4. Label(s), when required, *must* be printed on or affixed to the surface of the package near the proper shipping name. [Sec. 172.406(a)]

5. When two or more different labels are required, display them next to each other. [Sec. 172.406(c)]

6. When two or more packages containing compatible hazardous materials are packaged within the same overpack, the outside container *must* be labeled as required for each class of material contained therein. [Sec. 172.404(b)]

7. Material classed as an **Explosive A, Poison A,** or **Radioactive Material** also meeting the definition of another hazard class *must* be labeled for *each* class. [Sec. 172.402(a)]

8. Material classed as an **Oxidizer, Corrosive, Flammable Solid,** or **Flammable Liquid** that also meets the definition of a Poison B *must* be labeled POISON, in addition to the hazard class label. [Sec. 172.402(a)(3) and (5)]

9. Material classed as a **Flammable Solid** that also meets the definition of a water-reactive material *must* be labeled with FLAMMABLE SOLID and DANGEROUS WHEN WET labels. [Sec. 172.402(a)(4)]

10. Material classed as a **Poison B, Flammable Liquid, Flammable Solid,** or **Oxidizer** that also meets the definition of a Corrosive material *must* be labeled CORROSIVE in addition to the class label. [Sec. 172.402(a)(6) through (9)]

This Chart does not include all of the labeling requirements. For details, refer to the Code of Federal Regulations, Title 49, Part 172, Sec. 172.400 through 172.448.

U.S. Department of Transportation

Research and Special Programs Administration
Materials Transportation Bureau
Office of Operations and Enforcement
Washington, D.C. 20590

Chart 7   June 1981

Chart 7-2.

DOT Hazardous Materials Warning Placards

*Numbers in each square (illustration numbers) refer to TABLES 1 and 2.

1	2	3	4 POISON GAS	5 FLAMMABLE GAS	6 NON-FLAMMABLE GAS

7 CHLORINE	8 OXYGEN	9 FLAMMABLE	10 COMBUSTIBLE	11 FLAMMABLE SOLID	12 FLAMMABLE SOLID

13 OXIDIZER	14 ORGANIC PEROXIDE	15 POISON	16 RADIOACTIVE	17 CORROSIVE	18 DANGEROUS

Highway Shipments

RADIOACTIVE   GASOLINE   FUEL OIL

Cargo Tanks and Portable Tanks

RADIOACTIVE MATERIAL PLACARD (Square Background): Use on "large quantity" shipments of radioactive materials requiring *special routing.* NOTE: Required for use Feb. 1, 1982.

New Identification Numbers[1]

 1090   1541

ORANGE PANEL[2]

[1]**IDENTIFICATION NUMBER**—The four-digit ID number is found in the Hazardous Materials Table, Sec. 172.101, Column 3A and the Optional Hazardous Materials Table Sec. 172.102 Column 4. They are used on orange panels, certain placards and on shipping papers.

PLACARD—When ID numbers are used on placards, the ORANGE PANEL is not required.

[2]**ORANGE PANEL**—When ID numbers are used on ORANGE PANELS, the appropriate placard is also required.

NOTE: As of Nov. 1, 1981, ID numbers are required for use on cargo tanks, tank cars and portable tanks transporting hazardous materials. For details, see Sec. 172.332 through 172.338 and Sec. 172.519.

Rail Placards

EXPLOSIVES A   EMPTY FLAMMABLE

TABLE 1				

Hazard Classes .... *No.
Class A explosives 1
Class B explosives 2
Poison A 4
Flammable solid (DANGEROUS
  WHEN WET label only) 12
Radioactive material (YELLOW III
  label) 16
Radioactive material:
  Uranium hexafluoride, fissile (con-
    taining more than 0.7% U235 ..... 16 & 17
  Uranium hexafluoride, low-specific
    activity (containing 0.7% or less
    U235 .... 16 & 17

Guidelines

• Placard motor vehicles, freight containers, and rail cars containing *any quantity* of hazardous materials listed in TABLE 1.

• Placard motor vehicles and freight containers containing 1,000 pounds or more gross weight of hazardous materials classes listed in TABLE 2.

• Placard *any quantity* of hazardous materials classes listed in TABLES 1 and 2 when offered for transportation by air or water.

• Placard rail cars containing *any quantity* of hazardous materials classes listed in TABLE 2 except when less than 1,000 pounds gross weight of hazardous materials are transported in TOFC (Trailer on Flat Car) or COFC (Container on Flat Car) service.

TABLE 2				

Hazard Classes .... *No.
Class C explosives 18
Blasting agent 3
Nonflammable gas 6
Nonflammable gas (Chlorine) 7
Nonflammable gas (Fluorine) 15
Nonflammable gas
  (Oxygen, pressurized liquid) 8
Flammable gas 5
Combustible liquid 10
Flammable liquid 9
Flammable solid 11
Oxidizer 13
Organic peroxide 14
Poison B 15
Corrosive material 17
Irritating material 18

This Chart does not include all the placarding requirements. For details, refer to the Code of Federal Regulations, Title 49, Part 172, Sec. 172.500 through 172.558.

U.S. Department of Transportation
Research and Special Programs Administration
Materials Transportation Bureau
Office of Operations and Enforcement
Washington, D.C. 20590

Chart 7   June 1981

materials where each container is less than 640 cubic feet in size. For example, labels are required for metal or fiberboard drums, fiberboard boxes, plastic pails, and small, portable tanks, to name only a few. (See Fig. 7-1).

2. Placards

The placard size is a square 103/4 inches (27 cm) on a side set on its point similar to a label. (See Chart 7-2.) Placards are required to be placed on transporting vehicles, such as trucks, tank cars, or freight containers of 640 cubic feet or larger capacity.

As indicated previously, many hazardous materials have more than 1 hazard. That situation becomes even more dangerous when 2 or more hazardous materials are shipped together. Chart 7-3 summarizes the possible hazards indicated by a placard.

The labeling and placarding system developed for use within the United States is regulated by the DOT in 49 CFR. Those regulations require labels on shipping containers and placards on transportation vehicles containing specific, regulated hazardous materials. While Charts 7-1 and 7-2 show some of the labels and placards, a complete listing may be obtained in 49 CFR, Sec. 172.400 through 172.448 for labels and Sec. 172.500 through 172.558 for placards.

The label and placard requirements that have been described apply to the commercial transportation of hazardous materials interstate and within the borders of states. However, if the state in which you reside has not adopted the DOT regulations, then it is quite possible that some local carriers or a captive fleet of some business concern may be transporting hazardous materials without the use of placards.

3. Identification Numbers

In the past few years changes to 49 CFR have promulgated the use of a 4-digit number to assist in the identification of hazardous materials and to help the ER response team in identifying hazardous materials in the event of an incident. Since July 1, 1981, the 4-digit identification number must be placed on shipping papers. After January 1, 1982, it has to be visible on all portable tanks, over-the-road cargo tanks, tankers, and railroad tank cars. This identification number supplements the labels and placards because it makes the specific hazardous material known, and thus the multiple hazard factor becomes apparent to the ER team personnel, especially in handling an emergency response situation where every minute counts.

The 4-digit identification number is required on shipping papers and also on or near placards on bulk containers. Labels and placards are color coded, display a symbol and a 4-digit identification number on a placard. They may display a U.N. hazardous class number and may contain the name of the hazard class. The identification number may appear on the placard itself or on an orange panel placed adjacent to it. Labels and placards are similar in 3 ways: in color coding, in the display of the United Nations hazardous class number, and in the display of a symbol.

When required by DOT regulations, container labels are placed or printed near the name of the contents. Whenever labels cannot be placed directly on the container, it is adequate to place the label on a card or tag and attach it to the container or package.

Chart 7-3.   Possible Hazards Indicated By a Placard*

Placard	Hazardous Materials That May Be Present
BLASTING AGENT	1,000 pounds or more—Blasting agent(s)
CHLORINE	More than 110 gallons—Chlorine
COMBUSTIBLE	Over 110 gallons—a combustible liquid
CORROSIVE	1000 pounds or more—a corrosive material
DANGEROUS	1,000 pounds or more—materials with Explosive C labels 1,000 pounds or more—a *combination* of materials 1,000 pounds or more—an irritating material
EXPLOSIVE A	ANY QUANTITY—Explosive A ANY QUANTITY—Explosives A and B loaded together
EXPLOSIVE B	ANY QUANTITY—Explosive B
FLAMMABLE	1,000 pounds or more—flammable liquid 1,000 pounds or more—flammable solid
FLAMMABLE GAS	1,000 pounds or more—a flammable gas
FLAMMABLE SOLID	1,000 pounds or more—flammable solid
FLAMMABLE SOLID W	ANY QUANTITY—flammable solid which is dangerous WHEN WET
NONFLAMMABLE GAS	1,000 pounds or more—a nonflammable gas 110 gallons or *less*—Chlorine 1000 pounds or more—gaseous Oxygen
ORGANIC PEROXIDE	1,000 pounds or more—Organic Perioxide
OXIDIZER	1,000 pounds or more—Oxidizer
OXYGEN	1,000 pounds or more—liquefied Oxygen
POISON	1,000 pounds or more—class B poison OR Fluorine
POISON GAS	ANY QUANTITY—poison gas
RADIOACTIVE	ANY QUANTITY—Radioactive material with a radioactive class III label

*Note: (1) The above chart provides a general overview useful for the emergency response person. It is not intended to provide the precise information needed to comply with DOT regulations.

(2) Although hazardous material SHOULD be packaged, labeled, placarded, and shipped in accordance with the regulations . . . frequently, shipments are improperly placarded, labeled and/or marked. Also, shipping papers frequently contain errors. Expect the unexpected.

Chart from DOT's "Emergency Response to Hazardous Materials in Transportation."

As an example, although a metal drum would have a label placed or printed near the name of the contents, a gas cylinder would have its label attached on a tag wired around the neck of the cylinder.

The number that is printed in the lower corner of the label or placard corresponds to the United Nations hazard class, or IMCO code. IMCO is the abbreviation for the Intergovernmental Maritime Consultative Organization. International shipments of hazardous materials are subject to the requirements of the IMCO code. It contains both hazard class and division designations for materials. The abbreviations IMCO and U.N. are used interchangeably when one is referring to those classes and divisions.

The difference between the IMCO or U.N. system and the DOT hazard class system lies in the designations given to the various hazard classes. For example, instead of a name such as flammable liquid for a class of materials, IMCO (U.N.) assigns a class and a division number. U.N. class and division numbers may be printed at the bottom of placards. See Table 7-1, for a listing of IMCO (U.N.) classes and divisions including the basic definitions.

4. Labels and the Hazard Classes

In order to familiarize the reader with the accepted label requirements and to illustrate how the label requirements correspond to their usage, an abbreviated chart has been included in this chapter. Therefore, for a better understanding of the label requirements, refer to Chart 7-1 whenever the particular hazard class is mentioned in the text.

a. Explosives. All of the labels for Explosives on Chart 7-1 are colored "orange." As you can see, there is a label for each of the subdivisions of the hazard class of Explosives: Class A, Class B, Class C, and Blasting Agents. Also, please note that the IMCO (U.N.) Code of 1 appears in the lower corner. This numeral "1" is explained in Table 7-1 of the United Nations, IMCO Code on page 55.

The significant hazards for any orange color-coded DOT label would be as follows:

1. an explosion,
2. a shock wave,
3. the scattering of shrapnel-like fragments,
4. the starting of additional fires.

b. Compressed Gases. Compressed Gases have 2 labels, a red label for flammable gases and a green label for nonflammable gases. (See Item (6) for poison gases.)

If the reader is a member of an ER team, he or she would promptly associate the following hazards with red or green warning labels:

1. The materials may be toxic,
2. A BLEVE-like explosion may occur,
3. There is the possibility of frostbite on contact with a cryogenic substance,
4. The material may be highly corrosive,
5. The material may react with other combustibles.

c. Flammable Liquids. The DOT uses the color red to indicate a flammable hazard; therefore, like a Flammable Gas label, the label required for shipping containers of flammable liquids is red. The hazards associated with materials bearing a red flammable liquid label are very similar to the hazardous characteristics of flammable gases. Those hazards are the contamination of water supplies and fire. Combustible liquids, on the other hand, do not require labels, and usually there will not be labels on individual containers of the following liquids: fuel oil, stoddard solvent, and diesel oil.

d. Flammable Solids. Flammable solids sometimes require 1 or 2 labels, depending on their multiple hazards. For example, if the flammable solid is also water

Table 7-1 United Nations (IMCO Code) of Hazardous Classes with Definitions (DOT classes are in parentheses)

Class 1—Explosives

Division 1.1 — Substances and articles that have a mass explosion hazard (Explosive A)
Division 1.2 — Substances and articles that have a projection hazard but not a mass explosion hazard. (Explosive A or B)
Division 1.3 — Substances and articles that have a fire hazard and either a minor blast hazard or a minor projection hazard or both, but not a mass explosion hazard. (Explosive B)
Division 1.4 — Substances and articles that present no significant hazard. (Explosive C)
Division 1.5 — Very insensitive substances. (Blasting Agent)

Class 2—Gases (compressed, liquified or dissolved under pressure)

Division 2.1 — Flammable gases. (Flammable gas)
Division 2.2 — Nonflammable gases. (Nonflammable gas)
Division 2.3 — Poison gases. (Poison A)

Class 3—Flammable liquids

Division 3.1 — Low flash point group (liquids with flash points below 0°F). (Flammable liquid)
Division 3.2 — Intermediate flash point group (liquids with flash points of 0°F or above but less than 73°F). (Flammable liquid)
Division 3.3 — High flash point group (liquids with flash points of 73°F or above up to and including 141°F). (Flammable liquid or combustible liquid)

Class 4—Flammable solids or substances

Division 4.1 — Flammable solids. (Flammable solid)
Division 4.2 — Substances liable to spontaneous combustion. (Flammable solid or, for pyroforic liquids, flammable liquid)
Division 4.3 — Substances emitting flammable gases when wet. (Flammable solid)

Class 5—Oxidizing substances

Division 5.1 — Oxidizing substances or agents. (Oxidizer)
Division 5.2 — Organic peroxides. (Organic peroxide)

Class 6—Poisonous and infectious substances

Division 6.1 — Poisonous substances. (Poison B)
Division 6.2 — Infectious substances. (Etiologic agent)

Class 7—Radioactive substances. (Radioactive material)

Class 8—Corrosives. (Corrosive material)

Class 9—Miscellaneous dangerous substances. (Other regulated material)

reactive, then the "Dangerous-when-Wet" label must be used in addition to the flammable solid label.

If a flammable solid is not water-reactive, the label for the container is printed with alternating, vertical red and white stripes. If the material is water-reactive, then an additional blue-colored label is required. These labels are illustrated in Chart 7-1.

The blue label indicates 1 multiple hazard of the flammable solid class, that is, water-reactivity. Other multiple hazards of flammable solids are as follows:

1. They ignite easily and will burn with explosive force,
2. They will react with air,
3. They are susceptible to spontaneous combustion,
4. When in contact with water or air, they are toxic or corrosive, or they are capable of producing toxic or corrosive compounds.

e. Oxidizers and Organic Peroxides.   As shown in Chart 7-1, oxidizer and organic peroxide materials require yellow labels. That label indicates that those substances contribute significantly to the intensity of a fire.

The hazards associated with oxidizers are as follows:

1. They react with organic materials,
2. They form easily ignited mixtures,
3. They supply oxygen to increase the intensity of fires,
4. They are explosively sensitive to heat, shock, and friction.

Organic Peroxides have other multiple hazards worth noting:

1. They are highly flammable,
2. They are capable of releasing toxic, gaseous products,
3. Their reactivity inhibitors may be released by fire,
4. They are, like oxidizers, explosively sensitive to heat, shock, and friction.

f. Poisons.   Poisonous or Irritating substances may require 1 of 3 labels, as shown on Chart 7-1. Class A poisons require a poison gas label, Class B poisons require a poison label, and any material classed as irritating require labels having a white colored background.

ER team personnel should be warned by the white labels that the poisonous substances could enter the body by any of 3 means: inhalation, by breathing; ingestion, through the mouth; and by absorption, through skin contact.

g. Etiologic Agents.   Etiologic agents are biomedical products that carry a rectangular, red and white colored label rather than a square label on point.

The red and white warning label required on etiologic agent packages and containers is a necessary precaution, since those products may present hazards from inhalation, ingestion, or absorption of microorganisms that can cause human disease. In addition, the ER team confronted with such a label would be advised thereby of environmental contamination.

h. Radioactive Materials (RAMs).   Radioactive materials require 1 of 3 types of warning labels depending on the level of radiation emitted through the packaging.

Radioactive White-I identifies the lowest emission levels of radioactive substances; this is an all-white label.

Radioactive Yellow-II, a label that is yellow over white and indicates a midlevel of emitted radiation.

Radioactive Yellow-III is also yellow over white but indicates the highest permissible level of radiation emission.

i. Corrosive Materials. Corrosive materials are required to have a white and black label, as shown in Chart 7-1. That symbol warns the ER team members that the materials can be corrosive to both metals and to human tissue. It also warns that there are multiple hazards that exist for this substance:

1. It indicates that the substance could splatter upon contact with water;
2. The substance is toxic;
3. The substance can destroy human tissue by contact or inhalation;
4. It is thermally unstable;
5. The substance has oxidizing capability.

In a summation of this section, an alert observer may see the DOT warning labels on merchandise that is being transported. It is possible also that those labels may appear on shipping containers that are found in warehouses where goods are stored or in processing plants where products are being manufactured. Knowing the origin or destinations of hazardous materials may often help in determining what hazards are involved, what transportation routes may be used, and what packaging and modes of shipping must be used.

As we have indicated before, multiple hazards of certain materials present added problems to emergency response team personnel. The DOT now requires that certain hazardous materials must bear a label for each hazard classification for the material. That is beneficial to the ER team in response to an emergency situation; however, not at all times may those added requirements prove helpful. For example, a serious hazard may not be serious enough to fit the definition of a particular hazard class. Also, the ER team may not be able to see any of the labels on the containers because of smoke, fire, or some other problem. In addition, all of the labels may be damaged and may not be decipherable, thus increasing the multiple hazards of the situation.

5. Placards and Hazard Classes

When hazardous materials are transported in large quantities, the DOT requires placards for the bulk cargo tanks and the vehicles used to transport the materials. Since placards are placed on large bulk containers and large vehicles, they are similar to the warning labels used on the smaller containers of hazardous materials except that they are much larger in size in order to be fairly visible at a greater distance.

Placards have the same shape and color as the label, and they use the same hazard symbol; however, the placard is 10¾ inches, square, set on point, i.e., a larger, diamond shape. Placards must be placed on the front, rear, and both sides of vehicles. Those on trucks may be placed on the front of the cargo-carrying unit or the tank or trailer, or they may be placed on the cab of the tractor unit.

DOT placards generally indicate the major hazardous characteristic of a specific material by using the name of the hazard class. On the other hand, since the recent addition of the 4-digit identification numbers, the name of the hazard class may not necessarily appear on the placard. DOT labels, nevertheless, require the name of the hazard class. That particular and the size of the warning placards are the only differences between placards and labels except for the U.N. class number.

If a placard has the 4-digit identification number on it, then it must also carry the U.N. class number at the bottom of the diamond in 1¾ inch numerals. The

United Nations System uses both class numbers and class names, as indicated in Table 7-1.

A 4-digit identification number is not required for every hazardous material being transported. However, if a 4-digit identification number is required and is not printed on the placard itself, then it must appear in 4-inch (10 cm) high, black letters on an orange panel 6¼ inches (16 cm) high by 15¾ inches (40 cm) wide, next to the proper placard for the hazard class. See Figure 7-16, below.

The identification number that has been assigned to specific hazardous materials is intended to be of assistance during an emergency response situation. The ER team personnel should be able to read the number of the hazardous material in such a situation and be able to compare the number with a list of identification numbers and their respective hazardous materials. Therefore, in addition to the Hazardous Materials Table in Section 172.101 of 49 CFR, these numbers are listed in at least 2 other source books: "DOT Hazardous Materials—1980 Emergency Response Guidebook" and "1984 Emergency Response Guidebook, Guidebook for Hazardous Materials Incidents" (DOT P 5800.3).

If either of the above guidebooks are not available to the ER team, then one may obtain a listing by calling the chemical emergency response network, or CHEMTREC, whose telephone number is 1-800-424-9300.

We shall discuss how to obtain technical assistance in a hazardous materials incident in a later chapter.

a. Explosives.   The Class A explosive, the Class B explosive, and blasting agent placards are similar to the labels required for these materials except for size, as has been indicated. The color of the placards is orange. The Class C explosive must also be placarded with the red, dangerous placard. All of the placards are illustrated in Chart 7-2.

Explosives and blasting agents are included among those hazardous materials that do not have 4-digit identification numbers. Therefore, the placards for explosives and blasting agents will have the name of the hazard class but no number.

b. Compressed Gases.   Placards for compressed gases are the same as labels, except for their larger size. A flammable gas has a red placard, while a nonflammable Gas has a green placard. See Chart 7-2.

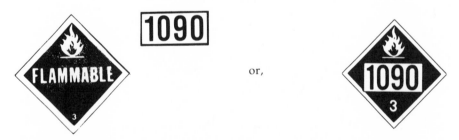

Fig. 7-16   In the above illustration either the placard plus identification number or the identification number in the placard is correct.

If there should be a 4-digit identification number on the placard instead of the name of the hazard class, viz. flammable gas or nonflammable gas, then the U.N. hazard class number for gases, 2, would be placed on the bottom of the placard. As we have indicated in the IMCO Code (U.N.) Table 7-1, the number "2" at the bottom of the placard indicates that the hazardous material is a compressed gas.

c. Flammable and Combustible Liquids. While a label is required only for a flammable liquid in "smaller containers," when a bulk, cargo container is involved, because of the quantities and the serious multiple hazard factor both flammable and combustible liquids are required to have placards.

When the hazard class names are used on the flammable and combustible placards, both placards are completely red. If the identification number should replace the hazard class name combustible liquid, then the bottom portion of the placard must be white with the U.N. class number 3. That separates a combustible liquid from a flammable liquid on a placard. See Chart 7-2, page 51 and the illustration Fig. 7-17, below.

Besides the exceptions noted above, there are 2 substitutions that may be used in these placards for highway transportation. For bulk cargo tanks of gasoline, the word *flammable* may be replaced by the word *gasoline*. For a fuel oil shipment, the word *combustible* may be replaced by the words *Fuel Oil*. That substitution of the hazard class name is permissible since both gasoline and fuel oil are common materials whose hazards are well known to ER personnel.

d. Flammable Solids. Flammable Solids have the use of 3 placards. The flammable placard may be used in place of the flammable solid placard, except for transportation by water. The "dangerous-when-wet" flammable solid placard is required for any flammable solid that is also water-reactive. The symbol **W** notifies ER team personnel to avoid using water on the material.

When the identification number is printed on a flammable solid placard instead of the hazard class name, then the U.N. class number must be used also. Should the solid red flammable placard be used for a flammable solid, then the U.N.

Combustible   Flammable

Fig. 7-17   When the hazard class name is replaced by the identification number, the lower portion of the combustible placard is colored white. Note: The U.N. class number 3 at the bottom of each placard.

class number 4 would be necessary to differentiate a flammable solid from either a flammable liquid or a flammable gas.

The two flammable solid placards have alternating, vertical red and white stripes. In the main, any hazardous material that has flammability as 1 of its primary hazards will have the color red on its label or placard. Examples of these hazardous materials are flammable liquids, flammable gases, flammable solids, and combustible liquids.

e. Oxidizers and Organic Peroxides. There is a similar correspondence between labels and placards for oxidizers and organic peroxides, as illustrated in the 2 Charts 7-1 and 7-2. Chart 7-2 also shows the special oxygen placard that is required when one is transporting oxygen that has been liquified by pressure. When the oxygen has been liquified by cooling and is being transported as a cryogenic material the placard is optional. All 3 of these labels, oxygen, oxidizer, and organic peroxide are colored yellow.

If the DOT hazard class name is replaced by the 4-digit identification number on the oxidizer or the organic peroxide placards, then the U.N. class number 5 would appear at the bottom of the placard. The number 5 and the yellow color on a placard, therefore, would indicate that the material is either an oxidizer or an organic peroxide.

As the reader will recall from Table 7-1, the U.N. system (the IMCO Code) uses both class and division numbers. Most placards use only the U.N. class number. Therefore, when a 4-digit identification number is used for oxidizer and organic peroxide placards, then in order to differentiate between the 2 hazardous materials the U.N. "division number" may appear next to the U.N. class number and we have the following:

5.1 Oxidizer; and,
5.2 Organic Peroxide.

f. Poisons. Placards for Class A and B poisons are the same labels for those chemicals, except for the larger size of the placard, with the exception that bulk shipments of irritating materials require the "dangerous" placard. Also, Class B poisons and irritating materials that carry the 4-digit identification number must also have the U.N. class number 6 at the bottom of the placard.

One of the 3 hazard classifications that is not allowed to display a 4-digit identification number on the placard itself is a Class A poison (poison gas). Radioactive materials and explosives are the 2 other hazard classes that are restricted in that manner. Explosives do not have assigned identification numbers. For Class A poisons and radioactive materials, any assigned 4-digit number must be on an orange panel placed alongside the placard.

Chlorine is classified as a nonflammable gas, but a special "chlorine" placard having a poison symbol (see Chart 7-2) is used for shipments of chlorine with packagings of more than 110 gallons per unit. The identification number that has been assigned is 1017. It should appear on the poison placard. However, the U.N. class number of poisons is replaced by the class number for a gas, which is 2.

Examples of the special treatment given to chlorine gas bulk shipments are shown in the illustrations below. Also check Chart 7-2, for the appropriate color of the placard.

g. Radioactive Material. It is necessary that all vehicles carrying any number of packages labeled radioactive yellow III and full-load shipments of other

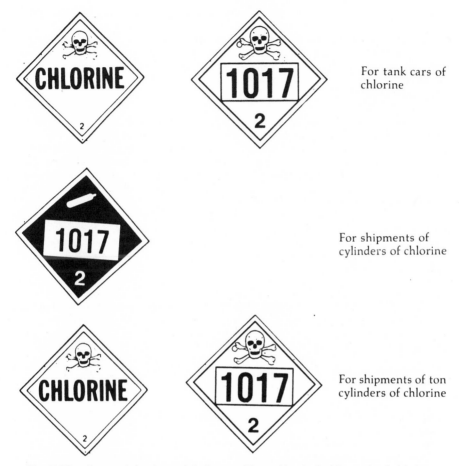

For tank cars of
chlorine

For shipments of
cylinders of chlorine

For shipments of ton
cylinders of chlorine

Fig. 7-18   Appropriate placards to be used in making bulk shipments of chlorine.

low-specific-activity (LSA) materials, such as radioactive I and II, be placarded with the "yellow over white" radioactive placard. For shipments thus placarded, the name of the hazard class is never replaced by a 4-digit ID number.

h. Corrosive Materials.   The placard for corrosive materials is the same as the label, and the background colors are black and white. As an example, 1830 is the 4-digit identification number assigned to the corrosive material called sulfuric acid. Since the U.N. class number for corrosives is 8, then it will be shown in the lower part of the diamond. See Fig. 7-19 on pg. 62.

6. Further Facts on Identification of Hazardous Materials

a. Placard Appropriateness.   Because the subject of hazardous material identification is so critical in the areas of emergency response, it may be seen from the following array of symbols that sometimes more than 1 placard may be

Flammable Solid

Chlorine
(under 110 gal.)

Oxidizer

Irritating Material

Explosive C

Water-Reactive
Flammable Solid

Organic Peroxide

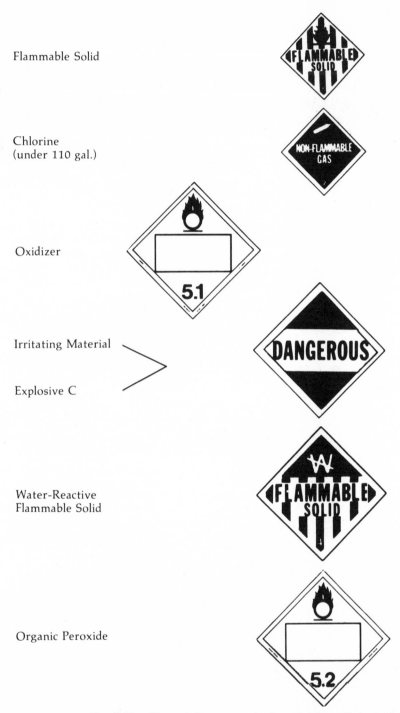

Fig. 7-19   Placards for a cargo tank containing sulfuric acid.

Fig. 7-19   Continued from page 62.

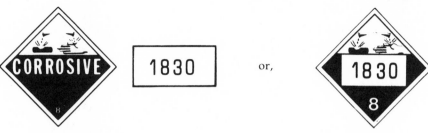

or,

appropriate for a chemical substance. The placards for "irritating material" and for "Explosive C" are similar.

b. Labels of Special Interest.   In addition to the labels that have been described earlier in this section, there are others that provide further information concerning hazardous materials in transportation. (See Fig. 7-20.)

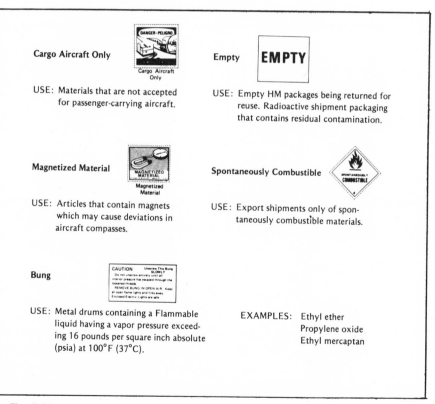

Cargo Aircraft Only

USE: Materials that are not accepted for passenger-carrying aircraft.

Empty

USE: Empty HM packages being returned for reuse. Radioactive shipment packaging that contains residual contamination.

Magnetized Material

USE: Articles that contain magnets which may cause deviations in aircraft compasses.

Spontaneously Combustible

USE: Export shipments only of spontaneously combustible materials.

Bung

USE: Metal drums containing a Flammable liquid having a vapor pressure exceeding 16 pounds per square inch absolute (psia) at 100°F (37°C).

EXAMPLES: Ethyl ether
Propylene oxide
Ethyl mercaptan

Fig. 7-20   Five additional labels that supply information about hazardous materials.

c. Special Placards. Special placards may be found on certain transportation vehicles that will be extremely useful in emergency response situations. A few examples follow. (See Fig. 7-21.)

d. Precautions Pertaining to the Regulations. While the DOT system of labeling, placarding, and classification has attempted to standardize the methodology of providing information concerning hazardous materials, it is not, in its present stage of development, a panacea for emergency response. All of the many Federal regulations only provide guidelines to shippers, carriers, and emergency response team personnel.

The DOT system is of decided advantage in providing information that ER personnel may use in the event of an accident. The system of labeling, placarding, and classification warns of the presence of types of hazardous materials involved in the incident. Yet the following caveat must be applied to certain

Empty

USE: Empty tank cars that have carried flammable materials. Note: those empty tank cars that have carried combustible liquids or have been thoroughly cleaned do not use this placard.

Dome

USE: Tank cars containing a flammable liquid having a vapor pressure exceeding 16 psia at 100°F (37°C).

EXAMPLES: Ethyl ether
Propylene oxide
Ethyl mercaptan

White Background

USE: On a rail car, each EXPLOSIVE A, POISON GAS, POISON GAS-EMPTY placard must be placed against a white square background.

Fumigation

USE: Any shipping container carrying freight that has been fumigated or treated with a poisonous liquid, solid, or gas, and will be shipped by rail.

Dangerous

USE: Explosives C, Irritating materials and mixed loads over 1,000 lbs (453.6 kg) total weight of two or more materials, which would require different placards.

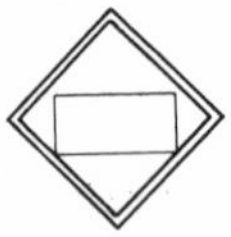

USE: for hazardous materials for which placards are not required, (ex. ORM-A, ORM-B, ORM-C, ORM-E). Identification numbers *may* be displayed on a plain white square-on-point configuration the size of a placard. Note: This is not considered a placard.

Fig. 7-21   Special placards that supply information about hazardous materials.

shipments of hazardous materials, and the reader should be aware of it: It is not always possible to depend upon labels and placards for information.

These examples will serve to illustrate the argument further.

Federal (DOT) regulations specify the use of the DANGEROUS placard for use with irritating materials, Class C EXPLOSIVES, and mixed loading of hazardous materials weighing more than certain amounts. The DANGEROUS placard does not indicate specific information about each of the hazardous materials. In other words, ER team personnel will not know immediately what control measures are to be employed when an incident occurs.

Also, it would be quite possible to have such materials as flammable liquids, flammable solids, and oxidizers in the same shipment, since only a DANGEROUS placard is necessary to conform to the regulations when one is transporting such a mixture of hazardous materials. In a fire emergency the ensuing results could be quite catastrophic, unless the ER crew were prepared for the multiple hazard factors in such a situation.

According to the regulations, many hazardous material shipments weighing less than specified amounts do not require placarding. Therefore, in an emergency, the ER personnel should carefully regard all materials as possible multiple hazards until they are properly identified.

It is not always possible to read placards and labels at the time of an accident or emergency situation. One reason is that the identifying devices may have been damaged or obliterated, or they may not be legible within a safe distance. It may be possible, however, to determine the color of the placards or labels, and that may be an indication of the specific hazards involved. On the other hand, some materials that the DOT does not classify as dangerous may not have either placards or labels, yet their properties may be hazardous enough to cause problems during an accident in transportation.

When substances are identified by class instead of by specific chemical, it suggests that the multiple-hazard problem is not recognized except in shipments where multiple labels are required. The system's limitation may be surmounted if the reader abides by the general rule that most hazardous materials have multiple, hazardous characteristics.

The effectiveness of labeling and placarding depends quite heavily upon human fallibility. Shipments may be improperly labeled or placarded as a result of human error. Moreover, the enforcement of the labeling and placarding regulations is difficult because of the large volume of hazardous materials that are transported and the large number of common carriers that are used in their transportation. It is advisable, therefore, that ER teams should exercise extreme care in all accidents involving tank trucks, tractor-trailers, or vans until their contents are positively identified.

When placards, 4-digit identification numbers, or labels may be seen and are legible when an emergency occurs, they are a good indication of the nature of the hazardous materials involved. Gasoline, fuel oil, chlorine, and oxygen are 4 hazardous materials that may be recognized by a name on a placard. When other chemicals are involved, however, the ER team may not be so fortunate. There may not be any labels or placards, and it will be necessary to rely upon other DOT regulations to provide the essential information. On certain shipping containers, package markings required by the DOT may help to identify the specific hazardous material in the shipment. For example, the markings may include the proper shipping name of the chemical, DOT-E exemption numbers, and the name and address of the shipper and consignee.

As indicated in an earlier chapter, the "proper shipping name" may be the chemical name, such as methanol, or it could be an appropriate common name, such as "refrigerant gas." Also, in addition to proper shipping names, the DOT has assigned 4-digit identification numbers for listed hazardous materials, and they have been referred to quite a few times in this text already. Other examples of proper shipping names are "ethyl ether" and "anhydrous ammonia."

A proper shipping name may be shown as "Flammable Liquid, n.o.s." or "Compressed Gas n.o.s." for certain materials. The abbreviation "n.o.s." means "not otherwise specified." Therefore, while "Flammable Liquid, n.o.s." is a proper shipping name, it is not a specific identifier of a particular chemical.

According to the DOT, there are more than 30 hazardous materials whose proper shipping names must be permanently stenciled on each side of a cargo tanker or tank car. (See Fig. 7-22 below.) The stenciled chemical name is a specific identifier of a hazardous material, inasmuch as the cargo tanker or tank car cannot legally be used to carry another material while that name remains on the sides of that transportation medium.

For the reader's information, a list of stenciled commodity names has been included in Table 7-2, below. This list is subject to frequent change as the DOT adds hazardous materials to the list when it learns about problems of incompatibility between materials. The stenciled commodity name on this list is the "proper shipping name" authorized by the DOT. Current information concerning commodities on this list may be obtained not only from the DOT but also from each state Department of Transportation, the Association of American Railroads (AAR), and the Operations Council of American Trucking Associations.

(1) Package Markings and Container Specification Numbers

When chemicals are packaged for shipment, each package must have its correct identity in the form of a "proper shipping name." Shipping containers must be clearly marked with a *container specification number* preceded by the letters "DOT." The container specification number indicates that the container has been manufactured according to the prescribed DOT specifications for the chemical

Fig. 7-22   A tank car of anhydrous ammonia with the stenciled commodity name on 1 side.

Table 7-2.   Stenciled Commodity Names

The following is a list of hazardous materials whose commodity names are required to be stenciled on the sides of tank cars by either the Department of Transportation or the Association of American Railroads. This list of commodities changes from time to time. Additions may be obtained by checking with AAR.

Acrolein
Anhydrous Ammonia
Bromine
Butadiene
Chlorine
Chloroprene (When transported in DOT 115A specification tank car.)
Difluoroethane*
Difluoromonochloromethane*
Dimethylamine, Anhydrous
Dimethyl Ether (Transported only in ton cylinders)
Ethylene Oxide
Formic Acid
Fused Potassium Nitrate and Sodium Nitrate
Hydrocyanic Acid
Hydrofluoric Acid
Hydrogen Chloride (By exemption from DOT)
Hydrogen Fluoride
Hydrogen Peroxide
Hydrogen Sulfide
Liquified Hydrogen
Liquified Hydrocarbon Gas   (May also be stenciled Propane, Butane, Propylene,
Liquified Petroleum Gas   Ethylene)
Methyl Acetylene Propadiene Stabilized
Methyl Chloride
Methyl Mercaptan
Methyl Chloride—Methylene Chloride Mixture
Monomethylamine, Anhydrous
Motor Fuel Antiknock Compound or Antiknock Compound
Nitric Acid
Nitrogen Tetroxide
Nitrogen Tetroxide—Nitric Oxide Mixture
Phosphorus
Sulfur Trioxide
Trifluorochloroethylene*
Trimethylamine, Anhydrous
Vinyl Chloride
Vinyl Fluoride Inhibited
Vinyl Methyl Ether Inhibited

*May be stenciled DISPERSANT GAS or REFRIGERANT GAS in lieu of name. Only *flammable* refrigerant or dispersant gases are stenciled."

substance; however, a container specification number will not be a specific identifier of a particular chemical. See Fig. 7-23, below.

It is also well to note that railroad tank cars, cargo tank trucks, tankers, and individual containers may sometimes carry hazardous chemicals in nonspecification containers. Under certain conditions, the DOT may allow an exemption. This "DOT-E" exemption number will be located on the side of the tank car, as shown in Fig. 7-24, below, and on the sides of cargo tank trucks or individual

Fig. 7-23 DOT container specification number on the side of a railroad tank car.

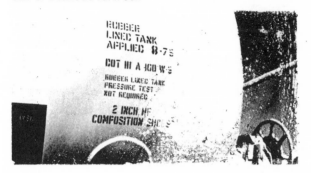

containers. Just as we have noted in the container specification number in the above paragraph, the DOT-E exemption number is not a specific identifier. However, it will help to name the shipper, who can identify the contents. See para. 6(2), below, on "Exemptions."

Package markings on individual containers should also include the name and address of the shipper or the consignee. Since the consignee is the individual or the company to which the merchandise is being shipped and the shipper is the one who has consigned the materials, it is probable that one or the other or both are capable of identifying the contents of the containers. Thus the hazardous characteristics of the substances involved in any incident that may occur will be revealed.

After July 1, 1983, individual packages containing hazardous materials must be marked with the DOT-assigned 4-digit ID number as well as with the "proper" shipping and the required label name. The 4-digit number has been designed to provide positive identification of the specific material being transported and to provide information to the ER team during an emergency.

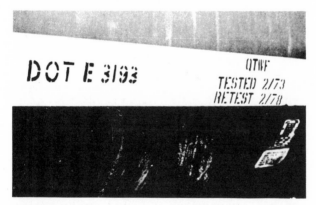

Fig. 7-24 DOT-E exemption number shown on the side of a railroad tank car.

When individual packages of hazardous materials are shipped, the ID number will be on the package and not on the label. If bulk containers are used in transporting hazardous materials, the ID number is to be placed on the placard or on an orange-colored panel.

(2) Exemptions and Exceptions

Section 107 of the Hazardous Materials Transportation Act (49 U.S.C. 1806) gives the Secretary of the Department of Transportation the authority to grant exemptions from the provisions of the act and regulations enacted under the authority of the act provided that certain conditions are met. Part 107 of 49 CFR starting at section 107.101 (49 CFR 107.101) sets forth the procedure to be used in applying for an exemption, the administrative procedure that will be used to process the application, and other items of concern to those persons seeking relief through the exemption process.

It must be noted from the outset that although there are certain similarities between "exemptions" and "exceptions," they are 2 different terms. An exception, like an exemption, provides relief from the hazardous materials regulations provided that certain conditions are complied with. Thereafter, however, the similarity ceases. Exceptions are specifically listed within 49 CFR and apply to all in accordance with the conditions stated. Exemptions, for the most part, however, are not specifically set forth in 49 CFR. They apply only to the specific person or class of persons to whom they are issued or who are parties to the exemptions, and they are issued only by the Materials Transportation Bureau, U.S. Dept. of Transportation, 400 7th St., S.W., Washington, D.C. 20590, in writing in response to a written application submitted in accordance with Subpart B of Part 107.

The process for obtaining a renewal of an exemption is basically the same as for applying for the original exemption. It should be pointed out, however, that the renewal application should be submitted at least 60 days before the expiration of the current exemption to permit timely consideration. In addition, if the application is submitted at least 60 days before the expiration of the current exemption, it will remain in effect until a final determination is made on the application even though the current exemption has expired.

Generally, applications for exemptions or renewals will be made available to the public for inspection. If confidential treatment is desired for any portion of the application, it must be requested at the time the documents are submitted along with a second copy of the documents from which the confidential material has been deleted. Moreover, the rest of the conditions in Sec. 107.5(a) must also be complied with.

Generally, only the holder of, or a party to, an exemption may operate under its authority. Section 173.22a, however, provides that if the exemption authorizes the use of a packaging by "any person" or a "class of persons" other than or in addition to the holder of the exemption, the packaging may be used by "any person" or a member of the class specified provided that they otherwise meet the requirements of 173.22a(b). A "class of persons" could be members of an association, customers of "X" company, etc. For example, a member of "X" association may use a packaging authorized in an exemption issued to the "ABC company" for use by members of "X" association, so long as he or she otherwise meets the requirements of 173.22a(b).

(3) Shipping Papers

Since it has been indicated in an earlier portion of the text that the required package markings may not always be legible or accessible to the ER team, an additional method of identifying the specific identity of the cargo is through its shipping papers. They are a necessary part of transportation regardless of the kind of equipment used. The law requires that all shipping papers used in the transportation of hazardous materials contain certain information.

Since there is no one specific U.S. government form required to be used in the shipment of hazardous materials, a shipper may use whatever form or shipping document appropriately fits his or her operation; what is important is the entries on the document.

a. The shipping paper entries must contain the following information:
   (1) the basic description of the hazardous material being shipped as required by 172.202(a), (b), and (c) plus
   (2) any applicable additional description requirements contained in 172.203,
   (3) which must be prepared and placed on the shipping paper in accordance with the requirements of 172.201(a) and (b).
b. The shipping paper must bear a signed shipper's certification prepared in accordance with 172.204(a), (b), and (d). (See Chapter 4, A.3.)

It should also be noted that in general hazardous material entries on shipping papers are not required for ORM—A, B, C, D, materials. The words "in general" are used because shipments of ORM, A, B, C subject to regulation by the air or water mode of transportation in 172.101 must have specific shipping paper entries when they are shipped by that mode. In addition, shipments of ORM-D by air must also have specific shipping paper entries. Shipments of hazardous substances, hazardous wastes, and some ORM-Es are also subject to specific shipping paper requirements for most modes of transportation.

The totally required hazardous materials description can be broken into 2 parts as follows:

Part 1—The basic description containing:
   a. The proper shipping name as listed in 172.101—Column #2. (172.202(a)(1))
   b. The prescribed hazard class as prescribed in 172.101—Column #3. (172.202(a)(2))
   c. The identification number as prescribed in 172.101—Column #3A. (172.202(a)(3))
   d. The total quantity of the material covered by the description entered either before or after or both before and after the required description. (172.202(a)(4) and (b)).

PLUS

Part 2—Additional description requirements—if applicable (172.203). It should be noted that in some cases there will be no additional description requirements. For those shipments where *no* additional description requirements are required by 172.203, the shipping description is complete after entering the basic description by 172.202(a).

In those instances where an additional description requirement *is* contained within 172.203, the required description is not complete unless the information there is entered as part of the required description in the manner prescribed. (See Fig. 7-25 for an example of a shipping paper.)

Fig. 7-25   An example of a shipping paper. The format of the document may vary from shipper to shipper; however, all of the required information must be provided.

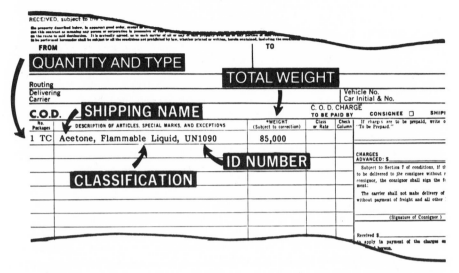

In addition to the required description of the hazardous material being shipped, the shipper must certify on the shipping paper that the materials as offered for transportation are properly classified, described, packaged, marked and labeled, and in proper condition for transportation according to applicable DOT regulations.

Except for air shipments, the certification must be as stated in 172.204(a). A shipper by air, even though a truck may be involved for a portion of the transportation, has the option of using the certification given in 172.204(a) or 172.204(c)(1). Regardless of which of the basic certifications is used for air shipments, either 172.204(a) or (c)(1), air shippers should be aware of the fact that they do not have a complete air certification unless they comply with all of the provisions of 172.204(c).

Particular emphasis should be placed on 172.204(c)(3), which requires the following statement: "This shipment is within the limitations prescribed for passenger/cargo only aircraft (delete not applicable)." At a minimum, an air certification requires: Either the basic certification contained in 172.204(a) or 172.204(c)(1) plus 172.204(c)(3) plus other applicable provisions of 172.204(c).

As a word of caution, no carrier, regardless of the transportation mode, may accept a shipment of hazardous materials unless a properly prepared shipping paper is offered with the shipment. Each mode of transportation has specific carrier requirements regarding information that must be shown on a document accompanying the shipment during transportation. Also, private carriers may not transport hazardous materials unless a properly prepared shipping paper accompanies the shipment.

In emergency response situations, the location of the shipping papers may vary with each mode of transportation. For over-the-road or highway transportation of hazardous materials, the driver of the truck should have in the cab of the truck some sort of driver packet containing the shipping papers. Usually, the papers will be in a holder on the inside of the door, on the driver's seat, or within his reach. If the driver leaves the vehicle, the papers should still be left in the cab.

In like manner, the conductor in the caboose of a railroad train carries a waybill. Either the conductor or the train's engineer may possess a "consist" or "wheel report," which lists the cars of the train in their order or a list of those cars in the train carrying hazardous materials together with their location in the train. If the ER team is looking for the train waybill, it should ordinarily search the caboose first.

On the consist some railroads list the cars beginning from the locomotive (engine) backward and others list them in the reverse order. The consist must indicate which cars are placarded, which cars contain hazardous materials, and each car's point of origin and destination. At the time of a derailment it is logical to look for, and at, the consist. Then it is important to find the waybills for each of the cars that have been derailed.

For domestic air shipments, the shipper prepares an "air bill." One copy is retained by the shipping agent, and another is attached to each package or container in the shipment. If any hazardous materials are being shipped, the shipper's hazardous materials declaration will also be attached to the shipment. A cargo manifest is also provided to the plane's captain or pilot by the airline, and it contains a listing of all hazardous materials being transported; therefore, it should be possible to obtain information on the shipment of all hazardous materials on board the aircraft.

For an ER team to identify the materials in a shipment in a marine terminal, it would be necessary to examine the bill of lading from the terminal shipping office. Once a cargo is on board the ship, however, a dangerous cargo manifest is prepared and given to the ship's master or the mate. It is customarily maintained in a special holder on the bridge of the vessel. The dangerous cargo manifest indicates the class of hazardous materials, the technical name of an n.o.s. (not otherwise specified) category, the quantity and type of containers, and the location of the hazardous materials on board.

For unmanned barges in a tow, the shipping papers or cargo manifest should be in the safekeeping of the tugboat operator or captain in the wheelhouse. Where manned barges are in tow, the shipping papers will normally be on the barge itself in the hands of the barge operator.

If shipping papers are not available, it may be possible to identify hazardous materials through members of the crew of the ship, the engineer or conductor of the train, the driver of the truck, or the pilot, tugboat operator, and so forth.

(4) Shippers' Information

In the unfortunate event that the involved personnel are killed, injured, or have been removed or have left the scene of the accident, there are other avenues to obtain the necessary information required by the ER team. While quickness of action is necessary in dealing with an incident, sometimes the only way to obtain clues about the contents of a shipment is to seek out the consignor or shipper.

The shipper can usually be discovered through the identifying names or numbers on the truck, airplane, or ship. A truck can readily be identified by its license number; however, it may be more advantageous to use the unit number assigned by the motor truck line, as shown on the end of the tanker in Fig. 7-26, below.

If an overseas shipping container is involved in an incident or if a piggybacked trailer is included, then the primary identification is a computer number. To

Fig. 7-26   Motor carrier's designation 515-4280.

identify the shipper and the contents, the computer number must be used. (See Fig. 7-27, below.)

Railroad boxcars, hopper cars, and flat cars, which are used for piggyback shipments, are usually owned by the individual railroad lines. Names and numbers serve to identify the shipper and the contents of a car. The cars are marked on both sides with serial numbers that correspond to the numbers used in the train consist. (See Fig. 7-28, below.)

Railroad tank cars may be leased by the tank car manufacturer to a shipper, or they may be owned by the shipper. An alpha-numeric ID number is stenciled on both sides and at the end of the tank car. It consists of the letters as an abbreviation of the car owner's name, followed by an "X" and a serial number. The tank car illustrated below in Fig. 7-29, belongs to the Pittsburgh Plate Glass Co.

Fig. 7-27   An overseas shipping container, with computer identification: ACTU 288427 GBX4300.

Fig. 7-28   A Conrail boxcar with Serial No. 166657.

Aircraft identification in the form of names and numbers appears on the tail section of the plane and beneath the cockpit, as shown in Fig. 7-30.

All marine vessels usually have identifying names on the bow and the stern. Barges usually have a vessel name or number that is legible on the stern and sides. If it is in tow, it is also advisable to obtain the towboat or tugboat name. An example of a marine vessel identifier is shown in Figs. 7-31 and 7-32.

Truck, tanker, aircraft, and vessel names and numbers can be used to identify the shipper of a hazardous material through a shipping agent or a truck or train dispatcher. In many instances that represents the ideal situation, and in reality it is sometimes more difficult and time consuming to do so. That is especially true when shipping papers or transportation personnel are not readily available, when the shipper cannot be identified, and when the ER team does not have the time to contact the shipper.

Fig. 7-29   A railroad tank car owned by the Pittsburgh Plate Glass Co. PPGX2643.

Fig. 7-30 Flying Tigers Cargo Plane No. N815FT.

Fig. 7-31 A marine vessel identifier.

Fig. 7-32   A barge showing its identifying name on the side.

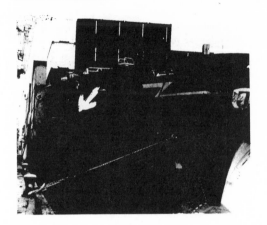

In any event it bears repeating that the identification of the contents of a shipping container may be a lifesaving factor in a transportation emergency. Specific identification of a hazardous material is necessary in order to determine the control actions necessary and to provide protective measures for the ER team personnel.

8

Responding to Hazardous Materials Incidents

A. Introduction

Since there are tremendous volumes and varieties of hazardous materials in use and in transit throughout the nation (and the world), there is a continuing need for technical assistance related to hazardous materials.

Thus it is that ER team personnel cannot be expected to know the properties and hazards of every material that may be involved in an incident. In addition, shipping papers or other means that are normally available for identifying cargoes of hazardous materials may have been lost or damaged during an accident, and valuable time must not be wasted whenever hazardous materials are concerned, when action and control measures need to be taken.

In many situations disaster lurks around every corner, and sometimes it can only be averted when the ER personnel know where and how to obtain assistance.

An example of just such an occurrence will serve to illustrate the objectives to be deduced in this chapter. It was early in the afternoon of February 9, 1976, when a freight train derailment in Chaseville, New York, sent a tank car containing a hazardous material into the Schenevus Creek. The state police contingent and the fire fighters from the nearby Schenevus volunteer fire department were fortunate enough to obtain the shipping papers for the tank car.

In a matter of minutes, a description of the hazardous chemical epichlorohydrin was obtained as well as the name of the shipper, which was the Shell Oil Company. The fire control center at Otsego was notified, and an evacuation of the surrounding area was begun. The danger area was roped off and otherwise secured from spectators as soon as it was learned that the tank car of epichlorohydrin was leaking.

The dispatcher at the Otsego fire control center immediately telephoned the Chemical Transportation Emergency Center, otherwise known as CHEMTREC, and gave it the name of the hazardous substance and the shipper of the tank car. CHEMTREC provided emergency information on epichlorohydrin and then proceeded to notify the Shell Oil Company. It in turn notified the hospital that might be receiving any of the exposure victims. Shell also provided the control center and the Schenevus fire chief with more detailed information concerning the chemical and also supplied suggestions for control actions.

After having been notified by CHEMTREC, the Shell personnel were in constant contact with the temporary command post that had been established as well as with all other personnel that were involved in controlling the incident. In addition, Shell sent a team of experts who began arriving at the site later that day. They were able to provide assistance to determine whether the tank car was still leaking and whether any environmental damage had occurred. During the ensuing 2 days, Shell Oil personnel worked with the local ER team to stabilize the situation and to return the community into some semblance of normality. The damaged tank car was removed from the Schenevus Creek after its chemical cargo was transferred to highway tank trucks.

The Otsego County Fire Department, the law enforcement authorities, and virtually all members of this community and surrounding ones were intensely relieved at the end result of a potentially dangerous accident.

Epichlorohydrin is a dangerous, hazardous chemical capable of causing death through vapor exposure. Since CHEMTREC had been promptly notified, it was able to place its response network into immediate action and as a result the incident was defused without a single death or injury.

The details of the story serve to delineate the importance of a common problem for ER team personnel, the desperate need for technical assistance when a disaster strikes, especially as it relates to hazardous materials.

That serious situation could have become worse, except for the fact that the emergency response team involved in the incident knew how and where to obtain assistance. The members were also aware of the procedures to be followed in the event of such an emergency and, therefore, because of their training and skills, the credit for resolving the emergency must go to them.

In the following chapter we shall explain the first steps involved in responding to hazardous materials incidents and the 2 main sources for obtaining technical assistance and emergency information in a time of crisis. Those sources are as follows:

1. CHEMTREC—The Chemical Transportation Emergency Center,
2. printed guides for handling emergency situations.

In addition, emphasis shall be placed upon the importance of learning how to obtain technical assistance in an emergency and how to use available assistance during the planning required to combat a hazardous materials problem.

B. The First Steps in Emergency Response

One of the difficulties that arise when one responds to an incident involving hazardous materials is that the ER team may be faced with a number of unknown chemical substances. It is important, therefore, that the immediate sources of technical assistance be understood. A brief review of those means is in order at this point.

The initial clues to chemical identity that may be observed at a safe distance are as follows:

1. labels,
2. placards,
3. types of shipping containers.

As soon as it is determined that hazardous materials might be present, then it is essential to identify the specific chemical. The several means available to make that more specific identification can be tabulated as follows:

1. shipping papers,
2. package markings,
3. placards,
4. four-digit identification numbers.

When the above means for identifying a hazardous material is not available, then the following methods should be investigated:

1. Interrogation of transportation or warehouse personnel,
2. Calling upon the shipper, or manufacturer,
3. Obtaining vehicle names or numbers when vehicles are involved.

In attempting to defuse a potentially dangerous incident involving hazardous materials there are 3 distinct areas of information that must be known. First, there is the name of the specific material or materials involved. Second, the hazardous characteristics of the materials must be understood. And, third, the necessary control actions must be taken.

The link between the information that the ER team has gathered and the information that is required through technical assistance is an important one in most emergency situations.

At the time of and during an emergency, technical assistance may be obtained from two reliable sources:

1. CHEMTREC, The Chemical Transportation Emergency Center,
2. DOT's Hazardous Materials—1984 Emergency Response Guidebook.

CHEMTREC is a public service of the Chemical Manufacturer's Association, CMA. It is provided by the CMA, which is a trade association of chemical manufacturers. It is also the single most available repository of all of the information known concerning commercially obtainable chemical substances. The agency was established in 1971. Its main objective is to assist ER personnel in responding to emergencies by promptly providing information on the best ways to control hazardous chemicals that is unfamiliar to the emergency crew.

C. How CHEMTREC Handles Emergencies

1. Getting Through to CHEMTREC

The CHEMTREC phone number is usually printed on the shipping documents of almost all of the manufacturers who belong to the Chemical Manufacturers' Association (CMA), in the following form:

"For help in chemical emergencies involving spills, leaks, fire, or exposure, call toll-free 800-424-9300, day or night."
(In Washington, D.C., the number is 483-7616; outside the U.S. call (202) 483-7616.)

There are a few manufacturers, such as Union Carbide, that list the company's own number. It is suggested that if a number is listed for the shipper, call the shipper directly rather than going through CHEMTREC. Also, if the railroad has stenciled its own emergency number on the tank car or the railroad's emergency number appears on the train waybills, then call the railroad directly; however, if no emergency phone numbers are available, then call CHEMTREC. Most of the manufacturers and shippers prefer that the ER team personnel reach them through the agency.

The CHEMTREC facility is manned on a 24-hour-per-day, seven-days-a-week basis. All of the personnel are ex-military and are used to handling emergency situations.

The CHEMTREC technician who will receive the call records the details of the emergency on tape and in writing. The technician will supply immediate technical assistance to the caller, based on information that has been supplied to the center by the chemical manufacturer.

The CHEMTREC technician promptly calls the shipper of the material or other experts for more detailed technical assistance and follow-up. The shipper or the manufacturer is then given the name and phone number of the response team member who made the call to the center. The shipper and the manufacturer are equally responsible for ensuring that their vital information gets to the person who has called the agency with the alert.

For example, if you had made a call to CHEMTREC concerning a shipment from the Dow Chemical Company, you would most assuredly receive a phone call directly from Dow very shortly.

A large number of chemical manufacturers have established their own response systems to provide support and assistance when accidents occur involving the chemicals they produce. The manufacturers also man their response activities on a round-the-clock basis. Therefore, when CHEMTREC alerts a large chemical shipper, the manufacturer's response staff may send a team of experts who have been especially and expertly trained for such emergencies. The team should be able to provide the best technical assistance available in both control and cleanup of a hazardous material. Before the team arrives at the scene, however, the manufacturer will have provided the best assistance possible over the phone until the factory's response team arrives.

2. When the Identity of the Shipper is Unknown.

At the time an emergency situation, when detailed assistance is needed as soon as possible and the identity of the shipper is unknown, then CHEMTREC turns to other valuable sources of information.

CHEMTREC has the following network of resources:

1. The DOT's National Response Center,
2. The Department of Defense,
3. Vinyl Chloride Monomer Emergency Response Program,
4. The Chlorine Institute,
5. The National Agricultural Chemical Association,
6. The Department of Energy,
7. The Bureau of Explosives of the Association of American Railroads.

CHEMTREC would immediately call upon 1 or more of the above agencies if the shipper's identity is unknown. Also, when the identity of the material is unknown, the center would help to identify the chemical through the DOT-E exemption number, placards, package markings, vehicle or vessel names or numbers, shipping papers, or carrier names.

3. Correct Information is Required

While CHEMTREC is capable of providing immediate technical assistance when the name of the hazardous material is known, it becomes time-consuming when insufficient data are supplied to the CHEMTREC technician. Having to identify an unknown substance becomes increasingly difficult if several leads have to be tracked down. Also, when calling the center, one should remember that the correct name of the substance is extremely critical; for example, 1 word or even 1 letter of a word can make a very large difference in reactivity. Some chemical substances sometimes have names that sound very much alike, but their hazardous characteristics are completely different.

Ethanol and ethanal sound alike but are spelled differently. In this example, ethanol is an abbreviated form for ethyl alcohol. It is a flammable liquid, and it will burn; however, its hazardous characteristics are only those hazards normally associated with the common flammable liquids. Ethanal, on the other hand, is a synonym for acetaldehyde. Acetaldehyde is also a flammable liquid and is water soluble. There the similarity ceases, because acetaldehyde is extremely volatile and must be stored in a pressurized tank. It is capable of producing serious eye burns, among other undesirable effects. Ethanal, therefore, is a good example of a hazardous material that has multiple hazardous characteristics.

The following procedures should be considered as recommendations that would help to provide CHEMTREC with the possibility of obtaining correct information in a timely fashion:

1. Have the fewest numbers of relays of information, as possible;
2. Use the shipping papers, if they are available;
3. Spell out the product name distinctly;
4. Use the four-digit identification number, if it is available.

(At the CHEMTREC center, chemical products are cross-referenced by synonyms and trade names, since various manufacturers may manufacture and distribute the same chemical under an entirely different brand name.)

After CHEMTEC has notified the shipper, the responsibility for further action belongs to the shipper or the manufacturer, whoever can place a response team in the field. The responsible entity will then call the person who has entered the alert; therefore, there should be no further requirement to remain in contact with CHEMTREC. Nevertheless, if the person who has initiated the alert does not receive a call from the agency or another responsible party within a half hour of calling it, then it is important to get in touch with it again.

Chart 8-1, below, clearly demonstrates the relationship between CHEMTREC and the ER team.

4. Guidelines for Calling CHEMTREC

CHEMTREC does not have to be called for all hazardous material incidents; thus, it is well to have some guidelines in mind that will serve to show what really necessitates such action.

Chart 8-1   The Relationship Between CHEMTREC and Emergency Response Personnel.

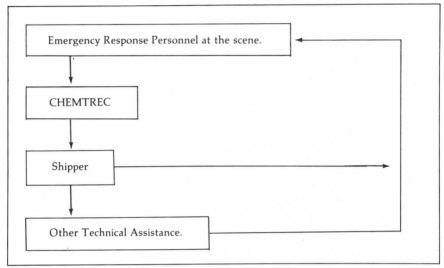

(1) CHEMTREC should be called when the substance involved is unfamiliar to the ER team crew. Most fire, police, and emergency medical departments or clinics are sufficiently well trained to cope with emergencies in which the more common flammable or combustible liquids, such as gasoline and home-heating fuel oil, are the hazardous materials. It is when the team is faced with an unfamiliar, hazardous material that CHEMTREC assistance should be sought. Its technical information may prevent the continuation of a spurious, control action that could produce disastrous results. It should be stressed, at this point, that the center does not relieve the local ER team of any responsibility. The agency can assist ER personnel by providing the latest technical information in order for the team to work safely and efficiently in performing its response function.

(2) CHEMTREC should be called when the material cannot be readily identified but has been determined to be potentially dangerous or hazardous. Some examples that may serve to illustrate when the agency should be called upon for assistance in the event of an accident are when a pressurized tank car is involved, when there are labels or placards placed on containers or vehicles, when there is a pungent odor emanating from damaged containers, when 4-digit identification numbers are visible, and when DOT container specification numbers are seen on metal drums.

(3) CHEMTREC should also be called when the chemical substance is unfamiliar or unidentified and the shipper cannot be readily identified. As indicated in the various passages above, the agency has the most effective resources for dealing with the problems of identification. It should be called immediately, for it can help trace the material.

(4) CHEMTREC should be called when only limited technical information is available from local sources and it becomes necessary to verify or substantiate

the data. It must be remembered that since the agency is an association of chemical manufacturers, it has the latest technical information available.

(5) CHEMTREC should be called when an incident involves any significant quantities of a hazardous chemical. For example, during a transportation or other accident where bulk transport containers, such as cargo tankers or railroad tank cars, have started to leak or there is an extensive spill or a fire is threatening. Also, the occasion may arise in a situation involving only a small quantity of an extremely hazardous substance such as a Class A poison. It is important to note that while CHEMTREC has a repository of technical assistance information, it does not provide general information on all chemicals, but it serves primarily as a hot line and source of *emergency* information for *transportation emergencies.*

5. A CHEMTREC Incident

In order to reinforce the general observations that have been made in the above paragraphs, the following descriptive narrative, which is completely fictitious, may serve to illustrate the CHEMTREC relationship to hazardous chemical incidents.

A police chief and his deputy are returning to their headquarters, which is located in a small city in the Midwest. Several miles outside of town they observe a freight train derailment in which 2 tank cars have rolled down a small embankment. One of the railroad tankers containing vegetable oil is leaking; the other tank car, a pressurized tanker, does not seem to be leaking.

Assured by the train's conductor that no injuries have been sustained by any of the crew members, the deputy examines the train's shipping papers obtained from the conductor. Since the police chief and deputy are not particularly concerned at the moment with the vegetable oil tanker, they find that a pressurized tanker contains a Flammable Liquid, n.o.s., U.N. 1993, and it is being shipped from a firm, the XYZ Company of Chicago. The police chief notes the serial number of the flammable liquid tank car. Since the two men are not familiar with the chemical substance, it is largely unidentified, and since it is a large and significant quantity of material evidently hazardous, they call CHEMTREC.

D. Emergency Response Guidebook

The "Hazardous Materials, Emergency Response Guidebook," DOT P 5800.3, was developed under the supervision of the Office of Hazardous Materials Regulation, Materials Transportation Bureau, Research and Special Programs Administration of the U.S. Department of Transportation. It was prepared by Andreas V. Jensen, senior staff chemist of the Chemical Propulsion Information Agency of the Johns Hopkins University's Applied Physics Laboratory at Laurel, Maryland 20707.

The guidebook was developed for use by fire fighters, police, and other emergency services personnel as a guide for initial actions to be taken to protect themselves and the public when they are called to handle incidents involving hazardous materials.

For your information, we have abstracted the following information from the Federal Register, Volume 45, Number 219, Monday, November 10, 1980, Rules and Regulations, page 74643. It explains the distribution process. (Italics added for emphasis.)

The Emergency Response Guidebook

The Emergency Response Guidebook (ERG) that is associated with the rule issued under Docket HM-126A has been completed. The MTB has entered into a contractual arrangement with the International Association of Fire Chiefs (IAFC) for distribution of the ERG to emergency services organizations. *MTB's principal objective is to have 1 ERG placed in each police, fire, civil defense, and rescue-squad vehicle in the United States* (estimated to be 450,000 vehicles) by November 1, 1981. Initial delivery of 200,000 copies of the ERG to the IAFC for distribution has been accomplished.

The IAFC is cooperating with the International Association of Chiefs of Police and the United States Civil Defense Council in the distribution of the ERG. The IAFC will make initial distribution to the members of those organizations before delivery to other emergency services entities. That will avoid duplication of requests from different levels of emergency services organizations. For example, the chief of a county fire and rescue department responds to the IAFC inquiry, which will be mailed before January 1, 1981, and requests ERGs for all fire and rescue squad vehicles in his county. If he handles the distribution for his county, it will not be necessary for individual departments in his county to submit separate requests for the guidebook.

The MTB requests the cooperation of all persons interested in the distribution of the ERG. It must be emphasized that the IAFC is distributing the guidebook only to emergency services organizations engaged in protecting the general public. The MTB has been informed that several private firms plan to reproduce and sell the ERG commercially. (See Appendix).

Commercial firms, organizations and private individuals should not contact the IAFC for copies of the ERG.

Persons representing emergency response entities that have not received copies of the ERG for their vehicles and have not been contacted by the IAFC may contact the IAFC after April 1 1981 by writing to:

International Association of Fire Chiefs
Attention: ERG, 1329 18th Street, N.W.,
Washington, D.C. 20036

The DOT P 5800.3 is a subsequent edition of the 1980 edition of the Emergency Response Guidebook (DOT P 5800.2) of which 750,000 copies were distributed free of charge between 1980 and 1983 by the DOT to emergency services organizations with the cooperation of the International Association of Fire Chiefs. The first printing of DOT P 5800.3 of December 1983 ran to 730,000 copies and was entitled, "1984 Emergency Response Guidebook."

The 1984 Guidebook has 5 sections:

(1) The first section is a numerical index of hazardous materials (yellow pages) that lists the chemical substances by their 4-digit identification number (see Fig. 8-1). This is followed by the corresponding "Guide No.," which is a separate section (Section 3 of the guidebook).

(2) The second section of the guidebook is cross-referenced to the first section since all of the chemical substances are arranged by "Name of Material" in alphabetical order. The chemical name is followed on the same line with the "Guide No." and the "ID No."

(3) The third section is a series of response guides organized according to single or multiple hazards. They describe in abbreviated form the potential

hazards of the substance and prescribe the initial, emergency action to be taken by the response team.

(4) The fourth section is a combined table of the minimum isolation and evacuation distances for removing unprotected people from hazardous areas.

(5) The fifth and last section shows some of the most commonly used placards and provides a Guide Number (see para. 3, above) below each placard to indicate what hazard is present and what responsive action is immediately required.

Since the book is an exceptionally fine piece of work, several of the pages from the DOT P 5800.3 Guidebook have been reproduced in the following pages so that the reader may observe how terse and objective the guidebook has been made in order to provide ready answers in an emergency situation.

KEEP THIS GUIDEBOOK IMMEDIATELY AVAILABLE IN EMERGENCY SERVICE VEHICLES

APPROACH INCIDENT FROM UPWIND, IF POSSIBLE
STAY CLEAR OF ALL SPILLS, VAPORS, FUMES, AND SMOKE

INSTRUCTIONS

HOW TO USE THIS GUIDEBOOK

DURING AN INCIDENT WITH HAZARDOUS MATERIALS

FIRST, IDENTIFY THE MATERIAL BY FINDING EITHER:

THE 4-DIGIT ID NUMBER ON A PLACARD OR ORANGE PANEL
or
THE 4-DIGIT ID NUMBER (after letters UN or NA), ON A SHIPPING PAPER
or
THE MATERIAL NAME ON A SHIPPING PAPER, PLACARD, LABEL OR PACKAGE

IF YOU FAIL TO FIND AN ID NUMBER OR NAME, SKIP TO THE NOTE BELOW.

SECOND, LOOK UP THE MATERIAL'S 2-DIGIT GUIDE NUMBER IN EITHER:

THE ID NUMBER INDEX (YELLOW Pages in FIRST Section of Guidebook)
or
THE MATERIAL NAME INDEX (BLUE Pages in SECOND Section of Guidebook)
or
THIS LIST FOR
ALL EXPLOSIVES

Explosives A - Use GUIDE 46
Explosives B - Use GUIDE 46
Explosives C - Use GUIDE 50
Blasting Agents - Use GUIDE 46

IF YOU FAIL TO FIND A GUIDE NUMBER IN EITHER INDEX:
CALL CHEMTREC TOLL-FREE (800) 424-9300

THIRD, TURN TO THE NUMBERED GUIDE PAGE: (WHITE Pages, ORANGE Tops)

READ THE GUIDE PAGE CAREFULLY BEFORE TAKING ACTION.

NOTE: IF YOU FAIL TO TURN TO A GUIDE PAGE IN THE STEPS ABOVE:

AND YOU FIND A PLACARD, LOOK IT UP IN THE TABLE OF PLACARDS
IN THE BACK OF THIS GUIDEBOOK AND USE THE GUIDE PAGE GIVEN
or
TURN TO AND USE GUIDE PAGE 11 NOW, UNTIL ADDITIONAL INFORMATION
OR HELP BECOMES AVAILABLE TO YOU.

Fig. 8-1   The First Page of the Guidebook.

WHEN APPROACHING THE SCENE OF AN ACCIDENT INVOLVING
ANY CARGO (NOT ONLY REGULATED HAZARDOUS MATERIALS):

- MOVE AND KEEP PEOPLE AWAY FROM INCIDENT SCENE
- DO NOT WALK INTO OR TOUCH ANY SPILLED MATERIAL
- AVOID INHALING FUMES, SMOKE AND VAPORS EVEN
  IF NO HAZARDOUS MATERIALS ARE INVOLVED
- DO NOT ASSUME THAT GASES OR VAPORS ARE
  HARMLESS BECAUSE OF LACK OF SMELL

FIRE EXTINGUISHING METHODS - USE OF WATER

The selection of the extinguishing method should be made with caution since there are many factors to be considered in any individual case. Water may be ineffective in fighting fires on some materials, but much will depend on the method of application. "Alcohol" foam is recommended for most of the materials covered under Guide 26 and ordinary or "Protein" foam for materials listed under Guide 27. It is impossible to recommend a choice of foam for flammable liquids with subsidiary corrosive or poison hazards although "Alcohol" foam may be effective with many of these materials. It should be emphasized that the final decision on the initial selection of the most effective extinguishing method must be left to the judgment of the firefighter. This judgment will depend on many factors such as location, exposure hazards, size and type of fire and the available extinguishing agent and equipment at his disposal.

Water is usually the most available and the most generally applicable fire extinguishing agent. It is also frequently used to flush spills and to control vapors in spill situations. However, a number of the hazardous materials indexed in this guidebook can react violently or even explosively with water. In these cases you should consider letting a fire burn until expert advice can be obtained. The guide pages clearly warn you against the use of water in controlling either small or large spills or fires in the event the material reacts vigorously with water. These materials require expert judgment during the attempted control of a large fire or spill since:

1. water getting inside the ruptured or leaking containers may cause an explosion;

2. water may be needed to cool adjoining containers to prevent rupturing (explosion) or further spread of the fire;

3. water may be effective in mitigating an incident involving a water-reactive material only if it can be applied at a sufficient "flooding" rate;

4. the products of the reaction with water may be more toxic, corrosive, or otherwise undesirable than the present products of the fire.

Incidents involving water-reactive chemicals must be handled by taking into account the existing conditions such as wind, precipitation, location and accessibility to the incident as well as the availability of materials to control the fire or spill. Because of the great number of possible variables, the decision to use water in large fires or spills involving certain materials should be made by an authoritative source, such as a manufacturer of that particular material who can be contacted through CHEMTREC.

ii

Fig. 8-2   Page ii of the Guidebook.

INTRODUCTION

This guidebook was developed for use by firefighters, police and other emergency services personnel as a guide for initial actions to be taken to protect themselves and the public when they are called to handle incidents involving hazardous materials. The information given is intended to provide guidance primarily during the initial phases following an incident. To obtain additional assistance for the most effective handling of an incident, it is important that a call be made as soon as possible to:

CHEMTREC TOLL-FREE (800) 424-9300

The purpose of this guidebook is to assist an individual in making decisions. It is not intended to serve as a substitute for his own knowledge or judgement. This distinction is important since the recommendations given are those most likely to be applicable in the majority of cases. It is not claimed that the recommendations are necessarily adequate or applicable in all cases. While this document was primarily designed for use at a hazardous materials incident occurring on a highway or a railroad, it will, with certain limitations, be of use in handling incidents in other modes of transportation and at facilities such as terminals and warehouses.

Each numbered response guide provides only the most vital information in the briefest practical form. It identifies the most significant potential hazards and gives information and guidance for initial actions to be taken. A numbered guide is assigned to each material listed in the indexes. Neither the numerical order of the guide presentation nor the guide number itself is of any significance. Since many materials represent similar types of hazards that call for similar initial emergency response actions, only a limited number of guide pages are required for all the materials listed. The guides cover hazardous materials taken one at a time and spilled or involved in a fire. Incidents involving more than one hazardous material at a time require that the on-scene leader get expert advice as soon as the scope of the incident can be determined. The materials involved in an accident may, by themselves, be nonhazardous; however, any combination of different materials or the involvement of a single commodity in a fire may produce serious health, fire or explosion hazards.

EXPLOSIVES and BLASTING AGENTS are not listed individually in this guidebook by ID number. If the shipping paper or placard identifies the material as Class A Explosives, Class B Explosives or as Blasting Agents (UN Class 1.1, 1.2, 1.3 or 1.5 for international shipments), follow the instructions given in Guide 46. If the shipping paper identifies the material as Class C Explosives (or UN Class 1.4 for international shipments), follow the instructions given in Guide 50. The instructions on the first page of the guidebook cover explosives as a special case.

Since the Identification Number assignments made in this guidebook, and the designation of hazardous materials are closely tied to the actions of regulatory authorities, each edition of this guidebook will have a limited useful life of two or three years from the date of publication.

iii

Fig. 8-3   Page iii of the Guidebook.

EXPLANATIONS OF WORDS AND TERMS

Full Protective Clothing

This means protection to prevent inhalation of, ingestion of, or skin contact with hazardous vapors, liquids and solids. It includes a helmet, self-contained breathing apparatus, coat, pants, rubber boots and gloves customarily worn by fire fighters. This turnout clothing may not provide protection from vapors, liquids or solids encountered during hazardous materials incidents. Full protective clothing should meet the OSHA Fire Brigades Standard (29 Code of Federal Regulations 1910.156). Chemical-cartridge respirators or gas masks are not acceptable substitutes for self-contained breathing apparatus. The demand-type self-contained breathing apparatus is being phased out of service since it does not meet the OSHA Fire Brigades Standard cited above.

Special Protective Clothing and Equipment

This category of clothing and equipment will protect the wearer against the specific hazard for which it was designed. The special clothing may afford protection only for certain chemicals and may be readily penetrated by chemicals for which it was not designed. Do not assume any protective clothing is fire resistant unless that is specifically stated by the manufacturer.

Isolate Hazard Area and Deny Entry

Keep everybody away from the hazard area if not directly involved with the emergency response or rescue operation. Do not let unprotected people into the area. Conduct any rescue operation as quickly as possible entering the scene from the upwind approach. This "isolate" step is the first to be taken even if "evacuation" is to follow.

Evacuate

Remove all people from area and buildings as far as recommended in the evacuation distance table presented in the back of this guidebook. Good judgment must be used in evacuation procedures to avoid placing people in greater danger. Topographic maps may assist you in the planning and execution of evacuations. You may obtain indexes of the topographic maps published for each state free of charge on request from the nearest office of The U.S. Geological Survey. Buy the maps you need to cover your area of responsibility. Preplanning and response team training is recommended.

Decontamination of Personnel and Equipment

Emergency services personnel should be decontaminated as soon as possible after contact occurs. Since the methods to be used differ from one chemical to another it is important to contact the shipper and medical authorities quickly to determine the most appropriate decontamination procedures. Contaminated protective clothing and equipment should be isolated to prevent further human contact, and should be stored in a restricted area (hot zone) at the incident site until appropriate decontamination procedures can be determined. In some cases, protective clothing and equipment cannot be decontaminated and will have to be disposed of according to appropriate state and federal guidelines.

Positive Pressure Breathing Apparatus

Positive pressure breathing apparatus is the best choice for complete protection during operations involving hazardous materials. Use apparatus certified by NIOSH and the Mine Safety and Health Administration in accordance with 30 Code of Federal Regulations Part II (30 CFR Part II) and used in accordance with the Respiratory Protection Standard (29 CFR 1910.134) and the OSHA Fire Brigades Standard (29 CFR 1910.156).

vi

Fig. 8-4   Page vi of the Guidebook.

ID No.	Guide No.	Name of Material	ID No.	Guide No.	Name of Material
1001	17	ACETYLENE	1027	22	CYCLOPROPANE, liquefied
1001	17	ACETYLENE, dissolved	1028	12	DICHLORODIFLUOROMETHANE
1002	12	AIR, compressed	1029	12	DICHLOROFLUOROMETHANE
1003	23	AIR, cryogenic liquid	1029	12	DICHLOROMONOFLUORO-METHANE
1003	23	AIR, liquid (refrigerated)			
1005	15	AMMONIA ANHYDROUS, liquefied *	1030	22	DIFLUOROETHANE
			1032	19	DIMETHYLAMINE, anhydrous *
1005	15	ANHYDROUS AMMONIA *	1033	22	DIMETHYL ETHER
1006	12	ARGON, compressed	1035	22	ETHANE, compressed
1008	15	BORON TRIFLUORIDE *	1036	68	ETHYLAMINE
1009	12	BROMOTRIFLUOROMETHANE	1036	68	MONOETHYLAMINE
1009	12	TRIFLUOROBROMOMETHANE	1037	27	ETHYL CHLORIDE
1010	17	BUTADIENE, inhibited	1038	22	ETHYLENE, cryogenic liquid
1011	22	BUTANE or BUTANE MIXTURE			
1012	22	BUTENE	1038	22	ETHYLENE, liquid (refrigerated)
1012	22	BUTYLENE			
1013	21	CARBON DIOXIDE	1039	26	ETHYL METHYL ETHER
1014	14	CARBON DIOXIDE-OXYGEN MIXTURE	1039	26	METHYL ETHYL ETHER
			1040	69	ETHYLENE OXIDE *
1015	12	CARBON DIOXIDE-NITROUS OXIDE MIXTURE	1041	17	CARBON DIOXIDE-ETHYLENE OXIDE MIXTURE, with more than 6% ETHYLENE OXIDE
1016	18	CARBON MONOXIDE			
1017	20	CHLORINE *	1041	17	ETHYLENE OXIDE-CARBON DIOXIDE MIXTURE, with more than 6% ETHYLENE OXIDE
1018	12	CHLORODIFLUOROMETHANE			
1018	12	MONOCHLORODIFLUORO-METHANE			
1020	12	CHLOROPENTAFLUORO-ETHANE	1043	16	FERTILIZER AMMONIATING SOLUTION, with more than 35% free ammonia
1021	12	CHLOROTETRAFLUORO-ETHANE	1043	16	NITROGEN FERTILIZER SOLUTION
1022	12	CHLOROTRIFLUOROMETHANE	1044	12	FIRE EXTINGUISHER, with compressed or liquefied gas
1022	12	TRIFLUOROCHLORO-METHANE	1045	20	FLUORINE, compressed
1023	18	COAL GAS	1046	12	HELIUM, compressed
1026	18	CYANOGEN	1048	15	HYDROGEN BROMIDE, anhydrous
1026	18	CYANOGEN, liquefied	1049	22	HYDROGEN, compressed
1027	22	CYCLOPROPANE	1050	15	HYDROCHLORIC ACID, anhydrous *

* Look for information next to this NAME in the TABLE OF EVACUATION DISTANCES in the back of this book. Use this in addition to the Guide Page if there is NO FIRE.

Fig. 8-5   The above represents the first page of the first section of the Guidebook's yellow pages. Note that the ID No. is in numerical sequence.

Guide 11

POTENTIAL HAZARDS

FIRE OR EXPLOSION

Flammable/combustible material; may be ignited by heat, sparks or flames.
May ignite other combustible materials (wood, paper, oil, etc.).
Container may explode in heat of fire.
Reaction with fuels may be violent.
Runoff to sewer may create fire or explosion hazard.

HEALTH HAZARDS

May be fatal if inhaled, swallowed or absorbed through skin.
Contact may cause burns to skin and eyes.
Fire may produce irritating or poisonous gases.
Runoff from fire control or dilution water may cause pollution.

EMERGENCY ACTION

Keep unnecessary people away; isolate hazard area and deny entry.
Stay upwind; keep out of low areas.
Wear self-contained (positive pressure if available) breathing apparatus and full protective clothing.
FOR EMERGENCY ASSISTANCE CALL CHEMTREC (800) 424-9300.
If water pollution occurs, notify appropriate authorities.

FIRE

Small Fires: Dry chemical, CO_2, water spray or foam.
Large Fires: Water spray, fog or foam.
Move container from fire area if you can do it without risk.
Cool containers that are exposed to flames with water from the side until well after fire is out.
For massive fire in cargo area, use unmanned hose holder or monitor nozzles; if this is impossible, withdraw from area and let fire burn.

SPILL OR LEAK

Shut off ignition sources; no flares, smoking or flames in hazard area.
Keep combustibles (wood, paper, oil, etc.) away from spilled material.
Do not touch spilled material.
Small Spills: Take up with sand or other noncombustible absorbent material and place into containers for later disposal.
Large Spills: Dike far ahead of spill for later disposal.

FIRST AID

Move victim to fresh air; call emergency medical care.
If not breathing, give artificial respiration.
If breathing is difficult, give oxygen.
In case of contact with material, immediately flush skin or eyes with running water for at least 15 minutes.
Remove and isolate contaminated clothing and shoes at the site.
Keep victim quiet and maintain normal body temperature.

Fig. 8-6   The above represents the first page of the guide numbers, the third section of the guidebook. The guides are numbered from 11 to 76. The 76th guide has been reserved for future use.

Name of Material	Guide No.	ID No.	Name of Material	Guide No.	ID No.
ACCUMULATORS, pressurized	12	1956	ACID MIXTURE, nitrating	73	1796
ACETAL	26	1088	ACID MIXTURE, spent, nitrating	60	1826
ACETALDEHYDE	26	1089	ACID SLUDGE	60	1906
ACETALDEHYDE AMMONIA	31	1841	ACRIDINE	32	2713
ACETALDEHYDE OXIME	26	2332	ACROLEIN, inhibited *	30	1092
ACETIC ACID, GLACIAL	29	2789	ACROLEIN DIMER, stabilized	26	2607
ACETIC ACID SOLUTION	29	1842	ACRYLAMIDE	55	2074
ACETIC ACID SOLUTION, more than 80% acid	29	2789	ACRYLIC ACID	29	2218
ACETIC ACID SOLUTION, more than 10% but not more than 80% acid	60	2790	ACRYLONITRILE, inhibited *	30	1093
ACETIC ANHYDRIDE	39	1715	ACTIVATED CARBON	32	1362
ACETONE	26	1090	ADHESIVE, containing flammable liquid, n.o.s.	26	1133
ACETONE CYANOHYDRIN	55	1541	ADHESIVE, n.o.s.	26	1133
ACETONE OIL	26	1091	ADIPIC ACID	31	9077
ACETONITRILE	28	1648	ADIPONITRILE	55	2205
ACETYL ACETONE PEROXIDE	48	2080	AEROSOLS	12	1950
ACETYL BENZOYL PEROXIDE	48	2081	AIR, compressed	12	1002
			AIR, cryogenic liquid	23	1003
ACETYL BROMIDE	60	1716	AIR, liquid (refrigerated)	23	1003
ACETYL CHLORIDE	29	1717	AIRCRAFT EVACUATION SLIDE	31	2990
ACETYL CYCLOHEXANE SULFONYL PEROXIDE	52	2082	AIRCRAFT ROCKET ENGINE	32	2791
ACETYL CYCLOHEXANE SULFONYL PEROXIDE	52	2083	AIRCRAFT ROCKET ENGINE IGNITER	32	2792
ACETYLENE	17	1001	AIRCRAFT SURVIVAL KIT	31	2990
ACETYLENE, dissolved	17	1001	AIRCRAFT THRUST DEVICE	32	2791
ACETYLENE TETRABROMIDE	58	2504	ALCOHOL (beverage)	26	1170
ACETYL IODIDE	60	1898	ALCOHOL, denatured	26	1987
ACETYL METHYL CARBINOL	26	2621	ALCOHOL, denatured (toxic)	28	1986
ACETYL PEROXIDE	49	2084	ALCOHOL (ethyl)	26	1170
ACID, liquid, n.o.s.	60	1760	ALCOHOL, nontoxic, n.o.s.	26	1987
ACID BUTYL PHOSPHATE	60	1718	ALCOHOL, n.o.s.	26	1987
ACID MIXTURE, hydrofluoric and sulfuric acids	59	1786	ALCOHOL, poisonous, n.o.s.	28	1986
			ALCOHOL, toxic, n.o.s.	28	1986
			ALCOHOLIC BEVERAGE	26	1170

* Look for information next to this **NAME** in the TABLE OF EVACUATION DISTANCES in the back of this book. Use this in addition to the Guide Page if there is NO FIRE.

Fig. 8-7   The above represents the first page of the second section of the guidebook's blue pages. Note that the chemical name is listed alphabetically in this section, thus providing a handy cross-reference.

I N T R O D U C T I O N   T O

T H E   E V A C U A T I O N   T A B L E S

F O R   S E L E C T E D   H A Z A R D O U S   M A T E R I A L S

The following tables give suggested distances for ISOLATING or EVACUATING unprotected people from spill areas involving the hazardous materials shown, if the materials are not on fire. These suggestions are only for the initial phase of an accident involving volatile hazardous liquids or gases shipped in bulk or multiple-container loads. Continuing reassessment of the situation will be necessary because there may be a change in circumstances, such as a change in wind direction. Good judgment must be used in evacuation procedures to avoid placing people in a more dangerous situation than is necessary.

If a hazardous material cloud goes between several multistory buildings or down a valley, the cloud may affect people much farther away than the distance specified in the tables and the evacuation distances should be increased for the downwind direction. It is important to note that the occupants of the upper floors of multistory buildings in the evacuation sector may be safer remaining where they are if the heating and air-handling equipment in the buildings can be shut down so that the hazardous vapors or gases will not be circulated within. A short-term spill cloud may be deflected or reflected by a multistory building and pass by without affecting the occupants or the equipment within the building.

For those materials listed in the tables, if a fire begins to burn the spilled material the health hazard may become less important and the evacuation distances may not have to be as great as they were with no fire involvement. It is important to notice that for some of these materials the potential fragmentation hazards from a tank car or truck involved in the fire may require isolation in all directions for at least one-half mile despite any shorter distance suggested in the tables. The guide page for the respective material clearly indicates if there is a one-half mile isolation requirement to handle the fragmentation hazard. Whatever number of feet or miles has been cleared, if unprotected people are being affected by one or more of the materials in the following tables—INCREASE THE DISTANCES and reassess the situation.

The isolation and evacuation distances presented in the tables are based on principles and calculations developed by the NASA Aerospace Safety Research and Data Institute, Lewis Research Center, Cleveland, Ohio; and under contract to The U.S. Department of Transportation by the Chemical Propulsion Information Agency, The John Hopkins University, Applied Physics Laboratory, Laurel, Maryland. The diffusion data and calculations were based on work contained in the publication "Workbook of Atmospheric Dispersion Estimates", by D. Bruce Turner, Environmental Protection Agency, Research Triangle Park, North Carolina.

Fig. 8-8   The above represents the first page of the fourth section of the guidebook and is self-explanatory; the tables follow on the next 2 pages.

T A B L E   O F   I S O L A T I O N   &   E V A C U A T I O N   D I S T A N C E S

NAME OF MATERIAL SPILLING OR LEAKING (ID No.)	INITIAL ISOLATION SPILL or LEAK FROM (drum, smaller container, or small leak from tank) ISOLATE in all Directions feet	INITIAL EVACUATION LARGE SPILL FROM A TANK (or from many containers, drums, etc.)		
		FIRST ISOLATE in all Directions feet	THEN EVACUATE IN A DOWNWIND DIRECTION Width miles	Length miles
Acrolein (1092)	300	620	1.5	2.4
Acrylonitrile (1093)	90	180	0.4	0.6
Ammonia, anhydrous liquified (1005) Ammonia solution, with more than 44% (2073) Anhydrous ammonia (1005)	80	160	0.4	0.6
Boron trifluoride (1008)	320	670	1.7	2.6
Bromine (1744)	170	350	0.8	1.3
Carbon bisulfide (1131) Carbon disulfide (1131)	50	100	0.2	0.3
Chloride of phosphorus (1809)	110	220	0.5	0.8
Chlorine (1017)	140	290	0.7	1.0
Dimethylamine, anhydrous (1032)	80	170	0.4	0.6
Dimethyl sulfate (1595)	80	170	1.4	2.2
Epichlorohydrin (2023)	40	80	0.2	0.3
Ethylene imine (1185)	270	570	1.4	2.2
Ethylene oxide (1040)	80	160	0.4	0.5
Fluorine, cryogenic liquid (9192)	460	980	2.5	3.9
Hydrochloric acid, anhydrous (1050) Hydrogen chloride, anhydrous (1050) Hydrogen chloride, liquid (refrigerated) (2186)	190	450	1.0	1.4
Hydrocyanic acid (1051) Hydrogen cyanide, anhydrous (1051)	90	190	0.5	0.7
Hydrofluoric acid, anhydrous (1052) Hydrogen fluoride, anhydrous (1052)	150	300	0.7	1.1

Fig. 8-9   The above table of isolation and evacuation distances and its continuation on the next page have been reproduced in their entirety since they may be of interest to the concerned reader.

TABLE   OF   ISOLATION   &   EVACUATION   DISTANCES

NAME OF MATERIAL SPILLING OR LEAKING (ID No.)	INITIAL ISOLATION SPILL or LEAK FROM (drum, smaller container, or small leak from tank) ISOLATE in all Directions feet	INITIAL EVACUATION LARGE SPILL FROM A TANK (or from many containers, drums, etc.)		
		FIRST ISOLATE in all Directions feet	THEN EVACUATE IN A DOWNWIND DIRECTION Width miles	Length miles
Hydrogen sulfide (1053)	120	240	0.6	0.9
Methylamine, anhydrous (1061) Monomethylamine, anhydrous (1061)	110	220	0.5	0.8
Methyl bromide (1062)	50	90	0.2	0.3
Methyl chloride (1063)	30	60	0.1	0.2
Methyl mercaptan (1064)	370	770	1.9	3.0
Methyl sulfate (1595)	80	170	1.4	2.2
Nitric acid, fuming (2032) Nitric acid, red fuming (2032)	100	210	0.5	0.7
Nitric oxide (1660) Nitric oxide and Nitrogen tetroxide mixture (1975)	90	180	0.4	0.6
Nitrogen dioxide (1067) Nitrogen peroxide (1067) Nitrogen tetroxide (1067)	110	220	0.5	0.8
Oleum (1831)	280	580	1.5	2.2
Perchloromethyl- mercaptan (1670)	220	450	1.1	1.6
Phosgene (1076)	600	1250	3.3	5.2
Phosphorus trichloride (1809)	110	220	0.5	0.8
Sulfur dioxide (1079)	120	250	0.6	0.9
Sulfuric acid, fuming (1831) Sulfuric anhydride (1829) Sulfur trioxide (1829)	280	580	1.5	2.2
Titanium tetrachloride (1838)	30	60	0.2	0.2
Trimethylamine, anhydrous (1083)	50	90	0.2	0.3

Fig. 8-10   A continuation of the table shown on the preceding page.

TABLE OF PLACARDS AND APPLICABLE RESPONSE GUIDE PAGES

USE **ONLY** IF MATERIALS CANNOT BE SPECIFICALLY IDENTIFIED

THROUGH SHIPPING PAPERS OR MARKINGS.

Fig. 8-11   Placards showing guide numbers that are explained in Section 3 of the guidebook. The placards represent the fifth section of the guidebook.

TABLE OF PLACARDS AND APPLICABLE RESPONSE GUIDE PAGES

USE <u>ONLY</u> IF MATERIALS CANNOT BE SPECIFICALLY IDENTIFIED
THROUGH SHIPPING PAPERS OR MARKINGS.

Fig. 8-12 A continuation of the placard guide numbers of the fifth section of the guidebook.

9

The Code of Federal Regulations—CFR Title 40

A. Introduction—Protection of the Environment

Title 40 is concerned with the protection of the environment and is composed of 9 volumes. The Environmental Protection Agency (EPA) is noted in all of the volumes. The purpose of this chapter is to inform the reader of the nature and philosophy of handling waste products. The following subjects, which are subchapters of 40 CFR, shall be discussed: radiation protection programs (Subchapter F), ocean dumping (Subchapter H), solid wastes (Subchapter I), and superfund programs (Subchapter J).

B. Radiation Protection Programs

Subchapter F of 40 CFR is mainly concerned with the quantity of radiation received by members of the public in the general environment and with radioactive materials introduced into the general environment as the result of operations that are part of a nuclear fuel cycle.

Before we begin a discussion of radiation effects, we must first define the terms used. They are as follows:

(a) "Nuclear fuel cycle" means the operations defined to be associated with the production of electrical power for public use by any fuel cycle through utilization of nuclear energy.

(b) "Uranium fuel cycle" means the operations of milling of uranium ore, chemical conversion of uranium, isotopic enrichment of uranium, fabrication of uranium fuel, generation of electricity by a light-water-cooled nuclear power plant using uranium fuel, and reprocessing of spent uranium fuel to the extent that they directly support the production of electrical power for public use utilizing nuclear energy, but the term excludes mining operations, operations at waste disposal sites, transportation of any radioactive material in support of those operations, and the reuse of recovered nonuranium special nuclear and by-product materials from the cycle.

(c) "General environment" means the total terrestrial, atmospheric, and aquatic environments outside sites upon which any operation that is part of a nuclear fuel cycle is conducted.

(d) "Site" means the area contained within the boundary of a location under the control of persons possessing or using radioactive material on which is conducted 1 or more operations covered by this part.

(e) "Radiation" means any or all of the following: Alpha, beta, gamma, or X rays; neutrons; and high-energy electrons, protons, or other atomic particles; but not sound or radio waves nor visible, infrared, or ultraviolet light.

(f) "Radioactive material" means any material that spontaneously emits radiation.

(g) "Curie" (Ci) means that quantity of radioactive material producing 37 billion nuclear transformations per second. (One millicurie (MCi)=0.001 (Ci.)

(h) "Dose equivalent" means the product of absorbed dose and appropriate factors to account for differences in biological effectiveness owing to the quality of radiation and its spatial distribution in the body. The unit of dose equivalent is the "rem." (One millirem (mrem)=0.001 rem.)

(i) "Organ" means any human organ exclusive of the dermis, the epidermis, or the cornea.

(j) "Gigawatt-year" refers to the quantity of electrical energy produced as the output of a generating station. A gigawatt is equal to 1 billion watts. A gigawatt-year is equivalent to the amount of energy output represented by an average electric power level of 1 gigawatt sustained for 1 year.

(k) "Member of the public" means any individual that can receive a radiation dose in the general environment, whether he may or may not also be exposed to radiation in an occupation associated with a nuclear fuel cycle. However, an individual is not considered a member of the public during any period in which he is engaged in carrying out any operation that is part of a nuclear fuel cycle.

(l) "Regulatory agency" means the government agency responsible for issuing regulations governing the use of sources of radiation or radioactive materials or emissions therefrom and carrying out inspection and enforcement activities to ensure compliance with such regulations.

1. Environmental Standards for the Uranium Fuel Cycle

Sec. 190.10 of 40 CFR covers the standards for normal operations in the uranium fuel cycle in order to provide assurances that dosages of radiation received by members of the public do not exceed specific levels.

The regulations state that:

(a) The annual dose equivalent does not exceed 25 millirems to the whole body, 75 millirems to the thyroid, and 25 millirems to any other organ of any member of the public as the result of exposures to planned discharges of radioactive materials, radon and its daughters excepted, to the general environment from uranium fuel cycle operations and to radiation from those operations.

(b) The total quantity of radioactive materials entering the general environment from the entire uranium fuel cycle, per gigawatt-year of electrical energy produced by the fuel cycle, contains less than 50,000 curies of krypton-85, 5 millicuries of iodine-129, and 0.5 millicuries combined of plutonium-239 and other alpha-emitting transuranic radionuclides with half-lives greater than 1 year.

It is to be noted, however, that the standards specified above in Sec. 190.10 of 40 CFR may be exceeded if the regulatory agency has granted a variance based upon its determination that a temporary and unusual operating condition exists and that continued operation of the facility is in the public interest.

Also, the above standards may be exceeded if information is promptly made a matter of public record explaining the nature of the unusual operating conditions, the degree to which this operation is expected to result in levels in excess of the standards, the basis of the variance, and the schedule for achieving conformance with the standards.

2. Health and Environmental Protection Standards

Part 192 of 40 CFR is concerned with the health and environmental standards for uranium and thorium mill tailings. Subpart A of Part 192 applies standards for the control of residual radioactive materials from inactive uranium-processing sites and for the restoration of such sites following any use of subsurface minerals.

In order to understand the standards, one must comprehend the terms used in defining them. Thus, the following definitions are offered for that purpose:

(a) Unless otherwise indicated in this subpart, all terms shall have the same meaning as in Title I of the Act.

(b) **Remedial** action means any action performed under Section 108 of the act.

(c) **Control** means any remedial action intended to stabilize, inhibit future misuse of, or reduce emissions or effluents from residual radioactive material.

(d) **Disposal site** means the region within the smallest perimeter of residual radioactive material (excluding cover materials) following completion of control activities.

(e) **Depository site** means a disposal site (other than a processing site) selected under Section 104(b) or 105(b) of the Act.

(f) **Curie** (Ci) means the amount of radioactive material that produces 37 billion nuclear transformations per second. One picocurie (pCi) = 10^{-12} Ci.

192.02 Standards

Control shall be designed[1] to:

(a) be effective for up to 1,000 years, to the extent reasonably achievable, and, in any case, for at least 200 years,

(b) provide reasonable assurance that releases of radon-222 from residual radioactive material to the atmosphere will not:

1. exceed an average[2] release rate of 20 picocuries per square meter per second, or

2. increase the annual average concentration of radon-222 in the air at or above any location outside the disposal site by more than one-half picocurie per liter.

---

[1]Because the standard applies to design, monitoring after disposal is not required to demonstrate compliance.

[2]This average shall apply over the entire surface of the disposal site and over at least a 1-year period. Radon will come from both residual radioactive materials and from materials covering them. Radon emissions from the covering materials should be estimated as part of developing a remedial action plan for each site. The standard, however, applies only to emissions from the residual radioactive materials to the atmosphere.

3. Standards for Cleanup of Land and Buildings Contaminated with Residual Materials from Inactive Uranium-Processing Sites

Subpart B, Sec. 192.10 applies to land and buildings that are a part of any processing site designated by the Secretary of Energy under Section 102 of the Uranium Mill Tailings Radiation Control Act of 1978. Under this Act, any site, including the mill, containing residual radioactive materials at which all or substantially all of the uranium was produced for sale to any Federal Agency before January 1, 1971, under a contract with any Federal Agency, except in the case of a site at or near Slick Rock, Colorado, unless such site was owned or controlled as of January 1, 1978, or is thereafter owned or controlled by any Federal Agency.

Also, some sites that were licensed by the Nuclear Regulatory Commission and its predecessor agency or by a state for the production of any uranium or thorium product or land near such sites that is determined to be contaminated with residual radioactive materials come under the regulatory standards of this section.

Again, it is necessary to define the terms used in this standard. There are 3 definitions: *land, working level (WL),* and *soil,* as follows:

(a) "Land" means any surface or subsurface land that is not part of a disposal site and is not covered by an occupiable building.

(b) "Working Level" (WL) means any combination of short-lived radon decay products in 1 liter of air that will result in the ultimate emission of alpha particles with a total energy of 130 billion electron volts.

(c) "Soil" means all unconsolidated materials normally found on or near the surface of the earth including, but not limited to, silts, clays, sands, gravel, and small rocks.

The standards state that remedial actions shall be conducted to provide reasonable assurance that the resultant residual radioactive materials from any designated processing site shall be such that:

(a) The concentration of radium-226 in land averaged over any area of 100 square meters shall not exceed the background level by more than:

(1) 5 pCi/g, averaged over the first 15 cm of soil below the surface, and

(2) 15 pCi/g, averaged over 15 cm thick layers of soil more than 15 cm below the surface.

(b) In any occupied or habitable building:

(1) The objective of remedial action shall be, and reasonable effort shall be made to achieve, an annual average (or equivalent radon decay product concentration (including background)) not to exceed 0.02 WL. In any case, the radon decay product concentration (including background) shall not exceed 0.03 WL, and

(2) The level of gamma radiation shall not exceed the background level by more than 20 microroentgens per hour.

4. Ground Water Protection

The radiation protection program of Subchapter F also contains provisions for protecting ground water supplies. Section 192.32(2) states that uranium by-product materials shall be managed so that they conform to the ground water

Table 9-1.   Table 1 of Sec. 264.94 - Maximum Concentration of Constituents for Ground-Water Protection

Constituent	Maximum concentration [1]
Arsenic	0.05
Barium	1.0
Cadmium	0.01
Chromium	0.05
Lead	0.05
Mercury	0.002
Selenium	0.01
Silver	0.05
Endrin (1,2,3,4,10,10-hexachloro-1,7-epoxy-1,4,4a,5,6,7,8,9a-octahydro-1, 4-endo, endo-5,8-dimethano naphthalene)	0.0002
Lindane (1,2,3,4,5,6-hexachlorocyclohexane, gamma isomer)	0.004
Methoxychlor (1,1,1-Trichloro-2,2-bis (p-methoxyphenylethane)	0.1
Toxaphene ($C_{10}H_{10}Cl_6$, Technical chlorinated camphene, 67–69 percent chlorine)	0.005
2,4-D (2,4-Dichlorophenoxyacetic acid)	0.1
2,4,5-TP Silvex (2,4,5-Trichlorophenoxypropionic acid)	0.01

[1] Milligrams per liter.

protection standards of Table 1 of 40 CFR Part 264, specifically Sec. 264.94. In addition, Table A of Sec. 192.34 has been added. Table 1 and Table A are given below as Table 9-1 and 9-2 of this text.

C. Ocean Dumping

Because of the dangers of the ecosystem and the belief that indiscriminate dumping in both coastal waters and the ocean can be damaging to future life on this planet, legislature has been enacted in an attempt to control such pollution. The United States is not alone in the endeavor, since other countries have controls that are similar but perhaps not so extensive.

While this text is primarily concerned with hazardous materials and waste, it is the latter aspect of hazardous materials that gives us cause for concern. If it

Table 9-2.   Table A of Sec. 192.34, 40 CFR

	pCi/liter
Combined radium-226 and radium-228	5
Gross alpha-particle activity (excluding radon and uranium)	15

were legally possible to dump any hazardous waste into the waters of the world, then rest assured that the profit motive would eventually turn our beautiful oceans and coastal waters into a vast cesspool.

It is for that reason that comments concerning Subchapter H, on Ocean Dumping of 40 CFR, have been included in this text.

1. Issuance of Permits

Both the EPA and the Corps of Engineers have certain authority in the statutes governing ocean dumping. The Corps of Engineers is concerned mainly in the review of activities involving the transportation of dredged material for the purpose of dumping it in ocean waters. There are, however, certain exceptions to those criteria, which shall be discussed later in the chapter.

The law states that no person shall transport from the United States any material for the purpose of dumping it into ocean waters. Also, no person shall dump any material transported from a location outside the United States into its territorial sea or into a zone contiguous to this country's territorial sea extending to a line 12 nautical miles seaward from the base line from which the breadth of the territorial sea is measured, to the extent that it may affect the territorial sea or the territory of the United States.

In accordance with international agreements the regulations and criteria included in Subchapter H apply the same standards and criteria binding upon the United States under the Convention on the Prevention of Marine Pollution by Dumping of Wastes and other matter.

There are certain exclusions to the regulations. For example, Subchapter H does not apply to *fish wastes,* and no permit is required for the transportation of those materials for dumping in the ocean. That does not mean, however, that fish wastes may be dumped in harbors or other protected or enclosed coastal waters, nor does it permit dumping where the result may reasonably be anticipated to endanger health, the environment, or ecological systems.

Oyster shells, which are used to propagate production of the bivalves, may be deposited on the seabed, usually in enclosed coastal waters and estuarial regions provided that the program is regulated by the state or Federal government when it is certified by the EPA or by the agency authorized to enforce the regulation. The National Oceanic and Atmospheric Administration, the U.S. Coast Guard, and the U.S. Army Corps of Engineers must concur in such placement as it may affect their responsibilities. Usually letters of concurrence may be obtained from those agencies to perform the seeding of oyster beds with shell deposits.

Permits are not required by vessels for the purpose of accomplishing the routine discharge of effluents incidental to propulsion or operating motor-driven equipment on board those vessels. It is not necessary or practical to obtain permits when emergency operations to safeguard life at sea require the dumping of materials into ocean waters from a vessel or aircraft. The owner or operator must, however, file timely reports of the incident in accordance with Sec. 224.2(b) of Subchapter H.

2. The Environmental Impact of Ocean Dumping

This Subpart B of Part 227 of 40 CFR sets specific environmental impact prohibitions, limits, and conditions for the dumping of materials into ocean waters. If the applicable prohibitions, limits, and conditions are satisfied, it is the determination of the EPA that the proposed disposal will not unduly degrade or endanger the marine environment and that the disposal will present no unacceptable adverse effects on human health and no significant damage to the resources of the marine environment; that it will present no unacceptable adverse effect on the marine ecosystem; that it will not present any unacceptable adverse persistent or permanent effects owing to the dumping of the particular volumes or concentrations of those materials; and that it will present no unacceptable adverse effect on the ocean for other uses as a result of direct environmental impact.

a. Prohibited Materials.   There are certain prohibited materials that cannot be issued permits under any circumstances, because the ocean dumping of the following materials will not be approved by the EPA or the Corps of Engineers:

(1) High-level radioactive wastes as defined in Sec. 227.30 of Subchapter H;

(2) Materials in whatever form (including without limitation, solids, liquids, semiliquids, gases, or organisms) produced or used for radiological, chemical or biological warfare;

(3) Materials insufficiently described by the applicant in terms of their compositions and properties to permit application of the environmental impact criteria;

(4) Persistent inert synthetic or natural materials that may float or remain in suspension in the ocean in such a manner that they may interfere materially with fishing, navigation, or other legitimate uses of the ocean.

b. Constituents Prohibited as Other than Trace Contaminants.   The ocean dumping or transportation for dumping of materials containing the following constituents as other than trace contaminants will not be approved on other than an emergency basis, with the exceptions included in paragraphs g, h, and i, of this section.

(1) Organohalogen compounds;

(2) Mercury and mercury compounds;

(3) Cadmium and cadmium compounds;

(4) Oil of any kind or in any form including but not limited to petroleum, oil sludge, oil refuse, crude oil, fuel oil, heavy diesel oil, lubricating oils, hydraulic fluids, and any mixtures containing them, transported for the purpose of dumping insofar ass these are not regulated under the FWPCA;

(5) Known carcinogens, mutagens, or teratogens or materials suspected to be carcinogens, mutagens, or teratogens by responsible scientific opinion.

c. Trace contaminants.   These constituents will be considered to be present as trace contaminants only when they occur in materials otherwise acceptable for ocean dumping in such forms and amounts in liquid, suspended particulate, and solid phases that the dumping of the materials will not cause significant undesirable effects, including the possibility of danger associated with their bioaccumulation in marine organisms.

d. Providing Constituent Acceptability.   The potential for significant undesirable effects caused by the presence of these constituents shall be determined by application of results of bioassays on liquid, suspended particulate, and solid phases of wastes according to procedures acceptable to the EPA, and for dredged material, acceptable to the EPA and the Corps of Engineers. Materials shall be deemed environmentally acceptable for ocean dumping only when the following conditions are met:

(1) The liquid phase does not contain any of these constituents in concentrations that will exceed applicable marine-water-quality criteria after allowance for initial mixing provided that mercury concentrations in the disposal site, after allowance for initial mixing, may exceed the average normal ambient concentrations of mercury in ocean waters at or near the dumping site that would be present in the absence of dumping by not more than 50 percent.

(2) Bioassay results on the suspended particulate phase of the waste do not indicate occurrence of significant mortality or significant adverse sublethal effects, including bioaccumulation resulting from the dumping of wastes containing the constituents listed in paragraph (a) of this section. The bioassays shall be conducted with appropriate sensitive marine organisms as defined in Sec. 227.27(c) of Part 227 of Subchapter H, using procedures for suspended particulate phase bioassays approved by the EPA or, for dredged material, approved by the EPA and the Corps of Engineers. Procedures approved for bioassays under this section will require exposure of organisms for a sufficient period of time and under appropriate conditions to provide reasonable assurance, based on consideration of the statistical significance of effects at the 95 percent confidence level, that, when the materials are dumped, no significant undesirable effects will occur because of either chronic toxicity or bioaccumulation of the constituents listed in paragraph b, above.

(3) Bioassay results on the solid phase of the wastes do not indicate occurrence of significant mortality or significant adverse sublethal effects attributable to the dumping of wastes containing the constituents listed in paragraph (b) of this section. The bioassays shall be conducted with the appropriate sensitive benthic marine organisms using benthic bioassay procedures approved by the EPA or, for dredged material, approved by the EPA and the Corps of Engineers. Procedures approved for bioassays under this section will require exposure of organisms for a sufficient period of time to provide reasonable assurance, based on considerations of statistical significance of effects at the 95 percent confidence level, that, when the materials are dumped, no significant undesirable effects will occur because of either chronic toxicity or bioaccumulation of the constituents listed in paragraph (b) of this section;

(4) For persistent organohalogens not included in the applicable marine water quality criteria, bioassay results on the liquid phase of the waste show that such compounds are not present in concentrations large enough to cause significant undesirable effects owing to either chronic toxicity or bioaccumulation in marine organisms after allowance for initial mixing.

e. Special Studies May Be Required. When the Administrator, Regional Administrator or District Engineer, as the case may be, has reasonable cause to believe that a material proposed for ocean dumping contains compounds identified as carcinogens, mutagens, or teratogens for which criteria have not been included in the applicable marine-water-quality criteria, he or she may require special studies to be done before issuance of a permit to determine the impact of disposal on human health and marine ecosystems.

f. The Validity of Interim Criteria. The criteria stated in paragraphs d(2) and (3) of this section will become mandatory as soon as announcement of the availability of acceptable procedures is made in the Federal Register. At that time the interim criteria contained in paragraph f of this section shall no longer be applicable. As interim measures, the criteria of paragraphs d(2) and (3) of this section may be applied on a case-by-case basis where interim guidance on acceptable bioassay procedures is provided by the Regional Administrator or, in the case of dredged material, by the District Engineer; in the absence of such guidance, permits may be issued for the dumping of any material only when the following conditions are met, except under an emergency permit:

(1) Mercury and its compounds are present in any solid phase of a material in concentrations less than 0.75 mg/kg, or less than 50 percent greater than the average total mercury content of natural sediments of similar lithologic characteristics as those at the disposal site;

(2) Cadmium and its compounds are present in any solid phase of a material in concentrations less than 0.6 mg/kg, or less than 50 percent greater than the average total cadmium content of natural sediments of similar lithologic characteristics as those at the disposal site;

(3) The total concentration of organohalogen constituents in the waste as transported for dumping is less than a concentration of such constituents known to be toxic to marine organisms. In calculating the concentration of organohalogens, the applicant shall consider that those constituents are all biologically available. The determination of the toxicity value will be based on existing scientific data or developed by the use of bioassays conducted in accordance with approved EPA procedures;

(4) The total amounts of oils and greases as identified in paragraph b(4) of this section do not produce a visible surface sheen in an undisturbed water sample when added at a ratio of 1 part waste material to 100 parts of water.

g. Nontoxic Compounds. The prohibitions and limitations of this section do not apply to the constituents identified in paragraph b of this section when the applicant can demonstrate that such constituents are (1) present in the material only as chemical compounds or forms (e.g., inert insoluble solid materials) nontoxic to marine life and nonbioaccumulative in the marine environment upon disposal and thereafter or (2) present in the material only as chemical compounds or forms that, at the time of dumping and thereafter, will be rapidly rendered nontoxic to marine life and nonbioaccumulative in the marine environment by chemical or biological degradation in the sea provided that they will not make edible marine organisms unpalatable or will not endanger human health or that of domestic animals, fish, shellfish, or wildlife.

h. Physical, Chemical, or Biological Changes to Constituents. The prohibitions and limitations of this section do not apply to the constituents identified in paragraph b of this section for the granting of research permits if the substances are rapidly rendered harmless by physical, chemical, or biological processes in the sea provided that they will not make edible marine organisms unpalatable and will not endanger human health or that of domestic animals.

i. Incineration at Sea. The prohibitions and limitations of this section do not apply to the constituents identified in paragraph b of this section for the granting of permits for the transport of these substances for the purpose of incineration at sea if the applicant can demonstrate that the stack emissions consist of substances that are rapidly rendered harmless by physical, chemical, or biological processes in the sea. Incinerator operations shall comply with requirements that will be established on a case-by-case basis.

3. Establishing Limits for Specific Wastes or Waste Constituents

Materials containing the following constituents must meet the additional limitations specified in this section to be deemed acceptable for ocean dumping:

(a) Liquid waste constituents immiscible with or slightly soluble in seawater, such as benzene, xylene, carbon disulfide, and toluene, may be dumped only when they are present in the waste in concentrations below their solubility limits in seawater. That provision does not apply to materials that may interact with ocean water to form insoluble materials;

(b) Radioactive materials, other than those prohibited by Sec. 227.5 of 40 CFR, must be contained in accordance with the provisions of Sec. 227.11 of 40 CFR to prevent their direct dispersion or dilution in ocean waters;

(c) Wastes containing living organisms may not be dumped if the organisms present would endanger human health or that of domestic animals, fish, shellfish, and wildlife by:

(1) Extending the range of biological pests, viruses, pathogenic microorganisms, or other agents capable of infesting, infecting, or extensively and permanently altering the normal populations of organisms;

(2) Degrading uninfected areas;

(3) Introducing viable species not indigenous to an area.

(d) In the dumping of wastes of highly acidic or alkaline nature into the ocean, consideration shall be given to:

(1) The effects of any change in acidity or alkalinity of the water at the disposal site;

(2) The potential for synergistic effects or for the formation of toxic compounds at or near the disposal site. Allowance may be made in the permit conditions for the capability of ocean waters to neutralize acid or alkaline wastes provided, however, that dumping conditions must be such that the average total alkalinity or total acidity of the ocean water after allowance for initial mixing, as defined in Sec. 227.29 of 40 CFR, may be changed, based on stoichiometric calculations, by no more than 10 percent during all dumping operations at a site to neutralize acid or alkaline wastes.

(e) Wastes containing biodegradable constituents or constituents that consume oxygen in any fashion may be dumped in the ocean only under conditions in which the dissolved oxygen after allowance for initial mixing, as defined in Sec. 227.29 of 40 CFR, will not be depressed by more than 25 percent below the normally anticipated ambient conditions in the disposal area at the time of dumping.

The government has indicated, through Subchapter H, that no wastes will be deemed acceptable for ocean dumping unless they can be dumped so as not to exceed certain permissible concentrations.

In addition, substances that may damage the ocean environment because of the quantities in which they are dumped or that may seriously reduce amenities may be dumped only when the quantities to be dumped at a single time and place are controlled to prevent long-term damage to the environment or to amenities. Wastes that may present a serious obstacle to fishing or navigation may be dumped only at disposal sites and under conditions that will ensure no unacceptable interference with fishing or navigation. Wastes that may present a hazard to shorelines or beaches may be dumped only at sites and under conditions that will ensure no unacceptable danger to shorelines or beaches.

4. Containerized Waste, Insoluble Waste, and Dredged Materials

a. Containerized Waste. Wastes containerized solely for transport to the dumping site and expected to rupture or leak on impact or shortly thereafter must meet the appropriate requirements. Other containerized wastes will be approved for dumping only under the following conditions:

(1) The materials to be disposed of decay, decompose, or radiodecay to environmentally innocuous materials within the life expectancy of the containers, and their inert matrix or both;

(2) Materials to be dumped are present in such quantities and are of such nature that only short-term localized adverse effects will occur should the containers rupture at any time;

(3) Containers are dumped at depths and locations where they will cause no threat to navigation, fishing, shorelines, or beaches.

b. Insoluble Wastes. Solid wastes consisting of inert natural minerals or materials compatible with the ocean environment may be generally approved for ocean dumping provided that they are insoluble above the applicable trace or limiting permissible concentrations and will settle rapidly and completely and are of a particle size and density that they would be deposited or rapidly dispersed without damage to benthic, demersal, or pelagic biota. (See the Glossary for a definition of those terms.)

Persistent inert synthetic or natural materials that may float or remain in suspension in the ocean are prohibited, except that they may be dumped in the ocean only when they have been processed in such a fashion that they will sink to the bottom and remain in place.

c. Dredged Materials. Dredged materials are bottom sediments or materials that have been dredged or excavated from the navigable waters of the United States and their disposal into ocean waters is regulated by the U.S. Army Corps of Engineers. Dredged material consists primarily of natural sediments or materials that may be contaminated by municipal or industrial wastes or by runoff from terrestrial sources, such as agricultural lands.

Dredged material that meets the criteria set forth in the following paragraphs (1), (2), or (3) of this section is environmentally acceptable for ocean dumping without further testing under this section:

(1) Dredged material is composed predominantly of sand, gravel, rock, or any other naturally occurring bottom material with particle sizes larger than silt, and the material is found in areas of high current or wave energy such as streams with large bed loads or coastal areas with shifting bars and channels;

(2) Dredged material is for beach nourishment or restoration and is composed predominantly of sand, gravel, or shell with particle sizes compatible with material on the receiving beaches;

(3) When the material proposed for dumping is substantially the same as the substrate at the proposed disposal site and the site from which the material proposed for dumping is to be taken is far removed from known existing and historical sources of pollution so as to provide reasonable assurance that such material has not been contaminated by such pollution.

5. Dumping Sites

A number of dumping sites have been tentatively approved for most of the coastal waters of the United States. The management authority for those sites has been delegated to the EPA organization in the region in which the site is located. The sites are listed under Subchapter H, Sec. 228.12 of 40 CFR and, as indicated, have been approved on an interim basis pending completion of baseline or trend assessment surveys and designation for continuation or termination.

D. Solid Waste and Hazardous Waste

1. Solid Waste

Of particular concern when one is dealing with the problems of the disposal of solid wastes in land-based sites as opposed to ocean dumping is the matter of ground water contamination. The EPA has developed specific criteria for various inorganic compounds that are best described in the following tabulations (Tables 9-3, 9-4, and 9-5); as follows:

In addition to the chemical contaminants noted, above, in Tables 9-3, 9-4, and 9-5, there are certain, prescribed maximum microbiological contaminant levels that have been established. Using the membrane filter technique, the maximum contaminant level for coliform bacteria from any one well is as follows:

(1) Four coliform bacteria per 100 milliliters if 1 sample is taken, or,
(2) Four coliform bacteria per 100 milliliters in more than 1 sample of all the samples analyzed in 1 month.

In the 5 tube most probable number procedure (the fermentation tube method) in accordance with the analytical recommendations set forth in "Standard Methods for Examination of Water and Waste Water," American Public Health Association, 13th Ed. pp. 662–688, and using a standard sample, each portion being one fifth of the sample, then:

Table 9-3. Maximum Contaminant Levels for Inorganic Chemicals (Maximum levels of inorganic chemicals other than fluoride.)

Contaminant	Level (milligrams per liter)
Arsenic	0.05
Barium	1
Cadmium	0.010
Chromium	0.05
Lead	0.05
Mercury	0.002
Nitrate (as N)	10
Selenium	0.01
Silver	0.05

Table 9-4. The Maximum Contaminant Levels for Fluoride.

Temperature [1] degrees Fahrenheit	Degrees Celsius	Level (milligrams per liter)
53.7 and below	12 and below	2.4
53.8 to 58.3	12.1 to 14.6	2.2
58.4 to 63.8	14.7 to 17.6	2.0
63.9 to 70.6	17.7 to 21.4	1.8
70.7 to 79.2	21.5 to 26.2	1.6
79.3 to 90.5	26.3 to 32.5	1.4

[1] Annual average of the maximum daily air temperature.

Table 9-5. Maximum Contaminant Levels for Organic Chemicals

	Level (milligrams per liter)
(a) Chlorinated hydrocarbons:	
Endrin (1,2,3,4,10,10-Hexachloro-6,7-epoxy-1,4,4a,5,6,7,8,8a-octahydro-1,4-endo, endo-5,8-dimethano naphthalene)	0.0002
Lindane (1,2,3,4,5,6-Hexachlorocyclohexane, gamma isomer	0.004
Methoxychlor (1,1,1-Trichloro-2,2-bis (p-methoxyphenyl) ethane)	0.1
Toxaphene (C$_{10}$H$_{10}$Cl$_8$-Technical chlorinated camphene, 67 to 69 percent chlorine)........	0.005
(b) Chlorophenoxys:	
2,4-D (2,4-Dichlorophenoxy-acetic acid)........	0.1
2,4,5-TP Silvex (2,4,5-Trichlorophen- oxy-propionic acid)	0.01

(1) If the standard portion is 10 milliliters, coliform in any 5 consecutive samples from a well shall not be present in 3 or more of the 25 portions, or

(2) If the standard portion is 100 milliliters, coliform in any 5 consecutive samples from a well shall not be present in 5 portions in any of 5 samples or in more than 15 of the 25 portions.

The maximum contaminant levels for radium-226, radium-228, and gross alpha particle radioactivity are as follows:

(a) Combined radium-226 and radium-228 is 5 pCi/1;

(b) Gross alpha particle radioactivity (including radium-226, but excluding radon and uranium) is 15 pCi/1.

In addition to ground water contamination, the use of sewage sludge and septic tank pumpings that has a viable impact on the economy in rejuvenating non-productive acreage must also be carefully controlled. Sewage sludge or septic tank pumpings that are applied to land surfaces or are incorporated into the soil for their fertilization value must be treated by a *process to further reduce pathogens* before application or incorporation if the crops for direct human consumption are grown within 18 months of fertilization. Such treatment is not required if there is no contact between the solid waste and the edible portion of the crop. In that event, the solid waste is treated by a *process to significantly reduce pathogens* before application, and public access to the area is controlled for at least 12 months, and grazing of animals whose products are consumed by humans is prevented for at least 1 month.

The methods to *significantly reduce* and to *further reduce* pathogens are given below:

(1) Processes to Significantly Reduce Pathogens

Aerobic digestion: The process is conducted by agitating sludge with air or oxygen to maintain aerobic conditions at residence times ranging from 60 days at 15° C to 40 days at 20° C, with volatile solids reduction at least 38 percent.

Air Drying: Liquid sludge is allowed to drain or dry on underdrained sand beds or paved or unpaved basins in which the sludge is at a depth of 9 inches. A minimum of 3 months is needed, 2 months of which temperatures average on a daily basis above 0° C.

Anaerobic digestion: The process is conducted in the absence of air at residence times ranging from 60 days at 20° C to 15 days at 35° C, with a volatile solids reduction of at least 38 percent.

Composting: Using the within-vessel, static aerated pile or windrow composting methods, the solid waste is maintained at minimum operating conditions of 40° C for 5 days. For 4 hours during this period the temperature exceeds 55° C.

Lime Stabilization: Sufficient lime is added to produce a pH of 12 after 2 hours of contact.

Other Methods: Other methods or operating conditions may be acceptable if pathogens and vector attraction of the waste (volatile solids) are reduced to an extent equivalent to the reduction achieved by any of the above methods.

(2) Processes to Further Reduce Pathogens

Composting: Using the within-vessel composting method, the solid waste is maintained at operating conditions of 55° C or greater for 3 days. Using the static aerated pile composting method, the solid waste is maintained at operating conditions of 55° C or greater for 3 days. Using the windrow composting method, the solid waste attains a temperature of 55° C or greater for at least 15 days during the composting period. Also, during the high temperature period, there will be a minimum of 5 turnings of the windrow.

Heat drying: Dewatered sludge cake is dried by direct or indirect contact with hot gases, and moisture content is reduced to 10 percent or lower. Sludge particles reach temperatures well in excess of 80° C, or the wet bulb temperature of the gas stream in contact with the sludge at the point where it leaves the dryer is in excess of 80° C.

Heat treatment: Liquid sludge is heated to temperatures of 180° C for 30 minutes.

Thermophilic Aerobic Digestion: Liquid sludge is agitated with air or oxygen to maintain aerobic conditions at residence times of 10 days at 55° to 60° C, with a volatile solids reduction of at least 38 percent.

Other methods: Other methods or operating conditions may be acceptable if pathogens and vector attraction of the waste (volatile solids) are reduced to an extent equivalent to the reduction achieved by any of the above methods.

Any of the processes listed below, if added to the processes described in Processes to Significantly Reduce Pathogens above, further reduce pathogens. Because the processes listed below, on their own, do not reduce the attraction of disease vectors[1] they are only add-on in nature.

Beta ray irradiation: Sludge is irradiated with beta rays from an accelerator at dosages of at least 1.0 megarad at room temperature (ca. 20° C).

Gamma ray irradiation: Sludge is irradiated with gamma rays from certain isotopes, such as ^{60}Cobalt and ^{137}Cesium, at dosages of at least 1.0 megarad at room temperature (ca. 20° C).

Pasteurization: Sludge is maintained for at least 30 minutes at a minimum temperature of 70°C.

Other methods: Other methods or operating conditions may be acceptable if pathogens are reduced to an extent equivalent to the reduction achieved by any of the above add-on methods.

2. Hazardous Waste

The definition of hazardous waste is involved and cumbersome. The EPA regulations that govern this subject field are legalistic and difficult to simplify,

---

[1] Disease vector means rodents, flies, and mosquitoes capable of transmitting diseases to humans.

yet for the purposes of this text the following paragraphs will provide a starting point that will serve as a basis for further definition and delination.

Hazardous wastes can be 1 or more of the following spent solvents, for example:

> carbon tetrachloride
> tetrachloroethylene
> trichloroethylene
> methylene chloride
> 1,1,1,-trichloroethane
> chlorobenzene
> o-dichlorobenzene
> cresols
> cresylic acid
> nitrobenzene
> toluene
> methyl ethyl ketone
> carbondisulfide
> isobutanol
> pyridine
> chlorofluorocarbons
> heat exchanger bundle cleaning sludge (from petroleum refining)
> waste water resulting from laboratory operations containing toxic wastes
> any solid waste generated from the treatment, storage, or disposal of a hazardous waste, including any sludge, spill residue, ash, emission control dust, or leachate
> waste pickle liquor sludge generated by lime stabilization of spent pickle liquor from the iron and steel industry

3. Special Requirements for Small Hazardous Waste Generators

A generator is a small-quantity generator if he or she generates less than 1,000 kilograms of hazardous waste in a month. As such, he or she is not subject to the full regulation of the laws governing hazardous wastes. In determining the quantity of hazardous waste generated, a generator need not include:

1. hazardous waste when it is removed from on-site storage;
2. hazardous waste produced by on-site treatment of hazardous waste.

The small-quantity generator should be careful not to mix solid wastes with hazardous wastes, because if the resultant mixture exceeds 1,000 kilograms he or she is subject to the full regulation of the law.

4. Residues of Hazardous Wastes in Empty Containers

Any hazardous waste remaining in an empty container or an inner lining from an empty container is not subject to regulation; however, any hazardous wastes remaining in containers that are not considered empty are considered hazardous wastes. The concept of "empty" is based upon having wastes removed by pouring, pumping, or aspirating, and no more than 2.5 centimeters (approximately 1 inch) of residue remains on the bottom of the container or inner liner. Other measures of "empty" are when no more than 3 percent by weight of the total capacity of the container remains in the container or inner liner if the container is less than or equal to 110 gallons in size or when a container that has held a hazardous compressed gas is empty when the pressure in the container is at or approaches atmospheric.

If the container or inner liner has been triple rinsed using a solvent capable of removing the commercial chemical or manufacturing chemical intermediate or if the container is cleaned by a method that is scientifically established as achieving equivalent removal or if the inner liner that prevented contact of the chemical with the container has been removed.

5. Criteria for Identifying Hazardous Wastes and HW Characteristics

Solid wastes that may cause or significantly contribute to an increase in mortality or an increase in a serious, an irreversible, or an incapacitating reversible illness are deemed to be hazardous. The same is true of any solid waste that poses a substantial present or potential threat to human health or the environment when it is improperly treated, stored, transported, disposed of, or otherwise managed, and when the characteristic may be measured by an available standardized test method that is reasonably within the capability of generators of solid waste or private sector laboratories that are available to serve generators of solid waste or may be reasonably detected by generators of solid waste through their knowledge of their waste.

The EPA lists a solid waste as a hazardous waste only upon determining that the solid waste meets 1 of the following criteria:

(1) It exhibits any of the characteristics of hazardous waste identified below.

(2) It has been found to be fatal to humans in low doses or, in the absence of data on human toxicity, it has been shown in studies to have an oral LD 50 toxicity (rat) of less than 50 milligrams per kilogram, an inhalation LC 50 toxicity (rat) of less than 2 milligrams per liter, or a dermal LD 50 toxicity (rabbit) of less than 200 milligrams per kilogram or is otherwise capable of causing or significantly contributing to an increase in serious irreversible or incapacitating reversible illness. (Waste listed in accordance with those criteria is designated Acute Hazardous Waste.)

(3) It contains any of the toxic constituents listed in Table 9-6, unless, after considering any of the following factors, the Administrator of the EPA concludes that the waste is not capable of posing a substantial present or potential hazard to human health or the environment when it is improperly treated, stored, transported or disposed of, or otherwise managed:

(i) The nature of the toxicity presented by the constituent.

(ii) The concentration of the constituent in the waste.

(iii) The potential of the constituent of any toxic degradation product of the constituent to migrate from the waste into the environment under the types of improper management considered in paragraph (a)(3)(vii) of this section.

(iv) The persistence of the constituent of any toxic degradation product of the constituent.

(v) The potential for the constituent or any toxic degradation product of the constituent to degrade into nonharmful constituents and the rate of degradation.

Table 9-6.   Hazardous Constituents (Appendix VIII of Part 261, 40 CFR)

Acetonitrile (Ethanenitrile)
Acetophenone (Ethanone, 1-phenyl)
3-(alpha-Acetonylbenzyl)-4-
hydroxycoumarin and salts (Warfarin)
2-Acetylaminofluorene (Acetamide, N-(9H-
fluoren-2-yl)-)
Acetyl chloride (Ethanoyl chloride)
1-Acetyl-2-thiourea (Acetamide, N-(amin-
othioxomethyl)-)
Acrolein (2-Propenal)
Acrylamide (2-Propenamide)
Acrylonitrile (2-Propenenitrile)
Aflatoxins
Aldrin   (1,2,3,4,10,10-Hexachloro-
1,4,4a,5,8,8a,8b-hexahydro-endo,exo-
1,4:5,8-Dimethanonaphthalene)
Allyl alcohol (2-Propen-1-ol)
Aluminum phosphide
4-Aminobiphenyl ([1,1'-Biphenyl]-4-amine)
6-Amino-1,1a,2,8,8a,8b-hexahydro-8-
(hydroxymethyl)-8a-methoxy-5-methyl-
carbamate   azirino[2',3':3,4]pyrrolo[1,2-
a]indole-4,7-dione, (ester) (Mitomycin C)
(Azirino[2'3':3,4]pyrrolo(1,2-a)indole-4,7-
dione,   6-amino-8-[((amino-
carbonyl)oxy)methyl]-1,1a,2,8,8a,8b-
hexahydro-8amethoxy-5-methy-)
5-(Aminomethyl)-3-isoxazolol (3(2H)-Isoxa-
zolone, 5-(aminomethyl)-) 4-Aminopyri-
dine (4-Pyridinamine)
Amitrole (1H-1,2,4-Triazol-3-amine)
Aniline (Benzenamine)
Antimony and compounds, N.O.S.*
Aramite (Sulfurous acid, 2-chloroethyl-, 2-
[4-(1,1-dimethylethyl)phenoxy]-1-
methylethyl ester)

Arsenic and compounds, N.O.S.*
Arsenic acid (Orthoarsenic acid)
Arsenic pentoxide (Arsenic (V) oxide)
Arsenic trioxide (Arsenic (III) oxide)
Auramine   (Benzenamine,   4,4'-
carbonimidoylbis[N,N-Dimethyl-,   mono-
hydrochloride)
Azaserine (L-Serine, diazoacetate (ester))
Barium and compounds, N.O.S.*
Barium cyanide
Benz[c]acridine (3,4-Benzacridine)
Benz[a]anthracene (1,2-Benzanthracene)
Benzene (Cyclohexatriene)
Benzenearsonic acid (Arsonic acid, phenyl-)
Benzene, dichloromethyl- (Benzal chloride)
Benzenethiol (Thiophenol)
Benzidine ([1,1'-Biphenyl]-4,4'diamine)
Benzo[b]fluoranthene (2,3-Benzofluoranth-
ene)

Benzo[j]fluoranthene   (7,8-Benzofluoranth-
ene)
Benzo[a]pyrene (3,4-Benzopyrene)
p-Benzoquinone (1,4-Cyclohexadienedione)
Benzotrichloride (Benzene, trichloro-
methyl)
Benzyl chloride (Benzene, (chloromethyl)-)
Beryllium and compounds, N.O.S.*
Bis(2-chloroethoxy)methane (Ethane, 1,1'-
[methylenebis(oxy)]bis[2-chloro-])
Bis(2-chloroethyl) ether (Ethane, 1,1'-
oxybis[2-chloro-])
N,N-Bis(2-chloroethyl)-2-naphthylamine
(Chlornaphazine)
Bis(2-chloroisopropyl) ether (Propane, 2,2'-
oxybis[2-chloro-])
Bis(chloromethyl)   ether   (Methane,
oxybis[chloro-])
Bis(2-ethylhexyl) phthalate (1,2-Benzenedi-
carboxylic acid, bis(2-ethylhexyl) ester)
Bromoacetone (2-Propanone, 1-bromo-)
Bromomethane (Methyl bromide)
4-Bromophenyl phenyl ether (Benzene, 1-
bromo-4-phenoxy-)
Brucine (Strychnidin-10-one, 2,3-dimethoxy-
)
2-Butanone peroxide (Methyl ethyl ketone,
peroxide)
Butyl benzyl phthalate (1,2-Benzenedicar-
boxylic acid, butyl phenylmethyl ester)
2-sec-Butyl-4,6-dinitrophenol   (DNBP)
(Phenol, 2,4-dinitro-6-(1-methylpropyl)-)
Cadmium and compounds, N.O.S.*
Calcium chromate (Chromic acid, calcium
salt)
Calcium cyanide
Carbon disulfide (Carbon bisulfide)
Carbon oxyfluoride (Carbonyl fluoride)
Chloral (Acetaldehyde, trichloro-)
Chlorambucil (Butanoic acid, 4-[bis(2-
chloroethyl)amino]benzene-)
Chlordane (alpha and gamma isomers) (4,7-
Methanoindan,   1,2,4,5,6,7,8,8-octachloro-
3,4,7,7a-tetrahydro-) (alpha and gamma
isomers)
Chlorinated benzenes, N.O.S.*
Chlorinated ethane, N.O.S.*
Chlorinated fluorocarbons, N.O.S.*

Chlorinated naphthalene, N.O.S.*
Chlorinated phenol, N.O.S.*
Chloroacetaldehyde (Acetaldehyde, chloro-)
Chloroalkyl ethers, N.O.S.*
p-Chloroaniline (Benzenamine, 4-chloro-)
Chlorobenzene (Benzene, chloro-)

---

*The abbreviation N.O.S. (not otherwise
specified) signifies those members of the
general class not specifically listed by name
in this appendix.

Table 9-6.   Hazardous Constituents (Continued)

Chlorobenzilate (Benzeneacetic acid, 4-chloro-alpha-(4-chlorophenyl)-alpha-hydroxy-, ethyl ester)

2-Chloro-1, 3-butadiene (chloroprene)

p-Chloro-m-cresol (Phenol, 4-chloro-3-methyl)

1-Chloro-2,3-epoxypropane (Oxirane, 2-(chloromethyl)-)

2-Chloroethyl vinyl ether (Ethene, (2-chloroethoxy)-)

Chloroform (Methane, trichloro-)

Chloromethane (Methyl chloride)

Chloromethyl methyl ether (Methane, chloromethoxy-)

2-Chloronaphthalene (Naphthalene, beta-chloro-)

2-Chlorophenol (Phenol, o-chloro-)

1-(o-Chlorophenyl)thiourea (Thiourea, (2-chlorophenyl)-)

3-Chloropropene (allyl chloride)

3-Chloropropionitrile (Propanenitrile, 3-chloro-)

Chromium and compounds, N.O.S.*

Chrysene (1,2-Benzphenanthrene)

Citrus red No. 2 (2-Naphthol, 1-[(2,5-dimethoxyphenyl)azo]-)

Coal tars

Copper cyanide

Creosote (Creosote, wood)

Cresols (Cresylic acid) (Phenol, methyl-)

Crotonaldehyde (2-Butenal)

Cyanides (soluble salts and complexes), N.O.S.*

Cyanogen (Ethanedinitrile)

Cyanogen bromide (Bromine cyanide)

Cyanogen chloride (Chlorine cyanide)

Cycasin (beta-D-Glucopyranoside, (methyl-ONN-azoxy)methyl-)

2-Cyclohexyl-4,6-dinitrophenol (Phenol, 2-cyclohexyl-4,6-dinitro-)

Cyclophosphamide (2H-1,3,2,-Oxazaphosphorine, [bis(2-chloroethyl)amino]-tetrahydro-, 2-oxide)

Daunomycin (5,12-Naphthacenedione, (8S-cis)-8-acetyl-10-[(3-amino-2,3,6-trideoxy)-alpha-L-lyxo-hexopyranosyl)oxy]-7,8,9,10-tetrahydro-6,8,11-trihydroxy-1-methoxy-)

DDD (Dichlorodiphenyldichloroethane) (Ethane, 1,1-dichloro-2,2-bis(p-chlorophenyl)-)

DDE (Ethylene, 1,1-dichloro-2,2-bis(4-chlorophenyl)-)

DDT (Dichlorodiphenyltrichloroethane) (Ethane, 1,1,1-trichloro-2,2-bis(p-chlorophenyl)-)

Diallate (S-(2,3-dichloroallyl) diisopropylthiocarbamate)

Dibenz[a,h]acridine (1,2,5,6-Dibenzacridine)

Dibenz[a,j]acridine (1,2,7,8-Dibenzacridine)

Dibenz[a,h]anthracene (1,2,5,6-Dibenzanthracene)

7H-Dibenzo[c,g]carbazole (3,4,5,6-Dibenzcarbazole)

Dibenzo[a,e]pyrene (1,2,4,5-Dibenzpyrene)

Dibenzo[a,h]pyrene (1,2,5,6-Dibenzpyrene)

Dibenzo[a,i]pyrene (1,2,7,8-Dibenzpyrene)

1,2-Dibromo-3-chloropropane (Propane, 1,2-dibromo-3-chloro-)

1,2-Dibromoethane (Ethylene dibromide)

Dibromomethane (Methylene bromide)

Di-n-butyl phthalate (1,2-Benzenedicarboxylic acid, dibutyl ester)

o-Dichlorobenzene (Benzene, 1,2-dichloro-)

m-Dichlorobenzene (Benzene, 1,3-dichloro-)

p-Dichlorobenzene (Benzene, 1,4-dichloro-)

Dichlorobenzene, N.O.S.* (Benzene, dichloro-, N.O.S.*)

3,3'-Dichlorobenzidine ([1,1'-Biphenyl]-4,4'-diamine, 3,3'-dichloro-)

1,4-Dichloro-2-butene (2-Butene, 1,4-dichloro-)

Dichlorodifluoromethane (Methane, dichlorodifluoro-)

1,1-Dichloroethane (Ethylidene dichloride)

1,2-Dichloroethane (Ethylene dichloride)

trans-1,2-Dichloroethene (1,2-Dichloroethylene)

Dichloroethylene, N.O.S.* (Ethene, dichloro-, N.O.S.*)

1,1-Dichloroethylene (Ethene, 1,1-dichloro-)

Dichloromethane (Methylene chloride)

2,4-Dichlorophenol (Phenol, 2,4-dichloro-)

2,6-Dichlorophenol (Phenol, 2,6-dichloro-)

2,4-Dichlorophenoxyacetic acid (2,4-D), salts and esters (Acetic acid, 2,4-dichlorophenoxy-, salts and esters)

Dichlorophenylarsine (Phenyl dichloroarsine)

Dichloropropane, N.O.S.* (Propane, dichloro-, N.O.S.*)

1,2-Dichloropropane (Propylene dichloride)

Dichloropropanol, N.O.S.* (Propanol, dichloro-, N.O.S.*)

Dichloropropene, N.O.S.* (Propene, dichloro-, N.O.S.*)

1,3-Dichloropropene (1-Propene, 1,3-dichloro-)

Dieldrin (1,2,3,4,10.10-hexachloro-6,7-epoxy-1,4,4a,5,6,7,8,8a-octa-hydro-endo,exo-1,4:5,8-Dimethanonaphthalene)

1,2:3,4-Diepoxybutane (2,2'-Bioxirane)

Diethylarsine (Arsine, diethyl-)

N,N-Diethylhydrazine (Hydrazine, 1,2-diethyl)

O,O-Diethyl S-methyl ester of phosphorodithioic acid (Phosphorodithioic acid, O,O-diethyl S-methyl ester)

O,O-Diethylphosphoric acid, O-p-nitrophenyl ester (Phosphoric acid, diethyl p-nitrophenyl ester)

Table 9-6.   Hazardous Constituents (Continued)

Diethyl phthalate (1,2-Benzenedicarboxylic acid, diethyl ester)

O,O-Diethyl O-2-pyrazinyl phosphorothioate (Phosphorothioic acid, O,O-diethyl O-pyrazinyl ester

Diethylstilbesterol   (4,4'-Stilbenediol, alpha,alpha-diethyl, bis(dihydrogen phosphate, (E)-)

Dihydrosafrole (Benzene, 1,2-methylene-dioxy-4-propyl-)

3,4-Dihydroxy-alpha-(methylamino)methyl benzyl alcohol (1,2-Benzenediol, 4-[1-hydroxy-2-(methylamino)ethyl]-)

Diisopropylfluorophosphate (DFP) (Phosphorofluoridic acid, bis(1-methylethyl) ester)

Dimethoate (Phosphorodithioic acid, O,O-dimethyl S-[2-(methylamino)-2-oxoethyl] ester

3,3'-Dimethoxybenzidine   ([1,1'-Biphenyl]-4,4'diamine, 3-3'-dimethoxy-)

p-Dimethylaminoazobenzene (Benzenamine, N,N-dimethyl-4-(phenylazo)-)

7,12-Dimethylbenz[a]anthracene   (1,2-Benzanthracene, 7,12-dimethyl-)

3,3'-Dimethylbenzidine ([1,1'-Biphenyl]-4,4'-diamine, 3,3'-dimethyl-) .).

Dimethylcarbamoyl chloride (Carbamoyl chloride, dimethyl-)

1,1-Dimethylhydrazine (Hydrazine, 1,1-dimethyl-)

1,2-Dimethylhydrazine (Hydrazine, 1,2-dimethyl-)

3,3-Dimethyl-1-(methylthio)-2-butanone, O-[(methylamino) carbonyl]oxime (Thiofanox)

alpha,alpha-Dimethylphenethylamine (Ethanamine, 1,1-dimethyl-2-phenyl-)

2,4-Dimethylphenol (Phenol, 2,4-dimethyl-)

Dimethyl phthalate (1,2-Benzenedicarboxylic acid, dimethyl ester)

Dimethyl sulfate (Sulfuric acid, dimethyl ester)

Dinitrobenzene, N.O.S.* (Benzene, dinitro-, N.O.S.*)

4,6-Dinitro-o-cresol and salts (Phenol, 2,4-dinitro-6-methyl-, and salts)

2,4-Dinitrophenol (Phenol, 2,4-dinitro-)

2,4-Dinitrotoluene (Benzene, 1-methyl-2,4-dinitro-)

2,6-Dinitrotoluene (Benzene, 1-methyl-2,6-dinitro-)

Di-n-octyl phthalate (1,2-Benzenedicarboxylic acid, dioctyl ester)

1,4-Dioxane (1,4-Diethylene oxide)

Diphenylamine (Benzenamine, N-phenyl-)

1,2-Diphenylhydrazine (Hydrazine, 1,2-diphenyl-)

Di-n-propylnitrosamine (N-Nitroso-di-n-propylamine)

Disulfoton   (O,O-diethyl   S-[2-(ethylthio)ethyl] phosphorodithioate)

2,4-Dithiobiuret (Thioimidodicarbonic diamide)

Endosulfan (5-Norbornene, 2,3-dimethanol, 1,4,5,6,7,7-hexachloro-, cyclic sulfite)

Endrin and metabolites (1,2,3,4,10,10-hexachloro-6,7-epoxy-1,4,4a,5,6,7,8,8a-octahydro-endo,endo-1,4:5,8-dimethanonaphthalene, and metabolites)

Ethyl carbamate (Urethan) (Carbamic acid, ethyl ester)

Ethyl cyanide (propanenitrile)

Ethylenebisdithiocarbamic acid, salts and esters   (1,2-Ethanediylbiscarbamodithioic acid, salts and esters

Ethyleneimine (Aziridine)

Ethylene oxide (Oxirane)

Ethylenethiourea (2-Imidazolidinethione)

Ethyl methacrylate (2-Propenoic acid, 2-methyl-, ethyl ester)

Ethyl methanesulfonate (Methanesulfonic acid, ethyl ester)

Fluoranthene (Benzo[j,k]fluorene)

Fluorine

2-Fluoroacetamide (Acetamide, 2-fluoro-)

Fluoroacetic acid, sodium salt (Acetic acid, fluoro-, sodium salt)

Formaldehyde (Methylene oxide)

Formic acid (Methanoic acid)

Glycidylaldehyde (1-Propanol-2,3-epoxy)

Halomethane, N.O.S.*

Heptachlor   (4,7-Methano-1H-indene, 1,4,5,6,7,8,8-heptachloro-3a,4,7,7a-tetrahydro-)

Heptachlor epoxide (alpha, beta, and gamma isomers) (4,7-Methano-1H-indene, 1,4,5,6,7,8,8-heptachloro-2,3-epoxy-3a,4,7,7-tetrahydro-, alpha, beta, and gamma isomers)

Hexachlorobenzene (Benzene, hexachloro-)

Hexachlorobutadiene   (1,3-Butadiene, 1,1,2,3,4,4-hexachloro-)

Hexachlorocyclohexane (all isomers) (Lindane and isomers)

Hexachlorocyclopentadiene   (1,3-Cyclopentadiene, 1,2,3,4,5,5-hexachloro-)

Hexachloroethane (Ethane, 1,1,1,2,2,2-hexachloro-)

1,2,3,4,10,10-Hexachloro-1,4,4a,5,8,8a-hexahydro-1,4:5,8-endo-endo-dimethanonaphthalene (Hexachlorohexahydro-endo,endo-dimethanonaphthalene)

Hexachlorophene (2,2'-Methylenebis(3,4,6-trichlorophenol))

Hexachloropropene (1-Propene, 1,1,2,3,3,3-hexachloro-)

Hexaethyl   tetraphosphate   (Tetraphosphoric acid, hexaethyl ester)

Hydrazine (Diamine)

Hydrocyanic acid (Hydrogen cyanide)

Table 9-6.   Hazardous Constituents (Continued)

Hydrofluoric acid (Hydrogen fluoride)

Hydrogen sulfide (Sulfur hydride)

Hydroxydimethylarsine oxide (Cacodylic acid)

Indeno(1,2,3-cd)pyrene   (1,10-(1,2-phenylene)pyrene)

Iodomethane (Methyl iodide)

Iron dextran (Ferric dextran)

Isocyanic acid, methyl ester (Methyl isocyanate)

Isobutyl alcohol (1-Propanol, 2-methyl-)

Isosafrole (Benzene, 1,2-methylenedioxy-4-allyl-)

Kepone (Decachlorooctahydro-1,3,4-Methano-2H-cyclobuta[cd]pentalen-2-one)

Lasiocarpine (2-Butenoic acid, 2-methyl-, 7-[(2,3-dihydroxy-2-(1-methoxyethyl)-3-methyl-1-oxobutoxy)methyl]-2,3,5,7a-tetrahydro-1H-pyrrolizin-1-yl ester)

Lead and compounds, N.O.S.*

Lead acetate (Acetic acid, lead salt)

Lead phosphate (Phosphoric acid, lead salt)

Lead   subacetate   (Lead,   bis(acetato-O)tetrahydroxytri-)

Maleic anhydride (2,5-Furandione)

Maleic hydrazide (1,2-Dihydro-3,6-pyridazinedione)

Malononitrile (Propanedinitrile)

Melphalan   (Alanine,   3-[p-bis(2-chloroethyl)amino]phenyl-, L-)

Mercury fulminate (Fulminic acid, mercury salt)

Mercury and compounds, N.O.S.*

Methacrylonitrile   (2-Propenenitrile,   2-methyl-)

Methanethiol (Thiomethanol)

Methapyrilene   (Pyridine,   2-[(2-dimethylamino)ethyl]-2-thenylamino-)

Metholmyl   (Acetimidic   acid,   N-[(methylcarbamoyl)oxy]thio-,   methyl ester)

Methoxychlor (Ethane, 1,1,1-trichloro-2,2'-bis(p-methoxyphenyl)-)

2-Methylaziridine (1,2-Propylenimine)

3-Methylcholanthrene

(Benz[j]aceanthrylene,   1,2-dihydro-3-methyl-)

Methyl chlorocarbonate (Carbonochloridic acid, methyl ester)

4,4'-Methylenebis(2-chloroaniline) (Benzenamine, 4,4'-methylenebis-(2-chloro-)

Methyl ethyl ketone (MEK) (2-Butanone)

Methyl hydrazine (Hydrazine, methyl-)

2-Methyllactonitrile (Propanenitrile, 2-hydroxy-2-methyl-)

Methyl methacrylate (2-Propenoic acid, 2-methyl-, methyl ester)

Methyl methanesulfonate (Methanesulfonic acid, methyl ester)

2-Methyl-2-(methylthio)propionaldehyde-o-(methylcarbonyl) oxime (Propanal, 2-methyl-2-(methylthio)-,   O-[(methylamino)carbonyl]oxime)

N-Methyl-N'-nitro-N-nitrosoguanidine (Guanidine, N-nitroso-N-methyl-N'-nitro-)

Methyl parathion (O,O-dimethyl O-(4-nitrophenyl) phosphorothioate)

Methylthiouracil (4-1H-Pyrimidinone, 2,3-dihydro-6-methyl-2-thioxo-)

Mustard gas (Sulfide, bis(2-chloroethyl)-)

Naphthalene

1,4-Naphthoquinone   (1,4-Naphthalenedione)

1-Naphthylamine (alpha-Naphthylamine)

2-Naphthylamine (beta-Naphthylamine)

1-Naphthyl-2-thiourea (Thiourea, 1-naphthalenyl-)

Nickel and compounds, N.O.S.*

Nickel carbonyl (Nickel tetracarbonyl)

Nickel cyanide (Nickel (II) cyanide)

Nicotine and salts (Pyridine, (S)-3-(1-methyl-2-pyrrolidinyl)-, and salts)

Nitric oxide (Nitrogen (II) oxide)

p-Nitroaniline (Benzenamine, 4-nitro-)

Nitrobenzine (Benzene, nitro-)

Nitrogen dioxide (Nitrogen (IV) oxide)

Nitrogen mustard and hydrochloride salt (Ethanamine, 2-chloro-, N-(2-chloroethyl)-N-methyl-, and hydrochloride salt)

Nitrogen mustard N-Oxide and hydrochloride salt (Ethanamine, 2-chloro-, N-(2-chloroethyl)-N-methyl-, and hydrochloride salt)

Nitroglycerine   (1,2,3-Propanetriol,   trinitrate)

4-Nitrophenol (Phenol, 4-nitro-)

4-Nitroquinoline-1-oxide (Quinoline, 4-nitro-1-oxide-)

Nitrosamine, N.O.S.*

N-Nitrosodi-n-butylamine   (1-Butanamine, N-butyl-N-nitroso-)

N-Nitrosodiethanolamine (Ethanol, 2,2'-(nitrosoimino)bis-)

N-Nitrosodiethylamine (Ethanamine, N-ethyl-N-nitroso-)

N-Nitrosodimethylamine (Dimethylnitrosamine)

N-Nitroso-N-ethylurea (Carbamide, N-ethyl-N-nitroso-)

N-Nitrosomethylethylamine (Ethanamine, N-methyl-N-nitroso-)

N-Nitroso-N-methylurea (Carbamide, N-methyl-N-nitroso-)

N-Nitroso-N-methylurethane (Carbamic acid, methylnitroso-, ethyl ester)

N-Nitrosomethylvinylamine (Ethenamine, N-methyl-N-nitroso-)

N-Nitrosomorpholine (Morpholine, N-nitroso-)

N-Nitrosonornicotine (Nornicotine, N-nitroso-)

N-Nitrosopiperidine (Pyridine, hexahydro-, N-nitroso-)

Nitrosopyrrolidine (Pyrrole, tetrahydro-, N-nitroso-)

N-Nitrososarcosine (Sarcosine, N-nitroso-)

Table 9-6.   Hazardous Constituents (Continued)

5-Nitro-o-toluidine (Benzenamine, 2-methyl-5-nitro-)

Octamethylpyrophosphoramide (Diphosphoramide, octamethyl-)

Osmium tetroxide (Osmium (VIII) oxide)

7-Oxabicyclo[2.2.1]heptane-2,3-dicarboxylic acid (Endothal)

Paraldehyde (1,3,5-Trioxane, 2,4,6-trimethyl-)

Parathion (Phosphorothioic acid, O,O-diethyl O-(p-nitrophenyl) ester

Pentachlorobenzene (Benzene, pentachloro-)

Pentachloroethane (Ethane, pentachloro-)

Pentachloronitrobenzene (PCNB) (Benzene, pentachloronitro-)

Pentachlorophenol (Phenol, pentachloro-)

Phenacetin (Acetamide, N-(4-ethoxyphenyl)-)

Phenol (Benzene, hydroxy-)

Phenylenediamine (Benzenediamine)

Phenylmercury acetate (Mercury, acetatophenyl-)

N-Phenylthiourea (Thiourea, phenyl-)

Phosgene (Carbonyl chloride)

Phosphine (Hydrogen phosphide)

Phosphorodithioic acid, O,O-diethyl S-[(ethylthio)methyl] ester (Phorate)

Phosphorothioic acid, O,O-dimethyl O-[p-((dimethylamino)sulfonyl)phenyl] ester (Famphur)

Phthalic acid esters, N.O.S.* (Benzene, 1,2-dicarboxylic acid, esters, N.O.S.*)

Phthalic anhydride (1,2-Benzenedicarboxylic acid anhydride)

2-Picoline (Pyridine, 2-methyl-)

Polychlorinated biphenyl, N.O.S.*

Potassium cyanide

Potassium silver cyanide (Argentate(1-), dicyano-, potassium)

Pronamide (3,5-Dichloro-N-(1,1-dimethyl-2-propynyl)benzamide)

1,3-Propane sultone (1,2-Oxathiolane, 2,2-dioxide)

n-Propylamine (1-Propanamine)

Propylthiouracil (Undecamethylenediamine, N,N'-bis(2-chlorobenzyl)-, dihydrochloride)

2-Propyn-1-ol (Propargyl alcohol)

Pyridine

Reserpine (Yohimban-16-carboxylic acid, 11,17-dimethoxy-18-[(3,4,5-trimethoxybenzoyl)oxy]-, methyl ester)

Resorcinol (1,3-Benzenediol)

Saccharin and salts (1,2-Benzoisothiazolin-3-one, 1,1-dioxide, and salts)

Safrole (Benzene, 1,2-methylenedioxy-4-allyl-)

Selenious acid (Selenium dioxide)

Selenium and compounds, N.O.S.*

Selenium sulfide (Sulfur selenide)

Selenourea (Carbamimidoselenoic acid)

Silver and compounds, N.O.S.*

Silver cyanide

Sodium cyanide

Streptozotocin (D-Glucopyranose, 2-deoxy-2-(3-methyl-3-nitrosoureido)-)

Strontium sulfide

Strychnine and salts (Strychnidin-10-one, and salts)

1,2,4,5-Tetrachlorobenzene (Benzene, 1,2,4,5-tetrachloro-)

2,3,7,8-Tetrachlorodibenzo-p-dioxin (TCDD) (Dibenzo-p-dioxin, 2,3,7,8-tetrachloro-)

Tetrachloroethane, N.O.S.* (Ethane, tetrachloro-, N.O.S.*)

1,1,1,2-Tetrachlorethane (Ethane, 1,1,1,2-tetrachloro-)

1,1,2,2-Tetrachlorethane (Ethane, 1,1,2,2-tetrachloro-)

Tetrachloroethane (Ethene, 1,1,2,2-tetrachloro-)

Tetrachloromethane (Carbon tetrachloride)

2,3,4,6,-Tetrachlorophenol (Phenol, 2,3,4,6-tetrachloro-)

Tetraethyldithiopyrophosphate (Dithiopyrophosphoric acid, tetraethyl-ester)

Tetraethyl lead (Plumbane, tetraethyl-)

Tetraethylpyrophosphate (Pyrophosphoric acide, tetraethyl ester)

Tetranitromethane (Methane, tetranitro-)

Thallium and compounds, N.O.S.*

Thallic oxide (Thallium (III) oxide)

Thallium (I) acetate (Acetic acid, thallium (I) salt)

Thallium (I) carbonate (Carbonic acid, dithallium (I) salt)

Thallium (I) chloride

Thallium (I) nitrate (Nitric acid, thallium (I) salt)

Thallium selenite

Thallium (I) sulfate (Sulfuric acid, thallium (I) salt)

Thioacetamide (Ethanethioamide)

Thiosemicarbazide (Hydrazinecarbothioamide)

Thiourea (Carbamide thio-)

Thiuram (Bis(dimethylthiocarbamoyl) disulfide)

Toluene (Benzene, methyl-)

Toluenediamine (Diaminotoluene)

o-Toluidine hydrochloride (Benzenamine, 2-methyl-, hydrochloride)

Tolylene diisocyanate (Benzene, 1,3-diisocyanatomethyl-)

Toxaphene (Camphene, octachloro-)

Tribromomethane (Bromoform)

1,2,4-Trichlorobenzene (Benzene, 1,2,4-trichloro-)

1,1,1-Trichloroethane (Methyl chloroform)

1,1,2-Trichloroethane (Ethane, 1,1,2-trichloro-)

Trichloroethene (Trichloroethylene)

Trichloromethanethiol (Methanethiol, trichloro-)

Table 9-6.   Hazardous Constituents (Continued)

Trichloromonofluoromethane   (Methane, trichlorofluoro-)

2,4,5-Trichlorophenol (Phenol, 2,4,5-trichloro-)

2,4,6-Trichlorophenol (Phenol, 2,4,6-trichloro-)

2,4,5-Trichlorophenoxyacetic acid (2,4,5-T) (Acetic acid, 2,4,5-trichlorophenoxy-)

2,4,5-Trichlorophenoxypropionic acid (2,(2,4,5-TP) (Silvex) (Propionoic acid, 2-(2,4,5-trichlorophenoxy)-)

Trichloropropane, N.O.S.* (Propane, trichloro-, N.O.S.*)

1,2,3-Trichloropropane (Propane, 1,2,3-trichloro-)

O,O,O-Triethyl phosphorothioate (Phosphorothioic acid, O,O,O-triethyl ester)

sym-Trinitrobenzene (Benzene, 1,3,5-trinitro-)

Tris(1-azridinyl) phosphine sulfide (Phosphine sulfide, tris(1-aziridinyl-)

Tris(2,3-dibromopropyl) phosphate (1-Propanol, 2,3-dibromo-, phosphate)

Trypan blue (2,7-Naphthalenedisulfonic acid, 3,3'-[(3,3'-dimethyl(1,1'-biphenyl)-4,4'-diyl)bis(azo)]bis(5-amino-4-hydroxy-, tetrasodium salt)

Uracil mustard (Uracil 5-[bis(2-chloroethyl)amino]-)

Vanadic acid, ammonium salt (ammonium vanadate)

Vanadium pentoxide (Vanadium (V) oxide)

Vinyl chloride (Ethene, chloro-)

Zinc cyanide

Zinc phosphide

(vi) The degree to which the constituent or any degradation product of the constituent bioaccumulates in ecosystems.

(vii) The plausible types of improper management to which the waste could be subjected.

(viii) The quantities of the waste generated at individual generation sites or on a regional or national basis.

(ix) The nature and severity of the human health and environmental damage that has occurred as a result of the improper management of wastes containing the constituent.

(x) Action taken by other governmental agencies or regulatory programs based on the health or environmental hazard posed by the waste or waste constituent.

(xi) Such other factors as may be appropriate.

Substances will be listed in Table 9-6 only if they have been shown in scientific studies to have toxic, carcinogenic, mutagenic, or teratogenic effects on humans or other life forms.

Wastes listed in accordance with those criteria will be designated toxic wastes.

It should be noted that the generator has the responsibility for determining whether his wastes exhibit 1 or more of the characteristics that have been described in the sections of this text that follow. The EPA has assigned EPA Hazardous Waste Numbers and certain record-keeping and reporting requirements making use of those numbers.

a. Characteristic of Ignitability. A solid waste exhibits the characteristics of ignitability if a representative sample of the waste has any of the following properties:

(1) It is a liquid, other than an aqueous solution containing less than 24 percent alcohol by volume and has a flash point less than 60°C (140°F), as determined by a Pensky-Martens Closed Cup Tester, using the test method specified in ASTM Standard D-93-79 or D-93-80, or a Setaflash Closed Cup Tester, using the test method specified in ASTM Standard D-3278-78, or as

determined by an equivalent test method approved by the Administrator of the EPA.

(2) It is not a liquid and is capable, under standard temperature and pressure, of causing fire through friction, absorption of moisture or spontaneous chemical changes and, when ignited, burns so vigorously and persistently that it creates a hazard.

(3) It is an ignitable compressed gas as defined in 49 CFR 173.300 and as determined by the test methods described in that regulation or equivalent test methods approved by the Administrator of the EPA.

(4) It is an oxidizer as defined in 49 CFR 173.151.

(5) A solid waste that exhibits the characteristic of ignitability but is not listed as hazardous waste in Subpart D of 40 CFR, which has been included in the following pages, and has the EPA Hazardous Waste Number of D001.

b. Characteristic of Corrosivity. A solid waste exhibits the characteristic of corrosivity if a representative sample of the waste has either of the following properties:

(1) It is aqueous and has a pH less than or equal to 2 or greater than or equal to 12.5, as determined by a pH meter using either an EPA test method or an equivalent test method approved by the Administrator under the procedures set forth in Sec. 260.20 and 260.21 of 40 CFR. The EPA test method for pH is specified as Method 5.2 in "Test Methods for the Evaluation of Solid Waste, Physical/Chemical Methods" (incorporated by reference, see Sec. 260.11 of 40 CFR).

(2) It is a liquid and corrodes steel (SAE 1020) at a rate greater than 6.35 mm (0.250 inch) per year at a test temperature of 550°C (130°F) as determined by the test method specified in the NACE (National Association of Corrosion Engineers) Standard TM-01-69 as standardized in "Test Methods for the Evaluation of Solid Waste, Physical/Chemical Methods" (incorporated by reference, see Sec. 260.11 of 40 CFR) or an equivalent test method approved by the Administrator under the procedures set forth in Sec. 260.20 and 260.21 (40 CFR).

(3) A solid waste that exhibits the characteristic of corrosivity but is not listed as a hazardous waste in Subpart D has the EPA Hazardous Waste Number of D002.

c. Characteristic of Reactivity. A solid waste exhibits the characteristic of reactivity if a representative sample of the waste has any of the following properties:

(1) It is normally unstable and readily undergoes violent change without detonating.

(2) It reacts violently with water.

(3) It forms potentially explosive mixtures with water.

(4) When mixed with water, it generates toxic gases, vapors or fumes in a quantity sufficient to present a danger to human health or the environment.

(5) It is a cyanide- or sulfide-bearing waste that, when exposed to pH conditions between 2 and 12.5, can generate toxic gases, vapors, or fumes in a quantity sufficient to present a danger to human health or the environment.

(6) It is capable of detonation or explosive reaction if it is subjected to a strong initiating source or if heated under confinement.

(7) It is readily capable of detonation or explosive decomposition or reaction at standard temperature and pressure.

(8) It is a forbidden explosive, as defined in 49 CFR 173.51, or a Class A explosive, as defined in 49 CFR 173.53, or a Class B explosive as defined in 49 CFR 173.88.

(9) A solid waste that exhibits the characteristic of reactivity but is not listed as a hazardous waste in Subpart D has the EPA Hazardous Waste Number of D003.

d. Characteristic of EP Toxicity.[1]   A solid waste exhibits the characteristic of EP toxicity if—when one uses the test methods described in Section 6d of this chapter, on EP Toxicity Test Procedures, page 124 or equivalent methods approved by the EPA administrator—the extract from a representative sample of the waste contains any of the contaminants listed in Table 9-7 at a concentration

Table 9-7.   Maximum Concentration of Contaminants for Characteristic of EP Toxicity (Table 1 of Part 261.24, CFR 40)

EPA hazardous waste number	Contaminant	Maximum concentration (milligrams per liter)
D004	Arsenic	5.0
D005	Barium	100.0
D006	Cadmium	1.0
D007	Chromium	5.0
D008	Lead	5.0
D009	Mercury	0.2
D010	Selenium	1.0
D011	Silver	5.0
D012	Endrin (1,2,3,4,10,10-hexachloro-1,7-epoxy-1,4,4a,5,6,7,8,8a-octahydro-1,4-endo, endo-5,8-dimethapo-naphthalene	0.02
D013	Lindane (1,2,3,4,5,6-hexachlorocyclohexane, gamma isomer.	0.4
D014	Methoxychlor (1,1,1-Trichloro-2,2-ba [p-methoxyphenyl]ethane)	10.0
D015	Toxaphene ($C_{10}H_{10}Cl_3$, Technical chlorinated camphene, 67–69 percent chlorine)	0.5
D016	2,4-D, (2,4-Dichlorophenoxyacetic acid	10.0
D017	2,4,5-TP Silvex (2,4,5-Trichlorophenoxypropionic acid)	1.0

[1]EP Toxicity refers to the extraction procedure (EP).

equal to or greater than the respective value given in that table. Where the waste contains less than 0.5 percent filterable solids, the waste itself, after filtering, is considered to be the extract for the purposes of this section.

A solid waste that exhibits the characteristic of EP toxicity but is not listed as a hazardous waste in Subpart D has the EPA Hazardous Waste Number specified in Table 9-7, which corresponds to the toxic contaminant causing it to be hazardous.

6. Lists of Hazardous Wastes

The classes or types of wastes that are listed in this section have been assigned Hazard Codes by EPA as follows:

Ignitable Waste	(I)
Corrosive Waste	(C)
Reactive Waste	(R)
EP Toxic Waste	(E)
Acute Hazardous Waste	(H)
Toxic Waste	(T)

In addition, each hazardous waste listed herein has been assigned the EPA Hazardous Number (noted in Section 5, above).

a. Hazardous Wastes from Discarded Commercial Chemical Products, Off-Specification Species, Container Residues, and Spill Residues. The following materials or items are hazardous wastes if and when they are discarded or intended to be discarded:

(a) Any commercial chemical product or manufacturing chemical intermediate having the generic name listed in paragraph (e) or (f) of this section.

(b) Any off-specification commercial chemical product or manufacturing chemical intermediate that, if it met specifications, would have the generic name listed in paragraph (e) or (f) of this section.

(c) Any container or inner liner removed from a container that has been used to hold any commercial chemical product or manufacturing chemical intermediate having the generic names listed in paragraph (e) of this section or any container or inner liner removed from a container that has been used to hold any off-specification chemical product and manufacturing chemical intermediate that, if it met specifications, would have the generic name listed in paragraph (e) of this section, unless the container is empty, as defined in para. C, 4, of this chapter.[1]

(d) Any residue or contaminated soil, water, or other debris resulting from the cleanup of a spill into or on any land or water of any commercial chemical product or manufacturing chemical product or manufacturing chemical inter-

---

[1]Unless the residue is being beneficially used or reused or legitimately recycled or reclaimed or being accumulated stored, transported, or treated before such use, reuse, recycling, or reclamation. The EPA considers the residue to be intended for discard and thus a hazardous waste. An example of a legitimate reuse of the residue would be where the residue remains in the container and the container is used to hold the same commercial chemical product or manufacturing chemical product or manufacturing chemical intermediate it previously held. An example of the discard of the residue would be where the drum is sent to a drum reconditioner who reconditions the drum but discards the residue.

mediate having the generic name listed[1] in paragraph (e) or (f) of this section or any residue or contaminated soil, water, or other debris resulting from the cleanup of a spill, into or on any land or water, of any off-specification chemical product and manufacturing chemical intermediate that, if it met specifications, would have the generic name listed in paragraph (e) or (f) of this section.

(e) The commercial chemical products, manufacturing chemical intermediates, or off-specification commercial chemical products or manufacturing chemical intermediates referred to in paragraphs (a) through (d) of this section are identified as acute hazardous[2] wastes (H) and are subject to the small-quantity exclusion defined in C, 3, of this chapter.

(f) The commercial chemical products, manufacturing chemical intermediates, or off-specification commercial chemical products referred to in paragraphs (a) through (d) of this section are identified as toxic wastes (T) unless otherwise designated and are subject to the small-quantity exclusion.[3]

Those wastes and their corresponding EPA Hazardous Waste Numbers are given in the following list:

Effective Date Note: At 49 FR 19923, May 10, 1984, the above listing was amended by revising 3 entries in the table in paragraph (e) and adding 3 entries to the table in paragraph (f) identified by hazardous waste numbers U248, (3-(alpha-acetonylbenzyl)-4-hydroxycoumarin and salts, when present at concentrations of 0.3 percent or less, and Warfarin, when present at concentrations of 0.3 percent or less), and U249 (Zinc phosphide, when present at concentrations of 10 percent or less), effective November 12, 1984. For the convenience of the user, the superseded entries from the listing are set out below in Table 9-12.

b. Representative Sampling Methods. The methods and equipment used for sampling waste materials will vary with the form and consistency of the waste materials to be sampled. Samples collected using the sampling protocols listed below, for sampling waste with properties similar to the indicated materials, will be considered by the agency to be representative of the waste.

Extremely viscous liquid—ASTM Standard D140-70 crushed or powdered material—ASTM Standard D346-75 soil or rocklike material—ASTM Standard D420-69 soillike material—ASTM Standard D1452-65.

---

[1] The phrase "commercial chemical product or manufacturing chemical product or manufacturing chemical intermediate having the generic name listed in . . ." refers to a chemical substance that is manufactured or formulated for commercial or manufacturing use, which consists of the commercially pure grade of the chemical, any technical grades of the chemical that are produced or marketed, and all formulations in which the chemical is the sole active ingredient. It does not refer to a material, such as a manufacturing process waste, that contains any of the substances listed in paragraphs (e) or (f). Where a manufacturing process waste is deemed to be a hazardous waste because it contains a substance listed in paragraphs (e) or (f), such waste will be listed in either or will be identified as a hazardous waste by the characteristics set forth in Subpart C, Sec. 261.20, of 40 CFR.

[2] For the convenience of the regulated community the primary hazardous properties of these materials have been indicated by the letters T (Toxicity) and R (Reactivity). Absence of a letter indicates that the compound is listed only for acute toxicity.

[3] For the convenience of the regulated community, the primary hazardous properties of these materials have been indicated by the letters T (Toxicity), R (Reactivity), I (Ignitability) and C (Corrosivity). Absence of a letter indicates that the compound is listed only for toxicity.

Fly ashlike material—ASTM Standard D2234-76 (ASTM Standards are available from ASTM, 1916 Race St., Philadelphia, PA 19103).

Containerized liquid wastes—"COLIWASA" described in "Test Methods for the Evaluation of Solid Waste, Physical/Chemical Methods," U.S. Environmental Protection Agency, Office of Solid Waste, Washington, D.C. 20460. (Copies may be obtained from Solid Waste Information, U.S. Environmental Protection Agency, 26 W. St. Clair St., Cincinnati, Ohio 45268.)

Liquid waste in pits, ponds, lagoons, and similar reservoirs—"Pond Sampler" described in "Test Methods for the Evaluation of Solid Waste, Physical/Chemical methods."[1] This manual also contains additional information on application of these protocols.

Since the EPA has not formally adopted the above sampling methods, a person who desires to employ an alternative sampling method is not required to demonstrate the equivalency of the method under the procedures set forth in Sec. 260.20 and 260.21 of 40 CFR.

c. EP Toxicity Test.   The Extraction Procedure (EP) is derived as follows:

1. A representative sample of the waste to be tested (minimum size 100 grams) shall be obtained using the methods specified in 6,c, above that corresponds to Appendix I of Sec. 261.33 of 40 CFR or any other method capable of yielding a representative sample within the meaning of Part 260. (For detailed guidance on conducting the various aspects of the EP, see "Test Methods for the Evaluation of Solid Waste, Physical/Chemical Methods.")

EPA Publication SW-846 (First Edition, 1980, as updated by Revision A (August 1980), B (July 1981), and C (February 1982) or (Second Edition, 1982). The first edition of SW-846 is no longer in print. Revisions A and B are available from EPA, Office of Solid Waste (WH565B) 401 M St., S.W., Washington, D.C. 20460. Revision C is available from NTIS, 5285 Port Royal Road, Springfield, VA 22161. The second edition of SW-846 includes material from the first edition and Revisions A, B, and C in a reorganized format. It is available from the Superintendent of Documents, U.S. Government Printing Office, Washington, D.C. 20402.

2. The sample shall be separated into its component liquid and solid phases using the method described in the "Separation Procedure" below. If the solid residue[1] obtained using this method totals less than 0.5 percent of the original weight of the waste, the residue can be discarded and the operator shall treat the liquid phase as the extract and proceed immediately to Step 8.

3. The solid material obtained from the Separation Procedure shall be evaluated for its particle size. If the solid material has a surface area per gram of material equal to or greater than 3.1 cm² or passes through a 9.5 mm (0.375 inch)

---

[1]The percent solids is determined by drying the filter pad at 80°C until it reaches constant weight and then calculating the percent solids using the following equation:

$$\text{Percent solids} = \frac{(\text{weight of pad} + \text{solid}) - (\text{tare weight of pad})}{\text{initial weight of sample}} \times 100$$

[1] These methods are also described in "Samplers and Sampling Procedures for Hazardous Waste Streams," EPA 600/2-80- 018, January 1980.

Table 9-8. Hazardous Wastes From Nonspecific Sources

Industry and EPA hazardous waste No.	Hazardous waste	Hazard code
Generic:		
F001	The following spent halogenated solvents used in degreasing: tetrachloroethylene, trichloroethylene, methylene chloride, 1,1,1-trichloroethane, carbon tetrachloride, and chlorinated fluorocarbons; and sludges from the recovery of these solvents in degreasing operations.	(T)
F002	The following spent halogenated solvents: tetrachloroethylene, methylene chloride, trichloroethylene, 1,1,1-trichloroethane, chlorobenzene, 1,1,2-trichloro-1,2,2-trifluoroethane, ortho-dichlorobenzene, and trichlorofluoromethane; and the still bottoms from the recovery of these solvents.	(T)
F003	The following spent non-halogenated solvents: xylene, acetone, ethyl acetate, ethyl benzene, ethyl ether, methyl isobutyl ketone, n-butyl alcohol, cyclohexanone, and methanol; and the still bottoms from the recovery of these solvents.	(I)
F004	The following spent non-halogenated solvents: cresols and cresylic acid, and nitrobenzene; and the still bottoms from the recovery of these solvents.	(T)
F005	The following spent non-halogenated solvents: toluene, methyl ethyl ketone, carbon disulfide, isobutanol, and pyridine; and the still bottoms from the recovery of these solvents.	(I, T)
F006	Wastewater treatment sludges from electroplating operations except from the following processes: (1) sulfuric acid anodizing of aluminum; (2) tin plating on carbon steel; (3) zinc plating (segregated basis) on carbon steel; (4) aluminum or zinc-aluminum plating on carbon steel; (5) cleaning/stripping associated with tin, zinc and aluminum plating on carbon steel; and (6) chemical etching and milling of aluminum.	(T)
F019	Wastewater treatment sludges from the chemical conversion coating of aluminum.	(T)
F007	Spent cyanide plating bath solutions from electroplating operations (except for precious metals electroplating spent cyanide plating bath solutions).	(R, T)
F008	Plating bath sludges from the bottom of plating baths from electroplating operations where cyanides are used in the process (except for precious metals electroplating plating bath sludges).	(R, T)
F009	Spent stripping and cleaning bath solutions from electroplating operations where cyanides are used in the process (except for precious metals electroplating spent stripping and cleaning bath solutions).	(R, T)
F010	Quenching bath sludge from oil baths from metal heat treating operations where cyanides are used in the process (except for precious metals heat-treating quenching bath sludges).	(R, T)
F011	Spent cyanide solutions from salt bath pot cleaning from metal heat treating operations (except for precious metals heat treating spent cyanide solutions from salt bath pot cleaning).	(R, T)
F012	Quenching wastewater treatment sludges from metal heat treating operations where cyanides are used in the process (except for precious metals heat treating quenching wastewater treatment sludges).	(T)
F024	Wastes, including but not limited to, distillation residues, heavy ends, tars, and reactor clean-out wastes from the production of chlorinated aliphatic hydrocarbons, having carbon content from one to five, utilizing free radical catalyzed processes. [This listing does not include light ends, spent filters and filter aids, spent dessicants, wastewater, wastewater treatment sludges, spent catalysts, and wastes listed in § 261.32.]	(T)

standard sieve, the operator shall proceed to Step 4. If the surface area is smaller or the particle size larger than specified above, the solid material shall be prepared for extraction by crushing, cutting, or grinding the material so that it passes through a 9.5 mm (0.375 in.) sieve or, if the material is in a single piece, by subjecting the material to the "Structural Integrity Procedure" described below.

4. The solid material obtained in Step 3 shall be weighed and placed in an extractor with 16 times its weight of deionized water. Do not allow the material to dry before weighing. For purposes of this test, an acceptable extractor is one that will impart sufficient agitation to the mixture not only to prevent stratification of the sample and extraction fluid but also to ensure that all sample surfaces are continuously brought into contact with well mixed extraction fluid.

Table 9-9.   Hazardous Wastes From Specific Sources.

Industry and EPA hazardous waste No.	Hazardous waste	Hazard code
Wood preservation: K001	Bottom sediment sludge from the treatment of wastewaters from wood preserving processes that use creosote and/or pentachlorophenol.	(T)
Inorganic pigments:		
K002	Wastewater treatment sludge from the production of chrome yellow and orange pigments.	(T)
K003	Wastewater treatment sludge from the production of molybdate orange pigments	(T)
K004	Wastewater treatment sludge from the production of zinc yellow pigments....................	(T)
K005	Wastewater treatment sludge from the production of chrome green pigments	(T)
K006	Wastewater treatment sludge from the production of chrome oxide green pigments (anhydrous and hydrated).	(T)
K007	Wastewater treatment sludge from the production of iron blue pigments	(T)
K008	Oven residue from the production of chrome oxide green pigments..............................	(T)
Organic chemicals:		
K009	Distillation bottoms from the production of acetaldehyde from ethylene	(T)
K010	Distillation side cuts from the production of acetaldehyde from ethylene......................	(T)
K011	Bottom stream from the wastewater stripper in the production of acrylonitrile...............	(R, T)
K013	Bottom stream from the acetonitrile column in the production of acrylonitrile................	(R, T)
K014	Bottoms from the acetonitrile purification column in the production of acrylonitrile	(T)
K015	Still bottoms from the distillation of benzyl chloride...	(T)
K016	Heavy ends or distillation residues from the production of carbon tetrachloride............	(T)
K017	Heavy ends (still bottoms) from the purification column in the production of epichlorohydrin.	(T)
K018	Heavy ends from the fractionation column in ethyl chloride production........................	(T)
K019	Heavy ends from the distillation of ethylene dichloride in ethylene dichloride production.	(T)
K020	Heavy ends from the distillation of vinyl chloride in vinyl chloride monomer production.	(T)
K021	Aqueous spent antimony catalyst waste from fluoromethanes production	(T)
K022	Distillation bottom tars from the production of phenol/acetone from cumene	(T)
K023	Distillation light ends from the production of phthalic anhydride from naphthalene.........	(T)
K024	Distillation bottoms from the production of phthalic anhydride from naphthalene...........	(T)
K093	Distillation light ends from the production of phthalic anhydride from ortho-xylene	(T)
K094	Distillation bottoms from the production of phthalic anhydride from ortho-xylene	(T)
K025	Distillation bottoms from the production of nitrobenzene by the nitration of benzene......	(T)
K026	Stripping still tails from the production of methy ethyl pyridines	(T)
K027	Centrifuge and distillation residues from toluene diisocyanate production.....................	(R, T)
K028	Spent catalyst from the hydrochlorinator reactor in the production of 1,1,1-trichloroethane.	(T)
K029	Waste from the product steam stripper in the production of 1,1,1-trichloroethane	(T)
K095	Distillation bottoms from the production of 1,1,1-trichloroethane	(T)
K096	Heavy ends from the heavy ends column from the production of 1,1,1-trichloroethane.	(T)
K030	Column bottoms or heavy ends from the combined production of trichloroethylene and perchloroethylene.	(T)
K083	Distillation bottoms from aniline production	(T)
K103	Process residues from aniline extraction from the production of aniline........................	(T)
K104	Combined wastewater streams generated from nitrobenzene/aniline production	(T)
K085	Distillation or fractionation column bottoms from the production of chlorobenzenes......	(T)
K105	Separated aqueous stream from the reactor product washing step in the production of chlorobenzenes.	(T)
Inorganic chemicals:		
K071	Brine purification muds from the mercury cell process in chlorine production, where separately prepurified brine is not used.	(T)
K073	Chlorinated hydrocarbon waste from the purification step of the diaphragm cell process using graphite anodes in chlorine production.	(T)
K106	Wastewater treatment sludge from the mercury cell process in chlorine production.......	(T)
Pesticides:		
K031	By-product salts generated in the production of MSMA and cacodylic acid	(T)
K032	Wastewater treatment sludge from the production of chlordane.................................	(T)
K033	Wastewater and scrub water from the chlorination of cyclopentadiene in the production of chlordane.	(T)
K034	Filter solids from the filtration of hexachlorocyclopentadiene in the production of chlordane.	(T)
K097	Vacuum stripper discharge from the chlordane chlorinator in the production of chlordane.	(T)
K035	Wastewater treatment sludges generated in the production of creosote........................	(T)
K036	Still bottoms from toluene reclamation distillation in the production of disulfoton...........	(T)
K037	Wastewater treatment sludges from the production of disulfoton.................................	(T)
K038	Wastewater from the washing and stripping of phorate production..............................	(T)

Table 9-9.   Hazardous Wastes From Specific Sources (Continued).

Industry and EPA hazardous waste No.	Hazardous waste	Hazard code
K039	Filter cake from the filtration of diethylphosphorodithioic acid in the production of phorate.	(T)
K040	Wastewater treatment sludge from the production of phorate	(T)
K041	Wastewater treatment sludge from the production of toxaphene	(T)
K098	Untreated process wastewater from the production of toxaphene	(T)
K042	Heavy ends or distillation residues from the distillation of tetrachlorobenzene in the production of 2,4,5-T.	(T)
K043	2,6-Dichlorophenol waste from the production of 2,4-D	(T)
K099	Untreated wastewater from the production of 2,4-D	(T)
Explosives:		
K044	Wastewater treatment sludges from the manufacturing and processing of explosives	(R)
K045	Spent carbon from the treatment of wastewater containing explosives	(R)
K046	Wastewater treatment sludges from the manufacturing, formulation and loading of lead-based initiating compounds.	(T)
K047	Pink/red water from TNT operations	(R)
Petroleum refining:		
K048	Dissolved air flotation (DAF) float from the petroleum refining industry	(T)
K049	Slop oil emulsion solids from the petroleum refining industry	(T)
K050	Heat exchanger bundle cleaning sludge from the petroleum refining industry	(T)
K051	API separator sludge from the petroleum refining industry	(T)
K052	Tank bottoms (leaded) from the petroleum refining industry	(T)
Iron and steel:		
K061	Emission control dust/sludge from the primary production of steel in electric furnaces.	(T)
K062	Spent pickle liquor from steel finishing operations	(C, T)
Secondary lead:		
K069	Emission control dust/sludge from secondary lead smelting	(T)
K100	Waste leaching solution from acid leaching of emission control dust/sludge from secondary lead smelting.	(T)
Veterinary pharmaceuticals:		
K084	Wastewater treatment sludges generated during the production of veterinary pharmaceuticals from arsenic or organo-arsenic compounds.	(T)
K101	Distillation tar residues from the distillation of aniline-based compounds in the production of veterinary pharmaceuticals from arsenic or organo-arsenic compounds.	(T)
K102	Residue from the use of activated carbon for decolorization in the production of veterinary pharmaceuticals from arsenic or organo-arsenic compounds.	(T)
Ink formulation: K086	Solvent washes and sludges, caustic washes and sludges, or water washes and sludges from cleaning tubs and equipment used in the formulation of ink from pigments, driers, soaps, and stabilizers containing chromium and lead.	(T)
Coking:		
K060	Ammonia still lime sludge from coking operations	(T)
K087	Decanter tank tar sludge from coking operations	(T)

5. After the solid material and deionized water are placed in the extractor, the operator shall begin agitation and measure the pH of the solution in the extractor. If the pH is greater than 5.0, the pH of the solution shall be decreased to 5.0 ± 0.2 by adding 0.5N acetic acid. If the pH is equal to or less than 5.0, no acetic acid should be added. The pH of the solution shall be monitored, as described below, during the course of the extraction, and if the pH rises above 5.2, 0.5N acetic acid shall be added to bring the pH down to 5.0 ± 0.2. However, in no event shall the aggregate amount of the acid added to the solution exceed 4 ml of acid per gram of solid. The mixture shall be agitated for 24 hours and maintained at 20° to 40°C (68° to 104°F) during this time. It is recommended that the operator monitor and adjust the pH during the course of the extraction with a device such as the Type 45-A pH Controller manufactured by Chemtrix, Inc., Hillsboro, Oregon 97123 or its equivalent, in conjunction with a metering

Table 9-10.   EPA Hazardous Waste Numbers.

Hazardous waste No.	Substance
P023	Acetaldehyde, chloro-
P002	Acetamide, N-(aminothioxomethyl)-
P057	Acetamide, 2-fluoro-
P058	Acetic acid, fluoro-, sodium salt
P066	Acetimidic acid, N-[(methylcarbamoyl)oxy]thio-, methyl ester
P001	3-(alpha-Acetonylbenzyl)-4-hydroxycoumarin and salts, when present at concentrations greater than 0.3%
P002	1-Acetyl-2-thiourea
P003	Acrolein
P070	Aldicarb
P004	Aldrin
P005	Allyl alcohol
P006	Aluminum phosphide
P007	5-(Aminomethyl)-3-isoxazolol
P008	4-aAminopyridine
P009	Ammonium picrate (R)
P119	Ammonium vanadate
P010	Arsenic acid
P012	Arsenic (III) oxide
P011	Arsenic (V) oxide
P011	Arsenic pentoxide
P012	Arsenic trioxide
P038	Arsine, diethyl-
P054	Aziridine
P013	Barium cyanide
P024	Benzenamine, 4-chloro-
P077	Benzenamine, 4-nitro-
P028	Benzene, (chloromethyl)
P042	1,2-Benzenediol, 4-[1-hydroxy-2-(methylamino)ethyl]-
P014	Benzenethiol
P028	Benzyl chloride
P015	Beryllium dust
P016	Bis(chloromethyl) ether
P017	Bromoacetone
P018	Brucine
P021	Calcium cyanide
P123	Camphene, octachloro-
P103	Carbamimidoselenoic acid
P022	Carbon bisulfide
P022	Carbon disulfide
P095	Carbonyl chloride
P033	Chlorine cyanide
P023	Chloroacetaldehyde
P024	p-Chloroaniline
P026	1-(o-Chlorophenyl)thiourea
P027	3-Chloropropionitrile
P029	Copper cyanides
P030	Cyanides (soluble cyanide salts), not elsewhere specified
P031	Cyanogen
P033	Cyanogen chloride
P036	Dichlorophenylarsine
P037	Dieldrin
P038	Diethylarsine
P039	O,O-Diethyl S-[2-(ethylthio)ethyl] phosphorodithioate
P041	Diethyl-p-nitrophenyl phosphate
P040	O,O-Diethyl O-pyrazinyl phosphorotbioate
P043	Diisopropyl fluorophosphate
P044	Dimethoate
P045	3,3-Dimethyl-1-(methylthio)-2-butanone, O-[(methylamino)carbonyl] oxime

Hazardous waste No.	Substance
P071	O,O-Dimethyl O-p-nitrophenyl phosphorothwoate
P082	Dimethylnitrosamine
P046	alpha, alpha-Dimethylphenethylamine
P047	4,6-Dinitro-o-cresol and salts
P034	4,6-Dinitro-o-cyclohexylphenol
P048	2,4-Dinitrophenol
P020	Dinoseb
P085	Diphosphoramide, octamethyl-
P039	Disulfoton
P049	2,4-Dithiobiuret
P109	Dithiopyrophosphoric acid, tetraethyl ester
P050	Endosulfan
P088	Endothall
P051	Endrin
P042	Epinephrine
P046	Ethanamine, 1,1-dimethyl-2-phenyl-
P084	Ethenamine, N-methyl-N-nitroso-
P101	Ethyl cyanide
P054	Ethylenimine
P097	Famphur
P056	Fluorine
P057	Fluoroacetamide
P058	Fluoroacetic acid, sodium salt
P065	Fulminic acid, mercury(II) salt (R,T)
P059	Heptachlor
P051	1,2,3,4,10,10-Hexachloro-6,7-epoxy-1,4,4a,5,6,7,8,8a-octahydro-endo,endo-1,4:5,8-dimethanonaphthalene
P037	1,2,3,4,10,10-Hexachloro-6,7-epoxy-1,4,4a,5,6,7,8,8a-octahydro-endo,exo-1,4:5,8-demethanonaphthalene
P060	1,2,3,4,10,10-Hexachloro-1,4,4a,5,8,8a-hexahydro-1,4:5,8-endo, endo-dimeth- anonaphthalene
P004	1,2,3,4,10,10-Hexachloro-1,4,4a,5,8,8a-hexahydro-1,4:5,8-endo,exo-dimethanonaphthalene
P060	Hexachlorohexahydro-exo,exo-dimethanonaphthalene
P062	Hexaethyl tetraphosphate
P116	Hydrazinecarbothio..mide
P068	Hydrazine, methyl-
P063	Hydrocyanic acid
P063	Hydrogen cyanide
P096	Hydrogen phosphide
P064	Isocyanic acid, methyl ester
P007	3(2H)-Isoxazolone, 5-(aminomethyl)-
P092	Mercury, (acetato-O)phenyl-
P065	Mercury fulminate (R,T)
P016	Methane, oxybis(chloro-
P112	Methane, tetranitro- (R)
P118	Methanethiol, trichloro-
P059	4,7-Methano-1H-indene, 1,4,5,6,7,8,8-heptachloro-3a,4,7,7a-tetrahydro-
P066	Methomyl
P067	2-Methylaziridine
P068	Methyl hydrazine
P064	Methyl isocyanate
P069	2-Methyllactonitrile
P071	Methyl parathion
P072	alpha-Naphthylthiourea
P073	Nickel carbonyl
P074	Nickel cyanide

Table 9-10.   EPA Hazardous Waste Numbers (Continued)

Hazardous waste No.	Substance	Hazardous waste No.	Substance
P074	Nickel(II) cyanide	P027	Propanenitrile, 3-chloro-
P073	Nickel tetracarbonyl	P069	Propanenitrile, 2-hydroxy-2-methyl-
P075	Nicotine and salts	P081	1,2,3-Propanetriol, trinitrate- (R)
P076	Nitric oxide	P017	2-Propanone, 1-bromo-
P077	p-Nitroaniline	P102	Propargyl alcohol
P078	Nitrogen dioxide	P003	2-Propenal
P076	Nitrogen(II) oxide	P005	2-Propen-1-ol
P078	Nitrogen(IV) oxide	P067	1,2-Propylenimine
P081	Nitroglycerine (R)	P102	2-Propyn-1-ol
P082	N-Nitrosodimethylamine	P008	4-Pyridinamine
P084	N-Nitrosomethylvinylamine	P075	Pyridine, (S)-3-(1-methyl-2-pyrrolidinyl)-, and salts
P050	5-Norbornene-2,3-dimethanol, 1,4,5,6,7,7-hexachloro, cyclic sulfite	P111	Pyrophosphoric acid, tetraethyl ester
P085	Octamethylpyrophosphoramide	P103	Selenourea
P087	Osmium oxide	P104	Silver cyanide
P087	Osmium tetroxide	P105	Sodium azide
P088	7-Oxabicyclo[2.2.1]heptane-2,3-dicarboxylic acid	P106	Sodium cyanide
		P107	Strontium sulfide
P089	Parathion	P108	Strychnidin-10-one, and salts
P034	Phenol, 2-cyclohexyl-4,6-dinitro-	P018	Strychnidin-10-one, 2,3-dimethoxy-
P048	Phenol, 2,4-dinitro-	P108	Strychnine and salts
P047	Phenol, 2,4-dinitro-6-methyl-	P115	Sulfuric acid, thallium(I) salt
P020	Phenol, 2,4-dinitro-6-(1-methylpropyl)-	P109	Tetraethyldithiopyrophosphate
P009	Phenol, 2,4,6-trinitro-, ammonium salt (R)	P110	Tetraethyl lead
P036	Phenyl dichloroarsine	P111	Tetraethylpyrophosphate
P092	Phenylmercuric acetate	P112	Tetranitromethane (R)
P093	N-Phenylthiourea	P062	Tetraphosphoric acid, hexaethyl ester
P094	Phorate	P113	Thallic oxide
P095	Phosgene	P113	Thallium(III) oxide
P096	Phosphine	P114	Thallium(I) selenite
P041	Phosphoric acid, diethyl p-nitrophenyl ester	P115	Thallium(I) sulfate
P044	Phosphorodithioic acid, O,O-dimethyl S-[2-(methylamino)-2-oxoethyl]ester	P045	Thiofanox
		P049	Thioimidodicarbonic diamide
P043	Phosphorofluoric acid, bis(1-methylethyl)-ester	P014	Thiophenol
		P116	Thiosemicarbazide
P094	Phosphorothioic acid, O,O-diethyl S-(ethylthio)methyl ester	P026	Thiourea, (2-chlorophenyl)-
		P072	Thiourea, 1-naphthalenyl-
P089	Phosphorothioci acid, O,O-diethyl O-(p-nitrophenyl) ester	P093	Thiourea, phenyl-
		P123	Toxaphene
P040	Phosphorothioic acid, O,O-diethyl O- pyrazinyl ester	P118	Trichloromethanethiol
		P119	Vanadic acid, ammonium salt
P097	Phosphorothioic acid, O,O-dimethyl O-[p-((dimethylamino)-sulfonyl)phenyl]ester	P120	Vanadium pentoxide
		P120	Vanadium(V) oxide
P110	Plumbane, tetraethyl-	P001	Warfarin, when present at concentrations greater than 0.3%
P098	Potassium cyanide		
P099	Potassium silver cyanide	P121	Zinc cyanide
P070	Propanal, 2-methyl-2-(methylthio)-, O-[(methylamino)carbonyl]oxime	P122	Zinc phosphide, when present at concentrations greater than 10%
P101	Propanenitrile		

pump and reservoir of 0.5N acetic acid. If such a system is not available, the following manual procedure shall be employed:

(a) A pH meter shall be calibrated in accordance with the manufacturer's specifications.

(b) The pH of the solution shall be checked and, if necessary, 0.5N acetic acid shall be manually added to the extractor until the pH reaches 5.0 ± 0.2. The pH of the solution shall be adjusted at 15-, 30-, and 60-minute intervals, moving to the next longer interval if the pH does not have to be adjusted more than 0.5N pH units.

(c) The adjustment procedure shall be continued for at least 6 hours.

(d) If at the end of the 24-hour extraction period, the pH of the solution is not below 5.2 and the maximum amount of acid (4 ml per gram of solids) has not been added, the pH shall be adjusted to 5.0 ± 0.2 and the extraction continued for an additional 4 hours, during which the pH shall be adjusted at 1-hour intervals.

6. At the end of the 24-hour extraction period, deionized water shall be added to the extractor in an amount determined by the following equation:

$V=(20)(W)—16(W)—A$
$V=$ml deionized water to be added
$W=$weight in grams of solid charged to extractor
$A=$ml of 0.5N acetic acid added during extraction

7. The material in the extractor shall be separated into its component liquid and solid phases as described under "Separation Procedure."

Table 9-11.   EPA Hazardous Waste Numbers.

Hazardous Waste No.	Substance	Hazardous Waste No.	Substance
U001	Acetaldehyde (I)	U019	Benzene (I,T)
U034	Acetaldehyde, trichloro-	U038	Benzeneacetic acid, 4-chloro-alpha-(4-chloro-phenyl)-alpha-hydroxy, ethyl ester
U187	Acetamide, N-(4-ethoxyphenyl)-		
U005	Acetamide, N-9H-fluoren-2-yl-	U030	Benzene, 1-bromo-4-phenoxy-
U112	Acetic acid, ethyl ester (I)	U037	Benzene, chloro-
U144	Acetic acid, lead salt	U190	1,2-Benzenedicarboxylic acid anhydride
U214	Acetic acid, thallium(I) salt	U028	1,2-Benzenedicarboxylic acid, [bis(2-ethyl-hexyl)] ester
U002	Acetone (I)		
U003	Acetonitrile (I,T)	U069	1,2-Benzenedicarboxylic acid, dibutyl ester
U248	3-(alpha-Acetonylbenzyl)-4-hydroxycoumarin and salts, when present at concentrations of 0.3% or less	U088	1,2-Benzenedicarboxylic acid, diethyl ester
		U102	1,2-Benzenedicarboxylic acid, dimethyl ester
		U107	1,2-Benzenedicarboxylic acid, di-n-octyl ester
U004	Acetophenone	U070	Benzene, 1,2-dichloro-
U005	2-Acetylaminofluorene	U071	Benzene, 1,3-dichloro-
U006	Acetyl chloride (C,R,T)	U072	Benzene, 1,4-dichloro-
U007	Acrylamide	U017	Benzene, (dichloromethyl)-
U008	Acrylic acid (I)	U223	Benzene, 1,3-diisocyanatomethyl- (R,T)
U009	Acrylonitrile	U239	Benzene, dimethyl-(I,T)
U150	Alanine, 3-[p-bis(2-chloroethyl)amino] phenyl-, L-	U201	1,3-Benzenediol
		U127	Benzene, hexachloro-
U011	Amitrole	U056	Benzene, hexahydro- (I)
U012	Aniline (I,T)	U188	Benzene, hydroxy-
U014	Auramine	U220	Benzene, methyl-
U015	Azaserine	U105	Benzene, 1-methyl-1-2,4-dinitro-
U010	Azirino(2',3':3,4)pyrrolo(1,2-a)indole-4,7-dione, 6-amino-8-[((aminocarbonyl) oxy)methyl]-1,1a,2,8,8a,8b-hexahydro-8a-methoxy-5-methyl-,	U106	Benzene, 1-methyl-2,6-dinitro-
		U203	Benzene, 1,2-methylenedioxy-4-allyl-
		U141	Benzene, 1,2-methylenedioxy-4-propenyl-
		U090	Benzene, 1,2-methylenedioxy-4-propyl-
U157	Benz[j]aceanthrylene, 1,2-dihydro-3-methyl-	U055	Benzene, (1-methylethyl)- (I)
U016	Benz[c]acridine	U169	Benzene, nitro- (I,T)
U016	3,4-Benzacridine	U183	Benzene, pentachloro-
U017	Benzal chloride	U185	Benzene, pentachloro-nitro-
U018	Benz[a]anthracene	U020	Benzenesulfonic acid chloride (C,R)
U018	1,2-Benzanthracene	U020	Benzenesulfonyl chloride (C,R)
U094	1,2-Benzanthracene, 7,12-dimethyl-	U207	Benzene, 1,2,4,5-tetrachloro-
U012	Benzenamine (I,T)	U023	Benzene, (trichloromethyl)-(C,R,T)
U014	Benzenamine, 4,4'-carbonimidoylbis(N,N-di-methyl-	0234	Benzene, 1,3,5-trinitro- (R,T)
		U021	Benzidine
U049	Benzenamine, 4-chloro-2-methyl-	U202	1,2-Benzisothiazolin-3-one, 1,1-dioxide
U093	Benzenamine, N,N'-dimethyl-4-phenylazo-	U120	Benzo[j,k]fluorene
U158	Benzenamine, 4,4'-methylenebis(2-chloro-	U022	Benzo[a]pyrene
U222	Benzenamine, 2-methyl-, hydrochloride	U022	3,4-Benzopyrene
U181	Benzenamine, 2-methyl-5-nitro	U197	p-Benzoquinone

Table 9-11.   EPA Hazardous Waste Numbers (Continued)

Hazardous Waste No.	Substance	Hazardous Waste No.	Substance
U023	Benzotrichloride (C,R,T)	U063	Dibenz[a,h]anthracene
U050	1,2-Benzphenanthrene	U063	1,2:5,6-Dibenzanthracene
U085	2,2'-Bioxirane (I,T)	U064	1,2:7,8-Dibenzopyrene
U021	(1,1'-Biphenyl)-4,4'-diamine	U064	Dibenz[a,i]pyrene
U073	(1,1'-Biphenyl)-4,4'-diamine, 3,3'-dichloro-	U066	1,2-Dibromo-3-chloropropane
U091	(1,1'-Biphenyl)-4,4'-diamine, 3,3'-dimethoxy-	U069	Dibutyl phthalate
U095	(1,1'-Biphenyl)-4,4'-diamine, 3,3'-dimethyl-	U062	S-(2,3-Dichloroallyl) diisopropylthiocarbamate
U024	Bis(2-chloroethoxy) methane	U070	o-Dichlorobenzene
U027	Bis(2-chloroisopropyl) ether	U071	m-Dichlorobenzene
U244	Bis(dimethylthiocarbamoyl) disulfide	U072	p-Dichlorobenzene
U028	Bis(2-ethylhexyl) phthalate	U073	3,3'-Dichlorobenzidine
U246	Bromine cyanide	U074	1,4-Dichloro-2-butene (I,T)
U225	Bromoform	U075	Dichlorodifluoromethane
U030	4-Bromophenyl phenyl ether	U192	3,5-Dichloro-N-(1,1-dimethyl-2-propynyl) benzamide
U128	1,3-Butadiene, 1,1,2,3,4,4-hexachloro-		
U172	1-Butanamine, N-butyl-N-nitroso-	U060	Dichloro diphenyl dichloroethane
U035	Butanoic acid, 4-[Bis(2-chloroethyl)amino] benzene-	U061	Dichloro diphenyl trichloroethane
		U078	1,1-Dichloroethylene
U031	1-Butanol (I)	U079	1,2-Dichloroethylene
U159	2-Butanone (I,T)	U025	Dichloroethyl ether
U160	2-Butanone peroxide (R,T)	U081	2,4-Dichlorophenol
U053	2-Butenal	U082	2,6-Dichlorophenol
U074	2-Butene, 1,4-dichloro- (I,T)	U240	2,4-Dichlorophenoxyacetic acid, salts and esters
U031	n-Butyl alchohol (I)		
U136	Cacodylic acid	U083	1,2-Dichloropropane
U032	Calcium chromate	U084	1,3-Dichloropropene
U238	Carbamic acid, ethyl ester	U085	1,2:3,4-Diepoxybutane (I,T)
U178	Carbamic acid, methylnitroso-, ethyl ester	U108	1,4-Diethylene dioxide
U176	Carbamide, N-ethyl-N-nitroso-	U086	N,N-Diethylhydrazine
U177	Carbamide, N-methyl-N-nitroso-	U087	O,O-Diethyl-S-methyl-dithiophosphate
U219	Carbamide, thio-	U088	Diethyl phthalate
U097	Carbamoyl chloride, dimethyl-	U089	Diethylstilbestrol
U215	Carbonic acid, dithallium(I) salt	U148	1,2-Dihydro-3,6-pyradizinedione
U156	Carbonochloridic acid, methyl ester (I,T)	U090	Dihydrosafrole
U033	Carbon oxyfluoride (R,T)	U091	3,3'-Dimethoxybenzidine
U211	Carbon tetrachloride	U092	Dimethylamine (I)
U033	Carbonyl fluoride (R,T)	U093	Dimethylaminoazobenzene
U034	Chloral	U094	7,12-Dimethylbenz[a]anthracene
U035	Chlorambucil	U095	3,3'-Dimethylbenzidine
U036	Chlordane, technical	U096	alpha,alpha-Dimethylbenzylhydroperoxide (R)
U026	Chlornaphazine	U097	Dimethylcarbamoyl chloride
U037	Chlorobenzene	U098	1,1-Dimethylhydrazine
U039	4-Chloro-m-cresol	U099	1,2-Dimethylhydrazine
U041	1-Chloro-2,3-epoxypropane	U101	2,4-Dimethylphenol
U042	2-Chloroethyl vinyl ether	U102	Dimethyl phthalate
U044	Chloroform	U103	Dimethyl sulfate
U046	Chloromethyl methyl ether	U105	2,4-Dinitrotoluene
U047	beta-Chloronaphthalene	U106	2,6-Dinitrotoluene
U048	o-Chlorophenol	U107	Di-n-octyl phthalate
U049	4-Chloro-o-toluidine, hydrochloride	U108	1,4-Dioxane
U032	Chromic acid, calcium salt	U109	1,2- Diphenylhydrazine
U050	Chrysene	U110	Dipropylamine (I)
U051	Creosote	U111	Di-N-propylnitrosamine
U052	Cresols	U001	Ethanal (I)
U052	Cresylic acid	U174	Ethanamine, N-ethyl-N-nitroso-
U053	Crotonaldehyde	U067	Ethane, 1,2-dibromo-
U055	Cumene (I)	U076	Ethane, 1,1-dichloro-
U246	Cyanogen bromide	U077	Ethane, 1,2-dichloro-
U197	1,4-Cyclohexadienedione	U114	1,2-Ethanediylbiscarbamodithioic acid
U056	Cyclohexane (I)	U131	Ethane, 1,1,1,2,2,2-hexachloro-
U057	Cyclohexanone (I)	U024	Ethane, 1,1'-[methylenebis(oxy)]bis[2-chloro-
U130	1,3-Cyclopentadiene, 1,2,3,4,5,5-hexa- chloro-	U003	Ethanenitrile (I, T)
U058	Cyclophosphamide	U117	Ethane,1,1'-oxybis- (I)
U240	2,44-D, salts and esters	U025	Ethane, 1,1'-oxybis[2-chloro-
U059	Daunomycin	U184	Ethane, pentachloro-
U060	DDD	U208	Ethane, 1,1,1,2-tetrachloro-
U061	DDT	U209	Ethane, 1,1,2,2-tetrachloro-
U142	Decachlorooctahydro-1,3,4-metheno-2H-cyclobuta[c,d]-pentalen-2-one	U218	Ethanethioamide
		U247	Ethane, 1,1,1,-trichloro-2,2-bis(p-methoxyphenyl).
U062	Diallate		
U133	Diamine (R,T)	U227	Ethane, 1,1,2-trichloro-
U221	Diaminotoluene	U043	Ethene, chloro-

Table 9-11.   EPA Hazardous Waste Numbers (Continued)

Hazardous Waste No.	Substance	Hazardous Waste No.	Substance
U042	Ethene, 2-chloroethoxy-	U068	Methane, dibromo-
U078	Ethene, 1,1-dichloro-	U080	Methane, dichloro-
U079	Ethene, trans-1,2-dichloro-	U075	Methane, dichlorodifluoro-
U210	Ethene, 1,1,2,2-tetrachloro-	U138	Methane, iodo-
U173	Ethanol, 2,2'-(nitrosoimino)bis-	U119	Methanesulfonic acid, ethyl ester
U004	Ethanone, 1-phenyl-	U211	Methane, tetrachloro-
U006	Ethanoyl chloride (C,R,T)	U121	Methane, trichlorofluoro-
U112	Ethyl acetate (I)	U153	Methanethiol (I,T)
U113	Ethyl acrylate (I)	U225	Methane, tribromo-
U238	Ethyl carbamate (urethan)	U044	Methane, trichloro-
U038	Ethyl 4,4'-dichlorobenzilate	U121	Methane, trichlorofluoro-
U114	Ethylenebis(dithiocarbamic acid)	U123	Methanoic acid (C,T)
U067	Etylene dibromide	U036	4,7-Methanoindan, 1,2,4,5,6,7,8,8-octa-chloro-3a,4,7,7a-tetrahydro-
U077	Ethylene dichloride		
U115	Ethlene oxide (I,T)	U154	Methanol (I)
U116	Ethylene thiourea	U155	Methapyrilene
U117	Ethyl ether (I)	U247	Methoxychlor.
U076	Ethylidene dichloride	U154	Methyl alcohol (I)
U118	Ethylmethacrylate	U029	Methyl bromide
U119	Ethyl methanesulfonate	U186	1-Methylbutadiene (I)
U139	Ferric dextran	U045	Methyl chloride (I,T)
U120	Fluoranthene	U156	Methyl chlorocarbonate (I,T)
U122	Formaldehyde	U226	Methylchloroform
U123	Formic acid (C,T)	U157	3-Methylcholanthrene
U124	Furan (I)	U158	4,4'-Methylenebis(2-chloroaniline)
U125	2-Furancarboxaldehyde (I)	U132	2,2'-Methylenebis(3,4,6-trichlorophenol)
U147	2,5-Furandione	U068	Methylene bromide
U213	Furan, tetrahydro- (I)	U080	Methylene chloride
U125	Furfural (I)	U122	Methylene oxide
U124	Furfuran (I)	U159	Methyl ethyl ketone (I,T)
U206	D-Glucopyranose, 2-deoxy-2(3-methyl-3-nitro-soureido)-	U160	Methyl ethyl ketone peroxide (R,T)
		U138	Methyl iodide
U126	Glycidylaldehyde	U161	Methyl isobutyl ketone (I)
U163	Guanidine, N-nitroso-N-methyl-N'nitro-	U162	Methyl methacrylate (I,T)
U127	Hexachlorobenzene	U163	N-Methyl-N'-nitro-N-nitrosoguanidine
U128	Hexachlorobutadiene	U161	4-Methyl-2-pentanone (I)
U129	Hexachlorocyclohexane (gamma isomer)	U164	Methylthiouracil
U130	Hexachlorocyclopentadiene	U010	Mitomycin C
U131	Hexachloroethane	U059	5,12-Naphthacenedione, (8S-cis)-8-acetyl-10-[(3-amino-2,3,6-trideoxy-alpha-L-lyxo-hexopyranosyl)oxyl]-7,8,9,10-tetrahydro-6,8,11-trihydroxy-1-methoxy-
U132	Hexachlorophene		
U243	Hexachloropropene		
U133	Hydrazine (R,T)		
U086	Hydrazine, 1,2-diethyl-	U165	Naphthalene
U098	Hydrazine, 1,1-dimethyl-	U047	Naphthalene, 2-chloro-
U099	Hydrazine, 1,2-dimethyl-	U166	1,4-Naphthalenedione
U109	Hydrazine, 1,2-diphenyl-	U236	2,7-Naphthalenedisulfonic acid, 3,3'-[(3,3'-di-methyl-(1,1'-biphenyl)-4,4'diyl)]-bis(azo)bis(5-amino-4-hydroxy)-,tetrasodium salt
U134	Hydrofluoric acid (C,T)		
U134	Hydrogen fluoride (C,T)		
U135	Hydrogen sulfide		
U096	Hydroperoxide, 1-methyl-1-phenylethyl- (R)		
U136	Hydroxydimethylarsine oxide	U166	1,4,Naphthaquinone
U116	2-Imidazolidinethione	U167	1-Naphthylamine
U137	Indeno[1,2,3-cd]pyrene	U168	2-Naphthylamine
U139	Iron dextran	U167	alpha-Naphthylamine
U140	Isobutyl alcohol (I,T)	U168	beta-Naphthylamine
U141	Isosafrole	U026	2-Naphthylamine, N,N'-bis(2-chloromethyl)-
U142	Kepone	U169	Nitrobenzene (I,T)
U143	Lasiocarpine	U170	p-Nitrophenol
U144	Lead acetate	U171	2-Nitropropane (I)
U145	Lead phosphate	U172	N-Nitrosodi-n-butylamine
U146	Lead subacetate	U173	N-Nitrosodiethanolamine
U129	Lindane	U174	N-Nitrosodiethylamine
U147	Maleic anhydride	U111	N-Nitroso-N-propylamine
U148	Maleic hydrazide	U176	N-Nitroso-N-ethylurea
U149	Malononitrile	U177	N-Nitroso-N-methylurea
U150	Melphalan	U178	N-Nitroso-N-methylurethane
U151	Mercury	U179	N-Nitrosopiperidine
U152	Methacrylonitrile (I,T)	U180	N-Nitrosopyrrolidine
U092	Methanamine, N-methyl- (I)	U181	5-Nitro-o-toluidine
U029	Methane, bromo-	U193	1,2-Oxathiolane, 2,2-dioxide
U045	Methane, chloro- (I,T)	U058	2H-1,3,2-Oxazaphosphorine, 2-[bis(2-chloro-ethyl)amino]tetrahydro-, oxide 2-
U046	Methane, chloromethoxy-		
		U115	Oxirane (I,T)

Table 9-11.   EPA Hazardous Waste Numbers (Continued)

Hazardous Waste No.	Substance	Hazardous Waste No.	Substance
U041	Oxirane, 2-(chloromethyl)-	U205	Sulfur selenide (R,T)
U182	Paraldehyde	U232	2,4,5-T
U183	Pentachlorobenzene	U207	1,2,4,5-Tetrachlorobenzene
U184	Pentachloroethane	U208	1,1,1,2-Tetrachloroethane
U185	Pentachloronitrobenzene	U209	1,1,2,2-Tetrachloroethane
U242	Pentachlorophenol	U210	Tetrachloroethylene
U186	1,3-Pentadiene (I)	U212	2,3,4,6-Tetrachlorophenol
U187	Phenacetin	U213	Tetrahydrofuran (I)
U188	Phenol	U214	Thallium(I) acetate
U048	Phenol, 2-chloro-	U215	Thallium(I) carbonate
U039	Phenol, 4-chloro-3-methyl-	U216	Thallium(I) chloride
U081	Phenol, 2,4-dichloro-	U217	Thallium(I) nitrate
U082	Phenol, 2,6-dichloro-	U218	Thioacetamide
U101	Phenol, 2,4-dimethyl-	U153	Thiomethanol (I,T)
U170	Phenol, 4-nitro-	U219	Thiourea
U242	Phenol, pentachloro-	U244	Thiram
U212	Phenol, 2,3,4,6-tetrachloro-	U220	Toluene
U230	Phenol, 2,4,5-trichloro-	U221	Toluenediamine
U231	Phenol, 2,4,6-trichloro-	U223	Toluene diisocyanate (R,T)
U137	1,10-(1,2-phenylene)pyrene	U222	O-Toluidine hydrochloride
U145	Phosphoric acid, Lead salt	U011	1H-1,2,4-Triazol-3-amine
U087	Phosphorodithioic acid, O,O-diethyl-, S-methyl-lester	U226	1,1,1-Trichloroethane
U189	Phosphorous sulfide (R)	U227	1,1,2-Trichloroethane
U190	Phthalic anhydride	U228	Trichloroethene
U191	2-Picoline	U228	Trichloroethylene
U192	Pronamide	U121	Trichloromonofluoromethane
U194	1-Propanamine (I,T)	U230	2,4,5-Trichlorophenol
U110	1-Propanamine, N-propyl- (I)	U231	2,4,6-Trichlorophenol
U066	Propane, 1,2-dibromo-3-chloro-	U232	2,4,5-Trichlorophenoxyacetic acid
U149	Propanedinitrile	U234	sym-Trinitrobenzene (R,T)
U171	Propane, 2-nitro- (I)	U182	1,3,5-Trioxane, 2,4,5-trimethyl-
U027	Propane, 2,2'oxybis[2-chloro-	U235	Tris(2,3-dibromopropyl) phosphate
U193	1,3-Propane sultone	U236	Trypan blue
U235	1-Propanol, 2,3-dibromo-, phosphate (3:1)	U237	Uracil, 5[bis(2-chloromethyl)amino]-
U126	1-Propanol, 2,3-epoxy-	U237	Uracil mustard
U140	1-Propanol, 2-methyl- (I,T)	U043	Vinyl chloride
U002	2-Propanone (I)	U248	Warfarin, when present at concentrations of 0.3% or less
U007	2-Propenamide		
U084	Propene, 1,3-dichloro-	U239	Xylene (I)
U243	1-Propene, 1,1,2,3,3,3-hexachloro-	U200	Yohimban-16-carboxylic acid, 11,17-dimeth-oxy-18-[(3,4,5-trimethoxy-benzoyl)oxy]-, methyl ester
U009	2-Propenenitrile		
U152	2-Propenenitrile, 2-methyl- (I,T)		
U008	2-Propenoic acid (I)		
U113	2-Propenoic acid, ethyl ester (I)	U249	Zinc phosphide, when present at concentra-tions of 10% or less
U118	2-Propenoic acid, 2-methyl-, ethyl ester		
U162	2-Propenoic acid, 2-methyl-, methyl ester (I,T)		
U233	Propionic acid, 2-(2,4,5-trichlorophenoxy)-		
U194	n-Propylamine (I,T)		
U083	Propylene dichloride		
U196	Pyridine		
U155	Pyridine, 2-[(2-(dimethylamino)-2-thenyla-mino]-		
U179	Pyridine, hexahydro-N-nitroso-		
U191	Pyridine, 2-methyl-		
U164	4(1H)-Pyrimidinone, 2,3-dihydro-6-methyl-2-thioxo-		
U180	Pyrrole, tetrahydro-N-nitroso-		
U200	Reserpine		
U201	Resorcinol		
U202	Saccharin and salts		
U203	Safrole		
U204	Selenious acid		
U204	Selenium dioxide		
U205	Selenium disulfide (R,T)		
U015	L-Serine, diazoacetate (ester)		
U233	Silvex		
U089	4,4'-Stilbenediol, alpha,alpha'-diethyl-		
U206	Streptozotocin		
U135	Sulfur hydride		
U103	Sulfuric acid, dimethyl ester		
U189	Sulfur phosphide (R)		

Table 9-12.   EPA Hazardous Waste Numbers.

Hazardous Waste No.	Substance
P001	3-(alpha-acetonylbenzyl)-4-hydroxycoumarin and salts
P001	Warfarin
P122	Zinc phosphide (R,T)

8. The liquids resulting from steps 2 and 7 shall be combined. The combined liquid (or the waste itself if it has less than 1/2 percent solids, as noted in step 2) is the extract and shall be analyzed for the presence of any of the contaminants specified in Table 9-7, using the analytical procedures described below.

Separation Procedure

Equipment: A filter holder, designed for filtration media having a nominal pore size of 0.45 micrometers and capable of applying a 5.3 kg/cm^2 (75 psi) hydrostatic pressure to the solution being filtered, shall be used. For mixtures containing nonabsorptive solids, where separation can be effected without imposing a 5.3 kg/cm^2 pressure differential, vacuum filters employing a 0.45 micrometers filter media can be used. (For further guidance on filtration equipment or procedures, see "Test Methods for Evaluating Solid Waste, Physical/Chemical Methods", referred to above.[1]

(i) Following the manufacturer's directions, the filter unit shall be assembled with a filter bed consisting of a 0.45 micrometer filter membrane. For difficult- or slow-to-filter mixtures a prefilter bed consisting of the following prefilters in increasing pore size (0.65 micrometer membrane, fine glass fiber prefilter, and coarse glass fiber prefilter) can be used.

(ii) The waste shall be poured into the filtration unit.

(iii) The reservoir shall be slowly pressurized until liquid begins to flow from the filtrate outlet at which point the pressure in the filter shall be immediately lowered to 10 psig to 15 psig. Filtration shall be continued until liquid flow ceases.

(iv) The pressure shall be increased stepwise in 10 psi increments to 75 psig and filtration continued until flow ceases or the pressurizing gas begins to exit from the filtrate outlet.

(v) The filter unit shall be depressurized and the solid material removed, weighed and then transferred to the extraction apparatus or, in the case of final

---

[1]This procedure is intended to result in separation of the "free" liquid portion of the waste from any solid matter having a particle size > 0.45 μm. If the sample will not filter, various other separation techniques can be used to aid in the filtration. As described above, pressure filtration is employed to speed up the filtration process. This does not alter the nature of the separation. If liquid does not separate during filtration, the waste can be centrifuged. If separation occurs during centrifugation, the liquid portion (centrifugate) is filtered through the 0.45 μm filter before becoming mixed with the liquid portion of the waste obtained from the initial filtration. Any material that will not pass through the filter after centrifugation is considered a solid and is extracted.

filtration before analysis, discarded. Do not allow the material retained on the filter pad to dry before weighing.

(vi) The liquid phase shall be stored at 4°C for subsequent use in Step 8.

Structural Integrity Procedure

Equipment: A Structural Integrity Tester having a 3.18 cm (1.25 in.) diameter hammer weighing 0.33 kg (0.73 lbs.) and having a free fall of 15.24 cm (6 in.) shall be used. The device is available from Associated Design and Manufacturing Company, Alexandria, VA 22314, as Part No. 125, or it may be fabricated to meet the specifications shown in Fig. 9-1.

Procedure

1. The sample holder shall be filled with the material to be tested. If the sample of waste is a large monolithic block, a portion shall be cut from the block having the dimensions of a 7.1 cm (2.8 in.) cylinder with a 3.3 cm (1.3 in.) diameter. For a fixated waste, samples may be cast in the form of 7.1 cm (2.8 in.) cylinder with a 3.3 cm (1.3 in.) diameter for purposes of conducting the test. In such cases the waste may be allowed to cure for 30 days before further testing.

2. The sample holder shall be placed into the Structural Integrity Tester and then the hammer shall be raised to its maximum height and dropped. That shall be repeated 15 times.

3. The material shall be removed from the sample holder, weighed, and transferred to the extraction apparatus for extraction.

Analytical Procedures for Analyzing Extract Contaminants

The test methods for analyzing the extract are as follows:

1. For arsenic, barium, cadmium, chromium, lead, mercury, selenium, silver, endrin, lindane, methoxychlor, toxaphene, 2,4-D(2,4-dichlorophenoxyacetic acid) or 2,4,5-TP (2,4,5-trichlorophenoxypropionic acid), see "Test Methods for the Evaluation of Solid Waste, Physical/Chemical Methods" above.

2. (Reserved)

For all analyses, the methods of standard addition shall be used for quantification of species concentration.

In summary, a solid waste exhibits the characteristic of EP toxicity if, when one uses the test methods described in the above paragraphs or equivalent methods approved by the EPA administrator, the extract from a representative sample of the waste contains any of the contaminants listed in Table 9-7 at a concentration equal to or greater than the respective value given in that table. Where the waste contains less than 0.5 percent filterable solids, the waste itself, after filtering, is considered to be the extract for the purposes of this section.

(b) A solid waste that exhibits the characteristic of EP toxicity but is not listed as a hazardous waste in Subpart D of CFR 40 has the EPA Hazardous Waste Number specified in Table 9-7, which corresponds to the toxic contaminant causing it to be hazardous.

(1) Chemical Analysis Test Methods

Table 9-13, 9-14, and 9-15 specify the appropriate analytical procedures described in "Test Methods for Evaluating Solid Waste, Physical/Chemical

Table 9-13. Analysis Methods for Organic Chemicals Contained in SW-846 (Table 1 of Appendix III, Part 261, CRF 40)

Compound	First edition method(s)	Second edition method(s)
Acetonitrile	8.03, 8.24	8030, 8240
Acrolein	8.03, 8.24	8030, 8240
Acrylamide	8.01, 8.24	8015, 8240
Acrylonitrile	8.03, 8.24	8030, 8240
Benzene	8.02, 8.24	8020, 8024
Benz(a)anthracene	8.10, 8.25	8100, 8250, 8310
Benzo(a)pyrene	8.10, 8.25	8100, 8250, 8310
Benzotrichloride	8.12, 8.25	8120, 8250
Benzyl chloride	8.01, 8.12, 8.24, 8.25	8120, 8250
Benzo(b)fluoanthene	8.10, 8.25	8100, 8250, 8310
Bis(2-chloroethoxymethane)	8.01, 8.24	8010, 8240
Bis(2-chloroethyl)ether	8.01, 8.24	8010, 8240
Bis(2-chloroisopropyl)ether	8.01, 8.24	8010, 8240
Carbon disulfide	8.01, 8.24	8015, 8240
Carbon tetrachloride	8.01, 8.24	8010, 8240
Chlordane	8.08, 8.25	8080, 8250
Chlorinated dibenzodioxins	8.08, 8.25	8080, 8250
Chlorinated biphenyls	8.08, 8.25	8080, 8250
Chloroacetaldehyde	8.01, 8.24	8010, 8240
Chlorobenzene	8.01, 8.02, 8.24	8020, 8240
Chloroform	8.01, 8.24	8010, 8240
Chloromethane	8.01, 8.24	8010, 8240
2-Chlorophenol	8.04, 8.25	8040, 8250
Chrysene	8.10, 8.25	8100, 8250, 8310
Creosote [1]	8.10, 8.25	8100, 8250
Cresol(s)	8.04, 8.25	8040, 8250
Cresylic Acid(s)	8.04, 8.25	8040, 8250
Dichlorobenzene(s)	8.01, 8.02, 8.12, 8.25	8010, 8120, 8250
Dichloroethane(s)	8.01, 8.24	8010, 8240
Dichloromethane	8.01, 8.24	8010, 8240
Dichlorophenoxyacetic acid	8.40, 8.25	8150, 8250
Dichloropropanol	8.12, 8.25	8120, 8250
2,4-Dimethylphenol	8.04, 8.25	8040, 8250
Dinitrobenzene	8.09, 8.25	8090, 8250
4,6-Dinitro-o-cresol	8.04, 8.25	8040, 8250
2,4-Dinitrotoluene	8.09, 8.25	8090, 8250
Endrin	8.08, 8.25	8080, 8250
Ethyl ether	8.01, 8.02, 8.24	8015, 8240
Formaldehyde	8.01, 8.24	8015, 8240
Formic acid	8.06, 8.25	8250
Heptachlor	8.06, 8.25	8080, 8250
Hexachlorobenzene	8.12, 8.25	8120, 8250
Hexachlorobutadiene	8.12, 8.25	8120, 8250
Hexachloroethane	8.12, 8.25	8010, 8240
Hexachlorocyclopentadiene	8.12, 8.25	8120, 8250
Lindane	8.08, 8.25	8080, 8250
Maleic anhydride	8.06, 8.25	8250
Methanol	8.01, 8.24	8010, 8240
Methomyl	8.32	8250
Methyl ethyl ketone	8.01, 8.02, 8.24	8015, 8240
Methyl isobutyl ketone	8.01, 8.02, 8.24	8015, 8240
Napthalene	8.10, 8.25	8100, 8250
Napthoquinone	8.06, 8.09, 8.25	8090, 8250
Nitrobenzene	8.09, 8.25	8090, 8250
4-Nitrophenol	8.04, 8.25	8040, 8240
Paraldehyde (trimer of acetaldehyde)	8.01, 8.24	8015, 8240
Pentachlorophenol	8.04, 8.25	8040, 8250
Phenol	8.04, 8.25	8040, 8250
Phorate	8.22	8140
Phosphorodithioic acid esters	8.06, 8.09, 8.22	8140
Phthalic anhydride	8.06, 8.09, 8.25	8090, 8250
2-Picoline	8.06, 8.09, 8.25	8090, 8250
Pyridine	8.06, 8.09, 8.25	8090, 8250
Tetrachlorobenzene(s)	8.12, 8.25	8120, 8250
Tetrachloroethane(s)	8.01, 8.24	8010, 8240
Tetrachloroethene	8.01, 8.24	8010, 8240
Tetrachlorophenol	8.04, 8.24	8040, 8250
Toluene	8.02, 8.24	8020, 8024
Toluenediamine	8.25	8250
Toluene diisocyanate(s)	8.06, 8.25	8250
Toxaphene	8.08, 8.25	8080, 8250
Trichloroethane	8.01, 8.24	8010, 8240
Trichloroethene(s)	8.01, 8.24	8010, 8240
Trichlorofluoromethane	8.01, 8.24	8010, 8240
Trichlorophenol(s)	8.04, 8.25	8040, 8250
2,4,5-Trichlorophenoxy propionic acid	8.40, 8.25	8150, 8250
Trichloropropane	8.01, 8.24	8010, 8240
Vinyl chloride	8.01, 8.24	8010, 8240
Vinylidene chloride	8.01, 8.24	8010, 8240
Xylene	8.02, 8.24	8020, 8240

[1] Analyne for phenanthrene and carbazole; if these are present in a ratio between 1.4:1 and 5:1 creosote should be considered present.

Table 9-14.   Analysis Methods for Inorganic Chemicals Contained in SW-846 (Table 2 of Appendix III, Part 261, CFR 40)

Compound	First edition method(s)	Second edition method(s)	Compound	First edition method(s)	Second edition method(s)
Antimony	8.50	7040, 7041	Mercury	8.57	7470, 7471
Arsenic	8.51	7060, 7061	Nickel	8.58	7520, 7521
Barium	8.52	7080, 7081	Selenium	8.59	7740, 7741
Cadmium	8.53	7090, 7091	Silver	8.60	7760, 7761
Chromium	8.54	7190, 7191	Cyanides	8.55	9010
Chromium: Hexavalent	8.545, 8.546, 8.547	7195, 7196, 7197	Total Organic Halogen	8.66	9020
Lead	8.56	7420, 7421	Sulfides	8.67	9030

Methods," SW-846, which are to be used to determine whether a sample contains a given toxic constituent listed in Table 9-6 or 9-8.

Table 9-13 identifies each Table 9-6 or 9-8 organic constituent along with the approved measurement method. Table 9-14 identifies the corresponding methods for inorganic species, and Table 9-15 summarizes the contents of SW-846 and supplies specific section and method numbers for sampling and analysis methods.

Before final sampling and analysis method selection, the analyst should consult the specific section or method described in SW-846 for additional guidance on which of the approved methods should be employed for a specific sample analysis situation.

E. Hazardous Waste Generators

1. Applicability

A generator who treats, stores, or disposes of hazardous waste on-site need only comply with certain sections of CFR 40 with respect to that waste. He has to determine whether or not the substance is a hazardous waste; that may be learned by consulting CFR 40 (Section 262.11). The generator has to obtain an EPA identification number from CFR (Section 262.12); accumulating hazardous waste has specific requirements that can be determined from CFR 40 (Section 262.34); the necessary record keeping is obtained from CFR 40 (Section 262.40)(c) and (d)); additional reporting is noted in CFR 40 (Section 262.43); if farmers are involved, see CFR 40 (Section 262.51).

A farmer who generates waste pesticides that are hazardous waste and who complies with all of the requirements of Sec. 262.51 is not required to comply with other standards in 40 CFR Parts 270, 264, or 265 with respect to such pesticides.

Any person who imports hazardous waste into the United States must comply with the standards applicable to generators established in this part.

2. Hazardous Waste Determination

A person who generates a solid waste, as defined in 40 CFR 261.2, must determine if that waste is a hazardous waste by means of the following method:

(a) He should first determine if the waste is excluded from regulation under 40 CFR 261.4.

(b) He must then determine if the waste is listed as a hazardous waste in Subpart D of 40 CFR Part 261.

Table 9-15. Sampling and Analysis Methods Contained in SW-846 (Table 3 of Appendix III, Part 261, CFR 40)

Title	First edition Section No.	First edition Method No.	Second edition Section No.	Second edition Method No.
Sampling of Solid Wastes	1.0		1.0	
Development of Appropriate Sampling Plans	1.0		1.1	
Regulatory and Scientific Objectives	1.0–2		1.1.1	
Fundamental Statistical Concepts	1.0–3		1.1.2	
Basic Statistical Strategies	1.0–7		1.1.3	
Simple Random Sampling			1.1.3.1	
Stratified Random Sampling			1.1.3.2	
Systematic Random Sampling			1.1.3.3	
Special Considerations	1.0–7		1.1.4	
Composite Sampling			1.1.4.1	
Subsampling			1.1.4.2	
Cost and Loss Functions			1.1.4.3	
Implementation of Sampling Plan	1.0–7		1.2	
Selection of Sampling Equipment			1.2.1	
Composite Liquid Waste Sampler	3.2.1		1.2.1.1	
Weighted Bottle	3.2.2		1.2.1.2	
Dipper	3.2.3		1.2.1.3	
Thief	3.2.4		1.2.1.4	
Trier	3.2.5		1.2.1.5	
Auger	3.2.6		1.2.1.6	
Scoop and Shovel	3.2.7		1.2.1.7	
Selection of Sample Containers	3.3		1.2.2	
Processing and Storage of Samples	3.3		1.2.3	
Documentation of Chain of Custody	2.0		1.3	
Sample Labels	2.0–1		1.3.1	
Sample Seals	2.0–3		1.3.2	
Field Log Book	2.0–5		1.3.3	
Chain-of-Custody Record	2.0–6		1.3.4	
Sample Analysis Request Sheet	2.0–9		1.3.5	
Sample Delivery to Laboratory	2.0–10		1.3.6	
Shipping of Samples	2.0–10		1.3.7	
Receipt and Logging of Sample	2.0–12		1.3.8	
Assignment of Sample for Analysis	2.0–13		1.3.9	
Sampling Methodology	3.0		1.4	
Containers	3.2–2		1.4.1	
Tanks	3.2–2		1.4.2	
Waste Piles	3.2–2		1.4.3	
Landfills and Lagoons	3.2–2		1.4.4	
Waste Evaluation Procedures			2.0	
Characteristics of Hazardous Waste			2.1	
Ignitability	4.0		2.1.1	
Pensky-Martens Closed-Cup Method	4.1		2.1.1	1010
Setaflash Closed-Cup Method	4.1		2.1.1	1020
Corrosivity	5.0		2.1.2	
Corrosivity Toward Steel	5.3		2.1.2	1110
Reactivity	6.0		2.1.3	
Extraction Procedure Toxicity	7.0		2.1.4	
Extraction Procedure Toxicity Test	7.1, 7.2, 7.5			
Method and Structural Integrity Test	7.4		2.1.4	1310
Sample Workup Techniques			4.0	
Inorganic Techniques	8.49		4.1	
Acid Digestion for Flame AAS	'		4.1	3010
Acid Digestion for Furnace AAS	'		4.1	3020
Acid Digestion of Oil, Grease, or Wax	8.49–9		4.1	3030
Dissolution Procedure for Oil, Grease or Wax	8.49–8		4.1	
Alkaline Digestion	8.0	8.458	4.1	3060
Organic Techniques	8.0		4.2	
Separatory Funnel Liquid-Liquid Extraction	9.0	9.1	4.2	3510
Continuous Liquid-Liquid Extraction	9.0	9.01	4.2	3520
Acid-Base Cleanup Extraction	8.0	8.84	4.2	3530
Soxhlet Extraction	8.0	8.86	4.2	3540
Sonication Extraction	8.0	8.85	4.2	3550

Table 9-15. Sampling and Analysis Methods Contained In SW-846 (Continued)

Title	First edition		Second edition	
	Section No.	Method No.	Section No.	Method No.
Sample Introduction Techniques			5.0	
Headspace	8.0	8.82	5.0	
Purge-and-Trap	8.0	8.83	5.0	5030
Inorganic Analytical Methods	8.0		7.0	
Antimony, Flame AAS	8.0	8.50	7.0	7470
Antimony, Furnace AAS	8.0	8.50	7.0	7471
Arsenic, Flame AAS	8.0	8.51	7.0	7060
Arsenic, Furnace AAS	8.0	8.51	7.0	7061
Barium, Flame AAS	8.0	8.52	7.0	7080
Barium, Furnace AAS	8.0	8.52	7.0	7081
Cadmium, Flame AAS	8.0	8.53	7.0	7130
Cadmium, Furnace AAS	8.0	8.53	7.0	7131
Chromium, Flame AAS	8.0	8.54	7.0	7090
Chromium, Furnace AAS	8.0	8.54	7.0	7191
Chromium, Hexavalent, Coprecipitation	8.0	8.545	7.0	7195
Chromium, Hexavalent, Colorimetric	8.0	8.546	7.0	7196
Chromium, Hexavalent, Chelation	8.0	8.547	7.0	7197
Lead, Flame AAS	8.0	8.56	7.0	7420
Lead, Furnace AAS	8.0	8.56	7.0	7421
Mercury, Cold Vapor, Liquid	8.0	8.57	7.0	7470
Mercury, Cold Vapor, Solid	8.0	8.57	7.0	7471
Nickel, Flame AAS	8.0	8.58	7.0	7520
Nickel, Furnace AAS	8.0	8.58	7.0	7521
Selenium, Flame AAS	8.0	8.59	7.0	7740
Selenium, Gaseous Hydride AAS	8.0	8.59	7.0	7741
Silver, Flame AAS	8.0	8.60	7.0	7760
Silver, Furnace AAS	8.0	8.60	7.0	7761
Organic Analytical Methods	8.0		8.0	
Gas Chromatographic Methods	8.0		8.1	
Halogenated Volatile Organics	8.0	8.01	8.1	8010
Nonhalogenated Volatile Organics	8.0	8.01	8.1	8015
Aromatic Volatile Organics	8.0	8.02	8.1	8020
Acrolein, Acrylonitrile, Acetonitrile	8.0	8.03	8.1	8030
Phenols	8.0	8.04	8.1	8040
Phthalate Esters	8.0	8.06	8.1	8060
Organochlorine Pesticides and PCBs	8.0	8.08	8.1	8080
Nitroaromatics and Cyclic Ketones	8.0	8.09	8.1	8090
Polynuclear Aromatic Hydrocarbons	8.0	8.10	8.1	8100
Chlorinated Hydrocarbons	8.0	8.12	8.1	8120
Organophosphorus Pesticides	8.0	8.22	8.1	8140
Chlorinated Herbicides	8.0	8.40	8.1	8150
Gas Chromatographic/Mass Spectroscopy Methods (GC/MS)	8.0		8.2	
GC/MS Volatiles	8.0	8.24	8.2	8240
GC/MS Semi-Volatiles, Packed Column	8.0	8.25	8.2	8250
GC/MS Semi-Volatiles, Capillary	8.0	8.27	8.2	8270
High Performance Liquid Chromatographic Methods (HPLC)	8.0		8.3	
Polynuclear Aromatic Hydrocarbons	8.0	8.10	8.3	8310
Miscellaneous Analytical Methods	8.0		9.0	
Cyanide; Total and Amenable to Chlorination	8.0	8.55	9.0	9010
Total Organic Halogen (TOX)	8.0	8.66	9.0	9020
Sulfides	8.0	8.67	9.0	9030
pH Measurement	5.0	5.2	9.0	9040
Quality Control/Quality Assurance	10.0		10.1	
Introduction	10.0		10.1	
Program Design	10.0		10.2	
Sampling	10.0		10.3	
Analysis	10.0		10.4	
Data Handling	10.0		10.5	

See specific metal.

Even if the waste is listed, the generator still has an opportunity under 40 CFR 260.22 to demonstrate to the administrator of the EPA that the waste from his particular facility or operation is not a hazardous waste.

(c) If the waste is not listed as a hazardous waste in Subpart D of 40 CFR Part 261, he must determine whether it is identified in Subpart C of 40 CFR 261 by either:

(1) Testing the waste according to the methods set forth in Subpart C of 40 CFR Part 261 or according to an equivalent method approved by the administrator under 40 CFR 260.21; or

(2) Applying knowledge of the hazard characteristic of the waste in light of the materials or the processes used.

3. EPA Identification Numbers

(a) A generator must not treat, store, dispose of, transport, or offer for transportation hazardous waste without having received an EPA identification number from the administrator of the EPA.

(b) A generator who has not received an EPA identification number may obtain one by applying to the administrator, using EPA form 8700-12. Upon receiving the request, the administrator will assign an EPA identification number to the generator.

(c) A generator must not offer his hazardous waste to transporters or to treatment, storage, or disposal facilities that have not received an EPA identification number.

4. The Manifest

A generator who transports or offers for transportation hazardous waste for off-site treatment, storage, or disposal must prepare a manifest on EPA form 8700-22 and, if necessary, EPA form 8700-22A, according to the instructions included in the Appendix to Part 262 of 40 CFR. A generator must designate on the manifest 1 facility that is permitted to handle the waste described on the manifest.

A generator may also designate on the manifest 1 alternate facility that is permitted to handle the waste in the event an emergency prevents its delivery to the primary designated facility. If the transporter is unable to deliver the hazardous waste to the designated or the alternate facility, the generator must either designate another facility or instruct the transporter to return the waste.

If the state to which the shipment is manifested (consignment state) supplies the manifest and requires its use, then the generator must use that manifest. If the consignment state does not supply the manifest, but the state in which the generator is located (generator state) supplies the manifest and requires its use, then the generator must use that state's manifest. When neither the generator state nor the consignment state supplies the manifest, then the generator may obtain the manifest from any source.

(a) *Required information* (Sec. 262.21 of CFR 40)

The manifest must contain all of the following information:

(1) A manifest document number,

(2) The generator's name, mailing address, telephone number, and EPA identification number,

(3) The name and EPA identification number of each transporter,

(4) The name, address, and EPA identification number of the designated facility and an alternate facility, if any,

(5) The description of the waste(s) (e.g., proper shipping name, etc.) required by regulations of the U.S. Department of Transportation in 49 CFR 172.101, 172.202, and 172.203,

(6) The total quantity of each hazardous waste by units of weight or volume and the type and number of containers as loaded into or onto the transport vehicle.

(b) *Certification*

The following certification must appear on the manifest: "This is to certify that the above-named materials are properly classified, described, packaged, marked, and are in proper condition for transportation according to the applicable regulations of the Department of Transportation and the EPA."

(c) *Number of copies* (Sec. 262.22 of CFR 40)

The manifest consists of at least the number of copies that will provide the generator, each transporter, and the owner or operator of the designated facility with 1 copy each for their records and another copy to be returned to the generator.

(d) *Use of the manifest* (Sec. 262.23 of CFR 40)

The generator must:

(1) Sign the manifest certification by hand,

(2) Obtain the handwritten signature of the initial transporter and date of acceptance on the manifest,

(3) Retain 1 copy, in accordance with Sec. 262.40(a) of CFR 40.

The generator must give the transporter the remaining copies of the manifest. For shipments of hazardous waste within the United States solely by water (bulk shipments only), the generator must send 3 copies of the manifest dated and signed in accordance with this section to the owner or operator of the designated facility or the last water (bulk shipment) transporter to handle the waste in the United States if it is exported by water. Copies of the manifest are not required for each transporter.

For all shipments of hazardous waste within the United States that originate at the site of generation, the generator must send at least 3 copies of the manifest dated and signed in accordance with this section to:

(i) The next nonrail transporter, if any,

(ii) The designated facility if transported solely by rail, or

(iii) The last rail transporter to handle the waste in the United States if it exported by rail.

(e) *Packaging* (Sec. 262.30 of CFR 40)

Before transporting hazardous waste or offering hazardous waste for transportation off-site, a generator must package the waste in accordance with the applicable DOT regulations on packaging under 49 CFR Parts 173, 178, and 179.

(f) *Labeling* (Sec. 262.31 of CFR 40)

Before transporting or offering hazardous waste for transportation off-site, a generator must label each package in accordance with the applicable DOT regulations on hazardous materials under 49 CFR Part 172.

(g) *Marking* (Sec. 262.32 of CFR 40)

Before transporting or offering hazardous waste for transportation off-site, a generator must mark each package of hazardous waste in accordance with the applicable DOT regulations on hazardous materials under 49 CFR Part 172.

Before transporting hazardous waste or offering hazardous waste for transportation off-site, a generator must mark each container of 110 gallons or less used in such transportation with the following words and information displayed in accordance with the requirements of 49 CFR 172.304:

> HAZARDOUS WASTE—Federal Law Prohibits Improper Disposal. If found, contact the nearest police or public safety authority or the U.S. Environmental Protection Agency.
> Generator's Name and Address _____.
> Manifest Document Number _____.

(h) *Placarding* (Sec. 262.33 of CFR 40)

Before transporting hazardous waste or offering hazardous waste for transportation off-site, a generator must placard or offer the initial transporter the appropriate placards according to DOT regulations for hazardous materials under 49 CFR Part 172, Subpart F.

(i) *Accumulation time* (Sec. 262.34 of CFR 40)

A generator may accumulate hazardous waste on-site for 90 days or less without a permit or without having interim status provided that:

(1) The waste is placed in containers and the generator complies with Subpart I of 40 CFR Part 265, or the waste is placed in tanks and the generator complies with Subpart J of 40 CFR Part 265 except Sec. 265.193;

(2) The date upon which each period of accumulation begins is clearly marked and visible for inspection on each container;

(3) While being accumulated on-site, each container and tank is labeled or marked clearly with the words "Hazardous Waste";

(4) The generator complies with the requirements for owners and operators in Subparts C and D in 40 CFR Part 265 and with Sec. 265.16.

A generator who accumulates hazardous waste for more than 90 days is an operator of a storage facility and is subject to the requirements of 40 CFR Parts 264 and 265 and the permit requirements of 40 CFR Part 270 unless he has been granted an extension to the 90-day period. The EPA may grant such an extension if hazardous wastes must remain on-site for longer than 90 days owing to unforeseen, temporary, or uncontrollable circumstances. An extension of up to 30 days may be granted at the discretion of the regional administrator of the EPA on a case-by-case basis.

(j) The EPA Form 8700-22 (Fig. 9-2) is the Uniform Hazardous Waste Manifest, and EPA Form 8700-22A is its continuation sheet, as shown in Fig. 9-3.

Federal regulations require generators and transporters of hazardous waste and owners or operators of hazardous waste treatment, storage, and disposal facilities to use this form (8700-22) and, if necessary, the continuation sheet (Form 8700-22A) for both interstate and intrastate transportation.

Federal regulations also require generators and transporters of hazardous waste and owners or operators of hazardous waste treatment, storage, and disposal facilities to complete the information on Form 8700-22 (Fig. 9-2).

5. Record Keeping and Reporting

(a) *Record keeping* (Sec. 262.40 of CFR 40)

A generator must keep a copy of each manifest signed in accordance with Sec. 262.23(a) of CFR 40 for 3 years or until he receives a signed copy from the designated facility that received the waste. The signed copy must be retained as a record for at least 3 years from the date the waste was accepted by the initial transporter. Also, a generator must keep a copy of each Biennial Report and Exception Report (see below) for a period of at least 3 years from the due date of the report.

A generator must keep records of any test results, waste analyses, or other determinations made in accordance with Sec. 262.11 of CFR 40 for at least 3 years from the date that the waste was last sent to on-site or off-site treatment, storage, or disposal.

The period or retention referred to in this section are extended automatically during the course of any unresolved enforcement action regarding the regulated activity or as requested by the Administrator of the EPA.

(b) *Biennial report* (Sec. 262.41 of CFR 40)

A generator who ships hazardous waste off-site must prepare and submit a single copy of a biennial report to the regional administrator of the EPA by March 1 of each even-numbered year. The report must be submitted on EPA Form 8700-13 A and must cover generator activities during the previous calendar year and must include the following information:

(1) The EPA identification number, name, and address of the generator;

(2) The calendar year covered by the report;

(3) The EPA identification number, name, and address for each off-site treatment, storage, or disposal facility to which waste was shipped during the year; for exported shipments, the report must give the name and address of the foreign facility;

(4) The name and EPA identification number of each transporter used during the reporting year;

(5) A description, EPA hazardous waste number (form 40 CFR Part 261, Subpart C or D), DOT hazard class, and quantity of each hazardous waste shipped off-site. That information must be listed by EPA identification number of each off-site facility to which waste was shipped;

(6) The certification signed by the generator or his authorized representative.

In addition, any generator who treats, stores, or disposes of hazardous waste on-site must submit a biennial report covering those wastes in accordance with the provisions of 40 CFR Parts 270, 264, 265, and 266.

(c) *Exception reporting* (Sec. 262.42 of CFR 40)

A generator who does not receive a copy of the manifest with the handwritten signature of the owner or operator of the designated facility within 35 days of the date the waste was accepted by the initial transporter must contact the transporter or the owner or operator of the designated facility or both to determine the status of the hazardous waste. Also, a generator must submit an Exception Report to the EPA regional administrator for the region in which the generator is located if he has not received a copy of the manifest with the handwritten signature of the owner or operator of the designated facility within 45 days of the date the waste was accepted by the initial transporter. The Exception Report must include:

(1) A legible copy of the manifest for which the generator does not have confirmation of delivery;

(2) A cover letter signed by the generator or his authorized representative explaining the efforts taken to locate the hazardous waste and the results of those efforts.

(d) *Additional reporting* (Sec.. 262.43 of CFR 40)

Fig. 9-1.   EPA Form 8700-22, Uniform Hazardous Waste Manifest

EPA Form 8700-22 (3-84)

Fig. 9-2. EPA Form 8700-22A. Continuation sheet for Uniform Hazardous Waste Manifest

Please print or type. (Form designed for use on elite (12-pitch) typewriter.) Form Approved. OMB No. 2000-0404. Expires 7-31-86

UNIFORM HAZARDOUS WASTE MANIFEST (Continuation Sheet)	21. Generator's US EPA ID No.	Manifest Document No.	22. Page	Information in the shaded areas is not required by Federal law.

GENERATOR

23. Generator's Name	L. State Manifest Document Number
	M. State Generator's ID
24. Transporter ___ Company Name / 25. US EPA ID Number	N. State Transporter's ID
	O. Transporter's Phone
26. Transporter ___ Company Name / 27. US EPA ID Number	P. State Transporter's ID
	Q. Transporter's Phone

28. US DOT Description (Including Proper Shipping Name, Hazard Class, and ID Number)	29. Containers No.	Type	30. Total Quantity	31. Unit Wt./Vol.	R. Waste No.
a.					
b.					
c.					
d.					
e.					
f.					
g.					
h.					
i.					

S. Additional Descriptions for Materials Listed Above	T. Handling Codes for Wastes Listed Above

32. Special Handling Instructions and Additional Information

TRANSPORTER

33. Transporter ___ Acknowledgement of Receipt of Materials		Date
Printed/Typed Name	Signature	Month Day Year
34. Transporter ___ Acknowledgement of Receipt of Materials		Date
Printed/Typed Name	Signature	Month Day Year

FACILITY

35. Discrepancy Indication Space

EPA Form 8700-22A (3-84)

The EPA administrator, as he deems necessary under section 2002(a) and section 3002(6) of the act, may require generators to furnish additional reports concerning the quantities and disposition of wastes identified or listed in 40 CFR Part 261.

6. Special Conditions

(a) *International shipments* (Sec. 262.50 of CFR 40)

Any person who exports hazardous waste to a foreign country or imports it from a foreign country into the United States must comply with the requirements of this part and with the special requirements of this section.

Also, when shipping hazardous waste outside the United States, the generator must:

(1) Notify the EPA Administrator in writing 4 weeks before the initial shipment of hazardous waste to each country in each calendar year; and

(i) The waste must be identified by its EPA hazardous waste identification number and its DOT shipping description;

(ii) The name and address of the foreign consignee must be included in this notice;

(iii) These notices must be sent to the Office of International Activities (A-106), United States Environmental Protection Agency, Washington, D.C. 20460.

Note: This requirement to notify will not be delegated to states authorized under 40 CFR Part 271. Therefore, all generators must notify the Administrator as stipulated above.

(2) Require that the foreign consignee confirm the delivery of the waste in the foreign country. A copy of the manifest signed by the foreign consignee may be used for this purpose;

(3) Meet the requirements under Sec. 262.20(a) for the manifest except that:

(i) In place of the name, address, and EPA identification number of the designated facility, the name and address of the foreign consignee must be used;

(ii) The generator must identify the point of departure from the United States through which the waste must travel before entering a foreign country.

(4) Obtain the manifest from the generator's state if that state supplies the manifest form and requires its use. If the generator's state does not supply the form, then the generator may obtain it from any source.

Also, a generator must file an Exception Report if:

(1) He has not received a copy of the manifest signed by the transporter, stating the date and place of departure from the United States within 45 days from the date is was accepted by the initial transporter; or

(2) Within 90 days from the date the waste was accepted by the initial transporter, the generator has not received written confirmation from the foreign consignee that the hazardous waste was received.

(b) *Importing Hazardous Waste*

When importing hazardous waste, a person must meet all the requirements of Sec. 262.20(a) of CFR 40, for the manifest except that:

(1) In place of the generator's name, address, and EPA identification number, the name and address of the foreign generator and the importer's name, address, and EPA identification number must be used.

(2) In place of the generator's signature on the certification statement, the U.S. importer or his agent must sign and date the certification and obtain the signature of the initial transporter.

A person who imports hazardous waste must obtain the manifest form from the consignment state if that state supplies the manifest and requires its use. If the consignment state does not supply the manifest form, then it may be obtained from any source.

7. Farm Disposal of Pesticides

A farmer disposing of waste pesticides that are from his own use and are hazardous wastes is not required to comply with the standards in this part or other standards in 40 CFR Parts 270, 264, or 265 for those wastes provided that he triple rinses each emptied pesticide container in accordance with Sec. 261.7(b)(3) of CFR 40 and disposes of the pesticide residues on his own farm in a manner consistent with the disposal instructions on the pesticide label.

F. Transporting Hazardous Materials and Wastes

The United States Congress in 1976, enacted the Resource Conservation and Recovery Act (RCRA or RECRA) for the primary purpose of encouraging its citizens to conserve the dwindling natural resources of the land. It also had a concomitant purpose, to stimulate the recovery of substances and materials that hitherto in ordinary circumstances would have been relegated to the scrap heap, garbage dump, or incinerator and would have been wasted.

This double-barreled approach was reinforced by government regulations that attempt to wring out of all materials by various processes, including recycling, a more extended and useful life until no further value can be obtained. There are more subtle and some not so subtle connotations with which RCRA was concerned, namely, that if specific materials or substances are not properly used or disposed of, they could have a harmful and negative impact on human and animal existence and, indeed, do irreparable damage to our environment and life-styles.

The EPA, therefore, in 1980 promulgated a series of regulations that directly concerned the treatment of hazardous wastes in order to provide direction to manage the billions of tons of solid waste that are generated each year in the United States. Those wastes include common household trash and garbage, industrial wastes, agricultural wastes, sewage, sludge, mine tailings, dredged materials, institutional wastes from hospitals and laboratories, and much more.

By 1980 approximately 60 million metric tons of the total wastes generated in the United States consisted of hazardous materials. For years before RCRA and the EPA, many hazardous materials that were simply discarded or dumped found their way back to haunt society by damaging local environmental settings. A catastrophic example was the Love Canal pollution, and there were others, some still to be discovered. The EPA has catalogued hundreds of incidents where indiscriminate dumping of hazardous wastes has wrought ecological damage and even threatened human life.

The RCRA has given the EPA the authority to perform specific functions to assess and manage hazardous wastes. Some of the regulations developed by the agency include, but are not limited to the following:

1. to identify hazardous wastes;
2. to set standards for the generators of hazardous wastes;
3. to set standards for the design, operation, and performance requirements for facilities that handle, store, and dispose of hazardous wastes;

4. to regulate and issue permits for facilities that handle hazardous wastes;
5. to set standards for the transportation of hazardous wastes;
6. to provide guidelines for the individual States to formulate their own internal hazardous waste management programs.

Before the RCCA and the EPA there were little, if any, restraints or regulations where hazardous materials were concerned. At the present time, Congress has mandated that anyone involved or in any way concerned with the handling of hazardous materials and waste must be trained regarding the regulations.

The new regulations include Treatment, Storage or Disposal Facilities (TSDF), which are issued either as interim or permanent permits to operate hazardous waste sites. A hazardous waste material may be transported only to a regulated TSDF. Therefore, it is only those facilities that have been duly inspected and have met the EPA requirements that can legally accept hazardous waste.

In Sec. D, Solid Waste and Hazardous Waste, of this chapter, which discussed Part 261 of CFR 40, the reader became acquainted with the several definitions of hazardous wastes and the criteria used in identifying those materials. The several hazardous material classes were divided into ignitable, corrosive, reactive, and EP toxic wastes. In addition, there are other classes of wastes that can be segregated into infectious, radioactive, and etiological or disease-carrying (pathological) wastes; fortunately, they account for only a small percentage of the transportation traffic, and they are rigidly controlled and managed.

In addition to the EPA identification number for each identifiable hazardous material, there is a code number for each generator, transporter, and TSDF that was derived from the Dun & Bradstreet's Data Universal Numbering System. That code numbering system, which was adapted by the EPA, is alphanumeric. There are 3 letters, XYZ, followed by 9 digits, 123456789. The identifying code number must appear on all shipping documents for hazardous waste materials.

A company may obtain a code number identification from the EPA by applying and filing a "Notification of Hazardous Waste Activity Report." Every facility of a company must also obtain an identification number if it has any hazardous waste. If the company is a shipper or transporter of hazardous wastes only, then the head office must obtain from the EPA an ID number, which will be the number used by all the terminals or shipping points.

Transporting companies that are also generators of hazardous wastes must obtain a separate ID number for each facility that is a hazardous waste generator.

Every generator of hazardous waste must prepare a manifest, which must accompany the load of waste to the TSDF. As you will recall, the manifest was discussed in 6,E, above. Since the characteristics of hazardous wastes are ignitability, corrosivity, reactivity, and EP toxicity, transporters, therefore, would do well to have a quick thumbnail image of each of the characteristics mentioned so that in making pickups they will also remember to have with them the manifest from the generating facility in order to comply with the regulations. The characteristics for hazardous wastes are as follows:

1. *Ignitability*—a liquid or solid waste with a flash point of less than 140°F (60°C).
2. *Corrosivity*—a liquid or solid waste that has a pH of 2 or less or 12.5 or greater or that will corrode steel at a rate exceeding 0.25 inches per year at a temperature of 130°F (55°C).
3. *Reactivity*—a liquid or solid waste that is normally unstable and readily under-

goes violent change without detonation or will react violently with water or forms an explosive mixture when in contact with water.

4. *EP toxicity*—a solid waste that, when subject to the Extraction Procedure, is found to be toxic.

5. *Infectious, etiological, and radioactive*—solid wastes that can spread disease or are radioactive.

Under the EPA regulations, hazardous waste shipments come under the DOT rules for hazardous materials that have been spelled out in Chapter IV. The EPA has imposed further restrictions, however, on hazardous wastes. Generators must carefully label each container of 110 gallons or less with the following information:

HAZARDOUS WASTE
Federal Law Prohibits Improper Disposal

Proper DOT Shipping Name: _____
 If N.O.S.[1] name is used, show
 EPA name, below:

_____

Appropriate EPA Name: _____
Generator's EPA I.D. Code No.: _____
Manifest Document No.: _____

1. Records and Reports

Another EPA-imposed requirement of the generator is that copies of the hazardous waste manifest be retained for at least a 3-year period from the time the shipment was released to the transporting company. The generator must also file an annual report to the EPA on form #8700-13 and 8700-13a. Those reports summarize all of the hazardous wastes that have been generated and shipped during the entire year. The annual report is due on March 1 of each year.

2. Exception Reports

If a generator of hazardous wastes does not receive a signed copy of his manifest from the TSDF within 35 days of tendering the shipment to the transporter, then the generator must trace the shipment. Ten days later the generator must send an Exception Report to his local regional EPA administrator if the shipment has not arrived at the TSDF and no signed manifest copy has been received.

The report itself is not an EPA form but merely a copy of the original manifest with an attached letter outlining the problem and the steps that were taken to trace the shipment.

3. Hazardous Waste Discharges

(a) *Immediate Action Required* (Sec. 263.30 of CFR 40)

In the event of a discharge of hazardous waste during transportation, the transporter must take appropriate immediate action to protect human health and the environment (e.g., notify local authorities or dike the area).

[1]N. O. S., not otherwise specified.

If hazardous waste is discharged during transportation and an official (with the state or local government or a Federal agency) acting within the scope of his official responsibilities determines that immediate removal of the waste is necessary to protect human health or the environment, he may authorize the removal of the waste by transporters who do not have EPA identification numbers and without the preparation of a manifest.

An air, rail, highway, or water transporter who has discharged hazardous waste must:

(1) Give notice, if required by 49 CFR 171.15, to the National Response Center (800-424-8802 or 202-426-2675);

(2) Report in writing, as required by 49 CFR 171.16, to the Director, Office of Hazardous Materials Regulations, Materials Transportation Bureau, Department of Transportation, Washington, D.C. 20590.

A water (bulk shipment) transporter who has discharged hazardous waste must give the same notice as required by 33 CFR 153.203 for oil and hazardous substances.

(b) *The Discharge Must Be Cleaned Up*

A transporter must clean up any hazardous waste discharge that occurs during transportation or take such action that may be required or approved by Federal, state, or local officials so that there is no longer any hazard to human health or the environment.

4. DOT Regulation

Transportation companies, both rail and trucking and air freight and water carriers, have been regulated by the DOT for many years, especially regarding hazardous materials and wastes. Historically, there has been a divergence in regulatory areas and in certain materials that have come under EPA control. Carriers, both private and common, are more likely to have their operations inspected by the DOT rather than by the EPA; however, that does not rule out the possibility of an EPA on-sight inspection, although this would occur only in an extreme case of a particular nature. Since the DOT has had years of experience in handling transportation matters, Congress has decided that the DOT will continue to enforce all of the regulations pertaining to carriers, including all of the new EPA rules pertaining to hazardous materials and waste.

5. Hazardous Waste Traffic

Whenever a common carrier, a private carrier (i.e., a carrier that is part of a company's operations) or a contract carrier enters onto a public highway or leaves a freight terminal, a railroad yard, an airport, or a boat dock, it is subject to both DOT and EPA regulations governing hazardous materials and wastes.

For example, only drivers who qualify under Federal Motor Carrier Safety Regulations may operate vehicles that carry hazardous materials, including hazardous wastes. Not only must the driver be qualified, but also the vehicle he is driving must meet Federal requirements, such as: the completion of DOT pre-trip vehicle inspections before it transports any hazardous materials or waste. The tires on such a vehicle must be examined at the beginning and at the end of each trip. Also tires must be scrutinized every 2 hours or 100 miles, whichever comes first.

As with all hazardous materials, placards must be placed on the vehicle before it may be moved. Before the pickup of hazardous wastes the trucking or vehicle dispatcher should know the consist of the load, including what markings should be on the containers or packages. The driver should be instructed about the required markings, such as: item name, durably marked on a contrasting background color in English, unobscured and having any special markings or labels. Special labels may say, "Store in cool place," or, "This end up," etc. Each container should have the appropriate hazardous waste label. Also, each container or package containing hazardous waste must be DOT authorized. Those packages should have a DOT specification, such as "DOT15B125." That indicates the box was made in accordance with DOT specification 15B fibreboard container, and the allowable gross weight is 125 pounds. (See CFR 49 Part 178 to 179.) The guidelines for hazardous materials packaging may be obtained from CFR 49 Part 172.01, as indicated in Chapter 3.

It is also the driver's responsibility to check the Hazardous Waste Manifest with the hazardous waste cargo in order to determine that he has the correct load and materials. he must keep shipping papers, including the Hazardous Waste Manifest, within reach, preferably in a pocket or container placed on the inside of the cab door.

When a driver is assigned to a vehicle carrying hazardous wastes, it is important that he be familiar with the routing. At certain times a dispatcher may prepare for the driver special instructions that specify the exact route that should be taken. A good driver and a capable dispatcher will avoid tunnels, ball-game crowds, and narrow streets and roads. When the vehicle is placarded, with a load of hazardous waste materials on board, full stops at all railroad crossings are required. The driver may not park the vehicle within 300 feet of an open fire, and it may be passed only when it is absolutely safe to do so. If a placarded vehicle must be refueled, the driver must turn the engine off, and the driver or the fuel attendant must remain at the fuel pump nozzle. A placarded vehicle should not be parked closer than 5 feet of the edge of the roadway of a public thoroughfare or highway. At no time shall a placarded vehicle be left unattended when stopped on a street or a roadway shoulder. As mentioned in F,3, above, if there is a spill, leak, or other discharge of the hazardous wastes being carried, the driver or other carrier personnel must notify the National Emergency Response Center as soon as it is possible to do so.

The above rules and regulations are fundamental to the transportation of hazardous wastes. It is very important, however, that all of the personnel involved in the hazardous waste network be properly informed and familiar with all of the regulations and concerns involving the substances.

Before the pickup driver accepts the load of hazardous materials or waste from the shipper, it is his responsibility to check each package or container to determine if it is properly marked. He should determine whether proper hazard waste labels are affixed to each unit, the consignee's name and address are correct on each package or container, and each unit is in adequate condition for transportation. He should also inspect each consigned unit for container damage, leaks, or the like so that the potential hazard in transit is minimized. It is also a wise precaution to advise the recipient, normally the receiving supervisor of the facility, if any type of specialized materials-handling equipment is to be required for the specific shipment. That responsibility should be delegated to the shipper's dispatcher.

It is through training and familiarization with all of the problems and potential hazards that employees and carrier personnel may learn how to comply with the many regulations that must be followed and to minimize the risks of handling hazardous materials.

6. Emergencies in Transit or Dockside

When a driver is hauling hazardous wastes and becomes involved in an incident that can be classified as an emergency, it is imperative that he follow a prescribed course of action. In the first place, it is necessary to ensure that no unauthorized persons get near the vehicle. If it is humanly possible, all open flames or fires are to be avoided. The driver should immediately set up warning flares or other advisory signals both in front of and behind the vehicle at reasonable distances. As soon as he has accomplished that, he should obtain police and fire personnel assistance. He might ask the aid of a passing motorist aid in placing the call or using a CB phone, emergency channels, etc., but we should not leave the vehicle unattended.

As far as possible, the driver should try to prevent leaking liquids from spilling onto the highway or into streams or sewers by attempting to dam up the liquids or by digging a drainage ditch around the flow. As soon as the public safety personnel arrive at the scene, the driver should make certain that the police or fire fighters have all of the information concerning the hazardous waste substances. The shipping papers and the Hazardous Waste Manifest will contain all of the pertinent information to assist the ER team at the scene. There is little difference between an emergency involving hazardous materials or hazardous wastes in terms of the above routine. If the driver is injured or otherwise unable to assist the response team and the transporter has complied with the regulations, the shipping documents are to be found in the vehicle's cab on the vehicle seat or driver's side door pocket.

The National Emergency Response Center at 1-800-424-8802 should be advised as soon as possible of any spill or discharge that may have occurred. Also, a DOT Hazardous Materials Incident Report must be filed within 15 days following the incident. When there is an incident involving a hazardous-waste-containing vehicle at dockside—whether fire, damaged or leaking containers, or the like—the area should immediately be cleared of personnel and the responsible supervisory personnel should be notified. If the problem is contained within the vehicle and it appears safe to approach the vehicle, it should be marked "out of service" and specific decontamination instructions or other information should be obtained from the shipper. If decontamination is not available from the shipper, CHEMTREC should be called immediately. (CHEMTREC was thoroughly discussed in Chapter 8.) Its toll-free number is 1-800-424-9300.

None of the facility personnel or bystanders should be permitted to collect, touch, or come in contact with broken packages, containers, or the spillage of the hazardous waste cargo. That also applies to any hazardous materials cargo that might be carried by any vehicle. In other words, no one should approach any part of the cargo until specific instructions are obtained from the shipper, manufacturer, or CHEMTREC.

Open flames, smokers, and smoking materials should be kept away from the dockside scene. Information such as shipping papers or the Hazardous Waste Manifest should be ready for inspection by police or fire officials when they arrive at the site. As in the roadside emergency, the hazardous waste liquids should be prevented from spilling out into other areas, sewers, and drains by being dammed up, impounded, or restricted from spreading. Hazardous waste substances should not be touched, gotten on one's person, shoes, or clothing.

As is always the case when an accident occurs involving hazardous materials or waste, it is necessary to advise the National Emergency Response Center of the Coast Guard at 1-800-424-8802, and within 15 days of the event, a Hazardous Materials Incident Report must be filed with the DOT.

7. State Regulations

The regulations that have been discussed in this section are, primarily, Federal regulations. A few states have their own hazardous wastes and solid wastes regulations that not only may differ but also may be more restrictive than the U.S. laws. For example, certain states require the use of their own coding system for identifying hazardous wastes rather than the use of the DOT "proper shipping name." Also, the "generator" and "transporter" under Federal regulations becomes the "producer" and "hauler" under several state regulatory codes.

The above caveats serve to indicate that, before transporting hazardous materials and hazardous wastes, it is advisable to be familiar with not only the Federal regulations but the state laws as well. A good place to obtain such information would be the appropriate state's Department of Transportation.

There is one final word on this subject, however. According to EPA regulations, the transporter of hazardous wastes is responsible for any costs that are incurred in property damage or environmental destruction while he has possession of the hazardous materials.

G. Handling Hazardous Materials and Waste

There is very little difference in the handling of hazardous materials or waste on the receiving dock or in warehouse or storage areas. As soon as the driver arrives at the consignee's facility, the receiving supervisor is notified of the shipment contents.

Each container or package that is loaded or off-loaded from a vehicle must be handled carefully with the proper equipment, and the unloading personnel should be familiar with the equipment and the methods to be used for particular containers.

Federal regulations are explicit concerning loading and unloading of hazardous materials. The following paragraphs parallel Subpart B of CFR 49 Sections 177.834 to 177.844.

1. General Requirements

(a) *Packages secured in a vehicle:* Any tank, barrel, drum, cylinder, or other packaging that is not permanently attached to a motor vehicle and contains any flammable liquid, compressed gas, corrosive material, poisonous material, or radioactive material must be secured against movement within the vehicle on which it is being transported, under conditions normally incident to transportation.

(b) *No hazardous material on pole trailers:* No hazardous materials may be loaded into or on or transported in or on any pole trailer.

(c) *No smoking during loading or unloading:* Smoking on or about any motor vehicle while loading or unloading any explosive, flammable liquid, flammable solid, oxidizing material, or flammable compressed gas is forbidden.

(d) *Keep fire away during loading and unloading:* Extreme care shall be taken in the loading or unloading of any explosive, flammable liquid, flammable solid, oxidizing material, or flammable compressed gas into or from any motor vehicle to keep fire away and to prevent persons in the vicinity from smoking, lighting matches, or carrying any flame or lighted cigar, pipe, or cigarette.

(e) *Handbrake set during loading and unloading:* No hazardous material shall be loaded into or on or unloaded from any motor vehicle unless the handbrake is securely set and all other reasonable precautions are taken to prevent motion of the motor vehicle during this time.

(f) *Use of tools during loading and unloading:* No tools that are likely to damage the effectiveness of the closure of any package or other container or are likely to affect such package or container adversely shall be used for the loading or unloading of any explosive or other dangerous article.

(g) *Prevent relative motion between containers:* Containers of explosives, flammable liquids, flammable solids, oxidizing materials, corrosive materials, compressed gases, and poisonous liquids or gases must braced so that they are stationary while the vehicle is in transit. Containers having valves or other fittings must be so loaded that there will be the minimum likelihood of damage during transportation.

(h) *Precautions concerning containers in transit; fueling road units:* Reasonable care should be taken to prevent undue rise in temperature of containers and their contents during transit. There must be no tampering with such containers or their contents nor any discharge of the contents of any container between the point of origin and the point of billed destination. Discharge of contents of any container, other than a cargo tank, must not be made before it is removed from the motor vehicle. Nothing contained in this paragraph shall be so construed as to prohibit the fueling of machinery or vehicles used in road construction or maintenance.

(i) *Attendance requirements:*

(1) Loading: A cargo tank must be attended by a qualified person at all times when it is being loaded. The person who is responsible for loading the cargo tank is also responsible for ensuring that it is so attended.

(2) Unloading: A motor carrier who transports hazardous materials by a cargo tank must ensure that the cargo tank is attended by a qualified person at all times during unloading. However, the carrier's obligation to ensure attendance during unloading ceases when:

(i) The carrier's obligation for transporting the materials is fulfilled;

(ii) The cargo tank has been placed upon the consignee's premises;

(iii) The motive power has been removed from the cargo tank and removed from the premises.

(3) A person "attends" the loading or unloading of a cargo tank if he is awake, has an unobstructed view of the cargo tank, and is within 7.62 meters (25 feet) of the cargo tank throughout the process.

(4) A person is "qualified," if he has been made aware of the nature of the hazardous material that is to be loaded or unloaded, if he has been instructed on the procedures to be followed in emergencies, and if he is authorized to move the cargo tank and has the means to do so.

(5) A delivery hose, when attached to the cargo tank, is considered a part of the vehicle.

(j) *Prohibited loading combinations:* In any single-drive motor vehicle or in any single unit of a combination of motor vehicles, hazardous materials shall not be loaded together if prohibited by loading and storage chart Sec. 177.848. That section shall not be so construed as to forbid the carrying of materials essential to safe operation of motor vehicles. (See Motor Carrier Safety Regulations Part 393 of this title.)

(k) *Access to mixed ladings:* Flammable solids, oxidizing materials, or corrosive liquids, when transported on a motor vehicle with other lading not otherwise forbidden, shall be so loaded as to provide ready access thereto for shifting or removal.

(l) *Use of cargo heaters when one is transporting certain hazardous material:* Transportation includes loading, carrying, and unloading.

(1) When one is transporting explosives: A motor vehicle equipped with a cargo heater of any type may transport explosives only if the cargo heater is rendered inoperable by: (i) draining or removing the cargo heater fuel tank and (ii) disconnecting the heater's power source.

(2) When one is transporting certain flammable material—(i) Use of combustion cargo heaters: A motor vehicle equipped with a combustion cargo heater may be used to transport flammable liquid or flammable gas only if each of the following requirements is met—

(A) It is a catalytic heater.

(B) The heater's surface temperature cannot exceed 130°F (54°C)—either on a thermostatic control when the outside or ambient temperature is 60°F (15.6°C) or less.

(C) The heater is not ignited in a loaded vehicle.

(D) There is no flame, either on the catalyst or anywhere in the heater.

(E) The manufacturer has certified that the heater meets the requirements under paragraph (1)(2)(i) of this section by permanently marking the heater "Meets DOT requirements for catalytic heaters used with flammable liquid and gas."

(F) The heater is also marked "Do not load into or use in cargo compartments containing flammable liquid or gas if flame is visible on catalyst or in heater."

(G) Heater requirements under Sec. 393.77 of this title are complied with.

(ii) Effective date for combustion heater requirements: The requirements under paragraph (1)(2)(i) of this section govern as follows—

(A) Use of a heater manufactured after November 4, 1975, is governed by every requirement under (1)(2)(i) of this section;

(B) Use of a heater manufactured before November 15, 1975, is governed only by the requirements under (1)(2)(i) (A), (C), (D), (F) and (G) of this section until October 1, 1976;

(C) Use of any heater after September 30, 1976, is governed by every requirement under paragraph (1)(2)(i) of this section.

(iii) Restrictions on automatic cargo-space-heating temperature control devices: Restrictions on these devices have 2 dimensions—restrictions upon use and those that apply when the device must not be used.

(A) Use restrictions: An automatic cargo-space-heating temperature control device may be used when one is transporting flammable liquid or flammable gas only if each of the following requirements is met:

(1) Electrical apparatus in the cargo compartment is nonsparking or explosion proof.

(2) There is no combustion apparatus in the cargo compartment.

(3) There is no connection for return of air from the cargo compartment to the combustion apparatus.

(4) The heating system will not heat any part of the cargo to more than 130°F (54°C).

(5) Heater requirements under Sec. 393.77 of this title are complied with.

(B) Protection against use: Flammable liquid or flammable gas may be transported by a vehicle, which is equipped with an automatic cargo-space-heating temperature control device that does not meet each requirement of paragraph (1)(2)(iii)(A) of this section, only if the device is first rendered inoperable, as follows—

(1) Each cargo heater fuel tank, if other than LPG, must be emptied or removed.

(2) Each LPG fuel tank for automatic temperature control equipment must have its discharge valve closed and its fuel feed line disconnected.

(m) Tanks constructed and maintained in compliance with spec. 106A or 110A (Sec. 179.300, 179.301 of this subchapter) that are authorized for the shipment of hazardous materials by highway in Part 173 of this subchapter must be carried in accordance with the following requirements:

(1) Tanks must be securely chocked or clamped on vehicles to prevent any shifting.

(2) Equipment suitable for handling a tank must be provided at any point where a tank is to be loaded upon or removed from a vehicle.

(3) No more than 2 cargo-carrying vehicles may be in the same combination of vehicles.

(4) Compliance with Sec. 174.200 and 174.204 of this subchapter for combination rail freight and highway shipments and for trailer-on-flat-car service is required.

(n) Specification 56, 57, IM 101, and IM 102 portable tanks, when loaded, may not be stacked on each other or placed under other freight during transportation by motor vehicle.

2. Explosives

(See also Sec. 177.834(a) to (k) of CFR 49.)

(a) *Engine stopped:* No explosives shall be loaded into or on or be unloaded from any motor vehicle with the engine running.

(b) *Care in loading, unloading, other handling of explosives:* No bale hooks or other metal tools shall be used for the loading, unloading, or other handling of explosives, nor shall any package or other container of explosives, except barrels or kegs, be rolled. No packages of explosives shall be thrown or dropped during the process of loading, unloading, or handling of explosives. Special care shall be exercised so that packages or other containers containing explosives shall not catch fire from sparks or hot gases from the exhaust tailpipe.

(1) Whenever tarpaulins are used for covering explosives, they shall be secured by means of rope or wire tiedowns. Explosives placards or markings required by Sec. 177.823 shall be secured in the appropriate locations directly on the equipment transporting the explosives. If the vehicle is provided with placard boards, the placards must be applied to them.

(c) *Explosives on vehicles in combination:* Class A explosives may not be loaded into or carried on any vehicle or a combination of vehicles if:

(1) More than 2 cargo-carrying vehicles are in the combination;

(2) Any full trailer in the combination has a wheel base of less than 184 in.

(3) Any vehicle in the combination is a tank motor vehicle that is required to be marked or placarded under Sec. 177.823;

(4) The other vehicle in the combination contains any:

(i) Initiating explosives,

(ii) Packages of radioactive materials bearing "Yellow III" labels,

(iii) Class A or B poisons,

(iv) Hazardous materials in a portable tank or a DOT specification 106A or 110A tank.

(e) *No sharp projections inside body of vehicles:* No motor vehicle transporting any kind of explosive shall have on the interior of the body in which the explosives are contained any inwardly projecting bolts, screws, nails, or other inwardly projecting parts likely to produce damage to any package or container of explosives during the loading or unloading process or in transit.

(f) *Explosives vehicles, floors tight and lined:* Motor vehicles transporting class A or class B explosives shall have tight floors and that portion of the interior that is in contact with the load shall be lined with either nonmetallic material or nonferrous metals, except that the lining is not required for truck load shipments loaded by the departments of the Army, Navy, or Air Force of the United States Government provided that the explosives are of such nature that they are not liable to leakage of dust, powder, or vapor that might become the cause of an explosion. The interior of the cargo space must be in good condition so that there will not be any likelihood of containers being damaged by exposed bolts, nuts, broken side panels, or floor boards, or any similar projections.

(g) *No detonating primer may be transported on the same motor vehicle with any Class A or Class B explosive (except detonating primers).* No detonator may be transported on the motor vehicle with any Class A or Class B explosive (except detonators) unless:

(1) It is packed in a specification MC 201 (Sec. 178.318 of this subchapter) container;

(2) It is packed and loaded in accordance with a method approved by the department. One method approved by the department is as follows:

(i) As is prescribed in Sec. 173.66 of this subchapter, the detonators are in packagings, which in turn are loaded into suitable containers or separate compartments. Both the detonators and the container or compartment must meet the requirements of the Institute of Makers of Explosives' Standard (IME Safety library Publication No. 22).

(h) *Lading within body or covered tailgate closed:* Except as provided in paragraphs (g), (k), and (m) of this section, dealing with the transportation of liquid nitroglycerin, desensitized liquid nitroglycerin or diethylene glycol dinitrate, other than as defined in Sec. 173.53(e) of this subchapter, all of that portion of the lading of any motor vehicle that consists of explosives shall be contained entirely within the body of the motor vehicle or within the horizontal outline thereof, without overhang or projection of any part of the load, and if such motor vehicle has a tailboard or tailgate, it shall be closed and secured in place during such transportation. Every motor vehicle transporting explosives must either have a closed body or have the body thereof covered with a tarpaulin, and in either event care must be taken to protect the load from moisture and sparks, except that subject to other provisions of these regulations, explosives other than black powder may be transported on flat-bed vehicles if the explosive portion of the load on each vehicle is packed in fire- and water-resistant containers or covered with a fire and water-resistant tarpaulin.

(i) *Explosives to be protected against damage by other lading:* No motor vehicle transporting any explosive may transport as a part of its load any metal or other articles or materials likely to damage such explosive or any package in which it is contained, unless the different parts of such load are so segregated or secured in place in or on the motor vehicle and separated by bulkheads or other suitable means as to prevent such damage.

(j) *Transfer of explosives en route:* No class A or class B explosive shall be transferred from 1 container to another or from 1 motor vehicle to another vehicle or from another vehicle to a motor vehicle on any public highway, street, or road except in case of emergency. In such cases red electric lanterns, red emergency reflectors, or red flags shall be set out in the manner prescribed for disabled or stopped motor vehicles. (See Motor Carrier Safety Regulations, part 392 of this title.) In any event, all practicable means, in addition to these prescribed, shall be taken to protect and warn other users of the highway against the hazard involved in any such transfer or against the hazard occasioned by the emergency making such transfer necessary.

(k) *Loading requirements for liquid nitroglycerin, desensitized liquid nitroglycerin, or diethylene glycol dinitrate:* Liquid nitroglycerin, desensitized liquid nitroglycerin, or diethylene glycol dinitrate, other than as defined in Sec. 173.53(e) of this chapter, may be accepted for transportation and transported only by motor carriers other than common carriers if it is loaded into or on a truck having the type of body specified in spec. MC200 (Sec. 178.315 of this subchapter). No liquid nitroglycerin, desensitized liquid nitroglycerin, or diethylene glycol dinitrate may be loaded directly above any other explosive or in any quantity in excess of 900 quarts on 1 motor vehicle or 10 quarts in any 1 individual container. Additional quantities of explosives, other than nitroglycerin, desensitized liquid nitroglycerin or diethylene glycol dinitrate, excepting any type of blasting or percussion cap or other detonating device, may be carried on such motor vehicle in a closed or covered bed or body that shall be firmly bolted or fastened above the lid of the compartment containing the nitroglycerin, desensitized liquid nitroglycerin, or diethylene glycol dinitrate. In no case shall the net load be more than 7,500 pounds. (See paragraph (m) of this section and spec. MC201 (Sec. 178.318) of this subchapter).

(l) *Separation of tools and supplies for preparing charges:* Motor vehicles transporting liquid nitroglycerin, desensitized liquid nitroglycerin, or diethylene glycol dinitrate may also transport the tools and supplies necessary for preparing and firing charges thereof provided that such tools and supplies are properly secured in place to prevent their coming in contact with the body above specified.

(m) *Detonators or other explosives:* Any explosive, including desensitized liquid explosives as defined in Sec. 173.53(e) of this subchapter, other than liquid nitroglycerin, desensitized nitroglycerin, or diethylene glycol dinitrate, transported on any motor vehicle transporting liquid nitroglycerin, desensitized liquid nitroglycerin, or diethylene glycol dinitrate, must be segregated, each kind from every other kind, and from tools or other supplies. Detonators must be packed in specification MC201 (Sec. 178.318 of this subchapter) containers.

3. Nonexplosive Material

(a) No restrictions are prescribed in Parts 170–189 of this subchapter for the packing, handling, and transportation of material relating to ammunition for cannon, but containing no explosive or other dangerous article, such as cartridge cases, "dummy" or "drill" cartridges, etc., sandloaded projectiles, sand-loaded bombs, empty projectiles, empty mines, empty bombs, solid projectiles, or empty torpedoes. Rotating bands should be protected against deformation by method of packing or loading.

4. Flammable Liquids

(See also Sec. 177.834(a) to (k) of CFR 49)

(a) *Engine stopped:* Unless the engine of the motor vehicle is to be used for the operation of a pump, no flammable liquid shall be loaded into, or on, or unloaded from any motor vehicle while the engine is running.

(b) *Bonding and grounding containers other than cargo tanks before and during transfer of lading:* For containers that are not in metallic contact with each other, either metallic bonds or ground conductors shall be provided for the neutralization of possible static charges before and during transfers of flammable liquids between such containers. Such bonding shall be made by first connecting an electric conductor to the container to be filled and subsequently connecting the conductor to the container from which the liquid is to come, and not in any other order. To provide ignition of vapors by discharge of static electricity, the latter connection shall be made at a point well removed from the opening from which the flammable liquid is to be discharged.

(c) *Bonding and grounding cargo tanks before and during transfer of lading:* (1) When a cargo tank is loaded through an open filling hole, one end of a bond wire shall be connected to the stationary system piping or integrally connected steel framing, and the other end to the shell of the cargo tank to provide a continuous electrical connection. (If bonding is to the framing, it is essential that piping and framing be electrically interconnected.) That connection must be made before any filling hole is opened and must remain in place until after the last filling hole has been closed. Additional bond wires are not needed around all-metal flexible or swivel joints but are required for nonmetallic flexible connections in the stationary system piping. When a cargo tank is unloaded by a suction-piping system through an open filling hole of the cargo tank, electrical continuity shall be maintained from cargo tank to receiving tank.

(2) When a cargo tank is loaded or unloaded through a vapor-tight (not open hole) top or bottom connection, so that there is no release of vapor at a point where a spark could occur, bonding or grounding is not required. Contact of the closed connection must be made before the flow starts and must not be broken until after the flow is completed.

(3) Bonding or grounding is not required when a cargo tank is unloaded through a nonvapor-tight connection into a stationary tank provided that the metallic filling connection is maintained in contact with the filling hole.

(d) *Pyroforic liquids in cylinders:* Cylinders containing pyroforic liquids, unless packed in a strong box or case and secured therein to protect valves, must be loaded with all valves and safety relief devices in the vapor space. All cylinders must be secured so that no shifting may occur in transit.

(e) *Manholes and valves closed:* A person shall not drive a tank motor vehicle and a motor carrier shall not require or permit a person to drive a tank motor vehicle containing flammable liquid (regardless of quantity) unless—

(1) All manhole closures on the cargo tank are closed and secured;

(2) All valves and other closures in liquid discharge systems are closed and free of leaks.

5. Flammable Solids and Oxidizing Materials

(See, also, Sec. 177.834 (a) to (k) of CFR 49.)

(a) *Lading within body or covered; tailgate closed; pickup and delivery:* All of that portion of the lading of any motor vehicle transporting flammable solids or oxidizing materials shall be contained entirely within the body of the motor vehicle and shall be covered by such body, tarpaulins, or other suitable means, and if such motor vehicle has a tailboard or tailgate, it shall be closed and secured in place during such transportation provided, however, that the provisions of this paragraph need not apply to "pickup and delivery" motor vehicles when they are used in no other transportation than in and about cities, towns, or villages. Shipment in water-tight bulk containers need not be covered by a tarpaulin or other means.

(b) *Articles to be kept dry:* In the loading of any motor vehicle with flammable solids or oxidizing materials that are likely to become hazardous to transport when wet, special care shall be taken to keep them from being wetted during the loading process and to keep them dry during transit. Special care shall also be taken in the loading of any motor vehicle with flammable solids or oxidizing materials, which are likely to become more hazardous to transport by wetting, to keep them from being wetted during the loading process and to keep them dry during transit. Examples of such dangerous materials are charcoal screenings; ground, crushed, or pulverized charcoal; and lump charcoal.

(c) *Lading ventilation, precautions against spontaneous combustion:* Whenever a motor carrier has knowledge concerning the hazards of spontaneous combustion or heating of any article to be loaded on a motor vehicle, the article shall be so loaded as to afford sufficient ventilation of the load to provide reasonable assurance against fire from this cause, and in such a case the motor vehicle shall be unloaded as soon as is practicable after reaching its destination. Charcoal screenings or ground, crushed, granulated, or pulverized charcoal in bags shall be so loaded that the bags are laid horizontally in the motor vehicle and so piled that there will be spaces for effective air circulation, which spaces shall not be less than 4 in. wide; and air spaces shall be maintained between rows of bags. Bags shall not be piled closer than 6 in. from the top of any motor vehicle with a closed body.

(d) *Loose or baled nitrate of soda bags:* Loose or baled unwashed, empty bags, having contained nitrate of soda, may be transported in truckload lots only in motor vehicles, and such motor vehicles must have closed or covered bodies lined with paper; such shipments are required to be loaded by the shipper and to be unloaded by the consignee.

(e) *Staying or blocking of packages of matches:* Special care shall be exercised in the loading of packages containing "strike-anywhere" matches to prevent the shifting or jamming of any such package during transit. To that end, the packages shall be compactly loaded with the strongest dimensions of each box or other container loaded lengthwise of the motor vehicle.

(1) Smooth vehicle interior for matches: Unless "strike-anywhere" matches are contained in wooden outside boxes, special care shall be taken to provide that the inside surfaces of any motor vehicle into which such matches are to be loaded and with which surfaces the containers might come in contact shall be smooth, without protrusions of any sort, such as bolts, nuts, sharp edges, or corners, etc., and smooth wooden inner linings shall be provided for this purpose if the interior of the motor vehicle is not otherwise smooth in accordance with this requirement.

(2) Flammable liquids: Matches must not be loaded next to a package bearing a flammable liquid label.

(f) *Nitrates, except ammonium nitrate having organic coating,* listed in Sec. 173.182(b) of this subchapter must be loaded in closed or open-type motor vehicles, which must be swept clean and be free of any projections capable of injuring bags when so packaged. When shipped in open-type motor vehicles, the lading must be suitably covered. Ammonium nitrate having organic coating must not be loaded in all-metal vehicles, other than those made of aluminum or aluminum alloys of the closed type.

(g) *Smokeless powder for small arms in quantities not exceeding 100 pounds net weight transported in 1 car or motor vehicle* may be classed as a flammable solid when examined for that classification by the Bureau of Explosives and approved by the Associate Director for OE (Office of Operations and Enforcement). Maximum quantity in any inside packaging must not exceed 8 lbs. and inside packagings must be arranged and protected to prevent simultaneous ignition of the contents. The complete package must be a type examined by the Bureau of Explosives and approved by the Associate Director for OE. Each outside packaging must bear a "flammable solid" label.

6. Corrosive Liquids

(See also Sec. 177.834 (a) to (k) of CFR 49)

(a) *Nitric acid:* In addition to the requirements set forth in paragraph (b) of this section, no carboy or other container of nitric acid shall be loaded above any container containing any other kind of material. The loading of carboys or other containers of nitric acid shall be limited to 2 tiers high.

(b) *Carboys and frangible containers:* In general, individual carboys and frangible containers of corrosive liquids, including charged electric storage batteries, must, when loaded by hand, be individually loaded into and unloaded from any motor vehicle in which they are to be or have been transported. All reasonable precautions must be taken to prevent, by all practicable means, the dropping of any such containers or batteries containing corrosive liquids. No such container or battery may be loaded into a motor vehicle having an uneven floor surface. It shall be permissible to load on or transport in any motor vehicle any authorized carboys or frangible shipping containers, containing corrosive liquids, more than 1 tier high above any floor only if such carboys or other containers are boxed or crated or are in barrels or kegs, as required by Parts 170–189 of this subchapter, and only if such containers are so stacked that the weight of each tier above the first is entirely supported by the boxes, crates, barrels, kegs, or other authorized means of enclosing the carboys or frangible containers.

Only so many tiers as may adequately be so supported without danger of crushing or breaking shall be permitted. Means must be provided to prevent by all practicable means in all cases the shifting of containers or batteries during transit. Nothing contained in this section shall be so construed as to prevent the use of cleats or other retaining means for the purpose of preventing shifting of containers or batteries. For the purposes of this section a false floor or platform, secured against relative motion within the body of the motor vehicle, shall be deemed to be a floor. (For recommendations for handling leaking or broken packages see Sec. 177.858(a) of CFR 49.)

(c) *Storage batteries:* In addition to the requirements set forth in paragraph (b) of this section, all storage batteries containing any electrolyte shall be so loaded, if loaded with other lading, that all such batteries will be protected against other lading falling onto or against them, and adequate means shall be provided in all cases for the protection and insulation of battery terminals against short circuits.

(d) *Corrosives in cargo tanks:* A person shall not drive a tank motor vehicle, and a motor carrier shall not require or permit a person to drive a tank motor vehicle containing corrosives (regardless of quantity) unless:

(1) All manhole closures on the cargo tank are closed and secured;

(2) All valves and other closures in liquid discharge systems are closed and free of leaks.

7. Compressed Gases

(See also Sec. 177.834 (a) to (k) of CFR 49.)

(a) *Floors or platforms essentially flat:* Cylinders containing compressed gases shall not be loaded onto any part of the floor or platform of any motor vehicle that is not essentially flat; cylinders containing compressed gases may be loaded onto any motor vehicle not having a floor or platform only if such motor vehicle is equipped with suitable racks having adequate means for securing such cylinders in place. Nothing contained in this section shall be so construed as to prohibit the loading of such cylinders on any motor vehicle having a floor or platform and racks as has been described.

(1) Cylinders: To prevent their overturning, cylinders containing compressed gases must be securely lashed in an upright position, loaded into racks securely attached to the motor vehicle, packed in boxes or crates of such dimensions as to prevent their overturning, or loaded in a horizontal position. Specification DOT-4L cylinders must be loaded in an upright position and securely braced.

(2) Cylinders for liquified hydrogen: Specification DOT-4L cylinders containing liquified hydrogen must be transported only on motor vehicles that have open bodies and are equipped with suitable racks or supports having clamps or securing bands capable of holding the cylinders upright when they are subjected to an acceleration of at least 2 gs in any horizontal direction.

(i) The combined total of the hydrogen venting rates as marked on the cylinders on 1 motor vehicle must not exceed 60 standard cubic feet per hour.

(ii) Motor vehicles loaded with cylinders containing liquified hydrogen may not be driven through tunnels.

(iii) Highway transportation is limited to private and contract motor carriers only and to direct movement from point of origin to destination.

(b) Portable tank containers containing compressed gases shall be loaded on motor vehicles only as follows:

(1) Onto a flat floor or platform of a motor vehicle.

(2) Onto a suitable frame of a motor vehicle.

(3) In either case, such containers shall be safely and securely blocked or held down to prevent movement relative to each other or to the supporting structure when in transit, particularly during sudden starts and stops and changes of direction of the vehicle.

(4) Requirements of paragraphs (1) and (2) of this paragraph (b) shall not be construed as prohibiting stacking of containers provided that the provisions of paragraph (3) of this paragraph (b) are fully complied with.

(d) *Engine to be stopped in tank motor vehicles, except for transfer pump:* No flammable compressed gas shall be loaded into or on or unloaded from any tank motor vehicle with the engine running unless the engine is used for the operation of the transfer pump of the vehicle. Unless the delivery hose is equipped with a shut-off valve at its discharge end, the engine of the motor vehicle shall be stopped at the finish of such loading or unloading operation while the filling or discharge connections are disconnected.

(e) *Chlorine cargo tanks* shall be shipped only when equipped (1) with a gas mask of a type approved by the U.S. Bureau of Mines for chlorine service; (2) with an emergency kit for controlling leaks in fittings on the dome cover plate.

(f) *No chlorine tank motor vehicle used for transportation of chlorine shall be moved,* coupled or uncoupled, when any loading or unloading connections are attached to the vehicle, nor shall any semitrailer or trailer be left without the power unit unless it is chocked or equivalent means are provided to prevent motion.

(g) *Each liquid discharge valve on a cargo tank,* other than an engine fuel line valve, must be closed during transportation, except during loading and unloading.

8. Poisons

(See also Sec. 177.834 (a) to (k) of CFR 49.)

(a) *Arsenical compounds in bulk:* Care shall be exercised in the loading and unloading of "arsenical dust," "arsenic trioxide," and "sodium arsenate," allowable to be loaded into sift-proof, steel hopper-type or dump-type motor vehicle bodies equipped with waterproof, dust-proof covers well secured in place on all openings, to accomplish such loading with the minimum spread of such compounds into the atmosphere by all means that are practicable, and no such loading or unloading shall be done near or adjacent to any place where there are or are likely to be, during the loading or unloading process, assemblages of persons other than those engaged in the loading or unloading process or upon any public highway or in any public place.

(1) The motor vehicles must be marked in accordance with Sec. 173.368(b) of this chapter.

(2) Before any motor vehicle may be used for transporting any other articles, all detectable traces of arsenical materials must be removed from it by flushing with water or by other appropriate methods, and the marking removed.

(b) *No Class A or irritating materials in cargo tanks:* No poison, Class A, or irritating material may be loaded into or transported in any cargo tank.

(c) *Class A poisons or irritating materials:* The transportation of a Class A poison or an irritating material is not permitted if there is any interconnection between packagings.

(d) *Poisons in cargo tanks:* A person shall not drive a tank motor vehicle, and a motor carrier shall not require or permit a person to drive a tank motor vehicle containing poisons (regardless of quantity) unless—

(1) All manhole closures on the cargo tank are closed and secured;

(2) All valves and other closures in the liquid discharge systems are closed and free of leaks.

(e) *A carrier may not transport a package bearing a poison label in the same transport vehicle with material that is marked as or is known to be foodstuff,* feed or any other edible material intended for consumption by humans or animals.

9. Radioactive Material

(a) The number of packages of radioactive materials in any motor vehicle, trailer, or storage location must be limited so that the total transport index number as defined in Sec. 173.389(i) of this subchapter and determined by adding together the transport index numbers on the labels of the individual packages, does not exceed 50. That provision does not apply to exclusive-use shipments described in Sec. 173.393(j), 173.396(f), or 173.392 of this subchapter of CFR 40.

(b) Packages of radioactive material bearing "radioactive yellow-II" or "radioactive yellow-III" labels must not be placed in a motor vehicle or in any other place closer than the distances shown in Table 9-16, below, to any areas that may be

Table 9-16.   Distance from Radioactive Materials to Undeveloped Film.

Total transport index	Minimum separation distances in feet to nearest undeveloped film for various times of transit					Minimum distance in feet to area of persons, or minimum distance in feet from dividing partition of cargo compartments
	Up to 2 hours	2–4 hours	4–8 hours	8–12 hours	Over 12 hours	
None	0	0	0	0	0	0
0.1 to 1.0	1	2	3	4	5	1
1.1 to 5.0	3	4	6	8	11	2
5.1 to 10.0	4	6	9	11	15	3
10.1 to 20.0	5	8	12	16	22	4
20.1 to 30.0	7	10	15	20	29	5
30.1 to 40.0	8	11	17	22	33	6
40.1 to 50.0	9	12	19	24	36	7

NOTE 1: The distance in the table must be measured from the nearest point on the packages of radioactive materials.

continuously occupied by passengers, employees, or shipments of animals, nor closer than the distances shown in Table 9-16, below, to any package containing undeveloped film (if so marked). If more than 1 of those packages is present, the distance shall be computed from Table 9-16 on the basis of the total transport index number (determined by adding together the transport index numbers on the labels of the individual packages) or packages in the vehicle or storeroom. Where more than 1 group of packages is present in any single storage location, a single group may not have a total transport index greater than 50. Each group of packages must be handled and stowed not closer than 6 meters (20 feet) (measured edge to edge) to any other group.

(c) Shipments of low-specific-activity materials, as defined in Sec. 173.389(c) of this subchapter, must be loaded to avoid spillage and scattering of loose materials. Loading restrictions are set forth in Sec. 173.392 of this subchapter of CFR 40.

(d) Packages must be so blocked and braced that they cannot change position during conditions normally incident to transportation.

(e) Persons should not remain unnecessarily in a vehicle containing radioactive materials.

(f) Each fissile class III radioactive material shipment (as defined in Sec. 173.389(a)(3) of this subchapter) must be transported in accordance with 1 of the methods prescribed in Sec. 173.396(g) of this subchapter. The transport controls must be adequate to ensure that no fissile class III shipment is transported in the same transport vehicle with any other fissile radioactive material shipment. In loading and storage areas each fissile class III shipment must be segregated by a distance of at least 20 feet from other packages required to bear one of the "Radioactive" labels described in Sec. 172.403 of this subchapter of CFR 40.

10. Contamination of Vehicles

(a) Each motor vehicle used for transporting low specific activity radioactive materials in truckload lots under the provisions of Sec. 173.392(d) of this subchapter must be surveyed with appropriate radiation detection instruments after each use. Carriers must not return such vehicles to service until the radiation dose rate at any accessible surface is not more than 0.5 millirem per hour, and there is no significant removable radioactive surface contamination. (See Sec. 173.397(a) of CFR 49.)

(b) This section does not apply to any vehicle used solely for transporting radioactive material if a survey of the interior surface shows that the radiation dose rate does not exceed 100 millirem per hour at the interior surface or 2 millirem per hour at 3 feet from any interior surface. These vehicles must be stenciled with the words "For Radioactive Materials Use Only" in lettering at least 3 in. high in a conspicuous place on both sides of the exterior of the vehicle. Those vehicles must be kept closed at all times other than loading and unloading.

(c) In case of fire, accident, breakage, or unusual delay involving shipments of radioactive material, see Sec. 177.861 of CFR 49.

11. Asbestos

Asbestos must be loaded, handled, and unloaded, and any asbestos contamination of transport vehicles removed in a manner that will minimize occupational exposure to airborne asbestos particles released incident to transportation. (See Sec. 173.1090 of CFR 49.)

Chart No. 9-4.   Loading and Storage Chart of Hazardous Materials

The following table shows the hazardous materials which must not be loaded or stored together.

The letter X at an intersection of horizontal and vertical columns shows that these articles must not be loaded or stored together, for example: Detonating fuzes, class A, with or without radioactive components k horizontal column must not be loaded or stored with high explosives or propellant explosives, class A vertical column.

Column / row key

No.	Description
15	Radioactive materials
14	Poisonous gases or liquids, in tank car tanks, cylinders, projectiles or bombs, poison gas label
13	Nonflammable gases: Nonflammable gas label.
12	Corrosive liquids: Corrosive label.
11	Flammable solids or oxidizing materials: Flammable solid, oxidizer, or organic peroxide label.
10	Flammable liquids or flammable gases: Flammable liquid or flammable gas label.
9	Fireworks, common.
8	Cordeau detonant fuse, safety squibs, fuse lighters, fuse igniters, delay electric igniters, electric squibs, instantaneous fuse or igniter cord
7	Time, combination or detonating fuzes, class C
6	Percussion fuzes, tracer fuzes or tracers
5	Primers for cannon or sm'll arms, empty cartridge bags—black powder igniters, empty cartridge cases, primed, combination primers or percussion caps, grenades, primed, combination primers or percussion caps, toy caps, explosive cable cutters, explosive rivets
4	Small arms ammunition, or cartridges, practice ammunition.
3	Fireworks special or railway torpedoes
2	Propellant explosives, class B; jet thrust units (Jato), class B; igniters, jet thrust, class B; rocket motors, class B; rocket engines (liquid), class B; igniters, rocket motor, class B; starter cartridges, jet engine, class B
1	Ammunition for cannon with empty, inert-loaded or solid projectiles, or without projectiles; rocket ammunition with empty, inert-loaded or solid projectiles
h	Detonating fuzes, class A, with or without radioactive components
g	Explosive projectiles; bombs; torpedoes; mines; rifle or hand grenades (explosive); jet thrust units (Jato), class A; igniters, Jet thrust, class A; rocket motors, class A; igniters, rocket motor, class A ●
f	Ammunition for cannon with explosive projectiles, gas projectiles or shell; ammunition for small arms with incendiary projectiles, smoke projectiles, incendiary projectiles, illuminating projectiles; ammunition for small arms with explosive projectiles, rocket ammunition with explosive projectiles, gas projectiles, smoke projectiles, incendiary projectiles, illuminating projectiles; boosters (explosive); bursters (explosive); projectiles, smoke projectiles, incendiary projectiles, illuminating projectiles, and supplementary charges (explosive); without detonators ● ●
e	Blasting caps, with or without safety fuse (including electric blasting caps), detonating primers
c	Initiating or priming explosives, wet: Diazodinitrophenol, fulminate of mercury, guanyl nitrosamino guanylidene hydrazine, lead azide, lead styphnate, nitro mannite, nitrosoguanidine, pentaerythrite tetranitrate, tetrazene, lead mononitroresorcinate
b	High explosives or propellant explosives, class A
a	Low explosives or black powder

CLASS A EXPLOSIVES

Material	15	14	13	12	11	10	9	8	7	6	5	4	3	2	1	h	g	f	e	d	c	b	a
																						(•)	(•)
Low explosives or black powder----	X	X	X	X	X	X	X					•	X								X		
High explosives or propellant explosive, class A----	X	X	X	X	X	X	X						X			X			X		X		
Initiating or priming explosives, wet: Diazodinitrophenol, fulminate of mercury, guanyl nitrosamino guanylidene hydrazine, lead azide, lead styphnate	X	X	X	X	X	X	X					•	X			X			X		X		

Chart No. 9-1.   Loading and Storage Chart of Hazardous Materials—Continued.

Material	Ref	1	2	3	4	5	6	7	8	9	10	11	12	13	14	15	16	17	18	19	20	21	22
styphnate, nitro mannite, nitrosog lanidine **pentaerythrite** tetranit ate, tetrazene, lead mononit orescorcinate	(ᵃ)	X	X	....	X	X	X	X	X	X	X	X	X	X	X	X	X	X	X	X	X	X	ˈX
Detonators	(ᵈ)	.....	•X	X	....	X	X	....	•X	....	....	....	•X	•X	••X	•X	•X	X	ˈX
Ammunition for cannon with explosive projectiles, gas projectiles, smoke projectiles, incendiary projectiles, illuminating projectiles or shell ammunition for small arms with explosive bullets, or ammunition for smal arms with explosive projectiles, or rocket ammunition with explosive projectiles, gas projectiles, smoke projectiles, incendiary projectiles, illuminating projectiles; and boosters (explosive), bursters (explosive), or supplementary charges (explosive) without detonators • ª	(ᵉ)	.....	....	X	X	....	X	X	....	X	....	X	X	X	X	X	X	ˈX
Explosive projectiles, bombs, torpedoes, or mines, rifle or hand grenades (explosive), jet thrust units (jato), explosive, class A, or igniters, jet thrust (jato), explosive, class A ª	(ᶠ)	.....	....	X	X	....	X	X	....	X	....	X	X	X	X	X	X	ˈX
Detonating fuzes, class A, with or without radioactive components	(ᵍ)	.....	X	X	....	X	X	....	X	....	X	X	X	X	X	X	ˈX
CLASS B EXPLOSIVES																							
Ammunition for cannon with empty, inert-loaded or solid projectiles, or without projectiles, or rocket ammunition with empty projectiles, inert-loaded or solid projectiles or without projectiles	1	.....	....	X	....	....	....	....	•X	....	X	....
Propellant explosives, class B, jet thrust units (jato), class B, igniters, jet thrust (jato), class B, or starter cartridges, jet engine, class B	2	.....	....	X	....	....	....	....	•X	X	....
Fireworks, special or railway torpedoes	3	X	X	X	•X	X	X	X	....	....	...	...	..	....	X	....

See **footnotes at end of table**

Chart No. 9-1. Loading and Storage Chart of Hazardous Materials—Continued.

The following table shows the hazardous materials which must not be loaded or stored together.

The letter X at an intersection of horizontal and vertical columns shows that these articles must not be loaded or stored together, for example: Detonating fuzes, class A, with or without radioactive components & horizontal column must not be loaded or stored with high explosives or propellant explosives, class A b vertical column.

	a	b	c	d	e	1	2	3	4	5	6	7	8	9	10	11	12	13	14	15
CLASS C EXPLOSIVES																				
4. Small arms ammunition, or cartridges, practice ammunition			X																	
5. Primers for cannon or small arms, empty cartridge bags—black powder igniters, empty cartridge cases, primed, empty primed, combination primers or percussion caps, toy caps, explosive cable cutters explosive rivets			X																	
6. Percussion fuzes, tracer fuzes or tracers			X																	
7. Time, combination or detonating fuzes, class C			X																	

Column headings:

a. Low explosives or black powder
b. High explosives or propellant explosives, class A
c. Initiating or priming explosives, wet: Diazodinitrophenol, fulminate of mercury, guanyl nitrosamino guanylidene hydrazine, lead azide, lead styphnate, nitro mannite, nitrosoguanidine, lead azide, lead styphnate, tetrazene, tetrazene, lead mononitroresorcinate
d. Detonators
e. Detonating fuzes, class A, with or without radioactive components
1. Ammunition for cannon with empty, inert-loaded or solid projectiles, inert-loaded or solid projectiles, or rocket ammunition with empty projectiles, inert-loaded or solid projectiles
2. Propellant explosives, class B, jet thrust units (jato), class B, igniters, jet thrust (jato), class B, or starter cartridges, jet engines, class B
3. Fireworks, special or railway torpedoes
4. Small arms ammunition, or cartridges, practice ammunition
5. Primers for cannon or small arms, empty cartridge bags—black powder igniters, empty cartridge cases, primed, empty powder igniters, primed, combination primers or percussion caps, toy caps, explosive cable cutters, explosive rivets
6. Percussion fuzes, tracer fuzes or tracers
7. Time, combination or detonating fuzes, class C
8. Cordeau detonant fuse, safety squibs, fuse lighters, fuse igniters, delay electric igniters, electric squibs, instantaneous fuse or igniter cord
9. Fireworks, common
10. Flammable liquids or compressed flammable gases, red label
11. Flammable solids or oxidizing materials, yellow label
12. Acids or corrosive liquids, white label
13. Compressed nonflammable gases, green label
14. Poisonous gases or liquids in tank car tanks, cylinders, projectiles or bombs, poison gas label
15. Radioactive materials

Chart No. 9-1.   Loading and Storage Chart of Hazardous Materials—Continued.

	8							X	
Fireworks, common	**9**	X	•X	X	X	X			
OTHER DANGEROUS ARTICLES									
Flammable liquids or flammable gases; Flammable liquid or flammable gas label	**10**	X	•X	X	X	X		X	
Flammable solids or oxidizing materials: Flammable solid, oxidizer, or organic peroxide label	**11**	X	•X	X	X	X	•X	X	
Corrosive liquids: Corrosive label	**12**	X	•X	X	X	'X	'X	X	X
Nonflammable gases: Nonflammable gas label	**13**	X	•X	X	X	X			
Poisonous gases or liquids in tank car tanks, cylinders, projectiles or bombs, poison gas label	**14**	X	X	X	X	X	X	X	
Radioactive materials	**15**	'X	'X	'X	'X	'X		'X	

ᵃ Detonators, Class C explosives, may also be loaded and transported with articles named in vertical and horizontal columns 3, 9, 10, 11, 12, and 13. In any quantity with articles named in vertical and horizontal columns, except as prescribed in §177.835, in any quantity with articles named in vertical or horizontal columns b, c, e, or f is prohibited.

ᵇ Corrosive liquids must not be loaded above or adjacent to flammable solids, oxidizing materials; ammunition for cannon with or without projectiles, or propellant explosives, except that shippers loading truckload shipments of corrosive liquids and flammable solids or oxidizing materials packages and who have obtained prior approval from the Department may load such materials together when it is known that the mixture of contents would not cause a dangerous evolution of heat or gas.

ᶜ Explosives, class A and explosives, class B must not be loaded or stored with chemical ammunition containing incendiary charges or white phosphorus either with or without bursting charges.

ᵈ Bursters (explosive), boosters (explosive), or supplementary charges (explosive) without (detonators when shipped by, to or for the Departments of the Army, Navy, and Air Force of the United States Government may be loaded with any of the articles named except those in columns c, d, 3, 9, 10, 11, 12, 13, 14, and 15.

ᵉ Does not include blasting agents, ammonium nitrate-fuel oil mixtures, or ammonium nitrate, fertilizer

grade, which may be loaded, transported or stored with high explosives, or with detonators, containing not more than 1 gram of explosive each, excluding ignition and delay charges.

ᶠ Normal uranium, depleted uranium, and thorium metal in solid form may also be loaded and transported with articles named in vertical and horizontal columns a, b, c, d, 6, f, and g.

NOTE 1: Charged electric storage batteries must not be loaded in the same vehicle with explosives, class A.

NOTE 2: Cyanides or cyanide mixtures must not be loaded or stored with acids or corrosive liquids.

NOTE 3: Gas identification sets may be loaded and transported with all articles named except those in column c.

NOTE 4: Nitric acid, when loaded in the same motor vehicle with other acids or other corrosive liquids in carboys, must be separated from the other carboys. A 2 by 6 inch plank, set on edge, should be nailed across the motor vehicle floor at least 12 inches from the nitric acid carboys, and the space between the plank and the carboys of nitric acid should be filled with sand, sifted ashes, or other incombustible absorbent material.

NOTE 5.—Smokeless powder for small arms in quantities not exceeding 100 pounds net weight in one motor vehicle shall be classed as a flammable solid for purposes of transportation when examined for this classification by the Bureau of Explosives and approved by the Associate Director for OE.

12. Loading and Storage of Hazardous Materials

In Subpart C of CFR 49, the DOT has prepared a chart that provides information to both transporters and warehousing entities concerning the compatibility of various hazardous materials, thus Sec. 177.848 of CFR 49 contains a chart that has been labeled in this text as "Chart No. 9-1—Loading and Storage of Hazardous Materials."

10

Polychlorinated Biphenyls-PCB's

A. Introduction

Polychlorinated biphenyls, or PCBs as they are often called, were manufactured in the United States from 1929 to 1977. Fortunately, because they are extremely toxic and hazardous to human health and the environment, they are no longer permitted to be manufactured in this country. Unfortunately, during the almost 50 years of their production many hundreds of million of pounds were produced and many pounds of PCBs are still in places where they were originally used.

B. The Composition of PCBs

PCBs are part of the large family of organic chemicals known as chlorinated hydrocarbons. They are produced by the combination of 1 or more chlorine atoms and a biphenyl molecule. Virtually all PCBs have been synthetically manufactured. (See Fig. 10-1.)

PCBs range in consistency from heavy, oily liquids weighing approximately 10 pounds per gallon to waxy solids. They are synthetic chemical compounds with a high boiling point and are very stable chemically. They have a low solubility in water, high solubility in fat, low flammability, and extremely low electrical conductivity. It is because of all of those properties that PCBs have been so widely used in many industrial applications. They were widely used in electric transformers, switches, voltage regulators, and capacitors because of their excellent cooling effects and low electrical conductivity. Much of the PCBs that were produced are still in service in such equipment. They have also been employed as hydraulic fluids and as heat transfer liquids, as dye carriers in carbonless copy paper, in paints, adhesives, and caulking compounds. In addition, they have been used as sealants in road coverings to control dust.

C. The Problem with PCBs

PCBs have been found to be extremely hazardous to living beings, especially humans, because they are toxic at very low concentrations. They are among the most stable chemical compounds known to man. They decompose very slowly over a long period of time, measured in decades after they are released into the environment. One of the problems is that they remain in the environment and are absorbed and stored in the fatty tissues of all organisms. The concentration

Fig. 10-1   The ring structure of a PCB module.

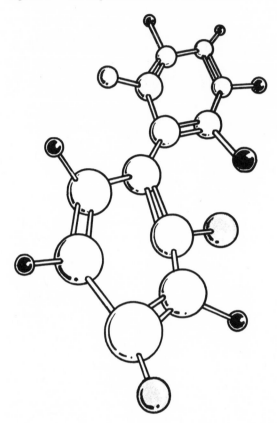

of the compounds in fatty tissue increases over the years, despite what may be very low exposure to them. Bioaccumulation and biomagnification are 2 biological terms used to describe the insidious presence of PCBs. The first term, bioaccumulation, has been used to indicate the process by which the concentration levels of PCBs are increased in fatty tissue; the second term, biomagnification, describes the process in which they build up their concentration in the food chain. As living organisms containing PCBs are consumed one by the other, the concentrations of PCBs consumed by each higher organism increases. The concentrations, therefore, that are consumed by human beings at the tail end of the food chain, can be very high.

PCBs can be concentrated in humans in several ways. They can enter the body through the lungs by inhaling their fumes, through the gastrointestinal tract by eating, and through the skin by contact with the substance. Once they are eaten, inhaled, or absorbed by contact, the PCBs are circulated throughout the body by means of the bloodstream, and they become stored in the fatty tissue of most of the organs of the body. Thus it is that PCBs may find resting places in the brain, heart, liver, kidneys, lungs, adrenal glands, and the skin. Once they have located themselves in those organs, they proceed to do their damage with disastrous results.

In tests using laboratory animals, PCBs have been found to cause birth defects, reproductive disorders, gastric disorders, skin lesions, cancers, swollen limbs, tumors, eye problems, liver disfunction, and a host of other diseases. The effects of PCBs on human beings was very vividly demonstrated quite by accident in 1968. In the Yusho area of Japan, a heat exchanger sprung a leak and contaminated a large quantity of rice oil. More than 1,000 people of Yusho who had used the rice oil developed a wide variety of complaints, such as abdominal pain, menstrual irregularity, fatigue, coughs, eye discharges, disorders of the nervous system, and hyperpigmentation of the skin, nails, and mucous membranes.

There is evidence also that the so-called Yusho disease increases the cancer rate among these victims of PCB ingestion. As a result of the Yusho health data, the Japanese government banned, almost entirely, the production and the importation or exportation of this chemical compound in 1972.

In the United States, PCB contamination has been discovered occurring in measurable amounts in soils, water, fish, milk, and human tissue. Hudson River and some Great Lakes fish as well as fish in other water sources have been found to have more PCB contamination than human consumption can permit. Also, there have been a number of accidents involving PCBs. For example, in North Carolina, as a result of a leaking heat exchanger, fishmeal used as animal fodder was found to be polluted. In Billings, Montana, PCBs that leaked from a transformer contaminated animal feed that was distributed and used in several states. Again, in Puerto Rico, fishmeal was contaminated when a fire broke out in a warehouse that also contained a number of stored, electrical transformers.

As recounted above, PCB has taken a toll of human disease, and environmental damage and, in addition, has caused vast numbers of animals to be put to death. There are no precise figures. However, large quantities of tainted foodstuffs and feed have had to be destroyed with large, economic losses for those companies that have had PCB problems.

D. Governmental Regulation of PCBs

Before 1976, the EPA could only regulate discharges of PCBs into waterways from companies that manufactured, processed, or used them. It was in 1976 that a milestone was passed, for it was in that year that Congress, in response to the growing and accumulating evidence that there were dangers inherent in PCBs and other toxic materials, enacted the "Toxic Substances Control Act (TSCA) and mandated that the EPA regulate all chemicals that present an unreasonable risk of injury to health or the environment. (See Chapter XX.)

The Toxic Substances and Control Act requires potentially toxic chemicals to be tested for safety before they are marketed to the public. If examinations show that a chemical compound is unsafe, the EPA can restrict the handling, use, or shipment of the substance and it may also prohibit its manufacture, processing, and use. Since PCB's toxicity has become known, the TSCA has a special section that prohibits the manufacture, processing, distribution, and use of PCBs, except for totally enclosed use, and it also requires the adequate labeling and safe disposal of PCBs now in use.

The EPA has taken several steps to fulfill its congressional mandate on the

PCB issue. EPA issued regulations of February 17, 1978, to establish requirements for marking and disposal of PCBs. On May 31, 1979, regulations were issued prohibiting the manufacture of PCBs after July 1 of that year, unless specifically exempted by the EPA, and prohibited further processing, distribution, and use of PCBs except in sealed systems. The May 31 regulations exempt totally enclosed systems, such as electrical equipment, from the standpoint that in the normal use of such products there is no human or environmental contamination, contact, or exposure to PCBs.

The TSCA permits some exceptions to the regulations if there does not appear to be unreasonable risk or danger to health or environment. Therefore, the EPA allowed several additional uses of PCBs until July 1, 1984. However, appropriate health and environmental safeguards were required to be taken.

PCBs are still with us today in sealed, electrical equipment, which is to be replaced over the next few years as it is overhauled or retired. But the EPA has published regulations for the final disposal of PCBs in high-temperature, environmentally safe incinerators and in hazardous waste landfills. In addition to the EPA, there are other government agencies that regulate PCBs. For example, the Food and Drug Administration (FDA) has established limits on the amount of PCBs permitted in foods and animal feeds. Products exceeding limits deemed to be safe may not be sold in interstate commerce. The FDA also prohibits the use of PCBs in food and feed-processing plants except in sealed transformers and capacitors. The EPA, FDA, and the U.S. Department of Agriculture may ban the use of PCBs in any kind of electrical equipment in food and feed processing or distribution channels.

E. An Example of PCB Proximity

As the news media take on various causes, they sometimes focus their spotlights of inquiry and investigation on subjects that have lain dormant for many years through neglect and inertia. It is only necessary to open the cupboard door under the kitchen sink to find all kinds of hazardous materials, cleansers, oven cleaners, solvents of various kinds, and so forth. It was not until very recently that some of them were not even properly labeled. Even now many of those products do not have child-proof fasteners or closures. Fortunately, progress is being made on that front, and so we shift our concern to the ubiquitous PCB, and the fluorescent light fixture used in homes, offices, and almost everywhere else that commercial lighting is used.

Light ballasts are the primary electric components of fluorescent light fixtures and are generally located within the fixture under a metal cover plate. (See Fig. 10-2.) The ballast units are generally composed of a **transformer** to reduce the incoming voltage, a small capacitor (which may contain PCBs), and possibly a **thermal cutoff switch** or **safety fuse** or both. Those components are surrounded by a tarlike substance that is designed to muffle the noise that is inherent in the operation of the ballast. The substance covers the small capacitor. When a ballast unit fails, excessive heat can be generated that will melt or burn the tar material, creating a characteristic foul odor.

In considering causes of ballast failure, some privately conducted tests have indicated that **operation of power-saving lamps with a standard ballast or standard lamp with a power-saving ballast tends to significantly increase the ballast operating temperature and decrease its normal life span.** It appears that

Fig. 10-2   The ballast portion of a typical fluorescent light fixture.

ballasts will fail less frequently if standard lamps are used only with standard ballasts and power-saving lamps with power-saving ballasts. **Fluorescent lamps should be changed in pairs; new lamps should not be used with old lamps.**

Before the EPA banned the manufacture of PCBs in 1978, they were used in the manufacture of fluorescent light ballasts. The EPA does not regulate the use of PCBs in ballasts manufactured before 1978. All light ballasts manufactured since 1978 that do not contain PCBs should be marked by the manufacturer with the statement **"No PCBs." For those manufactured before that time or for those ballasts that contain no statement regarding PCB content, one should assume that they do contain PCBs.**

If the ballast does not contain PCBs, they are located **inside the small capacitor.** There would be approximately 1 ounce to 1 1/2 ounces of PCB fluid in the capacitor itself. **If the ballast fails, the capacitor may break open, allowing the PCB oil to drip out of the fixture.** The capacitor does **not** always leak when the ballast fails, but when it does happen, measures should be taken to limit or avoid personal exposure.

The EPA has these recommendations for anyone with a fluorescent light ballast leaking PCBs:

1. Vacate the room or area immediately and open any windows to ventilate the room to the outside. If the incident occurred in a room that cannot be vented, the person replacing the failed ballast and cleaning up can reduce exposure by wearing a chemical cartridge respirator equipped with a organic vapor cartridge.
2. Turn off the light fixture at the switch and disconnect electricity at the fuse or breaker box. Let the ballast unit cool for 20 to 30 minutes before proceeding.

If the room is fully ventilated, the amount of PCB-contaminated particulate matter in the air should decrease significantly enough to make negligible any risk from breathing.

3. Wear rubber gloves that will not absorb PCBs (e.g., neoprene, butyl, or nitrile). Furthermore, if you will be working directly under the fixture, consider using additional protective gear, such as goggles or a face shield and a rubber apron to help guard against possible exposure from further leaking or cleanup activities. Exercise caution to avoid personal contamination, e.g., from touching your face with a contaminated glove. (See Fig. 10-3.) During the cleanup or removal period, smoking should be prohibited in the area because smoking increases the inhalation rate of contaminated air. In addition, you may be using a flammable solvent in the cleanup.

4. Remove the fluorescent lamps.

5. Recheck that the power is off at the fuse or breaker box. Remove the metal cover over the wiring and ballast unit and loosen the ballast unit by taking

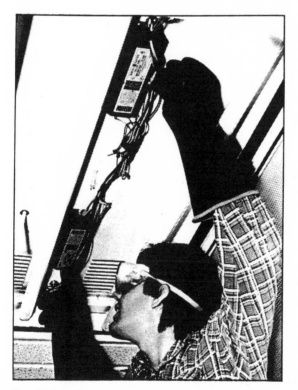

Fig. 10-3 After removal of the fluorescent tubes and the central cover from the light fixture, the ballasts may be seen. The use of gloves and goggles are recommended to prevent contact with PCBs from a leaking ballast.

out the metal screws that hold it to the end of the fixture. Cut the electrical wires going to the ballast and remove the ballast.
Note: Wire connectors can be used when one installs a new ballast.

6. Proceed to clean up leaks using the following guidelines:
PCBs that leak onto nonabsorbent surfaces such as table tops and uncarpeted floors should first be cleaned up by wiping with a rag or paper towel or by scraping with a putty knife if hardened. Avoid smearing the PCB around. That would only contaminate a larger area. Surfaces should then be thoroughly cleaned twice, using an appropriate solvent or detergent. Only certain solvents are effective in cleaning up spilled PCBs. They include mineral spirits, deodorized kerosene, turpentine, and rubbing alcohol. Certain detergents containing trisodium phosphate (such as "Soilex" or "Spic 'n Span") are also recommended. However, they should be used only at full strength and applied with a damp rag rather than diluted in a bucket. The solution would become contaminated and cannot legally be disposed of in the sewer system. Some of the other effective detergent products (which are commercially available) include: "Triton X-100" (Rohm-Haas), "Sterox" (Monsanto), and "Power Cleaner 155" (Penetone Corp.). The EPA does not endorse those particular products. Other effective products may also be available. For leaks onto absorbent materials, such as drapes and carpets, there is no reliable way to clean and decontaminate the material. In the case of rugs and fabrics, the material should be cut away in a 6-in. radius around the contamination point(s). In areas where foot traffic has spread pollution, the entire carpet should be disposed of. Proper disposal procedures for all such materials are described in the following section. Associated surfaces, such as flooring under tainted carpeting, should be thoroughly cleaned with a solvent or detergent as previously described.

7. Contaminated materials (ballasts, rags, clothing, gloves, drapes, carpets, etc.) should be packed into crumpled newspapers or other sorbent materials (sawdust, kitty litter, vermiculite, soil, etc.) and placed in a double-thickness plastic bag. It should be taken to one of the transporters indicated in item 10, below. There, the contaminated materials will be packed in a drum approved for PCBs by the DOT and finally disposed of at an EPA-approved site. One might consider discarding the entire light fixture instead of decontaminating the unit. That would eliminate the chance of skin coming into direct contact with the PCBs while one cleans inside the light fixture.

8. When one is completely through with the cleanup process and contaminated materials and protective clothing have been packed for disposal, one should wash one's hands thoroughly with detergent.

9. The room should continue to be ventilated for 24 hours before reuse.

10. To get rid of PCBs in the form of ballasts, PCB-soiled items, or fluorescent fixtures containing PCBs, one should check the telephone yellow pages for waste disposal operators to find an authorized transporter who will take the PCBs to an EPA-approved chemical waste-processing site. If one has difficulty in finding a transporter, one should call a local EPA regional office by looking under the U.S. Government section of the telephone directory. It may be possible, also, to obtain disposal information from your local health department office. In the Appendix of this text the reader will find a list of EPA regional offices.

11

Asbestos, The Curse and The Cure

A. Introduction

Since the early 1970s the EPA and the Occupational Safety and Health Administration (OSHA) have been concerned with the potential health hazards associated with exposure to asbestos. Their interest is based on medical evidence linking various types of cancer and noncancerous respiratory diseases with the exposure of asbestos workers and their wives to minute particles of airborne asbestos.

In recognition of those health hazards, the EPA has prepared a manual to provide guidance on how best to handle asbestos-containing materials during generation, transport, and final disposal. The number of the EPA Manual is EPA/530-SW-85-007, dated May 1985. Waste handling practices presented include not only those needed to meet current EPA and OSHA requirements but also additional recommendations reflecting practices needed to further minimize exposure to asbestos. In most cases the recommendations are consistent with the state-of-the-art procedures currently being followed by most knowledgeable asbestos waste-handling firms. State and local requirements may be more restrictive than Federal standards. Therefore, these agencies should be contacted before one handles asbestos-containing materials.

B. The Composition of Asbestos

Asbestos is a naturally occurring family of fibrous mineral substance. The typical size of asbestos fibers, as illustrated relative to other substances in Fig. 11-1, is 0.1μ to 10μ in length, a size that is not generally visible to the human eye. Somewhat longer fibers are used in making textile products. When disturbed, asbestos fibers are so small in particle size that they may become suspended in the air for many hours, thus increasing the extent of asbestos exposure for individuals within the area.

EPA regulations identify the following types of asbestos: chrysotile, amosite, crocidolite, anthophyllite, actinolite, and tremolite. Approximately 95 percent of all asbestos used in commercial products is chrysotile. Asbestos has been a useful, commercial product because it is noncombustible and resistant to corrosion and has a high tensile strength and low electrical conductivity. Asbestos had very little use until the early 1900s, when it was employed as thermal insulation for steam engines. Since then, asbestos fibers have been mixed with a number of different types of binding materials to create an estimated 3,000 commercial

Fig. 11-1   Asbestos particle size compared with other substances. Source: EPA 450/2-78-014, March 1978.

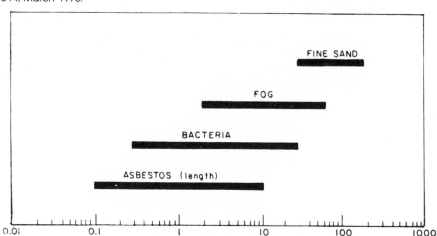

products. Asbestos has been used in brake linings, floor tile, sealants, plastics, cement pipe, cement sheet, paper products, textile products, and insulation. The amount of asbestos contained in those products varies significantly, from 1 percent to 100 percent, depending on the particular use.

The potential of an asbestos-containing product to release fibers is dependent upon how friable it is. Friability refers to the fact that the material can be crumbled with hand pressure and, therefore, is likely to shed fibers. The fibrous or fluffy spray-applied asbestos materials found in many buildings for fireproofing, insulation, soundproofing, or decorative purposes are generally considered friable. Pipe and boiler wrap are also friable and found in numerous buildings. Some materials, such as vinyl-asbestos floor tile, are considered nonfriable and generally do not emit airborne fibers unless sanded or sawed. Other materials, such as asbestos cement sheet and pipe, can emit asbestos fibers if the materials are broken or crushed in the demolition of structures that contain such materials. For that reason, such materials are considered friable under the National Emission Standards for Hazardous Air Pollutants (NESHAP) regulations for the demolition of structures.

Only on rare occasions can the asbestos content in a product be determined from the product labeling or by consulting the manufacturer. Most products that have been installed lack labels. A description of common asbestos-containing products is presented in the latter part of this chapter. Further information on the asbestos content of consumer products is available through the consumer Product Safety Commission Hotline:

Continental United States	1-800-638-2772
Maryland only	1-800-492-8363
Alaska, Hawaii, Puerto Rico, Virgin Islands	1-800-638-8333

Positive identification of asbestos requires laboratory examination of samples. Standard laboratory analysis using polarized light microscopy (PLM) may cost $30 to $60 per sample. For information on locating a laboratory capable of performing the test, contact any of the EPA's Regional Asbestos Coordinators listed in the Appendix or call EPA's toll-free number for assistance:

Continental United States　　　1-800-334-8571
ext. 6741

To acquire additional technical information and to obtain the EPA's publication regarding sampling and analysis of asbestos entitled "Guidance for Controlling Friable Asbestos-Containing Materials in Buildings" (EPA 560/5-83-002), contact any of the agency's Regional Asbestos Coordinators listed in the Appendix or call its toll-free TSCA[1] hotline:

Continental United States　　　1-800-424-9065
Washington, D.C. only　　　　　554-1404

C. Asbestos-Caused Diseases

Medical studies of asbestos-related diseases have revealed that the primary exposure stems from inhalation. They also suggest that there does not appear to be a safe level of exposure (e.g., a threshold) below which there would be no chance of disease. The exposure may be classified as "occupational exposure" of workers involved, for example, in mining, milling, manufacturing, fabricating, construction, spraying, or demolition activities. "Paraoccupational exposure" of workers' families occurs when asbestos on work clothes is taken home; of people living or working near such operations may experience "neighborhood exposure."

The following diseases can result from inhalation of airborne asbestos fibers:

Asbestosis—A noncancerous respiratory disease that consists of scarring of lung tissues. Symptoms of asbestosis include shortness of breath and rales, a dry crackling sound in the lungs during inhalation. Advanced asbestosis may produce cardiac failure and death. Asbestosis is rarely caused by neighborhood exposure.

Lung Cancer—Inhaled asbestos particles can produce lung cancer independent of the onset of asbestosis. In most lung cancer patients, a cough or a change in cough habit is found. A persistent chest pain unrelated to coughing is the second most common symptom.

Mesothelioma—This is a rare cancer of the thin membrane lining of the chest and abdomen. Most incidences of mesothelioma have been traced directly to a history of asbestos exposure. The earlier in life that one begins inhaling asbestos, the higher the likelihood of developing mesothelioma in later life. Thus there is concern over the exposure of schoolchildren to asbestos. The common symptoms are shortness of breath, pain in the walls of the chest, and abdominal pain. Mesothelioma is always fatal.

Other Cancers—Some medical studies have suggested that exposure to asbestos is responsible for some cancers of internal organs, including the esophagus, larynx, oral cavity, stomach, colon, and kidney. It is theorized that inhaled

---

[1]TSCA is the abbreviation for the Toxic Substance Control Act, which is discussed in Chapter 20.

asbestos fibers are absorbed into the bloodstream and carried to those other parts of the body.

Symptoms of asbestos respiratory disease generally do not appear for 20 or more years after the initial exposure to airborne asbestos. Early disease detection, however, is possible by a medical examination including a medical history, breathing capacity tests, and a chest X ray.

Most health risk data pertain to groups of asbestos workers with relatively high exposures. A study of mortality for 17,500 asbestos insulation workers is summarized in Fig. 11-2. The study compares death rates among insulation workers exposed to asbestos and other workers not exposed to asbestos. Citing that study and others, the National Institute for Occupational Safety and Health (NIOSH) has reported that persons exposed to asbestos may have 5 times the chance of developing an asbestos-related disease that persons who have not been exposed to asbestos have.

Studies have shown that exposure to asbestos and cigarette smoking combine to create a significantly higher risk of developing an asbestos-related disease. Statistics compiled by NIOSH indicate that a smoker exposed to asbestos may be 50 times more likely than a nonexposed nonsmoker to develop lung cancer. Some information suggests that quitting smoking can reduce the high risk.

There have been no conclusive studies to date stating that ingestion of asbestos in food or water may result in health hazards. Because of concern that there may be potential health impacts not yet identified, however, there are Federal regulations specifying asbestos limitations in ambient water and in products such as food-processing filters.

With regard to asbestos contact with the skin, there is currently no evidence to indicate that asbestos fibers can penetrate the skin tissue. Some workers have

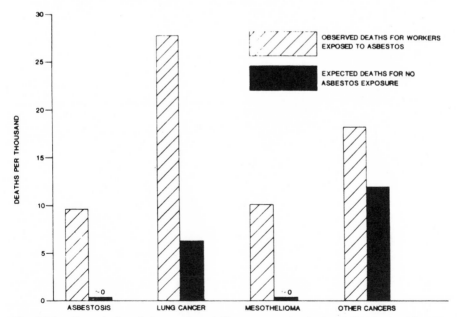

Fig. 11-2   Expected and observed mortality among asbestos insulation workers. Source: ASTM*834, PCN 04-834000-17, July 1984. *American Society for Testing and Materials.

indicated that asbestos fibers irritate the skin, resulting in a rash similar to that experienced with handling of other fibrous materials, such as fiberglass.

D. Regulatory Responsibility and Programs

Both the EPA and OSHA have a major responsibility for regulatory control over exposure to asbestos. Emissions of asbestos to the ambient air are controlled under Section 112 of the Clean Air Act, which established the National Emission Standards for Hazardous Air Pollutants (NESHAPs). The regulations specify control requirements for most asbestos emissions, including work practices to be followed to minimize the release of asbestos fibers during handling of asbestos waste materials; however, the regulations do not identify a safe threshold level for airborne asbestos fibers. For additional information about the NESHAPs regulations for asbestos, refer to the Code of Federal Regulations (40 CFR Part 61, Subpart M).

The OSHA rules have been established to protect workers handling asbestos or asbestos-containing products. The current OSHA precepts include a maximum workplace airborne asbestos concentration limit of 2 fibers/cc on an 8-hour time-weighted average basis and a ceiling limit of 10 fibers/cc in any 15-minute period. The standard includes requirements for respiratory protection and other safety equipment and established work practices to reduce indoor dust levels. For details regarding the OSHA regulations, refer to the Code of Federal Regulations (29 CFR Part 1910).

The EPA has implemented a separate regulation under the Toxic Substances Control Act (TSCA) to handle the problem of asbestos construction materials used in schools. In addition to requiring that the building be posted, the regulation prescribes that all schools be inspected to determine the presence and quantity of asbestos and that the local community be notified. Corrective actions, such as asbestos removal or encapsulation, are currently left to the discretion of the school administrators. The EPA provides technical assistance under this program through the Appendix contacts or the toll-free TSCA hotline: 1-800-424-9065 (554-1404 in Washington, D.C.). The specific details of the TSCA program are contained in the Code of Federal Regulations (40 CFR Part 763, Subpart F).

The Asbestos School Hazard Abatement Act of 1984 (ASHAA) establishes a $600 million grant and loan program to assist financially needy schools with asbestos abatement projects. The program also includes the compilation and distribution of information concerning asbestos and the establishment of standards for abatement projects and abatement contractors. Under this plan, centers for training contractors on asbestos handling and abatement have been established at the Georgia Institute of Technology, Atlanta, Georgia, and are scheduled to open in June 1985 at both Tufts University, Medford, Massachusetts and at the University of Kansas, Lawrence, Kansas. Additional information can be obtained through the toll-free ASHAA hotline: 1-800-835-6700 (554-1404 in Washington, D.C.).

Wastes containing asbestos are not hazardous wastes under the Resource Conservation and Recovery Act (RCRA). Because state regulations can be more restrictive than the Federal regulations under RCRA, however, some states may have listed asbestos-containing wastes as hazardous wastes. Since that will have a great impact on transportation and disposal of the waste, the state hazardous agency should be contacted. A list of state hazardous waste agencies may be obtained by calling the RCRA hotline: 1-800-424-9346 (382-3000 in Washington, D.C.). Current nonhazardous waste regulations under RCRA pertain to facility siting and general operation of disposal sites (including those that handle asbestos). Details concerning those RCRA requirements are contained in the Code of Federal Regulations (40 CFR Part 257).

Other Federal programs, authorities, and agencies controlling asbestos include the Clean Water Act, under which EPA has set standards for asbestos levels in effluents to navigable waters; the Mine Safety and Health Administration, which oversees the safety of workers involved in the mining of asbestos; the Consumer Product Safety Commission; the Food and Drug Administration; and the DOT.

State and local agencies may have more stringent standards than the federal requirements; therefore, those agencies should be contacted before asbestos removal or disposal operations.

E. The History and Use of Asbestos

Mining and commercially using asbestos has been with us since the early 1900s. U.S. consumption of asbestos increased to a peak of 800,000 tons per year in the early 1970s. Since then, consumption has dropped by more than 70 percent. The problem is, however, that much of the material originally installed in buildings may still be present.

The potential existence of asbestos in commercial products can be assessed first by understanding the physical and chemical characteristics of asbestos-containing products and their uses. This part of the chapter describes the appearance, composition, friability, and use of the most common asbestos-containing products.

Table 11-1 summarizes information on the products, many of which are still being manufactured. The list in Table 11-1 is not all-inclusive, because there are many more products made containing asbestos that are not on this list; however, the reader can readily see what a vast array of products contain this substance. Nevertheless, because of the recognized health risk, the manufacture of a few asbestos products has been banned. In addition, the concern of industry for exposure of their workers and the public and the increased availability of substitute products have rapidly reduced the use of asbestos.

Asbestos is used in brake linings for automobiles, buses, trucks, railcars, and industrial machinery and in vehicle or industrial clutch linings. Asbestos-containing brake linings include drum brake linings, disc brake pads, and brake blocks. In the past asbestos linings have accounted for up to 99 percent of this market. Friction materials are generally tough and nonfriable, but they release asbestos dust during fabrication operations. In addition, accumulated dust in a brake drum from lining wear contains high levels of asbestos. Brake installation facilities (e.g., city bus service centers and tire and brake shops) may generate significant quantities of asbestos waste. Substitute nonasbestos brake linings have been developed and are beginning to replace asbestos lining in some applications.

Asbestos has been used in combination with various plastics. Some of those plastic products include resilient vinyl and asphalt floor coverings, asphalt roof coatings, and traditional molded plastic products, such as cooking pot handles or plastic laboratory sinks. The products, in that category are usually tough and nonflexible. The asbestos in them is tightly bound and is not released under

Table 11-1.   A List of Products Containing Asbestos.

Product Type	Average Percent of Asbestos Content	Binders Used	Dates Used
Friction products	50	Various polymers	1910–present
Plastic products			
Floor tile and sheet	20	PVC, asphalt	1950–present
Coatings and sealants	10	Asphalt	1900–present
Rigid plastics	<50	Phenolic resin	?–present
Cement pipe and sheet	20	Portland cement	1930–present
Paper products			
Roofing felt	15	Asphalt	1910–present
Gaskets	80	Various polymers	?–present
Corrugated paper pipe wrap	80	Starches, sodium silicate	1910–present
Other paper	80	Polymers, starches, silicates	1910–present
Textile products	90	Cotton, wool	1910–present
Insulating and decorative products			
Sprayed coating	50	Portland cement, silicates, organic binders	1935–1978
Troweled coating	70	Portland cement, silicates	1935–1978
Preformed pipe wrap	50	Magnesium carbonate, calcium silicate	1926–1975
Insulation board	30	Silicates	Unknown
Boiler insulation	10	Magnesium carbonate, calcium silicate	1890–1978

typical conditions of use. Caution should be used in working with those products, however, because any sawing, drilling, or sanding during installation or removal would result in the release of asbestos dust.

Vinyl or linoleum and asphalt flooring are used in many types of buildings. Vinyl-asbestos flooring has about a 90 percent share of the resilient floor-covering market. The materials are not friable, and asbestos is released primarily through sawing or sanding operations during installation, remodeling, and removal. Asphalt-asbestos coatings, used primarily as roof sealants, generally remain flexible and nonfriable but may become friable or brittle as they age.

Asbestos-cement (A-C) pipe has been widely used for water and sewer mains, and it is occasionally used for electrical conduits, drainage pipe, and vent pipes. Also, asbestos-cement sheet, manufactured in flat or corrugated panels and shingles, has been used primarily for roofing and siding but also for cooling tower fill sheets, canal bulkheads, laboratory tables, and electrical switching gear panels. Asbestos-cement products are dense and rigid with gray coloration unless the material is lined or coated. The asbestos in them is tightly bound and would not be released to the air under typical conditions of use. Sawing, drilling, or sanding of the products during installation or renovation, however, would result in the release of asbestos dust. In addition, the normal breakage and crushing involved in the demolition of structures can release asbestos fibers. For that reason they are subject to the NESHAP's regulation during demolition operations. Also, normal use of A-C pipe for water or sewer mains has been shown to release asbestos fibers into the fluid being carried.

By the late 1970s, A-C pipe had a 40 percent share of the water main market and a 10 percent share of the sewer main market. However, since it has been in existence only for 50 years, it accounts for just a small fraction of the total pipe now in place in the United States. Other products, such as roofing felts, gaskets, and other paper materials, are manufactured on conventional paper-making equipment using asbestos fibers instead of cellulose. The raw asbestos paper produced in the process has a high asbestos content, about 85 percent, but it is typically coated or laminated with other materials in the final product. The asbestos fibers in most paper items are sufficiently bound to prevent their release during normal use. Cutting or tearing the material during installation, use, or removal would result in the release of asbestos dust.

Asbestos-containing roofing felt has been widely used for application of "built-up" roofs. Built-up roofing is used on a flat surface and consists of alternating layers of roofing felt and asphalt. The roofing felt consists of asbestos paper, saturated and coated with asphalt. Asphalt-asbestos roofing shingles for residential structures, made from roofing felt coated with asphalt, were reportedly used for only a short time between 1971 and 1974.

Other asbestos-containing paper products include pipeline wrap, millboard, rollboard, commercial insulating papers, and a variety of specialty papers. Pipeline wrap is used to protect underground pipes from corrosion, particularly in the oil and gas industry. Millboard and rollboard are laminated paper products used in commercial construction, such as walls and ceilings. Commercial insulating papers are used for high-temperature applications in the metals and ceramics industries, for low-grade electrical insulation, and for fireproofing steel decks in building construction.

Corrugated asbestos paper (see Fig. 11-3), was used for pipe coverings, block insulation, and specialty panel insulation. Although those uses have generally been discontinued, significant amounts are typically found in older structures. The products are generally considered friable.

Yarn, cloth, and other textiles are made from asbestos using conventional textile manufacturing equipment. Those materials are used to manufacture fire-resistant curtains or blankets, protective clothing, electrical insulation, thermal insulation, and packing seals that are used in machines. The raw textile products have a high asbestos content of approximately 85 percent. Asbestos is typically coated or impregnated with polymers before assembly into a final product, which is not required to be labeled as containing asbestos and usually is not so labeled. Those products may release asbestos dust if cut or torn, and some may release dust during normal use. A significant quantity of noncoated fabrics are still in use, especially in schools and fire departments.

Asbestos-containing thermal insulation generally refers to sprayed and troweled asbestos coatings and molded or wet-applied pipe coverings. Those materials generally have an asbestos content of 50 percent to 80 percent. The coatings were commonly applied to steel I-beams and decks (see Fig. 11-4), concrete ceilings and walls (see Fig. 11-5), and hot water tanks and boilers. The

Fig. 11-3   Corrugated asbestos paper used as a pipe wrapping.

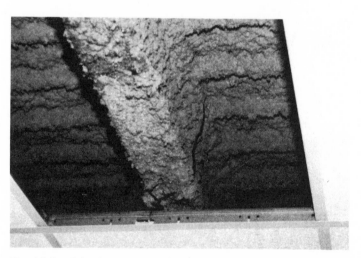

coatings were applied primarily for thermal insulation, although in many instances they also provided acoustical insulation and decorative finish. Sprayed coatings typically have a rough, fluffy appearance, while troweled coatings have a smooth finish and may be covered with a layer of plaster or other nonasbestos material. Both coatings are considered friable in most applications. Many spray-

Fig. 11-4   Asbestos insulation and fireproofing sprayed under steel deck and I-beam.

Fig. 11-5 Workers using safety clothing, engaged in removing sprayed-asbestos materials from a concrete ceiling.

applied asbestos coatings were banned for fireproofing and insulation in 1973 and for decorative purposes in 1978.

Asbestos insulating board was used as a thermal and fireproofing barrier in many types of walls, ceilings, ducts, and pipe enclosures. The material looks like A-C sheet but is less dense and much more friable. High asbestos dust levels

Fig. 11-6 Preformed asbestos pipe insulation with canvas wrap.

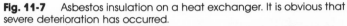

Fig. 11-7   Asbestos insulation on a heat exchanger. It is obvious that severe deterioration has occurred.

have been measured for many board-handling operations, including simple unloading of uncut sheets.

Preformed pipe coverings having an asbestos content of about 50 percent were used for thermal insulation on steam pipes in industrial, commercial, institutional, and residential applications. That product is usually white and chalky in appearance and was typically manufactured in 3-foot-long, half-round sections, joined around the pipe by means of plaster-saturated canvas or metal bands. A typical example of preformed pipe insulation is illustrated in Fig. 11-6. The covering was applied on straight pipe sections, while wet-applied coatings were used on elbows, flanges, and other irregular surfaces. The preformed pipe coverings may be slightly more dense than the insulating coatings but are still very friable. The installation of wet-applied and preformed asbestos insulations was banned in 1975. However, significant amounts are typically found in older structures.

Preformed block insulation was used as thermal insulation on boilers, hot water tanks, and heat exchangers, illustrated in Fig. 11-7, in industrial, commercial, institutional, and residential applications. The blocks are commonly chalky white, 2 inches thick, from 1 to 3 feet in length and held in place around the boiler by their metal wires or expanded metal lath or both. A plaster-saturated canvas was often utilized as a final covering or wrap. Asbestos block insulation is friable and rapidly deteriorates in a high-humidity environment or when exposed to water. In 1975 the EPA banned the installation of that type of asbestos insulation.

There have been many other uses of asbestos-containing materials: exterior siding shingles, shotgun shell base wads, asphalt paving mix, spackle and joint patching compounds, artificial fireplace logs for gas-burning fireplaces, and artificial snow. The use of asbestos as artificial logs in gas-burning fireplace installations was banned in 1977, and its use as a constituent of spackle and joint compounds was prohibited in 1978. Asbestos is still used in oil and gas drilling fluids and is added at a concentration of approximately 1 percent.

F. Removing and Handling Asbestos

Asbestos-containing wastes are generated by a variety of processes, including mining and milling asbestos ore, manufacturing and fabricating asbestos products, and removing asbestos building materials before demolition or renovation operations.

Asbestos is produced by mining ore deposits of the substance and separating the fibers from the nonasbestos rock. The United States currently has 4 active asbestos mines, at Copperopolis and Santa Rita, California, and two in North Carolina. Asbestos mines generate a large quantity of waste rock having insufficient asbestos content for additional, economical processing. This waste is typically piled in an area adjacent to the mine. (The Mine Safety Health Administration enforces asbestos exposure limits for mine workers. For additional information, refer to the Code of Federal Regulations, 30 CFR Parts 55-57 and 71).

The process of separating asbestos fibers from the mined ore and grading and packaging the fibers according to length is called milling. Asbestos mills are located at the mine sites in Copperopolis and Hyde Park, while the Santa Rita ore is hauled to a mill at King City, California. Asbestos mills generate a large quantity of waste rock, called tailings, that contain residual amounts of asbestos. Mills also generate asbestos-containing waste from air-cleaning control devices used to meet EPA and OSHA requirements. The EPA requires all asbestos-containing wastes from mills to be disposed without any visible emissions to the outside air, or certain wetting practices must be used to control emissions. Tailings are usually disposed by loading on a conveyor belt and dumping on a waste pile in the vicinity. Emission control during transport and dumping is usually achieved by wetting, although local exhaust ventilation may occasionally be used.

Asbestos products are manufactured by combining the milled asbestos fibers with binders, fillers, and other materials. The resultant mixture, which may be either dry or wet, is molded, formed, or sprayed, and then cured or dried. Some products require further machining or coating operations before being offered for sale. Manufactured products may then be fabricated by another manufacturer or by the installer or by the final consumer. Manufacturing and fabricating operations generate asbestos-containing wastes as follows:

1. Residual asbestos fibers left in empty asbestos shipping containers;
2. Process wastes such as cuttings, trimmings, and off-specification or rejected material;
3. Housekeeping waste from sweeping or vacuuming;
4. Pollution control device waste from recapture systems.

Process wastes and housekeeping wastes should be wetted before packaging using a mixture of surfactants, such as soap and water, in a fine mist. Empty shipping bags can be flattened and packaged under hoods exhausting to a pollution control device. Empty shipping drums are difficult to thoroughly clean and should be sealed and disposed of or used to contain other asbestos wastes for disposal. Air pollution control device waste is usually packaged directly by connecting a container to the waste hopper outlet. Vacuum bags or disposable paper filters should not be cleaned, but should be sprayed with a fine water mist and placed intact into a proper container.

A significant quantity of asbestos-containing waste may be generated during removal of friable asbestos materials from buildings. EPA regulations address the removal of friable asbestos materials before demolition or renovation of buildings in the Code of Federal Regulations, 40 CFR Part 61, Subpart M. Removal should also be considered for materials that may potentially become friable during the demolition or renovation activities. Currently, the Federal regulations apply to larger structures, i.e., structures with more than 4 apartments with certain minimum quantities of asbestos-containing materials. Some state and local health agencies, however, require removal of lesser quantities of asbestos from smaller buildings.

Regulatory requirements of the EPA and OSHA include written advance notice to the Regional NESHAPs office (see Appendix) of the planned removal, posting of warning signs, providing workers with protective equipment, wetting friable asbestos material to prevent emissions, monitoring indoor dust levels, and properly disposing of asbestos-containing wastes. It is also highly recommended that the work area be enclosed through the use of plastic barriers to prevent contamination of other parts of the structure. Guidelines for development of an asbestos removal contract are presented in a document entitled "Guide Specifications for the Abatement of Asbestos Release from Spray- or Trowel-Applied Materials in Buildings and Other Structures," published by the Foundation of the Wall and Ceiling Industry, 25 K Street N.E., Washington, D.C. 20002 (202-783-6580).

Asbestos removal contractors are encouraged to employ additional safety procedures beyond the minimum requirements of the EPA and OSHA. The use of a negative air pressure system, utilizing fans and filters to exhaust air from the room, and a shower or decontamination facility for anyone exiting the area, as illustrated in Fig. 11-8, is highly recommended. The air filters used in that system are high-efficiency particulate air (HEPA) filters, rated for 99.97 percent

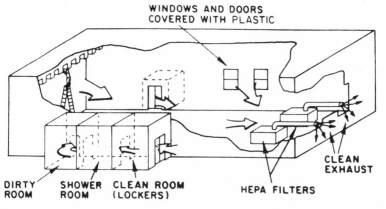

Fig. 11-8   A sketch showing a negative air pressure system.

Fig. 11-9   A plastic bag placed in a fiberboard container safeguards asbestos waste.

removal efficiency for asbestos-sized dust. Such safeguards provide better protection for workers and prevent contamination of the neighborhood. (For additional information, refer to the EPA document entitled "Guidance for Controlling Friable Asbestos-Containing Materials in Buildings," EPA publication No. 560/5-83-002, which is available from any of the EPA Regional Asbestos Coordinators, listed in the Appendix, or by calling EPA's toll-free TSCA hotline: 1-800-424-9065 (554-1404 in Washington, D.C.).

When the asbestos materials are prepared for removal, they are wetted with a water and surfactant mixture sprayed in a fine mist, allowing time between sprayings for complete penetration of the material. Once the thoroughly wetted asbestos material has been removed from a building component, the EPA and OSHA regulations require the wastes to be containerized as necessary to avoid creating dust during transport and disposal. The generally recommended containers are 6-mil thick, plastic bags, sealed in such a way as to make them leak-tight. When using plastic bags, one should minimize the amount of void space, or air, in the bag. That will help reduce any emissions should the bag burst under pressure. More thorough containerization may include double bagging and using plastic-lined cardboard containers, as illustrated in Fig. 11-9, or plastic-lined metal containers. Asbestos waste slurries can be packaged in leak-tight drums if they are too heavy for the plastic bag containers. Both the EPA and OSHA specify that the containers be tagged with a warning label. Either the EPA or the OSHA label must be used.

> "CAUTION
> CONTAINS ASBESTOS FIBERS
> AVOID OPENING OR BREAKING CONTAINER
> BREATHING ASBESTOS IS HAZARDOUS TO YOUR
> HEALTH"
> or
> "CAUTION
> CONTAINS ASBESTOS FIBERS
> AVOID CREATING DUST
> MAY CAUSE SERIOUS BODILY HARM"

In situations where pipes or other facility components containing asbestos materials are removed as sections without first removing the asbestos, 6-mil plastic can be used to wrap the section sufficiently to create a leak-tight container. There are currently no regulatory requirements that govern the time period that waste can remain on-site before being transported to a disposal site. It is well to recognize the health risk and potential liabilities associated with accidental exposure; therefore, waste should be guarded—i.e., protected against public access, such as by a fence or in a locked room or building—and transported as soon as possible.

After asbestos-containing materials have been removed, all plastic barriers should be removed and the facility should be thoroughly washed. The plastic used to line the walls, floors, etc., should be treated as asbestos waste and should be containerized appropriately. Cleanup of asbestos debris may be done with a HEPA vacuum cleaner. Any asbestos-containing waste collected by the HEPA vacuum cleaner must be bagged, labeled, and disposed of in accordance with regulatory methods described in this text.

All areas of the facility that might have been exposed to asbestos fibers should be washed down. Several washings should be performed along with air sampling and analysis to ensure a low, airborne asbestos fiber concentration. Various regulatory agencies have targeted asbestos fiber concentrations in the range of 0.001 to 0.0001 fibers/cc as a level desirable in the building air after cleanup. For example, the state of Arizona has specified 0.001 fibers/cc as a level above which additional cleanup is required, and British researchers have identified a level of 0.0001 fibers/cc to be attainable after cleanup. In some instances it may not be possible to remove all asbestos because of the irregularity or porosity of the subsurface materials. In those situations it may be necessary to spray an encapsulating paint over the surface to eliminate the potential for fiber release. For further information on encapsulants, contact any of the EPA Regional Asbestos Coordinators, listed in the Appendix, or call EPA's toll-free TSCA hotline: 1-800-424-9065 (554-1404 in Washington, D.C.).

If it is desired to use alternative techniques for removing asbestos materials from buildings, prior approval must be received from the EPA. At the present writing, the only alternate technique is by vacuum truck. The EPA will review vacuum trucks on a case-by-case basis. The 1 system found to be acceptable by the EPA has demonstrated the capability of removing asbestos materials in a wet condition. The asbestos material, contained within the truck as a slurry, is transported to the final disposal site. The air from the vacuum intake is dried and exhausted through a fabric filter located on the truck. Final filtration of exhaust air is through a HEPA filter.

G. Transporting Asbestos Wastes

Transportation of asbestos wastes, for the purpose of this discussion, is defined as all activities from receipt of the containerized asbestos waste at the generation site until it has been unloaded at the disposal site. Current EPA regulations state that there must be no visible emissions to the outside air during waste transport. It is well to recognize the potential hazards and subsequent liabilities associated with exposure to the release of asbestos particles. Therefore, the following additional precautions are recommended:

(1) Before accepting the asbestos wastes, a transporter should determine if the waste is properly wetted and containerized.

(2) The transporter should then require a chain-of-custody form signed by the generator. Such a form may include the name and address of the generator, the name and address of the pickup site, the estimated quantity of asbestos waste, types of containers used, and the destination of the waste.

(3) The chain-of-custody form should then be signed over to a disposal site operator to transfer responsibility for the asbestos waste.

(4) The transporter should maintain a copy of the form signed by the disposal site operator as evidence of receipt at the disposal site.

In further reinforcement of the precautions above, the transporter should ensure that the asbestos waste is properly contained in leak-tight containers with appropriate labels, and that the outside of the containers is not contaminated with asbestos debris. If there is reason to believe that the condition of the asbestos waste may allow fiber release, the transporter should not accept the waste. Improper containerization of waste is a violation of the NESHAP's regulation and should be reported to the EPA. A list of NESHAPs contacts is provided in the Appendix.

Once the transporter is satisfied with the condition of the asbestos waste and agrees to handle it, the containers should be loaded into the transport vehicle in a careful manner to prevent breakage. Similarly, at the disposal site, the asbestos waste containers should be transferred carefully to avoid fiber release.

Although there are no regulatory specifications regarding the transport vehicle, it is recommended that vehicles used for the transportation of containerized asbestos waste have an enclosed carrying compartment or utilize a canvas covering sufficient to contain the transported waste and prevent fiber release. Also, the vehicle or carrier must be such as to prevent damage to the containers. Transport of large quantities of asbestos waste is commonly conducted in a 20-cubic-yard "roll off" box, which should also be covered. Vehicles that use compactors to reduce waste volume should not be employed because the compactors will cause the waste containers to rupture. Vacuum trucks used to transport waste slurry must be inspected to ensure that water is not leaking from the truck.

H. Disposing of Asbestos

Disposing of asbestos involves the isolation of asbestos waste material in order to prevent fiber release to the air or water. Landfilling is recommended as an environmentally sound isolation method, because asbestos fibers are virtually immobile in soil. Other disposal techniques, such as incineration or chemical treatment, are not feasible owing to the unique properties of asbestos. For example, it is virtually noncombustible. The EPA has established asbestos disposal requirements for active and inactive disposal sites under NESHAPs; see 40 CFR Part 61, Subpart M, which specifies general requirements for solid waste disposal under RCRA, 40 CFR Part 257. NESHAP's regulations require the prior notification of the EPA about the intended disposal site.

An acceptable disposal facility for asbestos wastes must adhere to the EPA's prerequisites of no visible emissions to the air during disposal or minimizing emissions by covering the waste within 24 hours. The minimum essential cover is 6 inches of nonasbestos material, normally soil, or a dust-suppressing chemical. In addition to these Federal desiderata, many state or local governing agencies call for more stringent handling procedures. Those agencies usually supply a list of "approved" or licensed asbestos disposal sites upon request.

Solid-waste control agencies are listed in local telephone directories under state, county, or city headings. One may obtain a list of state solid-waste agencies by calling the RCRA hotline: 1-800-424-9346 (382-3000 in Washington, D.C.). Some landfill owners or operators place special requirements on asbestos waste, such as placing all bagged waste into 55-gallon metal drums. Therefore, asbestos removal contractors should contact the intended landfill before arriving with the waste.

A landfill approved for receipt of asbestos waste should normally require prior notification by the waste hauler that the load contains asbestos. The landfill operator should inspect the loads to verify that asbestos waste is properly contained in the leak-tight containers and is labeled appropriately. The EPA should be notified if the landfill operator believes that the asbestos waste is in a condition that may cause significant fiber release during disposal. A list of EPA Regional Asbestos Coordinators for disposal is provided in the Appendix. In situations when the wastes are not properly containerized, the landfill operator should thoroughly soak the asbestos with a water spray in advance of unloading, rinse out the truck, and immediately cover the wastes with nonasbestos material before compacting the waste in the landfill.

Since asbestos is hazardous to health and there are many dangers inherent in asbestos exposure, it is recommended that the current Federal regulations be further supplemented by the following procedures:

(1) It is advisable to designate a separate area for asbestos waste disposal. Provide a record for future landowners that asbestos waste has been buried there and that it would be hazardous to attempt to excavate that area. Future Federal, state, or local regulations may require property deeds to identify the location of any asbestos wastes and warn against excavation.

(2) Prepare a separate trench to receive asbestos wastes. The size will depend upon the quantity and frequency of asbestos waste delivered to the disposal site. The trenching technique allows application of soil cover without disturbing the asbestos waste containers. The ditch should be ramped to allow the transport vehicle to back into it, and it should be as narrow as possible to reduce the amount of cover required. If possible, the depression should be aligned perpendicular to prevailing winds.

(3) Place the asbestos waste containers into the trench carefully to avoid breaking them. Be particularly cautious with plastic bags, because when they break under pressure asbestos particles can be emitted.

(4) Completely cover the containerized waste within 24 hours with a minimum of 6 inches of nonasbestos material. Improperly containerized waste is a violation of the NESHAPs requirements, and the EPA should be notified; however, if improperly containerized waste is received at the disposal site, it should be covered immediately after unloading. Only after the wastes, including properly containerized wastes, are completely covered can they be compacted or other heavy equipment run over it. During compacting, avoid exposing wastes to the air or tracking asbestos material away from the trench.

(5) For final closure of an area containing asbestos waste, cover with at least an additional 30 inches of compacted nonasbestos material to provide a 36-in. final cover. To control erosion of the final cover, it should be properly graded and vegetated. In areas of the United States where excessive soil erosion may occur or the frost line exceeds 3 feet, additional cover is recommended. In desert areas where vegetation would be difficult to maintain, 3 in. to 6 in. of well-graded crushed rock is recommended for placement on top of the final cover.

Under the current NESHAPs regulation, the EPA does not require that a landfill used for asbestos disposal place warning signs or fencing if it meets the requirement to cover the asbestos wastes. Under the RCRA Act, however, the EPA requires that access be controlled to prevent exposure of the public to potential health and safety hazards at the disposal site. Therefore, for liability protection of operators of landfills that handle asbestos, fencing and warning signs are recommended to control public access when natural barriers do not exist. Access to a landfill should be limited to 1 or 2 entrances with gates that can be locked when left unattended. Fencing should be installed around the perimeter of the disposal site in a manner adequate to deter access by the general public. Chain-link fencing, 6 feet high and topped with a barbed wire guard, should be used. More specific fencing requirements may be specified by local regulations. Warning signs should be displayed at all entrances and at intervals of 330 feet or less along the property line of the landfill or perimeter of the sections where asbestos waste is deposited. The sign should read, as follows:

> "ASBESTOS WASTE DISPOSAL SITE
> BREATHING ASBESTOS DUST
> MAY CAUSE LUNG DISEASE AND CANCER"

For protection from liability and out of concern for possible future requirements for notification on disposal site deeds, a landfill owner should maintain documentation of the specific location and quantity of the buried asbestos wastes. In addition, the estimated depth of the waste below the surface should be recorded whenever a landfill section is closed. As mentioned previously, such information should be recorded in the land deed or other record along with a notice warning against excavation of the area.

The costs of handling asbestos waste are extremely variable. That variability is largely due to the range in handling practices, from those required to achieve minimal compliance with regulations to the use of extra safety precautions not required by law.

To avoid being charged inflated fees for asbestos handling, all cost data should be analyzed by comparing detailed descriptions of the work practices associated with each estimate.

Costs for disposal of containerized asbestos waste are not well documented. A few disposal sites using special handling techniques have quoted fees ranging from $5 to $50 per cubic yard. A cubic yard is about four 55-gallon containers. On the other hand, some sites may not even accept asbestos waste. The fee for hiring a waste hauler for transport of containerized asbestos waste depends on the quantity of waste and distance to an approved disposal site. Also, transportation charges may vary based on the degree of containerization, because rigid containers generally require less careful handling than do plastic bags.

The overall expense for removal of friable asbestos from buildings, including transportation and disposal, generally varies from $2 to $10 per square foot. About the same price range applies per linear foot for pipe insulation. Since this is a complicated handling operation, prices are highly dependent on each contractor's work practices. Higher ones are charged for safeguards that reduce the potential for exposure of building occupants, such as: (1) continuous fiber level monitoring; (2) use of "negative air pressure" systems (see Fig. 11-8 and the relevant discussion); (3) special cleanup and air testing at job completion; and (4) treating stripped porous surfaces with encapsulants. Other than dust control methods, the greatest factor affecting cost is usually the nature of the asbestos-coated surface. For example, a smooth concrete ceiling is much more easily stripped than a corrugated metal deck. The charges for asbestos removal generally include the price of waste hauling and disposal, but that should be confirmed during the negotiations with the contractors or clearly spelled out in the request for proposals.

12

Pesticides

A. Introduction

Pesticide is a word derived from the Latin *cida* (to kill) and *pestis* (a plague). It is a very apt, descriptive word, for that is exactly what pesticides are. They are intended to kill and therefore control insects, plants, fungi, mites, rodents, bacteria, and a host of other destructive nuisances. More than 34,000 pesticides that are derived from about 600 basic chemicals are registered by the EPA for use in this country.

Pesticides are and have been of enormous value to our economic well-being and agricultural productivity. Yet, on the other hand, when improperly or indiscriminately used, they are life threatening and detrimental to the environment and the ecological balance of this planet. Because of the complex nature of the problem that pesticides present, balancing the economic needs of mankind within nature's framework, the question of regulating those chemicals and their use is compounded by a general mistrust of government. That mistrust at times has been stimulated by the news media with words like *toxic, hazardous,* and *carcinogenic.* When the media use these words time and again, it is no small wonder that a significant number of our general public have developed a fear and a phobia about the very chemicals that have contributed so much to our agricultural productivity and the largesse that our Congress has seen fit to distribute to the underdeveloped nations of the world.

While pesticides contribute significantly to the production of food and other agricultural products, they are also important because of their ability to control disease-carrying pests. Since approximately 2 billion pounds of pesticides are used every year, they must be handled carefully, because the more toxic of the substances can present an immediate danger to the user and in some instances can prove fatal if the chemical is spilled on the skin, inhaled, or ingested in any way. Some of the pesticides that have been used in the past also have persisted in the environment over long periods of time. Those chemical substances can move up through the food chain from plankton in the seas and lakes, from insects and birds, to fish and animals, and eventually to humans through the foods that are eaten.

B. Regulation of Pesticides

Some pesticides exhibit evidence of long-term adverse health effects under some conditions. Thus, while pesticides are indeed a modern-day miracle, they must be viewed from another side as well, and that is the aspect of their toxicity and their threat to human health and the environment. Acquiescing to that judicious viewpoint, therefore, Congress has mandated that the EPA regulate the use of pesticides so that we may reap the many benefits of their use while minimizing the risks to health and the environment.

The EPA regulates pesticides under 2 laws: the Federal Insecticide, Fungicide, and Rodenticide Act (FIFRA) and the Pesticide Amendment to the Federal Food, Drug, and Cosmetic Act (FFDCA).

FIFRA was enacted originally in 1947 and was then administered by the U.S. Department of Agriculture. The EPA assumed that responsibility when it was established in 1970.

The 1947 law made it illegal to detach or destroy pesticide labels and provided for pesticide inspections, but it did not address actual pesticide use. In addition, it was not applicable to pesticides used solely within a single state. Despite the 1947 legislation, reports of pesticide-related health and environment problems became increasingly more frequent in subsequent years.

In 1972 Congress amended FIFRA to provide for a broader regulatory program, covering all pesticides used in the United States instead of only those involved in interstate commerce. Under that legislation, all pesticides must be registered by the EPA before they can be sold to the public, and the misuse of a registered pesticide is unlawful.

The Pesticide Amendment of the FFDCA provided an additional measure of public protection by authorizing the establishment of "tolerances." Those are amounts of pesticide residues that may safely remain on a treated food, or feed crop, after harvesting. Tolerance levels must be established for all pesticides used on food or feed crops.

The EPA registers specified uses of pesticide products on the basis of both safety and benefits. FIFRA requires that the EPA determine whether a pesticide can perform its intended function without causing "unreasonable adverse effects" upon human health or the environment while taking into account the potential benefits of the proposed use. This balancing of risks and benefits underlies all basic regulatory decisions under the act.

To make sound judgments, the EPA must have all pertinent information on every pesticide it evaluates. Manufacturers of pesticides are required to provide data on the potential for skin and eye irritation; hazards to nontarget organisms including fish and wildlife; the possibility of acute poisoning, tumor formation, birth defects, reproductive impairments, or other serious health effects; the behavior of the chemical in the environment after application; and the quantity and nature of residues likely to occur in food or feed crops.

Amassing health-related data is not limited to new, unregistered pesticides; for example, amendments to FIFRA enacted in 1972 require that the EPA reassess the safety of pesticide chemicals already in use. Most older pesticides do not meet the standards of testing required now. In the reregistration process, therefore, manufacturers of those older products must meet the same testing standards that new chemicals must meet. That normally requires performing and completing various tests that are then reviewed by the agency to determine whether the products may remain on the market.

If a pesticide ingredient poses a special concern because of a perceived health or environmental risk, the EPA can conduct a special review of the product's risks and benefits. The review process allows all interested parties—the general public, environmentalists, pesticide users, manufacturers, and scientists—to participate. At the conclusion of a special review, the EPA may decide to continue, restrict, or cancel the pesticide uses under consideration. A regulatory decision to cancel uses of a pesticide may be appealed to the EPA Administrator for an adjudicatory hearing.

C. Tolerance Levels of Pesticides

The amount of pesticide, or "tolerance level," that may remain on a crop after harvesting must be established under Federal law for all registered pesticide uses expected to result in residues in raw agricultural commodities, processed food, or feed. Tolerances are legal enforcement levels set well below—normally 100 times below—the level that might cause harm to people or the environment.

The FDA, which enforces tolerances, inspects agricultural commodities, both domestic and imported, to ensure that residues in food to be offered for sale in the marketplace do not exceed the limits established by the EPA. In addition, the U.S. Department of Agriculture inspects meat and poultry for such residues. Any food found to have residues in excess of the tolerance level is subject to seizure and destruction.

D. Enforcement of the Federal Insecticide, Fungicide, and Rodenticide Act

The responsibility for enforcing FIFRA is shared between the states and the EPA. Trained personnel continually check on marketed pesticide products. Samples are collected from manufacturers and distributors in all parts of the country. They are field tested and analyzed in government laboratories for verification of label claims concerning content, effectiveness, and safety. Labels are reviewed to determine that no claims are made other than those accepted by the EPA at the time of registration.

If a product is found to be ineffective or unsafe, the EPA may take 1 of several actions. If there is a minor violation, an informal notice to the company concerned is usually sufficient to guarantee the correction of a deficiency. More serious derelictions or deviations may result in a formal notice of violation, seizure of the company's goods, or initiation of civil or criminal proceedings against the violator.

When a question arises about the safety of a pesticide currently in use, the EPA may issue a formal notice of cancellation that becomes effective within 30 days, unless appealed. If the cancellation is challenged, the product registration remains valid and marketing may continue pending the completion of the appeal process.

If immediate action is warranted, agency can suspend the registration of a pesticide determined to be an imminent hazard to the public welfare. That action halts further sale, distribution, or shipment of the pesticide, regardless of any appeal by the manufacturer until a decision is made through the cancellation procedure.

E. The Classification and Certification of Pesticides

The EPA must classify all pesticide products for either "general" or "restricted" use. General-use pesticides are primarily those that will not cause unreasonable adverse effects to the user or the environment when they are used in accordance with label instructions. Such products usually are available to the public with no restrictions other than those specified on the label. Restricted-use pesticides are those that may cause adverse effects to the applicator or the environment unless applied by persons who have been specially trained in their use.

The law provides for government certification of applicators qualified to handle and apply restricted use pesticides without harming themselves, others, or the environment. Actual training of applicators as well as certification is carried out at the state level by the states and the Cooperative Extension Service. Virtually all states now have an active certification program. To qualify, applicators must demonstrate an understanding of labeling, safety requirements, environmental factors, consequences of pesticide misuse, hazards associated with residues, equipment use, and application techniques.

F. Label Information

One of the most important outcomes of the registration process is the product label, which must be written to exacting specifications and appear on every pesticide container. Because some of today's packaged products are so effective and toxic, users should read the label carefully in its entirety before use of a pesticide product. The user should be aware that it is illegal to use a pesticide in ways inconsistent with its labeling. Information on the label includes:

(1) *The EPA registration number.* This number assures the user that the product is legally registered and considered safe to use as directed.

(2) *Directions for use.* Always use a pesticide only on the sites specified and in the prescribed amounts. Do not think that twice that amount will double the effectiveness. It will not. It will only enhance the possibility of unintended harm, either to the user, houseplants, crops, etc. Some labels include directions for special applications, such as misting plants.

(3) *Precautions.* Pesticides with the highest degree of toxicity are marked with a skull and crossbones and state DANGER- POISON; the word WARNING appears on the labels of less toxic pesticides; and CAUTION is used on the least harmful products. Pay special attention to any instructions regarding precautions for children and pets or the need for protective clothing. Heed instructions about accidental spills on skin or clothes.

(4) *Read first-aid instructions.* Follow the instructions if an accident occurs. Call a poison control center and use the label to describe the chemical. If the accident results in a visit to a doctor or hospital, take the label with you.

(5) *Storage and disposal.* Store the product in the original container. Never transfer a pesticide to a soft-drink bottle or any other container that might be attractive to children. Dispose of empty containers as recommended by any special instructions on the label.

G. Dioxin

For the past several years, the reader has no doubt read or heard a good deal about a chemical called dioxin. Neither dioxin nor its associated problems are new. Questions about its effects have been raised since it was first synthesized in 1872, and many of those questions still need answers. Now new questions have been raised about dioxin, both as an environmental contaminant and a potential public-health problem.

The fact is that we still do not fully understand dioxin or how exposure to it affects human health. We do not fully understand how much dioxin is in the environment or where it is. We are not even sure how best to clean it up when we do find it. The only sure thing is that dioxin contamination has become an extremely complex and emotional issue.

Because of concerns about dioxin, the EPA initiated a "national dioxin strategy":

1. Study the nature of dioxin contamination throughout the United States and the risks to people and the environment;
2. Clean up dioxin-contaminated sites that threaten public health;
3. Find ways to prevent future contamination;
4. Find ways to destroy or dispose of existing dioxin.

As a first step, the EPA is sampling more than 1,000 sites all over the country. They range from locations were certain pesticides were produced and the agency most expects to find dioxin to places where it least expects to find the chemical. The places where the EPA least hopes to find dioxin include private property, and citizens are being asked to cooperate in the study by allowing field teams to take small samples of soil from their property. That sampling is extremely important, as it will help the EPA to learn if there are "background" levels of dioxin in the environment.

To begin with, the word dioxin is actually a generic term for a group of compounds known as polychlorinated dibenzo-p-dioxins (PCDDs), but in popular use it usually refers to the most toxic and carefully studied of these compounds—2,3,7,8 tetrachlorodibenzo-p-dioxin, or 2,3,7,8-TCDD, or simply TCDD. Whenever we discuss dioxin in this section, we are referring to 2,3,7,8-TCDD.

Nobody produces dioxin on purpose. It is an unwanted but almost unavoidable by-product that comes from manufacturing several commercial substances, chiefly the pesticide 2,4,5 trichlorophenol. That pesticide is then used as a basic ingredient in the manufacture of several other pesticides.

TCDD enters the environment in several ways, for example, through dioxin-contaminated chemical products, as a component of the wastes that are produced in manufacturing those products, and through the widespread use of the contaminated products. Certain types of combustion are other possible sources of dioxin contamination.

In general, it is the potential for exposure, either through ingestion or contact with contaminated soil or through eating contaminated fish, that presents the greatest possibilities for health risks. The Centers for Disease Control (CDC) considers 1 part per billion (ppb) of dioxin in the soil to be a level of concern in residential areas. The FDA recommends limiting consumption of fish with 25 parts per trillion (ppt) or greater of dioxin and not eating any fish with greater than 50 ppt of dioxin. The EPA, in conjunction with those other Federal agencies and state and local health agencies, will alert the reader that any precautions should be taken whenever dioxin is detected at levels that may adversely affect one's health. Those agencies will also decide what further actions are necessary.

Scientists disagree on the long-term health effects of exposure to dioxin; however, tests on laboratory animals indicate that it is 1 of the most toxic man-made chemicals known. Because information on effects to humans has come mostly from accidental exposures, the data are not definitive. Scientists do agree, however, that exposure to TCDD can cause a persistent skin rash called chloracne, as experienced by some workers exposed to TCDD in the workplace or through industrial accidents. Tests on laboratory animals indicate that exposure may result in a rare form of cancer called soft-tissue sarcoma in addition to nervousness and other problems.

In studying dioxin, the EPA is looking at 7 categories, or "tiers" of sites, ranging from those where no contamination is expected (Tier 7) to those sites most likely to be contaminated (Tiers 1 and 2).

Tier 7 includes background sites where the agency does not expect to find dioxin. The purpose of sampling those sites is to determine whether dioxin contamination is widespread and, if so, at what levels.

Tier 7 sampling consists of 2 phases. The first involves sampling soils from 30 urban and 200 rural locations across the country. Most of that sampling will take place on private property. No one whose property is selected is required to permit the sampling. Cooperation is encouraged but is strictly voluntary.

The second phase involves sampling fish or shellfish taken from 420 locations, including streams throughout the United States, waters of the Great Lakes, and estuarine and coastal waters. The sampling for that tier is a focus of EPA concern because dioxin in soil and in fish presents the greatest potential exposure to humans.

A tier by tier explanation is, as follows:

Tier 1—Production Sites: The EPA has already investigated and confirmed dioxin contamination at most of the 10 sites where 2,4,5-TCP was produced. At many of those locations companies are undertaking cleanup or are engaged in negotiations with the EPA. Additional investigations will be made where appropriate, and Superfund authority will be used to clean up the areas if removal or remedial actions are needed. The agency is still in the process of identifying places where wastes from these production facilities might have been disposed.

Tier 2—Precursor Sites: Tier 2 includes 9 sites where 2,4,5-TCP was used as a precursor to make other chemical products such as silvex, 2,4,5-T and hexachlorophene. Initial sampling has been completed at most sites. The waste disposal sites associated with those facilities will ultimately be included in this tier.

Tier 3—Formulation Sites: Tier 3 consists of sites and associated waste disposal sites where 2,4,5-TCP and its derivatives were formulated into herbicide products. Approximately 60 to 70 sites sampled between October 1983 and October 1985.

Tier 4—Combustion Sites: The EPA also will be investigating the possibility that various combustion processes produce dioxin. Examples of combustion sources to be studied include hazardous and municipal waste incinerators, internal combustion engines, and accidental fires involving PCB-transformers.

Tier 5—Commercial Use Sites: This tier includes sites where pesticides that may be contaminated with 2,3,7,8-TCDD have been used or are being used on a commercial basis for a variety of agricultural and silvicultural activities. Examples of those uses include clearing power line rights-of-ways of brush and vegetation and as a pesticide for rice and sugar cane fields in the southern United States and in forests of the Pacific Northwest. Approximately 20 to 30 sites are scheduled for testing.

Tier 6—Quality Control Problem Sites: In this tier the EPA will test certain organic chemical and pesticide manufacturing facilities where improper quality controls may have resulted in the production of 2,3,7,8-TCDD. Approximately 20 sites will be investigated.

Tier 7—Control Sites: These are sites where the EPA least expects to find dioxin. They have been included as part of study as "control" sites to determine if there are "background" levels of dioxin in the environment and, if so, how widespread they are. Soils at 500 randomly selected sites across the country, 200 in rural areas and 300 in urban areas, were sampled between July 1984 and July 1985. Fish will be sampled from more than 400 locations, including streams throughout the United States, the Great Lakes, and coastal and estuarine waters.

Soil sampling is done very simply. From the middle of each site, field crews take samples with a tulip bulb planter from a depth of about 3 ins. The sample is placed in a square mason jar, labeled, and sent to an EPA laboratory for analysis.

Fish sampling is more complicated. At freshwater sites, crews take samples both of bottom-feeding fish and game fish; at coastal and estuarine sites, only mussels and oysters are taken in most cases. Samples are taken very quickly to reduce damage and stress to the organisms and are wrapped in foil, placed in plastic bags, and labeled for analysis.

Because the levels of dioxin in the environment are so small, especially sensitive techniques have been developed for measuring it. The most common combines careful sample preparation with the use of gas chromatography and mass spectrometry. That technique, however, is expensive; a single test of each sample costs about $1,500. The EPA is working to develop analytical methods that are more sensitive, more rapid, and less costly.

The actual sampling takes very little time, but laboratory analysis may take as long as 2 to 4 months. The EPA will notify property owners of the sampling results as soon as possible.

It is important to note that the EPA and other Federal agencies have research under way to learn more about the extent of dioxin contamination and the risks of exposure. Since the movement and effects of dioxin in the environment are not fully understood at this time, the EPA is acting conservatively on the basis of current data.

The agency is evaluating methods of disposing of or destroying contaminated soils and wastes. Established technologies include incineration, chemical degradation, and biological treatment measures, but the EPA is working also to find other methods of disposal as well. One promising technique is to treat soil with a chemical compound and sunlight. That method holds promise for actually detoxifying the dioxin molecule. Some temporary methods to limit exposure include: excavating highly contaminated soil and removing it to a safe location, securing and capping the contaminated area, and using high-efficiency vacuums and liquid dust suppressants.

Because dioxin and other hazardous wastes have generated intense public interest and concern, each of the EPA's 10 regional offices has a dioxin information coordinator to answer questions and to provide the latest and most accurate information about dioxin studies in the community. See the Appendix for information concerning Regional Community Involvement Contacts for Dioxin.

H. Methyl Isocyanate (MIC)

Earlier in this text mention was made of the Bhopal, India, catastrophe, which occurred on December 3, 1984, and was one of the world's largest and most devastating chemical-industry disasters. The release into the atmosphere of a cloud of poisonous methyl isocyanate killed more than 2,000 persons and injured countless other thousands.

Although the details surrounding the disastrous incident are still unknown at this writing, the occurrence has stimulated government agencies and industry in the United States to review measures for preventing or responding to similar accidents in this country.

Methyl isocyanate (MIC) is a chemical used mainly in the process of manufacturing certain pesticides such as the brand names "Temik" and "Sevin." It is not contained in these finished products, and MIC is not a pesticide. As a chemical substance, it attacks any part of the body that is especially moist, such as the eyes and the respiratory tract. The immediate effect is eye irritation and shortness of breath, i.e., difficulty in breathing. Eventually, the symptoms are similar to those of pneumonia. The long-term effect may be permanent lung impairment and eye damage. MIC may also cause allergic reactions, such as asthma. If it is inhaled in high concentrations, as happened to many in the Bhopal incident, the result is death.

Fortunately, MIC is not known to linger in the environment. Moisture in the air breaks down or hydrolizes MIC, forming carbon dioxide and a largely inert compound. Thus it is the initial release of the gas that is very critical to human health and safety. Union Carbide, the only U.S. manufacturer of MIC for resale, has in Institute, West Virginia, near Charleston, a plant that makes MIC and sends it to other plants, where the chemical is used in pesticides production. Union Carbide has shut down the MIC part of the West Virginia plant until the accident investigation is completed at the Bhopal site.

Other facilities known to store and use MIC are located in Middleport, New York; LaPorte and Pasadena, Texas; Weeks Island, Louisiana; Muskegon, Michigan; and Woodbine, Georgia.

While accidental releases of toxic chemicals, including MIC, have taken place in this country, the probability of a disaster on the scale of Bhopal occurring here is low. Although no laws, no matter how well written or enforced, are ever fail-safe, the regulatory controls governing our chemical industries reduce the chance of such a large-scale release. In addition, public attention has been focused upon the possibility of a Bhopal-like occurrence in the United States, and many chemical companies are reviewing their operations and procedures in a rigorous manner. Also, the Federal Government has a National Contingency Plan to deal swiftly with releases of hazardous substances. Twelve agencies with environmental and public-health responsibilities form a coordinated national response team to provide leadership in responding to emergencies. Parallel regional teams work with state and local agencies to bring together the expertise, personnel, equipment and money necessary to counteract chemical and other accidents.

In addition, the EPA has an Environmental Response Team staffed by specially trained scientists and engineers. It provides information on chemicals and means of dealing with chemical emergencies throughout the country.

This national/regional/state/local network can act to evacuate an area, take measures to stop a release, and clean up any contamination. The system does work. For example, in November 1984, the FMC plant in Middleport, N.Y., experienced a release of MIC. Company, state, and local officials reacted rapidly, evacuating a school and neighboring residents. There were no serious injuries, and the incident was quickly resolved.

In large-scale or catastrophic events an accident may be declared by the President to be a "disaster" or an "emergency." Then the Federal Emergency Management Agency (FEMA) has the authority to respond and coordinate all Federal recovery efforts.

In general, the Occupational Safety and Health Administration (OSHA) is responsible for ensuring that every U.S. worker has a workplace that is safe, healthful, and free from hazards likely to cause death or serious injury. Specific health exposure standards for methyl isocyanate have been set at 0.02 parts per million over eight hours. That is 1,000 times below the level determined by OSHA to be immediately dangerous to life or health. OSHA recently adopted a "Hazard Communication Standard" commonly known as "Right-to-Know Law." The standard, scheduled to take effect in November 1985, will require chemical manufacturers and importers to notify employees of the use and hazard of chemicals in the workplace.

In addition, the EPA administers several laws that serve to prevent or respond to threats to human health and the environment:

1. The Comprehensive Environmental Response, Compensation and Liability Act, commonly known as "Superfund," gives the EPA authority and resources to deal with immediate threats to human health and the environment. It also provides a mechanism for drawing together other Federal resources and coordinating comprehensive responses to imminently hazardous situations. (See Chapter 18.)
2. The Toxic Substances Control Act gives the EPA board authority to obtain from industry information on what chemicals they produce, where they make and store them, and the potential dangers they pose to our society. (See Chapter 20.)
3. The Resource Conservation and Recovery Act (RCRA) deals with the safe management of hazardous wastes. Under RCRA, materials such as MIC must be managed in a way that protects human health and the environment when those substances are disposed of. New RCRA amendments signed into law by President Reagan on November 8, 1984, broaden the EPA's authority to prevent releases from underground storage tanks into subsurface soils, ground water, or surface water. (See Chapter 19.)
4. The Clean Air Act gives the EPA authority to set standards limiting emissions of hazardous air pollutants. The agency has generally established standards only for hazardous air pollutants known to be routinely emitted. Because MIC is not emitted into the air under normal circumstances, no standard has been established for the chemical. The EPA has emergency authority, however, that might be used to respond to a dangerous chemical release to protect against any "imminent and substantial" threat to human health due to air pollution.

The DOT has responsibility for the transportation of hazardous materials. MIC is now classified as a flammable liquid under DOT regulations, which means that certain protective shipping requirements must be met. The DOT recently proposed special packaging and more stringent labeling requirements for certain poisonous liquids, such as MIC, based on their potential inhalation hazards. Union Carbide, the only U.S. manufacturer of MIC, treats MIC as a "poison" and uses double-walled tankers for its transportation.

As a result of the Bhopal incident, the EPA is reviewing existing Federal authority and capabilities for preventing and responding to gas releases.

The Union Carbide facility at Institute, West Virginia, has been rigorously inspected to make sure the plant is in compliance with all Federal and state environmental laws. At EPA's request, Union Carbide has provided more data on releases of MIC that have occurred in past years. There have been many small releases of MIC at the facility. Fortunately, none have been on the scale of the Bhopal disaster. Additional studies will be carried out, if necessary, to ensure that the facility does not pose a danger to the community.

The agency is also compiling information about chemical releases nationally to reevaluate the potential threat to the environment and public health.

I. Suspended, Cancelled, and Restricted Pesticides

The following section contains a list of suspended, cancelled, and restricted pesticides prepared by the Compliance Monitoring Staff (EN-342) of the Office of Pesticides and Toxic Substances of the EPA. The term "restricted pesticide" as used in this section refers to pesticide uses that have been limited or "restricted" by the agency to require a label change such as "do not apply this pesticide within 25 ft. of any body of water." That term is not to be confused with pesticides whose uses have been restricted under Sec. 3(d) of the Federal Insecticide, Fungicide, and Rodenticide Act (FIFRA) to the application by or under the direct supervision of a certified applicator. The list was compiled by the above EPA staff as of June 1984 and is updated periodically; it is suggested, therefore, that if accurate information is required that the EPA Monitoring Staff (EN-342) be queried directly.

Pesticide/ Use Affected	Action	Reference
Aldrin	Cancelled, all uses *except* those in the following list: 1. subsurface ground insertion for termite control. 2. dipping of non-food roots and tops. 3. moth-proofing by manufacturing processes in a closed system.	PR Notice 71-4 March 18, 1971; Accelerated Decision of the Chief Administrative Law Judge May 27, 1975 and the order Declining Review of the Accelerated Decision of the Administrative Law Judge issued by the Chief Judicial Officer June 30, 1975; 39 FR 37246 October 18, 1974.
Amitraz	1. Conditional registration of amitraz for use on pears if the following requirements are fulfilled.*	44 FR 32736, June 7, 1979; 44 FR 59938, October 17, 1979

*In 1979 amitraz was conditionally registered for use on pears for four years. The registrant did submit the necessary benefits data and results of an oncogenic bioassay of amitraz in mice. Conditional registration of amitraz for use on pears has been extended to September 1984 while the Agency reviews the data.

a. Amitraz must be initially classified for restricted use on
pears and labeled as follows:

"Restricted Use Pesticide. For retail sale to and use only
by certified applicators or persons under their direct
supervision and only for those uses covered by the
certified applicator's certification.

General Precautions

1) Avoid getting in eyes, on skin or on clothing.

2) Avoid breathing vapors or spray mist.

3) In case of contact with skin, wash as soon as possible
·with soap and plenty of water.

4) If amitraz gets on clothing, remove contaminated
clothing and wash affected parts of body with soap and
water. If the extent of contamination is unknown, bathe
entire body thoroughly. Change to clean clothing.

5) Wash hands with soap and water each time before
eating, drinking, or smoking.

6) At the end of the work day, bathe entire body with
soap and plenty of water.

7) Wear clean clothes each day and launder before
reusing.

*Required clothing and equipment for mixing, loading, and
cleanup procedures:*

1) Long-sleeve shirt (fine weave).

2) Long pants (fine weave).

3) Rubber gloves.

4) Apron (only during cleanup).

5) Boots.

Required clothing for ground spray application:

1) Long-sleeve shirt (fine weave).

2) Long pants (fine weave).

3) Rubber gloves.

4) Boots.

Reentry Interval: Reentry into treated areas is prohibited
until the leaves are completely dry, and in any event, until
at least 24 hours after application.

Preharvest Interval: Harvest of treated pears is
prohibited until 7 days after application of amitraz."

b. The applicant for registration must agree to submit the
following information:

1) Repeat a mouse oncogenic bioassay conducted in
accordance with protocols approved by the Agency.

2) Additional economic benefits data.

3) Annual reports of progress and available test results.

2. Denial of application for registration of amitraz for use on
apples.

Pesticide/ Use Affected	Action	Reference
Arsenic Trioxide	In excess of 1.5%, labeling which bears directions for home use is unacceptable, and a warning against home use is required. The following statements must appear in a prominent position: "Do not use or store in or around the home" and "Do not allow domestic animals to graze treated area."	PR Notice 67-2 August 1, 1967 Interpretation No. 25 August 1968
Benomyl	Cancellations of registration and denial of future applications for registrations of benomyl products for uses allowing aerial application unless the conditions of registration are changed to include the following warning on the labels of products packaged in 5-pound or larger packages which do not prohibit aerial application: "Harmful if inhaled. Wear a cloth or disposable paper dust mask during handling and mixing."	47 FR 46747, October 20, 1982
BHC	All registered products have been eliminated. In some instances these products were voluntarily cancelled, in others lindane was substituted for BHC non-gamma isomers. These non-gamma isomers may not be sold, manufactured or distributed for use in the United States.	43 FR 31432 July 21, 1978
Bithionol	Cancelled, products intended for: 1. direct contact with the skin or can be expected to be in direct or continuous contact with the skin. 2. use in textiles or other materials likely to come in contact with the skin. 3. household use.	PR Notice 68-13 August 14, 1968
Chloranil	Voluntary cancellation, all products	42 FR 3702 January 19, 1977
Chlordane	Under the provisions of the Administrator's acceptance of the settlement plan to phase out certain uses of the pesticides chlordane and heptachlor, most registered products containing chlordane were effectively cancelled or their applications for registration denied by December 31, 1980. A summary of those uses not affected by this settlement, or a previous suspension follows: 1. subsurface ground insertion for termite control (clarified by 40 FR 30522, July 21, 1975, to apply to the use of emulsifiable or oil concentrate formulations for controlling subterranean termites on structural sites such as buildings, houses, barns, and sheds, using current control practices). 2. dipping of roots or tops of nonfood plants.	PR Notice 74-11 December 2, 1974 41 FR 7552 February 19, 1976 FIFRA Docket No. 336, et. al. March 6, 1978 PR Notice 78-2 March 28, 1978
Chloro-benzilate	Cancellation and denial of registrations of chlorobenzilate products for uses other than citrus uses in Florida, Texas, California, and Arizona. Notwithstanding the above, registration of chlorobenzilate products for citrus use in these four states will also be cancelled or denied unless registrants or applicants for new registrations modify the terms or conditions of registration as follows: 1. Classification of chlorobenzilate products for these citrus uses for restricted use, for use only by or under the supervision of certified applicators. 2. Modification of the labeling of chlorobenzilate products for these citrus uses to include the following: a. *Restricted Use Pesticide* - For retail sale to and use only by certified applicators or persons under their direct supervision and only for those uses covered by the certified applicator's certification.	44 FR 9548 February 13, 1979

Pesticide/ Use Affected	Action	Reference

b. *General Precautions -*

1) Take special care to avoid getting chlorobenzilate in eyes, on skin, or on clothing.

2) Avoid breathing vapors or spray mist.

3) In case of contact with skin, wash as soon as possible with soap and plenty of water.

4) If chlorobenzilate gets on clothing, remove contaminated clothing and wash affected parts of body with soap and water. If the extent of contamination is unknown, bathe entire body thoroughly. Change to clean clothing.

5) Wash hands with soap and water each time before eating, drinking, or smoking.

6) At the end of the work day, bathe entire body with soap and plenty of water.

7) Wear clean clothes each day and launder before reusing.

c. *Required Clothing and Equipment for Application -*

1) One-piece overalls which have long sleeves and long pants constructed of finely woven fabric as specified in the USDA/EPA *Guide for Commercial Applicators.*

2) Wide-brimmed hat.

3) Heavy-duty fabric work gloves.

4) Any article which has been worn while applying chlorobenzilate must be cleaned before reusing. Clothing which has been drenched or has otherwise absorbed concentrated pesticide must be buried or burned.

5) Facepiece respirator of the type approved for pesticide spray applications by the National Institute for Occupational Safety and Health.

6) Instead of the clothing and equipment specified above, the applicator can use enclosed tractor cab which provides a filtered air supply. Aerial application may be conducted without the specified clothing and equipment.

d. *Handling Precautions -* Heavy duty rubber or neoprene gloves and apron must be worn during loading, unloading, and equipment clean-up.

Pesticide/Use Affected	Action	Reference
Copper Arsenate (Basic)	Voluntary cancellation of the only product containing copper arsenate (basic).	42 FR 18422, April 7, 1977; PR Notice 83-1, February 17, 1983
DBCP	Voluntary cancellation of all registrations of end use products except for the use on pineapples in Hawaii.	FIFRA Docket Nos. 398, 399, and 400, October 27, 1977; 42 FR 57543, November 3, 1977; FIFRA Docket No. 435, October 29, 1979; 46 FR 19592, March 31, 1981
DDD(TDE)	Cancelled, all products containing DDD, a metabolite of DDT.	PR Notice 71-5 March 18, 1971
DDT	Cancelled, all products, *except* the following list of uses: 1. the U.S. Public Health Service and other Health Service Officials for control of vector diseases. 2. the USDA or military for health quarantine. 3. in drugs, for controlling body lice. (To be dispensed only by a physician.)	PR Notice 71-1 January 15, 1971 and 37 FR 13369 July 7, 1972

Pesticide/ Use Affected	Action	Reference
	4. in the formulation for prescription drugs for controlling body lice.	
2,4-D	Products bearing directions for use on small grains (barley, oats, rye, or wheat) must bear the following label precaution: "Do not forage or graze treated grain fields within 2 weeks after treatment with 2,4-D."	PR Notice 67-7 October 12, 1967 Reregistration Guidance package 23-4
Diallate	Cancellation of all existing registrations and denial of future registrations for products containing diallate unless registrants modify the terms of registration to include the following provisions on product labels:	47 FR 27109 June 23, 1982
	1. *Restricted Use Pesticide:* For retail sale to and use only by certified applicators or persons under their direct supervision and only for those uses covered by the applicator's certification. See FIFRA section 3(d).	
	2. *Protective Clothing Required:* The following items of clothing must be worn when mixing, loading, or applying Avadex:	
	a. Long trousers and long-sleeved shirt or jacket of close-knit material.	
	b. Gloves made of rubber or other similar impermeable material.	
	c. Leather or rubber boots high enough to cover the ankle.	
Dieldrin	Most uses cancelled. See Aldrin for uses allowed.	PR Notice 71-4 March 18, 1971; Accelerated Decision of the Chief Administrative Law Judge May 27, 1975 and the order Declining Review of the Accelerated Decision of the Administrative Law Judge issued by the Chief Judicial Officer June 30, 1975 37 FR 37246 October 18, 1974
Dimethoate	1. Unconditional denial of all applications for registration of dimethoate products for use in dust formulations. 2. Cancellation and denial of registrations of dimethoate products for all uses unless the registrants or applicants for registration modify the labeling of dimethoate products to include the following:	46 FR 5334, January 19, 1981

Required Clothing and Equipment for Application

All applicators, including homeowners and flaggers and personnel involved with the mixing, loading, and transferring operations, must wear the protective clothing and equipment enumerated below. Pilots are exempt from this requirement. The protective clothing and equipment to be worn is as follows:

a. Impermeable gloves (for example, rubber or plastic covered gloves).

b. Rubber or synthetic rubber boots or boot covers.

c. Long-sleeved shirt and long pants, made of closely woven fabric.

d. Wide-brimmed hat.

e. Respirators must be worn by flaggers and mixer/loaders.

Pesticide/ Use Affected	Action	Reference

3. Cancellation and denial of registration for all uses of dimethoate products labeled for aerial application unless the registrants or applicants for registration modify the labeling of these products to include the following statements:

AUTOMATIC FLAGGING DEVICES SHOULD BE USED WHENEVER FEASIBLE.

IF HUMAN FLAGGERS ARE EMPLOYED THEY MUST WEAR THE PROTECTIVE CLOTHING AND RESPIRATOR SPECIFIED ON THIS LABEL.

Disinfectants — Cancelled, products bearing labeling claims involving the terms "germ proofing," "germ proofs," and "germ proof."
PR Notice 69-13 August 8, 1969

EBDCs — Cancellation of registration and denial of future applications for registrations of all EBDC products unless registrants modify the conditions of registration to amend product labels as follows:
47 FR 47669, October 27, 1982

1. Add a wildlife warning for use of mancozeb on commercially grown wild rice:

"This product is toxic to fish. Discharge from treated areas may be hazardous to fish in neighboring areas. Do not contaminate water by cleaning of equipment or disposal of wastes."

2. Add a requirement that personnel involved in mixing and loading wear protective clothing (long pants, long sleeve shirt, gloves, hat, and boots).

3. Highlight preharvest intervals of non-commercial (homeowner) products.

EDB — 1. Cancellation of all registrations for products containing EDB *except* those to be used as described in 4.a.-e. below. The cancellation of products for use in the quarantine fumigation of fruits and vegetables (other than exported citrus and papaya) will be effective September 1, 1984.*
48 FR 46228, October 11, 1983; 49 FR 4452, February 6, 1984; 49 FR 14182, April 10, 1984

2. Denial of applications for new registrations for all cancelled uses of EDB.

3. Denial of applications for new registrations of EDB-containing products labeled for manufacturing use *unless* the products are labeled as follows:

"For use only to formulate pesticide products for end use as a post-harvest quarantine fumigant only until September 1, 1984 (except for exported citrus and papaya) or for end use for termite control, vault fumigation, the APHIS Japanese beetle

*Under the Federal Insecticide, Fungicide, and Rodenticide Act (FIFRA), a cancellation does not become final if an individual or company which would be adversely affected by such action requests a hearing to contest it. One registrant (Great Lakes Chemical Company) and several users (Florida Department of Citrus, Florida Citrus Mutual, Florida Citrus Packers, the Indian River Citrus League, Texas Citrus Mutual, the California Citrus Quality Council, and Hawaii Papaya Industry Association) contested cancellation of EDB in the quarantine fumigation of produce for use within the United States. Other companies have contested the cancellation of EDB-containing products for use in felled log fumigation. Regulation of the use of these products in domestic quarantine fumigation and felled log fumigation will not be final until the hearings are resolved.

control program, and beehive super and honeycomb fumigation. Such end use products must be labeled in accordance with the October 11, 1983 EPA Notice of Intent to Cancel EDB-containing pesticide products."

4. Cancellation of the following uses of EDB-containing products unless registrants modify the terms of registration as described below:

a. Use for termite control (must modify label to include statements 5.a., b., c., d., e., and f.).

b. Use in beehive supers and honeycombs (must modify label to include statements 5.b., c., d., g., h., and i.).

c. Use in vault fumigation (must modify label to include statements 5.c., j., and k.).

d. Use in APHIS Japanese beetle control program (must modify label to include statements 5.b., c., l., m., and n.).

e. Use as a quarantine fumigant for exported citrus and papaya (must modify label to include statements 5.h., o., p., q., r., s., t., u., and v.).

5. Labeling Statements:

a. *Pest for Which This Product May be Applied:* Use is restricted to control of subterranean termites in structures constructed on a concrete slab.

b. *Worker Restriction:* Fumigation may be conducted only by a certified applicator.

c. *Required Protection for Workers:* A full-face, black canister respirator approved for use with EDB by the Mine Safety Health Administration or the Occupational Safety and Health Administration (MSHA/OSHA), EDB-resistant gloves made of butyl rubber, and EDB-resistant boot covers and apron made of butyl rubber, nitrile or polyethylene must be worn by the applicator at all times during the treatment.

d. *Application Restriction:* Persons not involved with the fumigation for termites must leave the premises before fumigation, and may not return until after treatment and aeration are completed.

e. *Structural Requirements:* The label must specify the dosage per square foot, the hole spacing, and the volume of EDB which may be injected into each hole drilled through the concrete slab supporting the infested structure. The label must state that all treatment holes drilled in construction elements in commonly occupied areas of structures must be securely plugged. The label must state that use of EDB is not permitted under slabs which contain or cover heat and air conditioning ducts, electrical conduits, wells, water, sewer, gas, or steam pipes, or other similar structures which cause a potential for escape of EDB vapors and that application may not begin until the locations of such escape routes are known. Application must not be made in any manner to an area intended as a plenum air space.

f. *Aeration Requirements:* After fumigation, all doors, windows, and vents in the treatment area and connected buildings must be opened. Premises must be aerated 24 hours before reentry is permitted. Warning signs which prohibit entry during the aeration period must be posted at all entrances to the premises.

Pesticide/ Use Affected	Action	Reference

g. *Structural Requirements:* Supers and honeycombs must be placed in a gas-tight room or under a gas-tight covering such as a polyethylene tarpaulin held down with sand-filled "snakes." If the treatment is performed in a building, all windows, doors, and vents must be sealed. All persons not involved with the fumigation must be vacated from the building. If a tarpaulin is used, it must be completely aired for at least 24 hours before re-use.

h. *Required Protection During Aeration:* Following the treatment period, a full-face, black canister respirator approved by MSHA/OSHA must be worn when reentering during the 24 hours following treatment. All doors, windows and vents in the structure must be opened at this time. Aeration of the building and beehive supers or honeycombs for 24 hours is required before reentry is permitted. During fumigation and aeration, warning signs must be posted which contain these reentry restrictions.

i. *Application Restriction:* Treatments must be applied solely to clean supers or honeycombs mounted in frames in storage.

j. *Structural Requirements:* Vault fumigation with EDB formulations may be conducted only by certified commercial applicators in commercial fumigation vaults. Before fumigation, all doors and leaks in the vault must be sealed.

k. *Aeration Requirements:* Following the treatment period, a full-face, black canister respirator approved by MSHA/OSHA for EDB must be worn during reentry. All doors and vents in the treatment area and connected buildings must be opened at this time. Before workers may reenter the vault without a respirator, the vault must be aerated to reduce EDB air levels to the OSHA workplace standard. Warning signs must be posted stating these restrictions during fumigation and aeration.

l. *Disposal Requirement:* The pesticide spray mixture, rinse water, or excess treatment material that cannot be used must be disposed of according to the procedures established under Subchapter III of the Resource Conservation and Recovery Act.

m. *Application and Aeration Restrictions:* Treated sod, potting and bench soil, and soil for beds and other uses must be covered with heavy plastic sheeting or an impervious tarp during the treatment period to ensure an efficacious treatment and to minimize air exposures. The tarp or sheeting must be completely aired for at least 24 hours before re-use.

n. *Pests for Which This Product May be Applied:* Materials may be treated with EDB to control Japanese beetles only when beetle larvae and pupae are present.

o. *Pests for Which This Product May Be Applied:* The registration must be limited to use on fresh citrus fruit and papaya for quarantine control of fruit flies and must otherwise conform to the Agency's ordinary requirements for labeling as to dose, application directions, toxicity warnings, and related matters.

p. *Deadline for Treatment of U.S. Produce:* Treatment after September 1, 1984, of fresh citrus fruit or papaya for United States consumption must be expressly prohibited.

Pesticide/ Use Affected	Action	Reference

q. *Seasonal Restriction:* After September 1, 1984, citrus fruit for export outside the territorial United States may be treated only during the months of October, November, December, and January. The label may state that treatment of citrus fruit for export in any other month is permitted only after a written determination by the U.S. Environmental Protection Agency that a significant fruit fly outbreak has resulted in quarantine requirements for exported fruit which cannot be met by appropriate alternative quarantine treatments and that treatment with EDB for such purposes will not cause unreasonable adverse effects on the environment.

r. *Worker Restriction:* Fumigation may be conducted only by a certified applicator, and fumigation operations may be conducted only under the supervision of an authorized employee of a state or the United States government.

s. *Required Protection for Workers:* All persons actively engaged in the application or handling of EDB or in the cleanup of any EDB spills must wear self-contained breathing apparatus (SCBA) respirators approved by MSHA/NIOSH, and EDB-resistant gloves (butyl rubber only) and boot covers and apron (made of butyl rubber, nitrile, or polyethylene). All persons entering the fumigation chambers for any purpose within 24 hours following application must wear SCBA respirators approved by MSHA/NIOSH.

t. *Structural Requirements:* Fumigation chambers must have ventilation stacks which extend at least 30 feet above the rooftop of each fumigation chamber.

u. *Aeration Requirements:* When fumigated fruits are shipped by truck directly from the fumigation chambers without interim storage at the site of fumigation, trucks must be aerated in transit to warehouse storage or loading areas.

v. *Disposal Requirements:* Disposal of unused pesticide products containing EDB must be in accordance with the requirements of the Resource Conservation and Recovery Act and the regulations implementing it, and quantities exempted from those disposal requirements by virtue of small volume or the "farmer exemption" must be disposed of in an approved landfill or incinerator.

Electro-magnetic Pest Control Devices — Products ineffective in controlling rodents and insects. Regulatory actions have been taken to remove them from the marketplace.
Reference: EPA Publication No. EPA340102-80-001, October 1980, "Investigation of Efficacy and Enforcement Activities Relating to Electromagnetic Pest Control Devices"

Electronic (sonic) Mosquito Repelling Devices — Products ineffective in repelling mosquitoes. Regulatory actions have been taken to remove them from the marketplace.
Reference: EPA's Environmental News, October 13, 1976

Endrin —
1. Cancellation of uses on:
   a. Tobacco.
   b. Cotton in all areas east of Interstate Highway #35 (includes all states east of the Mississippi River, Arkansas, Louisiana, Missouri, and portions of Texas and Oklahoma).
   c. Small grains to control all pests other than the army cutworm, the pale western cutworm, and grasshoppers.
Reference: Unnumbered PR Notice May 20, 1964 44 FR 43632 July 25, 1979

d. Apple orchards in Eastern States to control meadow voles.

e. Sugarcane to control the sugarcane borer.

f. Ornamentals.

2. Denial of application for new registrations for the above uses (I. b.-e.), as well as for the use of endrin in unenclosed bird perch treatment.

3. Cancellation of the following registrations of endrin products *unless* registrants modify the terms and conditions of registration as specified below:

a. Use on cotton west of Interstate Highway #35 (must modify label to add statements 5. a., b., c., d., e., f., and g.).

b. Use on small grains to control army cutworms and pale western cutworms (must modify label to add statements 5. b., c., d., e., f., and h.).

c. Use on apple orchards in Eastern States to control the pine vole and in Western States to control meadow voles (must modify label to add statements 5. q., c., i., r., j., k., and l.).

d. Use on sugarcane to control the sugarcane beetle (must modify label to add statements 5. q., c., m., and n.).

e. Use for conifer seed treatment (must modify label to add statement 5. o.).

f. Use in enclosed bird perch treatments (must modify label to add statements 5. t., c., and p.).

4. Denial of applications for new registrations for any of the above endrin uses (3. a.-f.), as well as for the following endrin uses unless the applications are modified to meet the terms and conditions specified herein.

a. As a tree painting (in Texas) - must modify label to add statements 5. s. and c.

b. On alfalfa and clover seed crops (in Colorado) - must modify label to add statements 5. b., c., d., e., and f.

c. On small grains to control grasshopper (in Montana) - must modify label to add statements 5.b., c., d., e., f., and h.

5. Label Statements:

a. For use in areas west of Interstate Highway #35 only.

b. *Required Clothing for Female Workers* - Female ground applicators, mixers and loaders and flagpersons must wear long-sleeved shirts and long pants made of a closely woven fabric, and wide-brimmed hats. Mixers and loaders must also wear rubber or synthetic rubber boots and aprons.

c. *Warning to Female Workers* - The United States Environmental Protection Agency has determined that endrin causes birth defects in laboratory animals. Exposure to endrin during pregnancy should be avoided. Female workers must be sure to wear all protective clothing and use all protective equipment specified on this label. In case of accidental spills or other unusual exposure, cease work immediately and follow directions for contact with endrin.

d. *Equipment* -

1) Ground Application - For use with boom-nozzle ground equipment. Apply at not less than 5 gallons total mixture, water and chemical, per acre. Do not use nozzle liquid pressure at greater than 40 psi (pounds per square inch). Do not use cone nozzle size smaller than 0.16 gallons per minute (gpm), at 40 psi such as type D2-25 or TX-10, or any other atomizer or nozzle giving smaller drop size.

Pesticide/ Use Affected	Action	Reference

2) Aerial Application - Do not apply at less than 2 gallons total mixture of water and chemical per acre. Do not operate nozzle liquid pressure over 40 psi or with any fan nozzle smaller than 0.4 gpm or fan angle greater than 65 degrees such as type 6504. Do not use any cone type nozzles smaller than 0.4 gpm nor whirl plate smaller than #46 such as type D4-46 or any other atomizer or nozzle giving smaller drop size. Do not release this material at greater than 10 feet height above the crop.

e. *Application Restrictions* - Do not apply this product within 1/8 mile of human habitation. Do not apply this product by air within 1/4 mile or by ground within 1/8 mile of lakes, ponds, or streams. Application may be made at distances closer to ponds owned by the user but such application may result in excessive contamination and fish kills
Do not apply when rainfall is imminent. Apply only when wind velocity is between 2 and 10 mph.

f. *Procedures To Be Followed if Fish Kills Occur or if Ponds are Contaminated* - In case of fish kills, fish must be collected and disposed of by burial. Ponds in which fish kills have occurred, and user-owned ponds exposed to endrin by application at distances closer than otherwise prohibited, must be posted with signs stating: "Contaminated: No Fishing." Signs must remain for one year after a fish kill has occurred or for six months after lesser contamination unless laboratory analysis shows endrin residues in the edible portion of fish to be less than 0.3 parts per million (ppm).

g. *Prophylactic Use* - Unnecessary use of this product can lead to resistance in pest populations and subsequent lack of efficacy.

h. *Pests for Which This Product May be Applied* - This product may be applied to control the following pests only: army cutworm; pale western cutworm; grasshoppers. (Currently grasshoppers may only be included on endrin products for use in Montana).

i. *Application Restrictions* - Do not apply this product within 50 feet of lakes, ponds, or streams.
Do not apply this product within 50 feet of areas occupied by unprotected humans.
Do not apply when rainfall is imminent.

j. *Equipment* - Apply by ground equipment only.
Use a very coarse spray with minimum pressure necessary to penetrate ground cover. Do not apply as fine spray. Power air blast equipment must be modified to meet the above application restriction. Consult the State recommendations for acceptable methods of adapting equipment.

k. *Prophylactic Use* - Unnecessary use of this product can lead to resistance in the vole population and subsequent lack of efficacy.

l. *Pests for Which This Product May be Applied-* This product may be applied to control the following pests only:
Eastern United States - Pine Vole *(Microtus pinetorum)*
Western United States - Meadow Voles *(Microtus species)*.

m. *Application Restrictions* - Apply only with low-pressure ground equipment. Cover furrows with soil promptly after application.

n. *Pests for Which This Product May Be Applied* - This product may be applied only to control the sugarcane beetle.

Pesticide/ Use Affected	Action	Reference

o. *Application Restrictions* - Do not sow treated seed when large numbers of migratory birds are expected.

p. *Special Warning* - Do not use within one mile of roosting sites or within two miles of nesting sites of peregrine falcons, as identified by the United States Fish and Wildlife Service.

q. *Required Clothing for Female Workers* - Female applicators, mixers and loaders must wear long-sleeved shirts and long pants made of a closely woven fabric, and wide-brimmed hats. Mixers and loaders must also wear rubber or synthetic rubber boots and aprons.

r. *Procedures To Be Followed if Fish Kills Occur* - In case of fish kills, fish must be collected promptly and disposed of by burial. Ponds in which fish kills have occurred must be posted "Contaminated: No Fishing." Signs must remain for one year after a fish kill has occurred unless laboratory analysis shows endrin residues in the edible portion of fish to be less than 0.3 ppm.

s. *Required Clothing for Female Workers* - Female workers handling or applying this product must wear long-sleeved shirts and long pants made of a closely woven fabric, wide-brimmed hat, and rubber or synthetic rubber boots and aprons.

t. *Required Clothing for Female Workers* - Female workers handling this product must wear long-sleeved shirts and long pants made of a closely woven fabric, wide-brimmed hats, and rubber or synthetic rubber aprons.

EPN

1. Cancellations of all registrations of EPN-containing products for use as mosquito larvicides.

48 FR 39494, August 31, 1983

2. Cancellation of registrations and denial of future applications for registration of products for the following uses unless registrants modify the conditions of registration as specified:

a. Use on cotton (must modify registration to include 3.a., b., c., and d.).

b. Use on soybeans (must modify registration to include 3.a., b., c(1), and d.).

c. Use on field corn (must modify registration to include 3.a., b., d., and e.).

d. Use on sweet corn (must modify registration to include 3.a., b., d., and e.).

e. Use on pecans (must modify registration to include 3.b., c(1), and d.).

f. Use on other food crops (must modify registration to include 3.a., b., d., and f.).

3. Registration modifications:

a. Flagging in aerial applications must be accomplished by automated mechanical means or by humans working in completely enclosed vehicles.

b. Product labels must include the phrase "protective clothing required" in bold face type and the following statement:

"Protective Clothing Required: Wear clean protective clothing, goggles, and respirator approved by NIOSH or the American National Standards Institute when applying or handling, or when reentering fields within [at least 24] hours of treatment. The following protective clothing must be worn: lightweight, unlined, natural rubber gloves at least midforearm in length; a wide brimmed waterproof hat or waterproof hood; a protective suit or coveralls of a nonpermeable, non-cloth material covering the body from ankles to wrists; lightweight, unlined, natural rubber boots at least mid-calf in length; full-face respirators are recommended; half-face respirators and goggles are required. Aerial applicators in positive pressure cockpits and other applicators in comparable ground equipment with appropriate filters at all air intakes need not comply with these protective clothing requirements."

c. Labels must include the following instructions in the "Use Directions" section:

(1) "Do not apply this product when weather conditions favor drift from treated area."

(2) "[applies to WP and EC formulations only] Do not apply to blooming cotton if bees are visiting the treatment area."

d. The "Environmental Hazards" section of the label for WP and EC formulations must include the following statement:

"This product is highly toxic to bees exposed to direct treatment or residues on blooming crops and weeds. Do not apply this product or allow it to drift to blooming crops or weeds if bees are visiting the treatment area."

e. The "Use Directions" portion of the label must include the following statements:

(1) "Do not apply this product when weather conditions favor drift from treated area."

(2) "[applies only to WP and EC formulations] Do not apply to corn during the pollenshed period if bees are visiting the treatment area."

f. The "Use Directions" portion of the label must include the following statements:

(1) "Do not apply this product when weather conditions favor drift from treated area."

(2) "[applies only to WP and EC formulations labeled for use on stone fruits, pome fruits, and citrus] Do not apply when trees or a substantial number of weeds in the orchard/ grove are in bloom."

Fluoroace-tamide   Labeling amended to allow use *only* inside of sewers against the Norway and roof rat. This use is restricted and may be applied only by a certified applicator or a competent person acting under the instructions and control of a certified applicator.   Label amendment accepted by OPP November 2, 1979

Goal   Cancellation of registrations for oxyfluorfen-containing pesticide products (Goal) for use on nonbearing and bearing tree fruits and nuts, conifer seedbeds, transplants and out-plantings, soybeans, and field corn (in conjunction with the USDA Witchweed Eradication Program) *unless* registrants modify the conditions of registration as follows: 47 FR 27118, June 23, 1982

The perchloroethylene (PCE) contamination of oxyfluorfen products (Goal) must not exceed 200 ppm. This must be stated in the confidential statement of formula for each registered oxyfluorfen product.

Heptachlor   Under the provisions of the Administrator's acceptance of the settlement plan to phase out certain uses of the pesticides heptachlor and chlordane, most registed products containing heptachlor were effectively cancelled, or their application for registration denied by July 1, 1983. For a summary of those uses not affected by this settlement or a previous suspension, see chlordane.   PR Notice 74-11 December 2, 1974 41 FR 7552 February 19, 1976 FIFRA Docket No. 336, et. al. March 6, 1978 PR Notice 78-2 March 28, 1978

Pesticide/ Use Affected	Action	Reference
Kepone	All registered products containing Kepone were effectively cancelled by May 1, 1978. A summary of Kepone products, their registration numbers, effective cancellation dates, and dispostion of uses of existing stocks follows:	41 FR 24624 June 17, 1976; 42 FR 18885 April 11, 1977 42 FR 38205 July 27, 1977;

Kepone

All registered products containing Kepone were effectively cancelled by May 1, 1978. A summary of Kepone products, their registration numbers, effective cancellation dates, and dispostion of uses of existing stocks follows:

1. Inaccessible Products

a. Antrol Ant Trap, Reg. No 475-11; Black Flag Ant Trap, Reg. No. 475-82; Grant's Ant Trap, Reg. No. 1663-21; Grant's Roach Trap, Reg. No. 1663-22; Grant's Ant Control, Reg. No. 1663-24; and Dead Shot Ant Killer, Reg. No. 274-23 were cancelled as of May 11, 1977. Distribution, sale, and use of existing stocks formulated prior to May 11, 1977, was permitted until such stocks were exhausted.

b. Black Leaf Ant Trap, Reg. No. 5887-63; Hide Roach and Ant Trap, Reg. No. 3325-4; Lilly's Ant Trap With Kepone, Reg. No. 460-17; T.N.T. Roach and Ant Killer, Reg. No. 2095-2; Johnston's No-Roach Traps, Reg. No. 2019-19; Mysterious Ant Trap With Kepone, Reg. No. 395-19; Magikil Ant Trap With Kepone, Reg. No 395-21; Magikil Roach Trap with Kepone, Reg. No. 395-25; Ant-Not Ant Trap, Reg. No. 358-20; Nott Roach Trapp, Reg. No. 358-129; E-Z Ant Trap Contains Kepone, Reg. No. 506-109; Tat Ant Trap, Reg. No 506-126; and Ant Check Ant Trap, Reg. No. 506-129 were effectively cancelled on May 1, 1978. Distribution, sale, and use of kepone products formulated prior to May 1, 1978 was permitted until such stocks were exhausted.

2. Accessible Products: All of these products were cancelled as of December 13, 1977. Distribution, sale, and use of these products is now unlawful.

Note: the following definitions are included in order to distinguish between the two categories of Kepone products.

1. Inaccessible products: includes those enclosed Kepone traps made from metal or plastic as well as metal stakes containing enclosed Kepone bait which are hammered into the ground.

2. Accessible products: includes those which, in normal use, would be removed from their containers, as well as foil or cardboard covered traps.

Reference (Kepone):
41 FR 24624
June 17, 1976;
42 FR 18885
April 11, 1977
42 FR 38205
July 27, 1977;
FIFRA Dockets Nos. 392 et. al.
October 27, 1977;
and the affirmation of FIFRA Dockets Nos. 392 et. al. by the Judicial Officer December 13, 1977

Lindane

1. Cancellation of lindane-containing products for use in vaporizers.

2. Cancellation of lindane-containing products for indoor use in smoke fumigation devices.*

3. Cancellation of registrations and denial of applications for registration of lindane-containing products for all other uses unless the conditions of registration are modified as follows:**

a. Commercial Ornamentals, Avocados, Pecans, Livestock Sprays, Forestry, Christmas Trees, Structural Treatments, Dog Shampoos, and Dog Dusts

Reference (Lindane):
PR Notice 69-9,
April 28, 1969;
IF&R Docket No. 19,
December 2, 1974;
48 FR 48512,
October 19, 1983;
49 FR 26282,
June 27, 1984

---

*Under the Federal Insecticide, Fungicide, and Rodenticide Act (FIFRA), a cancellation is not final if an individual or company which would be adversely affected by such action requests a hearing to contest it. One registrant of a lindane-containing product used in smoke fumigation devices has contested the cancellation of this use. Regulation of the use of this product will not be final until the hearings are resolved.

**The cancelled uses include the direct application of lindane-containing products to aquatic environments; applications for registrations for direct application to aquatic environments will be denied.

Pesticide/ Use Affected	Action	Reference

These products must be classified for restricted use; their labels must include the following statement:

"Restricted Use Pesticide
For application only by or under the direct supervision of a certified applicator."

Products for the above uses (except dog shampoos) must also include the following labeling:

"Applicators must wear the following protective clothing during the application process: a light-weight protective suit or coveralls; water-resistant hat; unlined, waterproof gloves; and unlined, light-weight boots. Mixers and loaders must also wear goggles or face shield, waterproof gloves, and a waterproof apron."

1) *Additional requirements for dog dust use.* Labels of lindane-containing products for dog dust use must contain the following statement:

"This product should be applied in a well-ventilated area."

2) *Additional requirements for structural treatment. Labels of lindane products for structural treatment use must contain the following statement:*

"Applicators working in enclosed areas, such as crawl spaces, must wear a respirator approved by OSHA (29 CFR 1910.134)."

3) *Protective clothing requirements for dog shampoos.* Applicators of lindane-containing dog shampoos must wear waterproof, elbow-length gloves; a waterproof apron; and unlined, waterproof boots.

b. *Homeowner Ornamentals*
Lindane-containing products for use on homeowner ornamentals must be labeled with the following statement:

"Applicators must wear the following protective clothing during the application process: long-sleeved shirt, long pants, waterproof gloves, full foot covering, and a head covering."

c. *Hardwood Logs and Lumber*
Lindane-containing products for use on hardwood logs and lumber must be labeled with the following statement:

"Applicators must wear the following protective clothing during the application procss: lightweight protective suit or coveralls; unlined waterproof gloves; and unlined, lightweight boots."

d. *Dog Dips*
Lindane-containing products for use in dog dips must be labeled as follows***

1) The following statement must appear on the product label beneath the "MIX AS DIRECTED" statement under "CAUTION":

"AN INDIVIDUAL APPLICATOR MUST NOT APPLY THIS PRODUCT MORE THAN TWELVE TIMES PER YEAR."

2) The following statement shall be located on the front panel of the label beneath the product name and in the same size type as the signal word:

"FOR KENNEL, COMMERCIAL, FARM AND SPORT DOG USES ONLY."

***One registrant contested the initial cancellation of all lindane dog dip products to control pests other than mites. Consequently on June 27, 1984, the Agency agreed to continued registration of lindane dog dips to control fleas, ticks, lice, sarcoptic mange, and scabies provided certain additional protective measures were instituted.

3) The last two sentences under "DIRECTIONS FOR USE" shall state the following:

"An individual applicator must not use this product more than twelve times per year. Each treatment of three dogs or fewer should be considered one use."

4) Applicators must wear the following protective clothing during the treatment process: elbow-length, waterproof gloves; a waterproof apron; and unlined, waterproof boots.

5) The label shall state that "improper dilution could cause serious injury to your dog."

6) The label shall state that children under the age of thirteen should not be allowed to handle or apply this product.

7) The label shall be revised in accordance with the provisions of the Notice of Intent to Cancel dated September 30, 1983, regarding disposal of dips.

e. *Moth Sprays*
Lindane-containing products for use in moth sprays must include the following label statement:

"Applicators must wear MSHA/OSHA-approved cartridge respirators when applying this product."

f. *Seed Treatment*
Lindane-containing products for use in seed treatments must be labeled as follows:

"Applicators who apply this product manually or without the use of a closed system treatment procedure must wear the following protective clothing during the application process: long-sleeved shirt; long pants; gloves; and a disposable, paper dust mask which covers at least one-third of the face.

This product should be applied in a well-ventilated area."

Protective clothing for automated or closed-system treatment is not required.

g. *Other Household Uses (Flea Collars, Shelf Paper, and Household Sprays)*
Lindane-containing products for these uses must be labeled with the following warnings:

"Do not allow children to handle or apply this product."

"Children and pets should not be allowed in treated areas until sprayed surfaces are dry."

h. *Label Modifications Applicable to All Lindane Products Subject to This Notice*
Labels for all lindane-containing products must be modified to meet the standards of 40 CFR 162.10; labels must describe symptoms and proper practical treatment for poisoning, proper handling and disposal, and warnings appropriate for the product's toxicity category. Where applicable, labels must contain the following statement:

"Aerial application of lindane is prohibited."

Lindane products for residential use that contain more than 6.5 percent active ingredient must comply with the child-resistant packaging regulations described in 45 CFR 162.16.

Pesticide/ Use Affected	Action	Reference

i. *Disposal of Dips*

Lindane-containing products for dip uses (other than household uses) must be labeled with the following provisions:

"Used dip solutions must be disposed of in accordance with the Resource Conservation and Recovery Act (RCRA). If the applicator generates more than 1000 kg of used dip solution per month or more than 1000 kg used dip solution in combination with other hazardous waste, the material must be treated as a hazardous waste subject to subpart C of RCRA. Any user who wishes to treat, store or dispose of hazardous waste must obtain a permit to serve as a hazardous waste facility pursuant to RCRA."

Mercury

Cancelled, all uses *except* the following:

PR Notice 72-5
March 22, 1972
FIFRA Docket No 246 et.
al.
December 22, 1975
41 FR 16497
April 19, 1976
41 FR 26742
June 29, 1976
41 FR 36068
August 26, 1976

1. as a fungicide in the treatment of textiles and fabrics intended for continuous outdoor use.

2. as a fungicide to control brown mold on freshly sawn lumber.

3. as a fungicide treatment to control Dutch elm disease.

4. as an in-can preservative in water-based paints and coatings.

5. as a fungicide in water-based paints and coatings used for exterior application.

6. as a fungicide to control "winter turf diseases" such as *Sclerotinia boreales*, and gray and pink snow mold subject to the following

a. the use of these products shall be prohibited within 25 feet of any water body where fish are taken for human consumption.

b. these products can be applied only by or under the direct supervision of golf course superintendents.

c. the products will be classified as restricted use pesticides when they are reregistered and classified in accordance with section 4(c) of FEPCA.

Note: *For purposes of the settlement agreements, "winter diseases" refer to the forms of snow mold which can attach and damage the fine turf of greens, tees, and aprons.*

Metaldehyde

Labeling for metaldehyde snail and slug baits must have the following statement on the front panel of the product label:

PR Notice 74-7
July 1, 1974

"This pesticide may be fatal to children and dogs or other pets if eaten. Keep children and pets out of treated area."

Mirex

All registered products containing Mirex were effectively cancelled on December 1, 1977. (A technical Mirex product made by Hooker Chemical Company is unaffected by this Settlement Agreement. However since Mirex produced under this registration may be used only in the formulation of other pesticide products the registration was useless after December 1, 1977). All existing stocks of Mirex within the continental U.S. were not to be sold, distributed, or used after June 30, 1978.

FIFRA Docket No. 293
October 20, 1976.
41 FR 56694
December 29, 1976

Harvester Bait 300, Reg. No. 38962-5, may only be used for the control of the pheidole ant, Argentine ant, and fire ant on pineapples in Hawaii. The effective date of cancellation for these uses was December 1, 1977; existing stocks as of December 1, 1977 may not be applied aerially, but may be sold and used (other than aerially) indefinitely.

Pesticide/ Use Affected	Action	Reference

The application of Harvester Bait 300 is subject to the following restrictions:

1. Aerial Application: No longer permitted.

2. Ground Application

a. Permissible in all areas of infestation provided that there is no ground application to aquatic and heavily forested areas or areas where run-off or flooding will contaminate such areas.

b. Treatment shall be confined to areas where the imported fire ants are causing significant problems.

Note: *the following definition is included in order to clarify the ground application restrictions:*

Aquatic areas: encompasses without limitation estuaries, rivers, streams, wetlands (those land and water areas subject to inundation by tidal, riverine, or lacustrine flowage), lakes, ponds, and other bodies of water.

OMPA — Voluntary cancellation, all products — 41 FR 21859 May 28, 1976

10, 10'-Oxybisphenoxarsine — Product labels amended to eliminate use in windbreakers and baby pants. — Special Pesticide Review Division's Position Document on 10, 10'-Oxybisphenoxarsine Approved April 20, 1979

Parathion (Ethyl) — 1. Registration of ethyl parathion limited to those packed in one gallon containers or larger. — PR Notice 71-2 April 5, 1971

2. Manufacturers and formulators of registered ethyl parathion should be in compliance with the standardized safety label that was enclosed with PR 71-2.

Polychlorinated Biphenyls — Elimination (all use as active or inactive ingredients) — PR Notice 70-25 October 29, 1970

PCNB — Modification of the terms and conditions of all registrations of PCNB-containing products to meet the following conditions: — 47 FR 18177, April 28, 1982

1. Reduction of the hexachlorobenzene (HCB) level in technical PCNB products to 0.1 percent or less because of risks associated with the oncogenic effects of HCB.

2. Voluntary cancellation of the registrations of all dust base formulations except those used in planter box seed treatment.

3. Amendment to the labels of granular formulations used in parks and on golf courses to include the following precautionary statement:

"Do not apply directly adjacent to potable water supplies."

3. Amendment to the labels of homeowner products to include the following precautionary statement:

"Avoid contact with skin by wearing the following protective clothing: long-sleeved shirt, long pants, socks, and shoes. Wash hands thoroughly after using."

4. Amendment to the labels of professional applicator products to include the following protective clothing requirements during mixing/loading procedures:

"Granular formulations: gloves, longsleeved shirt, long pants, socks and shoes."

"Emulsifiable concentrate and liquid formulations: respirator, gloves, longsleeved shirt, long pants, socks and shoes."

Phenarzine Chloride — Voluntary cancellation, all uses. — 42 FR 59976 November 21, 1977

Pesticide/ Use Affected	Action	Reference
Polychlor- inated Terphenyls	Elimination (all use as active or inactive ingredients).	PR Notice 70-25 October 29, 1970
Predacides	Revocation of the requirement that agency heads prevent the use of chemical toxicants in Federal mammal/bird damage control programs or on any Federal lands where the field use of such toxicants is being employed to kill predatory mammals or birds, or where the field use of such toxicants causes any secondary poisoning effects on mammals, birds or reptiles.	Executive Order 11643; 37 FR 2876, February 9, 1972; Executive Order 11870; 40 FR 30611, July 22, 1975; Executive Order 11917; 41 FR 22239, June 2, 1976; Executive Order 12342 January 27, 1982
Pronamide	1. Cancellation and denial of registrations of hand spray application of pronamide for all uses except ornamentals and nursery stock.	44 FR 61640 October 26, 1979

2. Cancellation and denial of registrations of all pronamide products registered for use on lettuce, alfalfa and forage legumes and other uses unless the registrants or applicants for registration modify the terms and conditions of registration as follows:

a. Classification of pronamide wettable powder products for Restricted Use Only, for use only by or under the direct supervision of certified applicators and only for those uses covered by the certified applicator's certification.

b. Modification of the labeling of pronamide wettable powder products to include the following:

1) Restricted-use pesticide. For retail sale to and use only by certified applicators or persons under their direct supervision and only for those uses covered by the certified applicator's certification.

2) General precautions.

a) Take special care to avoid contact with eyes, skin, or clothing.

b) Wash clothing and gloves after use.

3) Protective clothing. The following items of clothing are required when mixing or applying pronamide:

a) Long-sleeved shirts and long pants, preferably one piece (overalls).

b) Hat with brim.

c) Heavy-duty fabric or rubber work gloves.

d) Hand-spray applications of pronamide will require the use of heavy-duty leather or rubber boots.

4) Water-soluble packaging. For all wettable-powder products introduced in commerce, the statement:

"Dilution Instructions

The enclosed pouches of this product are water-soluble. Do not allow pouches to become wet prior to adding to the spray tank. Do not handle the pouches with wet hands or gloves. Always reseal overwrap bag to protect remaining unused pouches. Do not remove except to add directly to the spray tank.

Add the required number of unopened pouches as determined by the dosage recommendations into the spray tank with agitation. Depending on the water temperature and the degree of agitation, the pouches should dissolve completely within approximately five minutes from the time they are added to the water."

c. Modification of the granular formulation pronamide labels to include the following for turf use:

"This product should be watered in within 24 hours."

Pesticide/ Use Affected	Action	Reference
Quaternary Ammonium Compounds	Cancelled; for use as a sanitizer in poultry drinking water.	PR Notice 73-5 August 29, 1973
Safrole	Voluntary cancellation, all uses. The distribution, sale, and use of existing stocks of Surf-Kote Pet Repellent, Reg. No. 1811-8, and Scram Dog Repellent Spray, Reg. No. 239-2057, are permitted only by persons other than the registrants.	42 FR 11039 February 25, 1977; 42 FR 16844 March 30, 1977 42 FR 29957 June 10, 1977
Seed Treatments	Cancelled, all registrations of these products not containing a dye or discoloring agent which will impart an unnatural color to the seed. Exceptions to this are products bearing directions for use solely as planter box treatments.	PR Notice 70-17 June 26, 1970 PR Notice 70-24 October 28, 1970
Silvex	Chlorodioxin contaminants not allowed.	PR Notice 70-22, September 28, 1970; 44 FR 15917, March 15, 1979; 44 FR 41536, July 17, 1979; 48 FR 48434, October 18, 1983.
	Cancellations of registrations of products which contain silvex as an active ingredient.* Except for those products whose registrations were suspended in 1979, all existing stocks which were packaged and labeled for non-suspended end use(s) and released for shipment before the receipt of the October 18, 1983 Federal Register notice may be distributed and sold for one year after the effective date of cancellation. Existing stocks of silvex which may be distributed and sold as described above include products for use on rice, rangeland, and sugarcane (field and stubble); for preharvest fruit drop of apples, prunes, and pears; for use on or around non-food crop areas and non-crop sites, including fencerows, hedgerows, fences (except for rights-of-way, pasture uses, and home and garden uses), industrial sites or buildings (except for rights-of-way, commercial/ornamental turf uses), storage areas, waste areas, vacant lots, and parking areas.	
	Range is non-pasture grazing land on which forage is produced through native species, or on which introduced species are managed as native species. This excludes land on which regular cultural practices of the nature contained in the pasture definition are followed.	
	Pasture is land producing forage for animal consumption, harvested by grazing, which has annual or more frequent cultivation, seeding, fertilization, irrigation, pesticide application or other similar practices applied to it. Fence rows enclosing pastures are included as part of the pasture.	
Sodium Arsenite	Unacceptable for home use if compound is in excess of 2.0%, and the following warning statements must appear on the label: "Do not use or store in or around the home" and "Do not allow domestic animals to graze treated area."	PR Notice 67-2 August 1, 1967 Interpretation No. 25 August 1968
Sodium Cyanide	Cancelled and suspended, all uses for mammalian predator control except; the registration of sodium cyanide capsules for use in the M-44 device is allowed for the purpose of controlling certain wild canid predators subject to the following 26 restrictions:	PR Notice 72-2, March 9, 1972 10th Circuit Court's Vacation of the Wyoming District Court's Predacide injunction, December 2, 1975, 40 FR 44726, September 29, 1975, 41 FR 21690, May 27, 1976 42 FR 8406, February 10, 1977
	1. Use of the M-44 device shall conform to all applicable Federal, State, and local laws and regulations.	
	2. The M-44 device shall be used only to take wild canids suspected of preying upon livestock and poultry.	
	3. The M-44 device shall not be used solely to take animals for the value of their fur.	

*Under the Federal Insecticide, Fungicide, and Rodenticide Act, an individual or organization which is adversely affected by a cancellation may contest such action in an adjudicatory hearing. Because some registrants of silvex products have pursued this course, regulatory action on these products will not be final until cancellation disputes have been resolved.

Pesticide/ Use Affected	Action	Reference

4. The M-44 device shall only be used in instances where actual livestock losses due to predation by wild canids are occurring. M-44 devices may also be used prior to recurrence of seasonal depredation, but only when a chronic problem exists in a specific area. In each case, full documentation of livestock depredation, including evidence that such losses were caused by wild canids, will be required before application of the M-44 is undertaken.

5. The M-44 device shall not be used in: (1) National or State Parks; (2) National or State Monuments; (3) Federally-designated Wilderness areas; (4) Wildlife refuge areas; (5) Prairie dog towns; (6) Areas where exposure to the public and family pets is probable.

6. The M-44 shall not be used in areas where threatened or endangered species might be adversely affected. Each applicator shall be issued a map which clearly indicates such areas.

7. The M-44 device shall not be placed within 200 feet of any lake, stream, or other body of water.

8. The M-44 device shall not be placed in areas where food crops are planted.

9. M-44 devices shall not be placed within 50 feet of public rights of way.

10. The maximum density of M-44's placed in any 100-acre pastureland area shall not exceed 10; and the density in any one square mile of open range shall not exceed 12.

11. The M-44 device may be placed in the vicinity of draw stations (livestock carcasses), provided that no M-44 device shall be placed within 30 feet of a carcass; no more than 4 M-44 devices shall be placed per draw station; and no more than 3 draw stations shall be operated per square mile.

12. M-44 devices shall be inspected at least once a week to check for interference or unusual conditions and shall be serviced as required.

13. Used sodium cyanide capsules shall be disposed of by deep burial or at a proper landfill site.

14. An M-44 device shall be removed from an area if, after 30 days, there is no sign that a target predator has visited the site.

15. Damaged or non-functional M-44 devices shall be removed from the field.

16. In all areas where the use of the M-44 device is anticipated, local hospitals, doctors, and clinics shall be notified of the intended use and informed of the antidotal and first-aid measures required for treatment of cyanide poisoning.

17. Bilingual warning signs in English and Spanish shall be used in all areas containing M-44 devices. All such signs shall be removed when M-44 devices are removed.

a. Main entrances or commonly used access points to areas in which M-44 devices are set shall be posted with warning signs to alert the public to the toxic nature of the cyanide and to the danger to pets. Signs shall be inspected weekly to insure their continued presence and insure that they are conspicuous and legible.

b. An elevated sign shall be placed within 6 feet of each individual M-44 device warning persons not to handle the device.

18. Registrations for sodium cyanide capsules to be used in the M-44 device may be granted to persons other than State

and Federal agencies; provided, that such persons shall be authorized to sell said capsules only to State and Federal registrants, except that Indian governing authorities on reservations not subject to State jurisdictions are also eligible to obtain registrations. Only State, Federal, and Indian registrants shall be permitted to sell, give or otherwise distribute sodium cyanide capsules to individual applicators. Such State, Federal, and Indian registrants of sodium cyanide capsules shall be responsible for insuring that the restrictions set forth herein are observed by individual applicators to whom such registrants sell or distribute such capsules and/or M-44 devices. State, Federal and Indian registrants shall train applicators, and such training shall include, but need not be limited to: (1) Training in safe handling of the capsules and placement of the device; (2) Training in the proper use of the antidote kit; (3) Instructions regarding proper placement of the device; and (4) Instructions in recordkeeping.

19. Each authorized M-44 applicator shall keep records dealing with the placement of the device and the results of each placement. Said records shall include, but need not be limited to: a. The number of devices placed. b. The location of each device placed. c. The date of each placement, as well as the date of each inspection. d. The number and location of devices which have been discharged and the apparent reason for each discharge. e. The species of animal taken. f. All accidents or injuries to humans or domestic animals.

20. M-44 devices and sodium cyanide capsules shall not be sold or transferred to, or entrusted to the care of, any person not authorized or licensed by, or under the supervision or control of a Federal, Indian, or State registrant.

21. All persons authorized to possess and use M-44 capsules and devices shall store said devices under lock and key.

22. Each authorized or licensed applicator shall carry an antidote kit on his person when placing and/or inspecting M-44 devices. The kit shall contain at least six pearls of amyl nitrate and instructions on their use. Each authorized or licensed applicator shall also carry on his person instructions for obtaining medical assistance in the event of accidental exposure to sodium cyanide.

23. One person other than the individual applicator must have knowledge of the exact placement location of all M-44 devices in the field.

24. Supervisors shall periodically check the records, signs, and devices of each applicator to verify that all applicable restrictions, laws, and regulations are being strictly followed.

25. In areas where more than one governmental agency is authorized to place M-44 devices, the agencies shall exchange placement information and other relevant facts to insure that the maximum number of M-44's allowed is not exceeded.

26. Registrants and applicators shall also be subject to such other restrictions as may be prescribed from time to time by the U.S. Environmental Protection Agency.

Pesticide/Use Affected	Action	Reference
Sodium Fluoride	Cancelled, for home use if the product contains more than 40% of this compound.	PR Notice 70-14, June 1, 1970.
Sodium Fluoroacetate	Cancelled and suspended for use in mammalia predator control. Label should have instructions for predator use blocked out. Nonetheless, the Agency has decided to accept applications for the registration of the toxic collar and to allow the use of single lethal dose baits by Federal or State employees if certain conditions are met.	PR Notice 72-2 March 9, 1972, 10th Circuit Court's vacation of the Wyoming District Court's Predacide Injunction, December 2, 1975 49 FR 4830 February 5, 1984

Pesticide/ Use Affected	Action	Reference
Sodium Hypo-chlorite	The following label requirements should be made in order to correct deficiencies in active ingredients: 1. Products bearing label claims for 7.0% sodium hypochlorite or less: no change requirements. 2. Products bearing label claims for more than 7.0% to 12.5% sodium hypochlorite, add this label statement. *"Degrades with age. Use a test kit and increase dosage as necessary to obtain required level of available chlorine."* Such products packaged in containers less than one gallon are not acceptable. 3. Products bearing label claims for more than 12.5% sodium hypochlorite are not acceptable for registration. In some cases overformulation to maintain claimed concentrations is necessary; however, such overformulation should not exceed 25% of the claimed concentrations.	PR Notice 70-16, June 25, 1970
Strobane	Voluntary cancellation, all products.	41 FR 26607, June 25, 1976
Strychnine	1. Cancellation and suspension of use in mammalian predator control. Label should have instructions for predator use blocked out. 2. Cancellation of registrations and denial of future applications for registration for strychnine-containing products for the following uses*: a. Control of prairie dogs on rangeland, pasture, cropland, and nonagricultural sites. b. Control of deer mice, meadow mice, and chipmunks on rangeland, pasture, cropland, and nonagricultural sites; marmots/ woodchucks on rangeland, cropland, and pasture; and cotton rats, kangaroo rats, mountain beavers, opossums, rabbits, and jackrabbits (except around airports) on nonagricultural sites. 3. Cancellation of registrations and denial of future applications for registration for strychnine-containing products for control of the following rodents unless registrants modify the conditions of registration as follows: a. *Ground squirrels on rangeland, pasture, cropland, and nonagricultural sites.* Add the following label statements: 1) "Do not expose baits in a manner which presents a likely hazard to pets, poultry or livestock." 2) "Do not place bait in piles." 3) "Pick up and burn or bury all visible carcasses of animals in or near treated areas." 4) "Do not use for ground squirrel control in areas occupied by the Utah prairie dog in Garfield, Iron, Kane, Piute, Sevier and Wayne Counties, Utah." 5) "Do not use for ground squirrel control within 200 yards of prairie dog colonies." 6) "Do not use within one mile of a prairie dog colony where the presence of a black-footed ferret is confirmed within a five-year period." 7) "The killing of an endangered species during strychnine baiting operations may result in a fine under the Endangered Species Act. Before baiting, the user is	PR Notice 72-2, March 9, 1972; 10th Circuit Court's vacation of the Wyoming District Court's Predacide Injunction, December 2, 1975; 48 FR 48522, October 19, 1983

---

*Under the Federal Insecticide, Fungicide, and Rodenticide Act (FIFRA), a cancellation is not final if an individual or company which would be adversely affected by such action requests a hearing to contest it. Several uses of strychnine-containing products were contested in Wyoming and South Dakota. Regulation of the use of these products will not be final until the hearings are resolved.

advised to contact the regional U.S. Fish and Wildlife
Service Endangered Species Specialist or the local Fish and
Game Office for specific information on endangered
species. Strychnine baits should not be used in geographic
ranges of the following species except under programs and
procedures approved by the USEPA: California Condor, San
Joaquin Kit Fox, Aleutian Canada Goose, Morro Bay
Kangaroo Rat, Gray Wolf, and Grizzly Bear."

8) "All registrations for products containing strychnine
which are not dyed in accordance with the standards
described by the State of California in the Vertebrate Pest
Control Handboook are cancelled; all registrants must
amend their statements of formula to provide that all
strychnine baits are dyed according to these guidelines."

b. *Other rodents and lagomorphs on rangeland, pastures, and
cropland.* Add the following label statements to products
used to control cotton rats, kangaroo rats, and jackrabbits:

1) "Do not expose baits in a manner which presents a
likely hazard to pets, poultry or livestock."

2) "Do not place bait in piles (for cotton and kangaroo
rats)."

3) "For jackrabbit control place the baits in piles."

4) "Pick up and burn or bury all visible carcasses of animals
found in or near treated areas."

5) "Do not use for cotton rat control in areas occupied by
the Mississippi sandhill crane in Jackson County,
Mississippi."

6) "Do not use for cotton rat control in areas occupied by
the Cape Sable sparrow in Collier, Dade and Monroe
Counties, Florida."

7) "Do not use for kangaroo rat or jackrabbit control in
areas occupied by the masked bobwhite quail in Pima and
Santa Cruz Counties, Arizona."

8) "Do not use for kangaroo rat or jackrabbit control in
areas occupied by the Utah prairie dog in Garfield, Iron,
Kane, Piute, Sevier, and Wayne Counties, Utah."

9) "Do not use for jackrabbit or cotton rat control in areas
occupied by Attwater's greater prairie chicken in the
following Texas counties: Arkansas, Austin, Brazoria,
Colorado, Fort Bend, Galveston, Goliad, Refugio, and
Victoria."

10) "The killing of an endangered species during strychnine
baiting operations may result in a fine under the
Endangered Species Act. Before baiting, the user is
advised to contact the Regional U.S. Fish and Wildlife
Service (Endangered Species Specialist) or the local Fish
and Game Office for specific information on endangered
species. Strychnine baits should not be used in geographic
ranges of the following species except under programs and
procedures approved by the USEPA: California Condor, San
Joaquin Kit Fox, Aleutian Canada Goose, Morro Bay
Kangaroo Rat, Gray Wolf, and Grizzly Bear."

11) "All registrations for products containing strychnine
which are not dyed in accordance with the standards
described by the State of California in the Vertebrate Pest
Control Handbook are cancelled; all registrants must
amend their statements of formula to provide that all
strychnine baits are dyed according to these guidelines."

c. *Other rodents and lagomorphs on nonagricultural sites.*

l) The following label changes are required for products
used for marmot/woodchuck control:

Pesticide/ Use Affected	Action	Reference

a) "Do not expose baits in a manner which presents a likely hazard to pets, poultry, or livestock."

b) "Do not place baits in piles."

c) "Pick up and burn or bury all visible carcasses of animals found in or near treated areas."

d) "The use of strychnine to control marmot/ woodchucks .is prohibited in all areas except where dens are located in rocky areas where fumigants cannot be used."

2) The following label changes are required for products used for marmot/woodchuck control:

a) "Treated salt blocks will be nailed at least ten feet above the snowline."

b) "Use is prohibited in areas known to be occupied by the gray wolf or grizzly bear."

3) "The use of strychnine to control jackrabbits is prohibited except around airports."

d. *Birds on croplands.* The following label changes are required:

l) To control blackbirds, cowbirds, crowned sparrows, horned larks, house finch (linnets), and meadow larks in orchards and vineyards:

a) "Bait must be placed in troughs no less than 3 inches deep (V-shaped, with ends blocked to avoid spillage)."

b) "Troughs must be removable so that they can be cleaned out after each change of bait."

c) "Troughs must be at least four feet from ground level."

d) "Place troughs within orchards or vineyards where damage exists or is about to occur."

e) "Expose strychnine bait sparingly—one-half inch deep in each trough."

f) "Pick up and dispose of any bait spilled daily."

g) "Pick up and burn or bury all visible bird carcasses at the end of each day."

2) To control horned larks on crops:

a) "Expose bait sparsely in a depression between bedded crops."

b) "Pick up and dispose of any bait spilled outside the depression daily."

c) "Pick up and burn or bury all visible carcasses of birds at the end of each day."

e. *Birds on nonagricultural sites.* The following label changes are required for products used to control pigeons and house sparrows:

1) "Pick up and dispose of uneaten pigeon bait after three days of exposure per site."

2) "Where uneaten bait is not easily retrievable, place in trays or V-shaped troughs."

3) "Inspect each site for nontarget species hazard daily."

4) "Pick up and burn or bury all visible dead birds daily."

5) "The use to control pigeons and sparrows will be prohibited in Puerto Rico for the protection of the Yellow-Shouldered Blackbird and the Puerto Rican Plain Pigeon."

6) "The use to control pigeons will be prohibited in the following areas for the protection of the Peregrine Falcon: within five miles of aeries or critical habitats."

Pesticide/ Use Affected	Action	Reference

| | 7) "Do not use when migratory falcons may be present. Consult with local fish and game or regional Fish and Wildlife Service." | |
| **2,4,5-T** | Chlorodioxin contaminants not allowed. | PR Notice 70-22, September 28, 1970; PR Notice 70-11, April 20, 1970; PR Notice 70-13, May 1, 1970; 44 FR 15874, March 15, 1979; 44 FR 41531, July 17, 1979; 48 FR 48434, October 18, 1983. |

Cancellations of registrations of products which contain 2,4,5-T as an active ingredient.[*] Except for those products whose registrations were suspended in 1979, all existing stocks which were packaged and labeled for non-suspended end use(s) and released for shipment before the receipt of the October 18, 1983 Federal Register notice may be distributed and sold for one year after the effective date of cancellation. Existing stocks of 2,4,5-T which may be distributed and sold as described above include products for use on rice, rangeland, and non-crop areas, including airports, fences, hedgerows (except for rights-of-way, pasture uses), lumber yards, refineries, non-food crop areas, storage areas, wastelands (except for forestry uses), vacant lots, tank farms, and industrial sites and areas (except for rights-of-way uses).

Note: For definitions of range and pasture see silvex.

Thallium Sulfate	Cancelled and suspended, all products.	PR Notice 72-3, March 9, 1972
TOK	Voluntary cancellation, all products.	49 FR 2151, January 18, 1984; conversation with product manager
Toxaphene	1. Cancellation of registrations of all pesticide products containing toxaphene except those which meet the conditions described in 2-3.	PR Notice 69-5, February 14, 1969; 47 FR 53784, November 29, 1982.

2. Registrations of manufacturing use products or technicals amended to require labels that limit use to formulation of pesticide products which are registered according to the Notice of Intent to Cancel or Restrict Registration of Pesticide Products Containing Toxaphene (47 FR 53784).

3. Registrations of toxaphene products where registrants have requested modification of the conditions of registration as described below:

a. Use to control scabies on beef cattle or sheep (must modify label to include provisions 5.a., b., c., d., e., f., g., and h.).

b. Minor use for armyworm/cutworm/grasshopper control (must modify label to include provisions 5.c., h., i., j., k., l., m., and n.).

c. Minor use for mealybug and pineapple gummosis moth control on pineapples and weevil control on bananas (must modify label to include provisions 5.c., h., k., l., o., and p.)

4. Existing Stocks:

a. Existing stocks of cancelled toxaphene products must be disposed of by registrants within 90 days after receipt of 47 FR 53784 and in accordance with the Resource Conservation and Recovery Act unless a registrant holding existing stocks provided the Agency with a request to qualify for the provisions of existing stocks within 30 days after receipt of the above FR notice. Annual inventory reports are required of qualifying registrants. Existing stocks which qualify for this disposal exemption may be distributed, sold, and used only for the retained registered

---

[*]Under the Federal Insecticide, Fungicide, and Rodenticide Act, an individual or organization which is adversely affected by a cancellation may contest such action in an adjudicatory hearing. Because some registrants of 2,4,5-T products have pursued this course, regulatory action on these products will not be final until cancellation disputes have been resolved.

uses or with the following requirements, which must be stated on the supplemental labeling:

1) Stocks which can be made into formulations for the following purposes must be applied only to these uses: control of sicklepod in soybeans and peanuts in states which have 24(c) registrations or make a special local needs finding within the 90-day period of receipt of 47 FR 53784; control of insects in corn cultivated without tillage and in dry and southern peas; emergency use until December 31, 1986, following a statement by a Federal or State agency and finding by EPA that emergency conditions warrant the requested use.

2) Stocks which cannot be applied to the above uses may be sold, distributed, and used according to current label instructions until December 31, 1983.

3) In aerial applications, flagging must be accomplished by automated mechanical means or by humans working in completely enclosed vehicles.

4) Protective clothing (described in 5.1. and m.) must be worn by all who handle the pesticide.

5) Labels must include this warning:

"This product is toxic to fish, birds and other wildlife. Use of this product may be fatal to birds and other wildlife in treated areas. Do not spray over lakes, streams, ponds, tidal marshes and estuaries. Do not apply where runoff is likely to occur. Do not apply when weather conditions favor drift from areas treated. Do not contaminate water by cleansing of equipment or disposal of wastes. Overspray of this product into water at application rates recommended on this label may be fatal to shrimp and crab; do not apply where these are important resources. Apply this product only as specified on this label."

b. The above required supplemental labeling may either accompany the present labeling of the product or may be incorporated into the present labeling to include all necessary provisions and to mark the product as permitted for sale, distribution, and use until December 31, 1986.

c. Existing stocks not within the possession of the registrant (e.g., possessed by dealers and retailers) nor in the facilities owned or leased by the registrant may be distributed or sold under current labeling until December 31, 1983. Existing stocks obtained by users on or before December 31, 1983, may be used according to the product label.

d. Existing stocks of cancelled toxaphene products may not be distributed or sold in a manner which would violate FIFRA section 12 if the product were still registered. Except as described in 4.c., products will be used only by or under the supervision of certified applicators; the labels will indicate this restriction.

e. After December 1, 1986, all existing stocks may not be distributed, sold, or used; they must be disposed of according to the Resource Conservation and Recovery Act (RCRA) or the product label.

f. Formulations of toxaphene for registered uses must be produced from material designated as existing stocks under the terms and conditions of 47 FR 53784. This requirement shall be in effect until existing stocks are used up or until December 31, 1986, whichever is sooner.

Pesticide/ Use Affected	Action	Reference

5. Label Provisions:

a. The product may be used only to treat scabies on beef cattle or sheep.

b. Application may be made only by dipping exposed or infected beef cattle or sheep in toxaphene vat solutions, or by using a spray dip machine.

c. The products are classified as Restricted Use pesticides (for use only by certified applicators).

d. Vat cleaning and mixing directions are included to assure the safe handling of materials and and the correct percentages of toxaphene in mixtures.

e. Disposal or dispersal to ponds, lakes, streams, and other water bodies is prohibited.

f. Every applicator must wear boots, extended impermeable gloves, a head covering, an apron, long pants, a long-sleeved shirt, and a respirator.

g. Used dip solutions must be disposed of in accordance with the RCRA. If an owner generates more than 1000 kg used dip solution per month or more than 1000 kg used dip solution with other hazardous wastes the material must be treated as a hazardous waste subject to Subpart C of RCRA. Any user who wishes to manage hazardous waste must have a permit to operate a hazardous waste facility. A farmer or rancher who only uses the product to treat sheep and beef cattle owned by family members may be exempted from disposal requirements.

h. Toxaphene products for registered uses must be formulated from existing stocks until existing stocks are used up or until December 31, 1986, whichever is sooner.

i. The product may only be used in the treatment of armyworms, cutworms, or grasshoppers on cotton, corn, or small grain.

j. The sale, distribution, and use of the product are limited to emergency situations. A description of the emergency, including a description of the infestation and proposed treatment, must be presented by a Federal or State agency to EPA for EPA determination of emergency conditions. An agency may anticipate an upcoming emergency and submit the appropriate notification to EPA and obtain advance approval.

k. Flagging must be accomplished by automated mechanical means or by humans in completely enclosed vehicles.

l. Workers who mix, load, transfer, or otherwise handle toxaphene products must wear hats, impermeable gloves, rubber or synthetic boots or boot covers, long-sleeved shirts, and long pants. Full-face respirators are recommended; half-face respirators are required.

m. Instead of wearing protective clothing workers may use closed system methods of mixing, loading, and transferring; mixers and loaders are encouraged to wear waterproof gloves. Aerial applicators in positive pressure cockpits and applicators in comparable ground equipment with appropriate filters at all air intakes are exempted from these clothing requirements.

n. Product labels must include the warning listed in 4.a.5.

o. The product may be used only for mealybug and pineapple gummosis moth control on pineapples and weevil control in bananas.

Pesticide/ Use Affected	Action	Reference
	p. The product must be labeled "For Use Only in Puerto Rico or the Virgin Islands."	
	Note: Existing stocks of cancelled toxaphene products *include any toxaphene formulated for end·use and any technical toxaphene which cannot be diverted for any other use by the registrant or for sale on the open market for any purpose other than manufacture of toxaphene.*	
Trifluralin	Cancellation of all registrations and denial of all applications for registration of trifluralin-containing products *unless* registrants and applicants amend the terms of registration as follows:	47 FR 33777, August 4, 1982
	1. *Amendment to the confidential statement of formula*	
	Registrants must amend the inert ingredients statement in the confidential statement of formula for each technical registered product to read as follows:	
	"Total N-nitrosamine contamination: no greater than 0.5 ppm."	
	Registrations for formulated registered products must be amended to include the following statement in the confidential statement of formula:	
	"Total N-nitrosamine contamination: no greater than (number to be calculated as follows: 0.5 ppm total N-nitrosamine contamination allowed in technical trifluralin x X% technical trifluralin in the end-use product X 2 to allow for possible generation of nitrosamines during formulation)."	
	Distribution or sale of a trifluralin-containing pesticide product whose chemical composition differs from that stated in the confidential statement of formula is prohibited.	
	2. *Quality and records* Registrants of trifluralin-containing pesticides must maintain accurate records of their quality control efforts which will be subject to EPA review and must inform EPA of the quality control procedures which they will adopt to ensure that the N-nitrosamine contamination limit listed in the confidential statement of formula will not be exceeded.	
Vinyl Chloride	Cancelled and suspended, all pesticide products containing this compound, whether an active or inert ingredient, for uses in the home, food handling establishments, hospitals or in enclosed areas.	PR Notice 74-5, April 30, 1974, 40 FR 3494, January 2, 1975

J. Forest Chemicals

With one-half billion acres of commercial forest lands in the United States there is a continuous flow of chemical application to large areas of growing timber. It is estimated that, by and large, as many as 2 million acres receive 1 or more chemical treatments every year. In those applications 3 general groups of chemicals are used, all of which might result in significant water pollution, if application procedures are not carefully controlled. Included in these groups are (1) fertilizers for increased tree growth, (2) herbicides to cull out unwanted species, and (3) insecticides and rodenticides for pest control.

1. Application Methods

Specially equipped airplanes and helicopters are the most prevalently used equipment for applying the chemicals used in the above groupings. They are, also, a source of lake and stream pollution when not carefully controlled.

The ground application of all categories of forest chemicals, where a worker with a tank of spray on a backpack and a spray wand applicator has never involved more than small amounts of chemicals. The chemical applicator can pinpoint his applications and never get near open water. Also, the small quantities that are used do not appear to affect run-off waters and the impact on lakes and streams is less severe than aerial applications.

2. Toxic Chemicals

Quite a few of the forest chemicals are toxic at low concentrations. The minimum dosage necessary to produce a measurable effect, or "threshold dosage," in an organism varies with the organism and the chemical used. The toxicity or hazard of a chemical depends not only on the dosage used but also on the level of exposure, the duration of this exposure, the method of contact with the chemical, and the absolute toxicity of the chemical. The concentration of a toxic chemical has a direct effect on aquatic organisms in relation to the amount absorbed and the period of retention.

3. Fertilizers

The toxic impact of chemical fertilizers generally requires extremely large application quantities. The primary concern, when those chemicals are used, is that surface waters or runoffs or even direct application to lakes and streams will result in heavy aquatic growth of weeds or other aquatic organisms. This undesirable growth could lead to oxygen depletion, fish kills, and other serious problems of an ecological nature.

4. The Effects of Climate and Physical Characteristics

The usual application of forest chemicals within a watershed area is usually fairly well controlled. It is only when nature steps in that there are problems with even the most carefully controlled application procedures. For example:

(a) Soon after application of the chemicals, there are torrential rains that dissolve the water-soluble chemicals and an enormous quantity of the substance is contained in the runoff. If, however, there is slight rainfall, even intermittent rainfall weeks after the application, then very little pollution will result.

(b) The nature of the terrain is important to the effects of pollution. For where the slopes are steep there may be greater runoff; however, it is usually fortunate that ground cover in some areas may retain runoffs, especially if rains are light or intermittent. Where there are steep slopes and light ground cover, then greater runoffs will, of course, occur.

(c) Ground and Surface Water

Forest waters will often test at higher levels of pollution where there are meadows and plains and the local water tables are relatively close to the surface and where streams and swamps are numerous. Often contamination occurs because it is difficult to avoid these areas when one is spraying chemicals from airplanes or helicopters.

(d) Soils and Other Influences

Chemical pesticides that persist for fairly long periods may be washed from vegetation by rains and through the materials from which forest floors are composed. Depending on the type of chemicals that have been used, the substance may be adsorbed onto the surfaces of the smaller soil particles, where a large variety of microorganisms is capable of decomposing many of them. If groundwaters, i.e., the water table, are not close to the surface, high soil porosity can add to the effectiveness of the process. On the other hand, water that moves through highly porous soils rapidly may give little opportunity for the chemicals to become adsorbed or to degrade microbially.

There are a number of other factors that tend to affect chemical pollution in the forests. There is the extent of streamside cover vegetation that may stop some aerial sprays; there is also the water depth of streams, velocity of current, water temperature, quantity of water to dilute the chemicals, and the sensitivity of the aquatic community to the chemicals. The degree of suspended sedimentation and stream or lake bottom characteristics may have important bearing on the degree of pollution that may result from chemical applications.

5. Applying Forest Chemicals

Insecticide applications require greater control for forest safety than do herbicides primarily because acute insecticide toxicity tends to cause direct and immediate injury to aquatic and other nontarget organisms. Also, chronically toxic materials, for instance, the persistent insecticides, require greater safety measures to ensure against long-term effects.

Small streams, i.e., those with less than 10 cubic feet per second of velocity, will clean more rapidly than will a larger stream. A smaller stream will tolerate a given peak level of contamination more readily and without as much major harm as a large stream. Therefore, since large streams clean themselves more slowly than small streams and more environmental damage may be done to them, it is necessary to use greater care in protecting them from chemical exposures.

Certain state forest practice regulations prescribe the isolation of treatment areas from open water. The effectiveness of the buffer strip between open water and the spray area depends on a number of factors. These are application droplet size, basic spray path width, and the degree of precision in operating the aircraft. Therefore, it is necessary to maintain control of the closeness of the application to any lakes and streams, to maintain control over the selection of chemicals to be used in order that the path of application may not cause damage, and to maintain control of the dispersion rate from the application equipment.

Since fine droplets of chemicals may be widely dispersed in winds of only 1 to 10 miles an hour, the applicator may control that overspray in a variety of ways: (1) by reducing the boom pressure; (2) by increasing the orifice size through which the chemical is ejected into the air; (3) by orienting the nozzle into the air stream of the aircraft; (4) by using special boom and nozzle designs; (5) by minimizing straight oil in spray mixtures; and (6) by using spray adjuvants to thicken the spray mixture.

Other considerations for using forest chemicals more judiciously consist of substituting less harmful spraying practices when one is operating close to streams or lakes and providing buffer strips or using hand-spraying methods. It would be well, also, to restrict the use of organiochlorine insecticides and to eliminate picloram and dicamba chemical residues in water used for irrigation. Forest chemical applications should be closely monitored so that recommended dosages and application procedures are being followed in order that water quality may be maintained.

In addition to the above program, all chemical applicators should be duly trained and licensed. Further information on silviculture chemicals and procedures will be found in the Appendix.

13

Heavy Metal Wastes and Other Toxic Pollutants

A. Introduction

Heavy metals, such as lead, mercury, cadmium, zinc, and chromium, are natural components in the environment. They become potentially lethal hazards, however, when their concentrations build up to higher than normal levels. The major sources of heavy-metal pollution are, primarily, the metals-processing industries, the mineral processors, the manufacturers of inorganic products, and the large-scale users of coal, such as the public utilities.

Of the 10,000 tons of mercury that are mined annually, about 5,000 tons are introduced into the environment in the United States. Also mercury poisoning is not unknown in this country and in other countries of the world. The most disastrous incident involving mercury occurred in the 1950s and is known as the Minamata Bay poisonings. This unfortunate occurrence happened when a Japanese plastics manufacturing company kept dumping methyl mercury into the bay as an effluent of its production processing. The fish, of course, ingested the methyl mercury, and since fish are a staple of the Japanese diet, many people of the Minamata Bay communities were afflicted.

Large quantities of cadmium are used in metal plating and in the manufacture of batteries. When the smelting process produces cadmium, some of it is vaporized into the atmosphere. Higher than normal concentrations of the metal from chronic exposure is the probable cause of kidney and liver damage as well as of bone-marrow diseases. In addition, overexposure has been associated with hypertension, and heavy smoking appears to increase the risks of emphysema and other toxic effects. Studies on animals have demonstrated that cadmium can also produce tumors and birth defects.

Lead from gasoline-burning automobiles and factory emissions becomes airborne and a threat to our environment. It has been found that at the present time the average blood lead levels in an urban adult is anywhere from 75 to 100 times the natural lead levels that constitute normalcy. Overexposure to lead, which is particularly dangerous to younger children, may result in anorexia, and severe lead ingesting, such as in the form of paint chips, may result in permanent brain damage.

Chromium is used in electroplating, in photography, and in the manufacture of paint pigments. Acute ingestion of some forms of chromium cause hemorrhages of the gastrointestinal tract. Airborne chromium has been known to cause lung and other respiratory cancers in workers who are constantly exposed to it on the job.

Then there are other toxic pollutants, such as the polycyclic aromatic hydrocarbons. These PAHs are formed by some industrial processes, the burning of gasoline, and incinerating certain materials. Chemical substances in that category are among the most carcinogenic of all compounds. Benzene, which is a common component of gasoline and is widely used in industrial processing operations, has been thought to cause leukemia and disorders of the central nervous system. Also, chromosome damage has been known to occur in workers exposed to very low levels of benzene compounds.

Organic solvents that are used in many industrial processes also are oftentimes toxic and are even combined with other equally toxic substances. One of the problems associated with those substances is that they are sold, stored, and disposed of in 55-gallon steel barrels, which eventually corrode, leak, or spill onto the ground and sometimes find their way into underground water supplies.

In some of the worst examples, those spent solvents may find their way into local water sources directly as effluent from poorly managed manufacturing operations. Such effluvia have been responsible, in large part, for polluting such European waterways as the beautiful Rhine River as well as the Mediterranean Sea. Although such pollution is due not only to effluents, such as organic solvents, the list of pollutants is a lengthy one.

Many aromatic hydrocarbons, such as benzene, toluene, and xylene, together with the organochlorine compounds, such as carbon tetrachloride and trichlorethylene, are used as organic solvents. When the writer was a product engineer for a company, he used trichlorethylene to pressure-test fire extinguisher cylinders at 600 pounds per square inch. The reason for using that solvent was its extreme volatility. In other words, if water were used to test the steel cylinder, it would have to be air-dried before the cylinder could be filled with the powdered fire extinguishant. Since the trichlorethylene was so volitile, it would dry the instant it was voided from the cylinder. That is one of the many industrial uses of organic solvents.

In the organo-halogen group we have compounds of chlorine, bromine, and fluorine, which are widely used as solvents, pesticides, herbicides, and fire retardants. From those basic ingredients many plastics are also made. Some of the compounds have extremely toxic side effects. They are disseminated virtually all over the globe and persist in the environment for long periods of time, even years. Kepone was a pesticide that polluted the James River of Virginia to such an extent that both the river and parts of the Chesapeake Bay were closed to fishing. Polychlorinated biphenyls, discussed in Chapter 10, which were widely used in electrical appliances, were dumped directly into the Hudson River in New York and the Housatonic River in Massachusetts for many years. Fishing is virtually impossible in parts of both of those rivers, and the bottom mud may retain that toxic material for a very long time. So prevalent and pervasive is the contamination of those substances that trace quantities have been found in the milk of human mothers.

One of the most poisonous of the organo-halogen group of compounds is tetrachlorodioxin (TCDD) dioxin, or Agent Orange, as it was known as in Vietnam, or 2,4,5-T. This substance, which was discussed in Chapter 12, was used by the U.S. military to defoliate large areas of the jungle so that it would destroy the enemy cover. It is believed that as a result of such large-scale use of the herbicide many animal deaths occurred and there was a considerable increase in prenatal deaths and deformed births of many Vietnamese children. Some of the surviving U.S. veterans of the Vietnam episode believe that cancers and other illnesses may have been caused by the dioxin to which they had been exposed.

The EPA has since placed a restriction on the production and use of 2,4,5-T, because it was found that there was a high incidence of human miscarriage in areas where herbicidal spraying using the substance was being carried out. To the writer's knowledge, the use of that herbicide is still permitted on certain range lands and in some rice fields.

Arsenic, which is a by-product of the production of lead, copper, and zinc in the smeltering process, is a toxic pollutant of considerable importance and effect. It is also widely known that the large-scale burning of coal, such as in the production of power, leads to many thousands of tons of arsenic being spewed into the environment every year.

Fluorides, which are not as harmless as the toothpaste variety, enter the atmosphere by way of the production of aluminum and inorganic fertilizers. The presence of that substance on forage causes fluorosis, or fluoride poisoning, in cattle.

B. The Clean Water Act

The Federal Water Pollution Control Act Amendments of 1972 established a comprehensive program to "restore and maintain the chemical, physical, and biological integrity of the Nation's waters," Section 101(a). By July 1, 1977, existing industrial waste dischargers were required to achieve "effluent limitations requiring the application of the *best practical* control *technology* currently available" (BPT), Section 301(b)(1)(A), and by July 1, 1983, those dischargers were required to achieve "effluent limitations requiring the application of the *best available technology* economically achievable . . . which will result in reasonable further programs toward the national goal of eliminating the discharge of all pollutants" (BAT), Sec. 301(b)(2)(A). New industrial dischargers were required to comply with Section 306, *new source performance standards* (NSPS), based upon the best available demonstrated technology, and new and existing dischargers to *publicly owned treatment works* (POTWs) were subject to pretreatment standards under Sections 307(b) and (c) of the act.

While the requirements for direct discharges were to be incorporated into the National Pollutant Discharge Elimination System (NPDES) permits issued under Section 402 of the act, pretreatment standards were made enforceable directly against dischargers to POTWs (indirect dischargers).

Although section 402(a)(1) of the 1972 act authorized the setting of requirements for direct dischargers on a case-by-case basis, Congress intended that, for the most part, control requirements would be based on regulations promulgated by the administrator of the EPA. Section 304(b) of the act required the administrator to promulgate regulations providing guidelines for effluent limitations setting forth the degree of effluent reduction attainable through the application of BPT and BAT. Moreover, Sections 304(c) and Sections 304(f), 307(b), and 307(c) required promulgation of regulations for pretreatment standards.

In addition to those regulations for designated industry categories, Section 307(a) of the act required the administrator to promulgate effluent standards applicable to all dischargers of toxic pollutants. Finally, Section 501(a) of the act authorized the administrator to prescribe any additional regulations "necessary to carry out his functions" under the act.

The EPA was unable to promulgate many of the regulations by the dates contained in the act. In 1976, the EPA was sued by several environmental groups, and in settlement of this lawsuit the agency and the plaintiffs executed a "Settlement Agreement" that was approved by the court. It required the EPA to develop a program and adhere to a schedule for promulgating for 21 major industries BAT effluent limitations guidelines, pretreatment standards, and new source performance standards for 65 "priority" pollutants and classes of pollutants. See *Natural Resources Defense Council, Inc. et al. v. Train,* 8 ERC 2120 (D.D.C. 1976), modified March 9, 1979.

On December 27, 1977, the President signed into law the Clean Water Act of 1977. Although that law makes several important changes in the Federal water pollution control program, its most significant feature is its incorporation into the act of several of the basic elements of the Settlement Agreement program for toxic pollution control. Sections 301(b)(2)(A) and 301(b)(2)(C) of the act required the achievement by July 1, 1984, of effluent limitations requiring application of BAT for "toxic" pollutants, including the 65 "priority" pollutants and classes of pollutants that Congress declared "toxic" under Section 307(a) of the act. Likewise, the EPA's programs for new source performance standards and pretreatment standards are now aimed principally at toxic pollutant controls. Moreover, to strengthen the toxics control program, Section 304(e) of the act authorizes the administrator to prescribe "best management practices" (BMP's) to prevent the release of toxic and hazardous pollutants from plant site runoff, spillage or leaks, sludge or waste disposal, and drainage from raw material storage associated with, or ancillary to, the manufacturing or treatment process.

In keeping with the emphasis on toxic pollutants, the Clean Water Act of 1977 also revises the control program for nontoxic pollutants. Instead of BAT for "conventional" pollutants identified under Section 304(a)(4) (including biochemical oxygen demand, suspended solids, fecal coliform, and pH), the new Section 301(b)(2)(F) required achievement by July 1, 1984, of "effluent limitations requiring the application of the best conventional pollutant control technology" (BCT). The factors considered in assessing BCT for an industry include the costs of attaining a reduction in effluents and the effluent reduction benefits derived, compared with the costs and effluent reduction benefits from the discharge of publicly owned treatment works, Section 304(b)(4)(B). For nontoxic, nonconventional pollutants, Sections 301(b)(2)(A) and (b)(2)(F) required achievement of BAT effluent limitations within 3 years after their establishment or July 1, 1984, whichever was later, but not later than July 1, 1987.

C. Development of Industry Guidelines

The metal finishing industry encompasses 46 unit operations involved in the machining, fabrication, and finishing of products primarily associated with Standard Industrial Classification (SIC) groups 34 through 39 (see Table 13-1). The effluent guidelines for the metal finishing industry were developed from data obtained from previous EPA studies, literature searches, plant surveys and evaluations, and long-term self-monitoring data supplied by industry. Initially, all existing information from EPA records and data from literature searches were collected. That information was then compiled in a format that summarized the

Table 13-1. Industries Within the Metal Finishing Industry

Major Group 34 Fabricated Metal Products Except Machinery and Transportation Equipment

341 Metal Cans and Shipping Containers.
342 Cutlery, Hand Tools, and General Hardware.
343 Heating Equipment (except Electric and Warm Air, Plumbing Fixtures).
344 Fabricated Structural Metal Products.
345 Screw Machine Products and Bolts, Nuts, Screws, Rivets, and Washers.
346 Metal Forgings and Stampings.
347 Coating, Engraving, and Allied Services.
348 Ordnance and Accessories, except Vehicles and Guided Missiles.
349 Miscellaneous Fabricated Metal Products.

Major Group 35 Machinery, Except Electrical

351 Engines and Turbines.
352 Farm and Garden Machinery and Equipment.
353 Construction, Mining and Materials Handling Machinery and Equipment.
354 Metalworking Machinery and Equipment.
355 Special Industry Machinery, except Metalworking Machinery.
356 General Industrial Machinery and Equipment.
357 Office, Computing, and Accounting Machines.
358 Refrigeration and Service Industry Machinery.
359 Miscellaneous Machinery, except Electrical.

Major Group 36 Electrical and Electronic Machinery, Equipment and Supplies

361 Electric Transmission and Distribution Equipment.
362 Electrical Industrial Apparatus.
363 Household Appliances.
364 Electric Lighting and Wiring Equipment.
365 Radio and Television Receiving Equipment, except Communication Types.
366 Communication Equipment.
367 Electronic Components and Accessories.
369 Miscellaneous Electrical Machinery, Equipment, and Supplies.

Major Group 37 Transportation Equipment

371 Motor Vehicles and Motor Vehicle Equipment.
372 Aircraft and Parts.
373 Ship and Boat Building and Repairing.
374 Railroad Equipment.
375 Motorcycles, Bicycles, and Parts.
376 Guided Missiles and Space Vehicles and Parts.
379 Miscellaneous Transportation Equipment.

Major Group 38 Measuring, Analyzing, and Controlling Instruments, Photographic, Medical, and Optical Goods; Watches and Clocks

381 Engineering, Laboratory, Scientific, and Research Instruments and Associated Equipment.
382 Measuring and Controlling Instruments.
383 Optical Instruments and Lenses.
384 Surgical, Medical, and Dental Instruments and Supplies.
385 Opthalmic Goods.
386 Photographic Equipment and Supplies.
387 Watches, Clocks, Clockwork-Operated Devices, and Parts.

Major Group 39 Miscellaneous Manufacturing Industries

391 Jewelry, Silverware, and Plated Ware.
393 Musical Instruments.
394 Dolls.
395 Pens, Pencils, and Other Office and Artists' Materials
396 Costume Jewelry, Costume Novelties, Buttons and Miscellaneous Notions, Except Precious Metal.
399 Miscellaneous Manufacturing Industries.

individual plant descriptions from the following information: manufacturing unit operations performed, water usage, process water discharges, waste water treatment practices, and waste water constituents.

In addition to providing a quantitative description of the metal finishing industry, the existing information was used to determine if the waste water characteristics of the industry as a whole were uniform and thus amenable to 1 set of discharge standards. The discharge characteristics of all plants in the existing data base were not uniform; however, it was found that the discharge from those plants lent itself to a common end-of-pipe treatment technology. Therefore, the entire metal finishing industry is represented by a single subcategory and is subject to 1 set of effluent discharge limitations. Seven classifications of raw waste are present and were studied to establish treatment requirements. These 7 waste types are:

1. Common Metals
2. Precious Metals
3. Complexed Metals
4. Hexavalent Chromium
5. Cyanide
6. Oils
7. Toxic Organics

To supplement existing data, data collection portfolios (DCPs) under the authority of Section 308 of the Federal Water Pollution Control Act, as amended, were transmitted by the EPA to a large number of manufacturing facilities in the metal finishing industry. In addition to the existing data base and the plant-supplied information, in accordance with the completed DCPs, a sampling program was conducted at selected plant locations. It was used to establish the sources and quantities of pollutant parameters in the raw process waste water and the treated effluent. The sites visited were chosen on the basis of the specific manufacturing operations performed or the particular waste treatment technology employed. The EPA requested historical effluent information in the form of long-term self-monitoring data, and nearly 100 plants responded. All of the data collected were analyzed to correlate the pollutants generated with the manufacturing processes performed by each facility.

In addition to evaluating pollutant constituents and discharge rates, the full range of control and treatment technologies within the metal finishing industry was identified and examined. That was done by considering the pollutants to be treated and their chemical, physical, and biological characteristics. Special attention was paid to in-process technology, such as the recovery and reuse of process solutions, the recycling of process water, and the reduction of water use.

The information was then evaluated in order to determine the levels of technology appropriate as bases for effluent limitations for existing sources after July 1, 1977 ("Best Practicable Control Technology Currently Available"), and after July 1, 1984 ("Best Available Technology Economically Achievable"). Levels of technology appropriate for direct discharge and pretreatment of waste water POTWs from both new and existing sources were also identified, as were the demonstrated control technology, processes, operating methods, or other alternatives. In the evaluation of those technologies various factors were considered. They included demonstrated effluent performance of treatment technologies, the total cost of application of the technology in relation to the pollution reduction benefits to be achieved, the production processes employed, the engineering aspects of the application of various types of control techniques and process changes, and nonwater quality environmental impact (including energy requirements).

There are approximately 13,500 manufacturing facilities in the United States that are covered by the metal finishing industry. Those plants are engaged in the manufacture of a variety of products that are constructed primarily by using metals. The operations performed (see Table 13-2) usually begin with materials in the form of raw stock—as, for example, rods, bars, sheet, castings, forgings, etc.—and can progress to the most sophisticated surface-finishing operations. The functions are performed in plants that vary greatly in size, age, number of employees, and number and type of operations performed. They range from very small job shops with less than 10 employees to large factories employing thousands of production workers. Because of the differences in size and processes, production facilities are custom-tailored to the specific needs of individual places. Fig. 13-1 illustrates the variation in number of unit operations that can be performed depending upon the complexity of the product. The possible operational variations within the metal finishing industry are extensive.

The process (and their sequence) presented in Fig. 13-1 are not actual plants but are representative of possible manufacturers within the metal finishing industry. Some complex products could require the use of nearly all 46 unit operations, while a simple product might require only a single one.

Many different raw materials are used by the plants in the metal finishing category. Basic materials are almost exclusively metals, which range from common copper and steel to extremely expensive high-grade alloys and precious metals. The chemical solutions utilized in the various unit operations can contain acids, bases, cyanide, metals, complexing agents, organic additives, oils, and

Table 13-2.   Metal Finishing Operations.

(1)	Electroplating	(24)	Electrochemical machining
(2)	Electroless plating	(25)	Electron beam machining
(3)	Anodizing	(26)	Laser beam machining
(4)	Conversion coating	(27)	Plasma arc machining
(5)	Etching (chemical milling)	(28)	Ultrasound machining
(6)	Cleaning	(29)	Sintering
(7)	Machining	(30)	Laminating
(8)	Grinding	(31)	Hot dip coating
(9)	Polishing	(32)	Sputtering
(10)	Tumbling (barrel finishing)	(33)	Vapor plating
(11)	Burnishing	(34)	Thermal infusion
(12)	Impact deformation	(35)	Salt bath descaling
(13)	Pressure deformation	(36)	Solvent degreasing
(14)	Shearing	(37)	Paint stripping
(15)	Heat treating	(38)	Painting
(16)	Thermal cutting	(39)	Electrostatic painting
(17)	Welding	(40)	Electropainting
(18)	Brazing	(41)	Vacuum metalizing
(19)	Soldering	(42)	Assembly
(20)	Flame spraying	(43)	Calibration
(21)	Sand blasting	(44)	Testing
(22)	Other abrasive jet machining	(45)	Mechanical plating
(23)	Electric discharge machining	(46)	Printed circuit board manufacturing

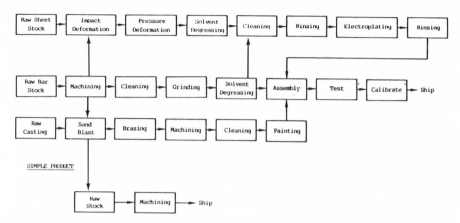

Fig. 13-1 Metal finishing processes. Source: Effluent Guidelines Division, EPA 440/1-83/091, June 1983.

detergents. All of those raw materials can potentially enter waste water streams during the production sequence.

Plating and cleaning operations are typically the biggest water users. While the majority of metal-finishing operations use water, some of them are completely dry. The type of rinsing utilized can have a marked effect on water usage, as can the flow rates within the particular rinse types. Product quality requirements often dictate the amount of rinsing needed for specific parts. Parts requiring extensive surface preparation will generally necessitate the use of larger amounts of water.

D. Descriptions of Metal-Finishing Operations

The list of operations that are performed in the metal-finishing industry have been tabulated, as indicated above, in Table 13-2. This section describes the operations in brief and will serve to give the reader at least a nodding acquaintance with the many processes that are the backbone of industry as practiced not only in the United States but also in many countries all over the world. Unfortunately, while these industrial processes contribute in large measure to a higher living standard, their negative aspects, i.e., their pollutant effects, are only now receiving serious attention from the major powers and the lesser countries.

(1) *Electroplating* is the production of a thin surface coating of 1 metal upon another by electrodeposition. The surface coating is applied to provide corrosion protection, wear- or erosion-resistance, antifrictional characteristics or for decorative purposes. The electroplating of common metals includes the processes in which ferrous or nonferrous basis material is electroplated with copper, nickel, chromium, brass, bronze, zinc, tin, lead, cadmium, iron, aluminum, or in combinations of those metals. Precious metals electroplating includes the processes in which a ferrous or nonferrous basis material is plated with gold, silver,

palladium, platinum, rhodium, indium, ruthenium, iridium, osmium, or combinations thereof.

In electroplating, metal ions in either acid, alkaline, or neutral solutions are reduced on cathodic surfaces. Those surfaces are the workpieces being plated. The metal ions in solution are usually replenished by the dissolution of metal from anodes or small pieces contained in inert wire or metal baskets. Replenishment with metal salts is also practiced, especially for chromium plating. In that method an inert material must be selected for the anodes. Hundreds of different electroplating solutions have been adopted commercially, but only 2 or 3 types are utilized widely for a particular metal or alloy. For example, cyanide solutions are popular for copper, zinc, brass, cadmium, silver, and gold. Noncyanide alkaline solutions containing pyrophosphate, however, have come into use recently for zinc and copper. Zinc, copper, tin, and nickel are plated with acid sulfate solutions, especially for plating relatively simple shapes. Cadmium and zinc are sometimes electroplated from neutral or slightly acidic chloride solutions. The most common methods of plating are in barrels, on racks, and continuously from a spool or coil.

(2) *Electroless Plating* is a chemical reduction process dependent upon the catalytic reduction of a metallic ion in an aqueous solution containing a reducing agent and upon the subsequent deposition of metal without the use of external energy. It has found widespread use in industry owing to several unique advantages it has over conventional electroplating. Electroless plating provides a uniform plating thickness on all areas of the part regardless of the configuration of the part. An electroless plate on a properly prepared surface is dense and virtually nonporous. Copper and nickel electroless plating is the most common. The basic ingredients in an electroless plating solution are as follows:

1. A source of metal, usually a salt;
2. A reducer to reduce the metal to its base states;
3. A complexing agent to hold the metal in solution so that the metal will not plate out indiscriminately; and,
4. Various buffers and other chemicals designed to maintain bath stability and increase the bath life.

Electroless plating is an autocatalytic process where catalysis is promoted from 1 of the products of a chemical reaction. The chemistry of electroless plating is best demonstrated by examining electroless nickel plating. The source of nickel is a salt, such as nickel chloride or nickel sulfate, and the reducer is sodium hypophosphite. There are several complexing agents that can be used, the most common of which is citric and glycolic acid. Hypophosphite anions in the presence of water are dehydrogenated by the solid catalytic surface provided by nickel to form acid orthophosphite anions. Active hydrogen atoms are bonded on the catalyst forming a hydride. Nickel ions are reduced to metallic nickel by the active hydrogen atoms that are in turn oxidized to hydrogen ions. Simultaneously, a portion of the hypophosphite anions are reduced by the active hydrogen and adsorbed on the catalytic surface producing elemental phosphorus, water, and hydroxyl ions. Elemental phosphorus is bonded to or dissolved in the nickel, making the reaction irreversible. At the same time hypophosphite anions are catalytically oxidized to acid orthophosphite anions, evolving gaseous hydrogen. The basic plating reactions proceed as follows:

The nickel salt is ionized in water

$$NiSO_4 = Ni^{+2} + SO_4^{-2}$$

There is then a reduction-oxidation reaction with nickel and sodium hypophosphite.

$$Ni^{+2} + SO_4^{-2} + 2NaH_2PO_2 + 2 H_2O = Ni + 2NaH_2PO_3 + H_2 + H_2SO_4$$

The sodium hypophosphite also reacts in the following manner:

$$2NaH_2PO_2 + H_2 = 2P + 2NaOH + 2H_2O$$

As can be seen in the equations above, both nickel and phosphorus are produced, and the actual metal deposited is a nickel-phosphorus alloy. The phosphorus content can be varied to produce different characteristics in the nickel plate.

When electroless plating is done on a plastic basis material, catalyst application and acceleration steps are necessary as surface preparation operations. Those steps are considered part of the electroless plating unit operation.

Immersion plating is a chemical plating process in which a thin metal deposit is obtained by chemical displacement of the basis metal. Unlike electroless plating, it is not an autocatalytic process. In immersion plating a metal will displace from solution any other metal that is below it in the electromotive series of elements.

The lower, more noble metal will be deposited from solution while the more active metal, higher in the electromotive series, will be dissolved. A common example of immersion plating is the deposition of copper on steel from an acid copper solution. Because of the similarity of the wastes produced and the materials involved, immersion plating is considered part of the electroless plating unit operation.

(3) *Anodizing* is an electrolytic oxidation process that converts the surface of the metal to an insoluble oxide. Those oxide coatings provide corrosion protection, decorative surfaces, a base for painting and other coating processes, and special electrical and mechanical properties. Aluminum is the most frequently anodized material, while some magnesium and limited amounts of zinc and titanium are also treated.

Although the majority of anodizing is carried out by immersion of racked parts in tanks, continuous anodizing is done on large coils of aluminum in a manner similar to continuous electroplating. For aluminum parts, the formation of the oxide course when the segments are made anodic in dilute sulfuric acid or dilute chromic acid solutions. The oxide layer begins formation at the extreme outer surface, and as the reaction proceeds, the oxide grows into the metal. The last-formed oxide, known as the boundary layer, is located at the interface between the base metal and the oxide. The border is extremely thin and nonporous. The sulfuric acid process is typically used for all parts fabricated from aluminum alloys except for those subject to stress or containing recesses in which the sulfuric acid solution may be retained and attack the aluminum. Chromic acid anodic coatings are more protective than sulfuric acid coatings and have a relatively thick boundary layer. For those reasons, a chromic acid bath is used if a complete rinsing of the part cannot be achieved.

(4) *Coating or Conversion Coating*—This manufacturing operation includes chromating, phosphating, metal coloring, and passivating (see below). The coatings are applied to previously deposited metal or basis material for increased corrosion

protection, lubricity, preparation of the surface for additional coatings, or formulation of a special surface appearance. In chromating a portion of the base metal is converted to 1 of the components of the protective film formed by the coating solution. That occurs by reaction with aqueous solutions containing hexavalent chromium and active organic or inorganic compounds. Chromate coatings are most frequently applied to zinc, cadmium, aluminum, magnesium, copper, brass, bronze, and silver. Most of the coatings are applied by chemical immersion, although a spray or brush treatment can be used. Changes in the solutions can impart a wide range of colors to the coatings from colorless to iridescent yellow, brass, brown, and olive drab. Additional coloring of the coatings can be achieved by dipping the parts in organic dye baths to produce red, green, blue, and other colors.

Phosphate coatings are used to provide a good base for paints and other organic coatings, to condition the surfaces for cold-forming operations by providing a base for drawing compounds and lubricants, and to impart corrosion resistance to the metal surface by the coating itself or by providing a suitable base for rust-preventive oils or waxes. Phosphate conversion coatings are formed by the immersion of iron, steel, or zinc-plated steel in a dilute solution of phosphoric acid plus other reagents. The method of applying the phosphate coating is dependent upon the size and shape of the part to be coated. Small parts are coated in barrels immersed in the phosphating solution. Large parts, such as steel sheet and strip, are spray coated or continuously passed through the phosphating solution. Supplemental oil or wax coatings are usually applied after phosphating unless the part is to be painted.

Metal coloring by chemical conversion methods produces a large group of decorative finishes. That operation covers only chemical methods of coloring in which the metal surface is converted into an oxide or similar metallic compound. The most common colored finishes are used on copper, steel, zinc, and cadmium.

Application of the color to the cleaned basis metal involves only a brief immersion in a dilute aqueous solution. The colored films produced on the metal surface are extremely thin and delicate. Consequently, they lack resistance to handling and the atmosphere. A clear lacquer is often used to protect the colored metal surface. A large quantity of copper and brass is colored to yield a wide variety of shades and colors. Shades of black, brown, gray, green, and patina can be obtained on copper and brass by use of appropriate coloring solutions. The most widely used colors for ferrous metals are based on oxides that yield black, brown, or blue colors. A number of colors can be developed on zinc depending on the length of immersion in the coloring solution. Yellow, bronze, dark green, black, and brown colors can be produced on cadmium. Silver, tin, and aluminum are also colored commercially. Silver is given a gray color by immersion in a polysulfide solution, such as ammonium polysulfide. Tin can be darkened to produce an antique finish of pewter by immersion in a solution of nitric acid and copper sulfate.

Passivation refers to forming a protective film on metals, particularly stainless steel and copper, by immersion in an acid solution. Stainless steel is passivated in order to dissolve any imbedded iron particles and to form a thin oxide film on the surface of the metal. Typical solutions for passivating stainless steel include nitric acid and nitric acid with sodium dichromate. Copper is passivated with a solution of ammonium sulfate and copper sulfate forming a blue-green patina on the surface of the metal.

(5) *Etching and Chemical Milling*—Those processes are used to produce specific

design configurations and tolerances or surface appearances on parts or metal-cladding plastic in the case of printed circuit boards by the controlled dissolution of the surface using chemical reagents or etchants. Included in that classification are the processes of chemical milling, chemical etching, and bright dipping. Chemical etching is the same process as chemical milling except that the rates and depths of metal removal are usually much greater in the latter. Typical solutions for chemical milling and etching include ferric chloride, nitric acid, ammonium persulfate, chromic acid, cupric chloride, hydrochloric acid, and combinations of those reagents. Bright dipping is a specialized form of etching and is used to remove oxide and tarnish from ferrous and nonferrous materials and is frequently performed just before anodizing. Bright dipping can produce a range of surface appearances from bright clean to brilliant depending on the surface smoothness desired for the finished part. Bright dipping solutions usually involve mixtures of 2 or more of sulfuric, chromic, phosphoric, nitric, and hydrochloric acids. Also included in the processing operation is the stripping of metallic coatings.

(6) *Cleaning* involves the removal of oil, grease, and dirt from the surface of the basis material using water with or without a detergent or other dispersing agent. Alkaline cleaning (both electrolytic and nonelectrolytic) and acid cleaning are both included.

Alkaline cleaning is used to remove oily dirt or solid soils from workpieces. The detergent nature of the cleaning solution provides most of the cleaning action with agitation of the solution and movement of the workpiece being of secondary importance. Alkaline cleaners are classified into 3 types: soak, spray, and electrolytic. Soak cleaners are used on easily removed soil. Such a cleaner is less efficient than spray or electrolytic cleaners. Spray cleaners combine the detergent properties of the solution with the impact force of the spray, which mechanically loosens the soil. Electrolytic cleaning produces the cleanest surface available from conventional methods of alkaline cleaning. The effectiveness of that method results from the strong agitation of the solution, by gas evolution, and oxidation-reduction reactions that occur during electrolysis. Also, certain dirt particles become electrically charged and are repelled from the surface. Direct current (cathodic) cleaning uses the workpiece as the cathode, while for reverse current (anodic) cleaning the workpiece is the anode. In periodic reverse current cleaning, the current is periodically reversed from direct current to reverse current. Periodic reverse cleaning gives improved smut removal, accelerated cleaning and a more active surface for any subsequent surface-finishing operation.

Acidic cleaning is a process in which a solution of an inorganic (mineral) acid, organic acid, or an acid salt, in combination with a wetting agent or detergent, is employed to remove oil, dirt, or oxide from metal surfaces. Acidic cleaning is done with various acid concentrations and can be referred to as pickling, acid dipping, descaling, or desmutting. The solution may or may not be heated and can be an immersion or spray operation. Agitation is normally required with soaking, and spray is usually used with complex shapes. An acid dip operation may also follow alkaline cleaning before plating. Phosphoric acid mixtures are in common use to remove oils and light rust while leaving a phosphate coating that provides a paint base or temporary resistance to rusting. Strong acidic solutions are used to remove rust and scale before surface finishing.

(7) *Machining* is the general process of removing stock from a workpiece by forcing a cutting tool through the workpiece, thereby removing a chip of basis

material. Machining operations, such as turning, milling, drilling, boring, tapping, planing, broaching, sawing and cutoff, shaving, threading, reaming, shaping, slotting, hobbing, filing, and chamfering, are included in that definition.

(8) *Grinding* is the process of removing stock from a workpiece by the use of a tool consisting of abrasive grains held by a rigid or semirigid binder. The tool is usually in the form of a disk, for example, a grinding wheel, but may also be in the form of a cylinder, ring, cup, stick, strip, or belt. The most commonly used abrasiveness are aluminum oxide, silicon carbide, and diamond. The processes included in this unit operation are sanding or cleaning to remove rough edges or excess material, surface finishing, and separating—as in cutoff or slicing operations.

(9) *Polishing* is an abrading operation used to remove or smooth out surface defects, scratches, pits, tool marks, and so forth, that adversely affect the appearance or function of a part. Polishing is usually performed with either a belt or a wheel to which an abrasive, such as aluminum oxide or silicone carbide, is bonded. Both wheels and belts are flexible and will conform to irregular or rounded areas where necessary. The operation usually referred to as "buffing" is included in the polishing operation.

(10) *Tumbling, or Barrel Finishing,* is a controlled method of processing parts to remove burrs, scale, flash, and oxides as well as to improve surface finish. Widely used as a finishing operation for many parts, it obtains a uniformity of surface finish not possible by hand finishing. For large quantities of small parts it is generally the most economical method of cleaning and surface conditioning.

Parts to be finished are placed in a rotating barrel or vibrating unit with an abrasive media, water or oil, and usually some chemical compound to assist in the operation. As the barrel rotates slowly, the upper layer of the work is given a sliding movement toward the lower side of the barrel, causing the abrading or polishing action to occur. The same results may also be accomplished in a vibrating unit, in which the entire contents of the container are in constant motion.

(11) *Burnishing* is the process of finish sizing or smooth-finishing a workpiece, which has been previously machined or ground, by displacement, rather than by removal, of minute surface irregularities. It is accomplished with a smooth point or line-contact and fixed or rotating tools.

(12) *Impact Deformation* is the process of applying an impact force to a workpiece so that it is permanently deformed or shaped. Impact deformation operations include shot peening, peening, forging, high energy forming, heading, and stamping.

(13) *Pressure Deformation* is the process of applying force, at a slower rate than an impact force, to permanently deform or shape a workpiece. Pressure deformation includes such operations as rolling, drawing, bending, embossing, coining, swaging, sizing, extruding, squeezing, spinning, seaming, staking, piercing, necking, reducing, forming, crimping, coiling, twisting, winding, flaring, or weaving.

(14) *Shearing* is the process of severing or cutting a workpiece by forcing a sharp edge or opposed sharp edges into the workpiece, thereby stressing the material to the point of shear failure and separation.

(15) *Heat Treating* is the modification of the physical properties of a workpiece through the application of controlled heating and cooling cycles. Such operations as tempering, carburizing, cyaniding, nitriding, annealing, normalizing, austenizing, quenching, austempering, siliconizing, martempering, and malleabilizing are included in this definition.

(16) *Thermal Cutting* is the process of cutting, slotting, or piercing a workpiece using an oxyacetylene oxygen lance or electric-arc cutting tool.

(17) *Welding* is the process of joining 2 or more pieces of material by applying heat, pressure, or both, with or without filler material, to produce a localized union through fusion or recrystallization across the interface. Included in the process are gas welding, resistance welding, arc welding, cold welding, electron beam welding, and laser beam welding.

(18) *Brazing* is the process of joining metals by flowing a thin, capillary thickness layer of nonferrous filler metal into the space between them. Bonding results from the intimate contact produced by the dissolution of a small amount of base metal in the molten filler metal, without fusion of the base metal. The term brazing is used where the temperature exceeds 425°C (800°F).

(19) *Soldering* is the process of joining metals by flowing a thin, capillary-thickness layer of nonferrous filler metal into the space between them. Bonding results from the intimate contact produced by the dissolution of a small amount of base metal in the molten filler metal, without fusion of the base metal. The term soldering is used where the temperature range falls below 425°C (800°F).

(20) *Flame spraying* is the process of applying a metallic coating to a workpiece using finely powdered fragments of wire, together with suitable fluxes; the fragments are projected through a cone of flame onto the workpiece.

(21) *Sand Blasting* is the process of removing stock, including surface films, from a workpiece by the use of abrasive grains that are blown against the workpiece at high pressures. The abrasive grains include sand, metal shot, slag, silica, pumice, or natural materials, such as walnut shells.

(22) *Abrasive Jet Machining* is a mechanical process for cutting hard, brittle materials. It is similar to sand blasting but uses much finer abrasives carried at high velocities (500 fps to 3000 fps) by a liquid or gas stream. Its uses include frosting glass, removing metal oxides, deburring, and drilling and cutting thin sections of metal.

(23) *Electrical Discharge Machining* is a process that can remove metal, using good dimensional control, from any metal. It cannot be used for machining glass, ceramics, or other nonconducting materials. The machining action is caused by the formation of an electrical spark between an electrode, shaped to the required contour, and the workpiece. Since the cutting tool has no contact with the workpiece, it can be made from a soft, easily worked material, such as brass. The tool works in conjunction with a fluid such as mineral oil or kerosene, which is fed to the work under pressure. The function of that coolant is to serve as a dielectric, to wash away particles of eroded metal from the workpiece or tool, and to maintain a uniform resistance to the flow of current.

Electrical discharge machining is also known as spark machining or electronic erosion. The operation was developed primarily for machining carbides, hard nonferrous alloys, and other hard-to-machine materials.

(24) *Electrochemical Machining* is a process based on the same principles used in electroplating except the workpiece is the anode and the tool is the cathode. Electrolyte is pumped between the electrodes and as an electric current is applied, metal is rapidly removed from the part (or workpiece).

In the process electrode accuracy is important since the surface finish of the electrode tool will be reproduced in the surface of the workpiece. While copper is frequently used as the electrode, brass, graphics, and copper-tungsten are also used. The tool must be an electrical conductor, easy to machine, corrosion-resistant, and able to conduct the quantity of current needed. Although there is no standard electrolyte, sodium chloride is more generally used than other electrolytes.

(25) *Electron Beam Machining* is a thermoelectric process. In electron beam machining heat is generated by high-velocity electrons impinging on part of the workpiece. At the point where the energy of the electrons is focused, it is transformed into sufficient thermal energy to vaporize the material locally. The process is generally carried out in a vacuum. While the metal-removal rate of electron beam machining is approximately 0.01 milligrams per second, the tool is accurate and is especially adapted for micro-machining. There is no heat-affected zone or pressure on the workpiece, and extremely close tolerances can be maintained. The process results in X-ray emission, which requires that the work area be shielded to absorb radiation. At present the process is used for drilling holes as small as 0.0508 mm (0.002 in.) in any known material, cutting slots, shaping small parts, and machining sapphire-jewel bearings.

(26) *Laser Beam Machining* is the process whereby a highly focused monochromatic collimated beam of light is used to remove material at the point of impingement on a workpiece. Laser beam machining is a thermoelectric process, and material removal is largely accomplished by evaporation, although some material is removed in the liquid state at high velocity. Since the metal removal rate is very small, it is used for such jobs as drilling microscopic holes in carbides or diamond, wire-drawing dies and for removing metal in the balancing of high-speed rotating machinery.

Lasers can vaporize any known material. They have small heat-affected zones and work easily with nonmetallic, hard materials.

(27) *Plasma Arc Machining* is the process of material removal or shaping of a workpiece by a high-velocity jet of high temperature ionized gas. A gas (nitrogen, argon, or hydrogen) is passed through an electric arc, causing it to become ionized and raised to temperatures in excess of 16,649° C (30,000° F). The relatively narrow plasma jet melts and displaces the workpiece material in its path. Because plasma machining does not depend on a chemical reaction between the gas and the work material and because plasma temperatures are extremely high, the process can be used on almost all metals, including those that are resistant to oxygen-fuel gas cutting. The method is of commercial importance mainly for profile cutting of stainless steel and aluminum alloys.

(28) *Ultrasonic Machining* is a mechanical process designed to effectively machine hard, brittle materials. It removes particles by the use of abrasive grains, which are carried in the work surface at high velocity. This action gradually chips away minute pieces of material in a pattern controlled by the tool shape and contour. A transducer causes an attached tool to oscillate linearly at a frequency of 20,000 to 30,000 times per second at an amplitude of 0.0254 mm to 0.127 mm (0.001 in to 0.005 in). The tool motion is produced by being part of a sound wave energy transmission line that causes the tool material to change its normal length by contraction and expansion. The tool holder is threaded to the transducer and oscillates linearly at ultrasonic frequencies, thus driving the grit particles into the workpiece. The cutting particles, boron carbide and similar materials, are of a 280-mesh size or finer, depending upon the accuracy and the finish desired. Operations that can be performed include drilling, tapping, coining, and the making of openings in all types of dies. Ultrasonic machining is used principally for machining materials such as carbides, tool steels, ceramics, glass, gemstones, and synthetic crystals.

(29) *Sintering* is the process of forming a mechanical part from a powdered

metal by fusing the particles together under pressure and heat. The temperature is maintained below the melting point of the basis metal.

(30) *Laminating* is the process of adhesive bonding layers of metal, plastic, or wood to form a part.

(31) *Hot Dip Coating* is the process of coating a metallic workpiece with another metal by immersion in a molten bath to provide a protective film. Galvanizing (hot dip zinc) is the most common hot dip coating.

(32) *Sputtering* is the process of covering a metallic or nonmetallic workpiece with thin films of metal. The surface to be coated is bombarded with positive ions in a gas discharge tube, which is evacuated to a low pressure.

(33) *Vapor Plating* is the process of decomposition of a volatile compound at a temperature below the melting point of either the deposit or the basis material.

(34) *Thermal Infusion* is the process of applying a fused zinc, cadmium, or other metal coating to a ferrous workpiece by placing metal powder or dust on the surface of the workpiece in the presence of heat.

(35) *Salt Bath Descaling* is the process of removing surface oxides or scale from a workpiece by immersing the workpiece in a molten salt bath or a hot salt solution. Molten salt baths are used in a salt bath—water quench—acid dip sequence to clean hard-to-remove oxides from stainless steels and other corrosion-resistant alloys. The work is immersed in a molten salt, with a temperature range from 400° C to 540° C, quenched with water, and then dipped in acid. Oxidizing, reducing, and electrolytic baths are available, and the particular type needed is dependent on the oxide to be removed.

(36) *Solvent Degreasing* is a process for removing oils and grease from the surfaces of a workpiece by the use of organic solvents, such as aliphatic petroleums, i.e., kerosene or naptha aromatics, i.e., benzene or toluene; oxygenated hydrocarbons, i.e., ketones, alcohol, or ether, halogenated hydrocarbons, i.e., 1,1,1-trichloroethane, trichloroethylene, or methylene chloride, and combinations of those classes of solvents. Solvent cleaning can be accomplished by either the liquid or vapor phase. Solvent vapor degreasing is normally quicker than solvent liquid degreasing. Sometimes, however, ultrasonic vibration is used with a liquid solvent to decrease the required immersion time with complex shapes. Solvent cleaning is often used as a precleaning operation, such as before the alkaline cleaning that precedes plating, as a final cleaning of precision parts, or as a surface preparation for some painting operations.

Emulsion cleaning is a type of solvent degreasing that uses common organic solvents, for example, kerosene, mineral oil, glycols, and benzene, dispersed in an aqueous medium with the aid of an emulsifying agent. Depending on the solvent used, cleaning is done at temperatures from room temperature to 82° C (180° F). This operation uses less chemical than solvent degreasing because of the lower solvent concentration employed. The process is used for rapid superficial cleaning and is usually performed as emulsion spray cleaning.

(37) *Paint Stripping* is the process of removing an organic coating from a workpiece. The stripping of such coatings is usually performed with caustic, acid, solvent, or molten salt.

(38) *Painting* is the process of applying an organic coating to a workpiece. The application of coatings such as paint, varnish, lacquer, shellac, and plastics by processes such as spraying, dipping, brushing, roll coating, lithographing, and wiping are included. Spray painting is by far the most common and can be used with nearly all varieties of paint. It can be sprayed manually or automatically, hot or cold, and it may be atomized with or without compressed air to force the

paint through an orifice. Other processes included under this unit operation are printing, silk screening, and stenciling.

(39) *Electrostatic Painting* is the application of electrostatically charged paint particles to an oppositely charged workpiece followed by thermal fusing of the paint particles to form a cohesive paint film. Usually the paint is applied in spray form and may be applied manually or automatically, hot or cold, and with or without compressed air atomization. Both waterborne and solvent-borne coatings can be sprayed electrostatically.

(40) *Electropainting* is the process of coating a workpiece by either making it anodic or cathodic in a bath that is generally an aqueous emulsion of the coating material. The electrodeposition bath contains stabilized resin, dispersed pigment, surfactants, and sometimes organic solvents in water. Electropainting is used primarily for primer coats, because it gives a fairly thick, highly uniform, corrosion-resistant coating in a relatively short time.

(41) *Vacuum Metalizing* is the process of coating a workpiece with metal by flash heating metal vapor in a high-vacuum chamber containing the workpiece. The vapor condenses on all exposed surfaces.

(42) *Assembly* is the fitting together of previously manufactured parts or components into a complete machine, unit of a machine, or structure.

(43) *Calibration* is the application of thermal, electrical, or mechanical energy to set or establish reference points for a component or complete assembly.

(44) *Testing* is the application of thermal, electrical, or mechanical energy to determine the suitability or functionality of a component or complete assembly.

(45) *Mechanical Plating* is the process of depositing metal coatings on a workpiece via the use of a tumbling barrel, metal powder, and, usually, glass beads for the impaction media. The operation is subject to the same cleaning and rinsing operations that are applied before and after the electroplating operation.

(46) *Printed Circuit Board Manufacturing* involves the formation of a circuit pattern of conductive metal (usually copper) on nonconductive board materials, such as plastic or glass. There are 5 basic steps involved in the manufacture of printed circuit boards: cleaning and surface preparation, catalyst and electroless plating, pattern printing and masking, electroplating, and etching.

After the initial cutting, drilling, and sanding of the boards, the board surface is prepared for plating electroless copper. This surface preparation involves an etchback, i.e., removal of built-up plastic around holes, and an acid and alkaline cleaning to remove grime, oils, and fingerprints. The board is then etched and rinsed. Following etching, the catalyst is applied, and rinsing operations follow the catalyst application. The entire board is then electroless copper plated and rinsed.

Following electroless copper plating, a plating resist is applied in noncircuit areas. Following application of a resist, a series of electroplates are applied; first, the circuit is copper plated. Then solder electroplate is applied; that is followed by a rinse. For copper removal in noncircuit areas, an etching step is next. After the etching operation, a variety of tab plating processes can be utilized depending on the board design requirements. Those include nickel electroplating, gold electroplating, rhodium electroplating, and tin immersion plating.

There are currently 3 main production methods in the manufacture of circuit boards: (1) additive, (2) semiadditive, and (3) subtractive. The additive method uses presensitized, unclad material as the starting board; the semiadditive method uses unclad, unsensitized material as the starting board; and the subtractive method begins with the copper-clad, unsensitized material.

E. Wastes from Metal-Finishing Operations

There are 7 distinct types of raw wastes present in metal finishing waste waters. It is possible to divide these raw wastes into 2 major categories, as, for example, (1) inorganic and (2) organic wastes. The 2 major constituents can be subdivided further into the specific types of wastes that occur in each of the 2 major categories, and we have the following classification:

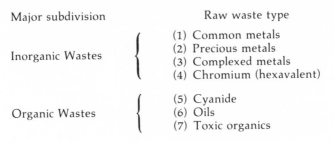

Major subdivision	Raw waste type
Inorganic Wastes	(1) Common metals (2) Precious metals (3) Complexed metals (4) Chromium (hexavalent)
Organic Wastes	(5) Cyanide (6) Oils (7) Toxic organics

In Fig. 13-2 there is shown a schematic of the waste treatment requirements for the metal-finishing industry in general. All of the process raw wastes

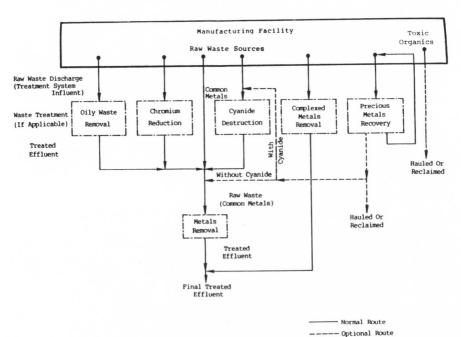

Fig. 13-2   Waste treatment schematic. Source: Effluent Guidelines Division, EPA 440/1-83/091, June 1983.

resulting from each of the 46 operations listed in Table 13-2, Metal Finishing Operations, and described earlier in this chapter are included in 1 or more of the raw waste types.

The types of raw wastes that each of the 46 operational processes may potentially generate are shown in Table 13-3. Thus it is that a direct relationship exists between the treatment system requirements and the processing operations at a given manufacturing facility. It may be seen, upon examining Table 13-3, that many plant operations generate the same waste constituents.

F. How Waste Water Is Generated

In the metal-finishing industries water is used for rinsing workpieces, washing away spills, air scrubbing, process fluid replenishment, cooling and lubrication, washing of equipment and workpieces, quenching, spray booths, and assembly and testing. Descriptions of those uses are as follows:

1. Rinsing

A large proportion of the water usage in metal finishing is for rinsing. That water is used to remove the film, both fluid and solid, that is deposited on the surfaces of the workpieces during the preceding process. As a result of the rinsing, the water becomes contaminated with the constituents of the film. Rinsing can be used in some capacity after virtually all of the operations covered by the metal finishing processes listed in Table 13-2 and is considered to be an integral part of the process operation that it follows.

2. Spills and Air Scrubbing

Water is used for washing away floor spills and for scrubbing ventilation exhaust air. In both cases those waste waters are contaminated with constituents of process materials and dirt.

3. Process Fluid Replenishment

As process fluids, i.e., cleaning solutions, plating solutions, paint formulations, etc., become exhausted or spent, new solutions have to be made up, with water a major constituent. When a fluid is used at high temperature, water must be added periodically to make up for evaporative losses. Exhausted or spent process solutions to be dumped are either collected in sumps for batch processing or are slowly metered into discharged rinse water before treatment.

4. Cooling and Lubrication

Coolants and lubricants in the form of free oils, emulsified oils, and grease are required by many metal removal operations. The films and residues from those fluids are removed during cleaning, washing, or rinsing operations, and the constituents contaminate other fluids. In addition, spent fluids in the sumps represent a further waste contribution that is processed either batchwise, i.e., segregated, or is discharged to other waste streams.

Table 13-3. Waste Characteristics of Metal Finishing Operations.

Unit Operation \ Waste Characteristics	INORGANICS					ORGANICS	
	Common Metals	Precious Metals	Complexed Metals	Chromium (Hexavalent)	Cyanide	Oils	Toxic Organics
1. Electroplating	×	×		×	×		
2. Electroless Plating	×	×	×		×		
3. Anodizing	×			×	×		
4. Conversion Coating	×	×		×			
5. Etching (Chem. Milling)	×	×	×	×			
6. Cleaning	×	×	×	×	×	×	×
7. Machining	×					×	
8. Grinding	×					×	
9. Polishing	×	×				×	
10. Tumbling	×			×	×	×	
11. Burnishing	×	×			×	×	
12. Impact Deformation	×					×	
13. Pressure Deformation	×					×	
14. Shearing	×					×	
15. Heat Treating	×				×	×	×
16. Thermal Cutting	×						
17. Welding	×						
18. Brazing	×						
19. Soldering	×		×				
20. Flame Spraying	×						
21. Sand Blasting	×						
22. Other Abr. Jet Machining	×						
23. Elec. Discharge Machining	×					×	
24. Electrochemical Machining	×				×	×	
25. Electron Beam Machining	×					×	
26. Laser Beam Machining	×						
27. Plasma Arc Machining	×						
28. Ultrasonic Machining	×						
29. Sintering	×						
30. Laminating	×						×
31. Hot Dip Coating	×						

32. Sputtering
33. Vapor Plating
34. Thermal Infusion
35. Salt Bath Descaling
36. Solvent Degreasing
37. Paint Stripping
38. Painting
39. Electrostatic Painting
40. Electropainting
41. Vacuum Metalizing
42. Assembly
43. Calibration
44. Testing
45. Mechanical Plating
46. Printed Circuit Board
    Manufacturing

Source: Effluent Guidelines Division, EPA 440/1-83/091, June, 1983

5. Water from Auxiliary Operations

Auxiliary operations, such as stripping of plating or painting racks, are essential to plant operations; waters used in these processes do become contaminated and require treatment.

6. Washing

Water used for washing workpieces or for washing equipment, such as filters, pumps, and tanks, picks up residues of concentrated process solutions, salts, or oils and is routed to an appropriate waste water stream for treatment.

7. Quenching

Workpieces that have undergone an operation involving intense heat, such as heat treating, welding, or hot dip coating, are frequently quenched or cooled in aqueous solutions to achieve the desired properties or to facilitate subsequent handling of the part. The solutions become contaminated and require treatment.

8. Spray Booths

Plants that employ spray-painting processes use spray booths to capture over-sprayed paint in a particular medium. Many of the booths use water curtains to capture the paint overspray. The paint is directed against a flowing stream of water, which scrubs the air so that paint and solvents are not released into the outside atmosphere. The paint collected in the water is removed by skimming or by using an ultrafilter and the water is reused in the curtain. That water will periodically be dumped.

9. Testing and Calibration

Many types of testing, such as leak, pressure, and performance testing, make use of large quantities of water that becomes contaminated.

G. Treatment Technologies

1. Introduction

The remainder of this chapter identifies the component technologies that are applicable for the treatment of raw wastes that are generated by the 46 metal-finishing industry operations previously described in this part. Table 13-4 lists the component technologies and indicates their specific application to the specific metal-finishing operation. Table 13-5 indicates the applicability of each technology to the particular type of waste.

2. Treatment of Common Metals Wastes

Common metals wastes can be generated in the metal-finishing industries by the processing operations that have previously been described. The methods used for treating those wastes are discussed in this section and fall into 2 group-ings—recovery techniques and solids-removal techniques. Recovery techniques are treatment methods used for recovering or regenerating process constituents that would otherwise be lost in the waste water or discarded. Included in that

Table 13-4.   Specific Application of Treatment Technologies.

Technology	Application or Potential Application to Metal Finishing
Aerobic Decomposition	Oil breakdown and organics removal
Carbon Adsorption	Removal of trace metals and organics
Centrifugation	Sludge dewatering, oil removal
Chemical Reduction	Treatment of chromic acid and chromates
Chemical Reduction-Precipitation/ Sedimentation	Removal of Complexed Metals
Coalescing	Oil removal
Diatomaceous Earth Filtration	Metal hydroxides and suspended solids removal
Electrochemical Oxidation	Destruction of free cyanide and cyanates
Electrochemical Reduction	Reduction of chromium from metal finishing and cooling tower blowdowns
Electrochemical Regeneration	Conversion of trivalent chromium to hexavalent valence
Electrolytic Recovery	Recovery of precious and common metals
Emulsion Breaking	Breakdown of emulsified oil mixtures
Evaporation	Concentration and recovery of process chemicals
Ferrous Sulfate ($FeSO_4$)-Precipitation/ Sedimentation	Removal of complexed metals and cyanides
Flotation	Suspended solids and oil removal
Granular Bed Filtration	Solids polishing of settling tank effluent
Gravity Sludge Thickening	Dewatering of clarifier underflow
High pH Precipitation/Sedimentation	Removal of complexed metals
Hydroxide Precipitation	Dissolved metals removal
Insoluble Starch Xanthate	Dissolved metals removal
Integrated Adsorption	Emulsified oils and paints removal
Ion Exchange	Recovery or removal of dissolved metals
Membrane Filtration	Dissolved metals and suspended solids removal
Oxidation by Chlorine	Destruction of cyanides and cyanates
Oxidation by Hydrogen Peroxide	Cyanide destruction and metals removal
Oxidation by Ozone	Destruction of cyanides and cyanates
Oxidation by Ozone w/UV Radiation	Destruction of cyanides and cyanates
Peat Adsorption	Dissolved metals removal
Pressure Filtration	Sludge dewatering or suspended solids removal
Resin Adsorption	Removal of organics
Reverse Osmosis	Removal of dissolved salts for water reuse
Sedimentation	Suspended solids and metals removal
Skimming	Free oil removal
Sludge Bed Drying	Sludge dewatering
Sulfide Precipitation	Dissolved metals removal
Ultrafiltration	Oil and suspended solids removal and paint purification
Vacuum Filtration	Sludge dewatering

methodology are evaporation, ion exchange, electrolytic recovery, electrodialysis, and reverse osmosis. Solids-removal techniques are employed to remove metals and other pollutants from process waste waters to make them suitable for reuse or discharge. Those methods include hydroxide and sulfide precipitation, sedimentation, diatomaceous earth filtration, membrane filtration, granular bed

Table 13-5.   Applicability of Treatment Technologies to Raw Waste Types
Source: Effluent Guidelines Division, EPA 440/1-83/091, June, 1983

Technology	Common Metals	Precious Metals	Complexed Metals	Chromium Bearing	Cyanide Bearing	Toxic Organics	Oily Wastes	Sludge	In-Process
Aerobic Decomposition						×	×		
Carbon Adsorption	×	×				×	×		
Centrifugation			×				×	×	
Chemical Reduction				×					
Coalescing							×		
Diatomaceous Earth Filtration	×	×	×			×	×		
Electrochemical Oxidation					×				
Electrochemical Reduction				×					
Electrochemical Regeneration				×					×
Electrodialysis				×					×
Electrolytic Recovery	×	×							×
Emulsion Breaking							×		×
Evaporation	×	×	×	×	×				
Flotation	×	×	×				×		
Granular Bed Filtration	×	×	×						
Gravity Sludge Thickening								×	
High pH Precipitation			×						
Hydroxide Precipitation	×	×	×		×				
Insoluble Starch Xanthate	×	×	×						
Ion Exchange	×	×	×	×	×				
Membrane Filtration	×	×	×						×
Oxidation by Chlorine					×				
Oxidation by Hydrogen Peroxide					×				
Oxidation by Ozone					×				
Oxidation by Ozone with UV Radiation					×				
Peat Adsorption	×	×	×			×	×		
Pressure Filtration	×	×	×				×		
Resin Adsorption						×			×
Reverse Osmosis	×	×	×			×	×		
Sedimentation	×	×	×						

Skimming								
Sludge Bed Drying	×	×	×		×		×	
Sulfide Precipitation								
Ultrafiltration				×				×
Vacuum Filtration	×	×	×		×	×		

filtration, sedimentation, peat adsorption, insoluble starch xanthate treatment, and flotation.

This section presents the treatment systems that are applicable to common metals removal for treatment Options 1, 2, and 3; describes the treatment techniques applicable to each Option; and defines the effluent performance levels for the Options. Option 1 common metals removal incorporates hydroxide precipitation and sedimentation. Option 2 for common metals removal consists of the addition of filtration devices to the Option 1 system. The Option 3 treatment system for common metals wastes consists of the Option 1 end-of-pipe treatment system with the addition of in-plant controls for cadmium. Alternative treatment techniques that can be applied to provide Option 1, 2, or 3 system performance are described following the Option 3 discussion.

a. Treatment of Common Metals Wastes, Option 1

The Option 1 system for the treatment of common metals wastes consists of hydroxide precipitation followed by sedimentation, as is shown in Fig. 13-3. That system accomplishes the end-of-pipe metals removal from all common metals bearing waste water streams that are present at a facility. The recovery of

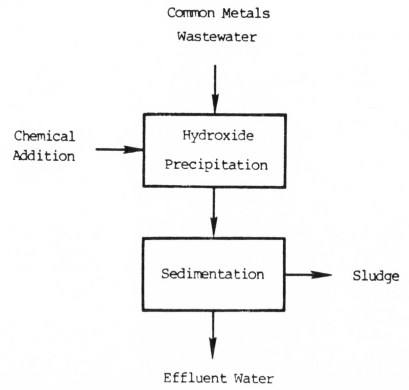

Fig. 13-3 Treatment of common metals wastes, option 1, hydroxide precipitation followed by sedimentation. Source: Effluent Guidelines Division, EPA 440/1-83/091, June 1983.

precious metals, the reduction of hexavalent chromium, the removal of oily wastes, and the destruction of cyanide must be accomplished before common metals removal, as was shown in Fig. 13-2.

Cyanide-bearing wastes must undergo oxidation so that the cyanide in the waste water is destroyed. Cyanide, as well as being a highly toxic pollutant, will complex metals such as copper, cadmium, and zinc and prevent efficient removal of them in the solids-removal device. Similarly, complexed metal wastes must be kept segregated and treated separately to avoid complexing metals in the primary solids removal device. Complexed metal wastes should be treated in a separate solids removal device, such as a membrane filter or a high pH clarifier. The specific techniques for the treatment of all other waste types, a description of the 3 levels of treatment Options for each waste type, and the performance for all levels of those options are presented in a later portion of this test.

The treatment techniques incorporated in the option 1 common metals waste treatment system include pH adjustment, hydroxide precipitation, flocculation, and sedimentation. Sedimentation may be carried out with equipment such as clarifiers, tube settlers, settling tanks, and sedimentation lagoons, or it may be replaced by various filtration devices preceded by hydroxide precipitation. The following paragraphs describe the hydroxide precipitation and sedimentation techniques that are employed for the Option 1 common metals treatment system.

1. Hydroxide Precipitation

Dissolved heavy metal ions are often chemically precipitated as hydroxides so that they may be removed by physical means, such as sedimentation, filtration, or centrifugation. Reagents commonly used to effect that precipitation include alkaline compounds, such as lime and sodium hydroxide. Calcium hydroxide precipitates trivalent chromium and other metals as metal hydroxides and phosphates as insoluble calcium phosphate. Those treatment chemicals may be added to a flash mixer or rapid mix tank or directly to the sedimentation device. Because metal hydroxides tend to be colloidal in nature, coagulating agents may also be added to facilitate settling. Fig. 13-4 illustrates typical chemical precipitation equipment as well as the accompanying sedimentation device.

After the solids have been removed, final pH adjustment may be required to reduce the high pH created by the alkaline chemicals used in the treatment process.

Hydroxide precipitation is used in metal finishing for precipitation of dissolved metals and phosphates. It can be utilized in conjunction with a solids removal device, such as a clarifier, or filter, for removal of metal ions, such as iron, lead, tin, copper, zinc, cadmium, aluminum, mercury, manganese, cobalt, antimony, arsenic, beryllium, and trivalent chromium. The process is also applicable to any substance that can be transformed into an insoluble form like soaps, phosphates, fluorides, and a variety of compounds.

Hydroxide precipitation has proved to be an effective technique for removing many pollutants from industrial waste water. Hydroxide precipitation operates at ambient conditions and is well suited to automatic control. Lime is usually added as a slurry when used in hydroxide precipitation. The slurry must be kept well mixed, and the addition lines periodically checked to prevent blocking, which results from a buildup of solids. The use of hydroxide precipitation produces large quantities of sludge requiring disposal, following precipitation and settling. Treatment chemicals should be used with caution because of the potentially

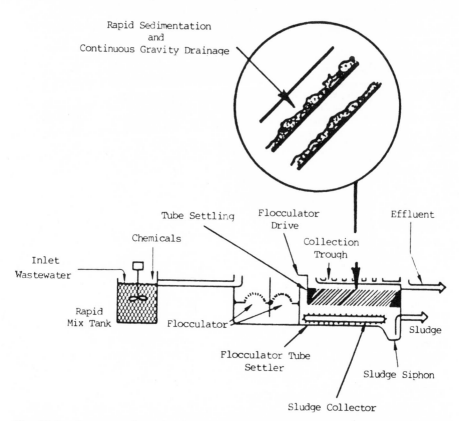

Fig. 13-4 A cross-section of a typical precipitation and sedimentation equipment arrangement. Source: Effluent Guidelines Division, EPA 440/1-83/091, June 1983.

hazardous situation involving storage and handling. Recovery of the precipitated species is sometimes difficult because of the homogeneous nature of most hydroxide sludges, where no single metal hydroxide is present in high concentrations, and because of the difficulty in smelting that results from the interference of calcium compounds.

The performance of hydroxide precipitation depends on several variables. The most important factors affecting precipitation effectiveness are as follows:

(a) The addition of sufficient excess anions to drive the precipitation reaction to completion;

(b) the maintenance of an alkaline pH throughout the precipitation reaction and subsequent settling (Fig. 13-5 illustrates the solubilities of various metal hydroxides as a function of pH);

(c) the effective removal of precipitated solids; see appropriate solids-removal technologies.

Hydroxide precipitation of metals is a classic waste treatment technology used in most industrial waste treatment systems. As noted earlier, sedimentation to

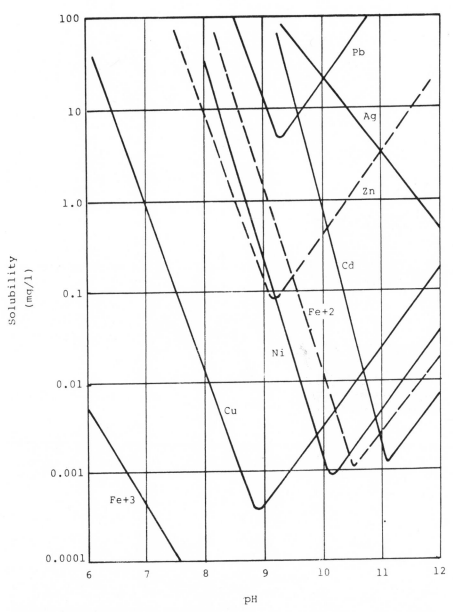

Fig. 13-5 Solubilities of metal hydroxides as a function of pH. Source: Effluent Guidelines Division, EPA 440/1-83/091, June 1983.

remove precipitates is discussed separately; however, both techniques have been illustrated in Fig. 13-4.

2. Sedimentation

Sedimentation is a process that removes solid particles from a liquid waste stream by gravitational settling. The operation is effected by reducing the velocity of the feed stream in a large volume tank or lagoon so that gravitational settling can occur. Fig. 13-6 shows cross-sections of 2 typical sedimentation devices.

For the Option 1 system, sedimentation is preceded by hydroxide precipitation, which converts dissolved metallic pollutants to solid forms and coagulates suspended precipitates into larger, faster-settling particles. Waste water is fed into a high-volume tank or lagoon, where it loses velocity and the suspended solids are allowed to settle. High retention times are generally required. Plants use retention times ranging from 1 to 48 hours or longer. Accumulated sludge can be collected and removed either periodically or continuously and either manually or mechanically.

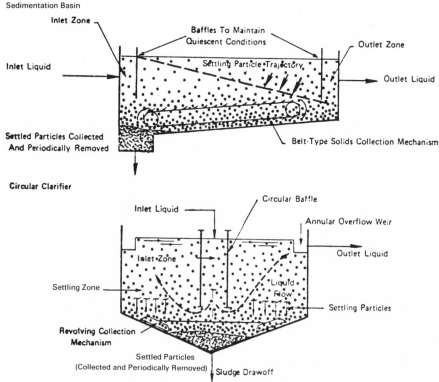

Fig. 13-6   Representative types of sedimentation. Source: Effluent Guidelines Division, EPA 440/1-83/091, June 1983.

Inorganic coagulants or polyelectrolytic flocculants are added to enhance coagulation. Common inorganic coagulants include sodium sulfate, sodium aluminate, ferrous or ferric sulfate, and ferric chloride. Organic polyelectrolytes vary in structure, but all usually form larger floccules than coagulants used alone.

The use of a clarifier for sedimentation reduces space requirements, shortens retention time, and increases solids removal efficiency. Conventional clarifiers generally consist of a circular or rectangular tank with a mechanical sludge-collecting device or a sloping funnel-shaped bottom designed for sludge collection. In advanced clarifiers, inclined plates, slanted tubes, or a lamellar network may be included within the clarifier tank in order to increase the effective settling area. A more recently developed "clarifier" utilizes centrifugal force rather than gravity to effect the separation of solids from a liquid. The precipitates are forced outward and accumulate against an outer wall, where they can later be collected. A fraction of the sludge stream is often recirculated to the clarifier inlet, promoting formation of a denser sludge.

Sedimentation is used in metal finishing to remove precipitated metals, phosphates, and suspended solids. Because most metal ion pollutants are easily converted to solid metal hydroxide precipitates, sedimentation is of particular use in industries associated with metal finishing and in other industries with high concentrations of metal ions in their wastes. In addition to heavy metals, suitably precipitated materials effectively removed by sedimentation/clarification include aluminum, manganese, cobalt, arsenic, antimony, beryllium, molybdenum, fluoride, and phosphate.

The major advantage of simple sedimentation is the simplicity of the process itself, the gravitational settling of solid particulate waste in a holding tank or lagoon. The major disadvantage of sedimentation involves the long retention times necessary for achieving complete settling, especially if the specific gravity of the suspended matter is close to that of water.

A clarifier is more effective in removing slow-settling suspended matter in a shorter time and in less space than a simple sedimentation system. Also, effluent quality from a clarifier is often better. The cost of installing and maintaining a clarifier is, however, substantially greater than those associated with sedimentation lagoons.

The inclined plate, slant tube, and lamellar clarifiers have even higher removal efficiencies than do conventional clarifiers, and greater capacities per unit area are possible. Installation costs for these advanced clarification systems are claimed to be one-half that of conventional systems of similar capacities.

A properly operating sedimentation system is capable of efficient removal of suspended solids, precipitated metal hydroxides, and other impurities from waste water. The performance of the process depends on a variety of factors, including the effective charge on the suspended particles; however, adjustments can be made in the type and dosage of flocculant or coagulant and in the types of chemicals used in prior treatment. It has been found that the site of flocculant or coagulant addition may significantly influence the effectiveness of sedimentation. If the flocculant is subjected to too much mixing before entering the settling device, the agglomerated complexes may be broken up and the settling effectiveness diminished. At the same time the flocculant must have sufficient mixing in order for effective setup and settling to occur. Most plant personnel select the line or trough leading into the clarifier as the most efficient site for flocculant addition. The performance of sedimentation is a function of the retention time, particle size and density, and the surface area of the sedimentation catchment.

Sedimentation in conjunction with hydroxide precipitation, the option 1 system, represents the typical method of solids removal and is employed extensively in industrial waste treatment. The advanced clarifiers are just beginning to appear in significant numbers in commercial applications, while the centrifugal force "clarifier" has yet to be used commercially.

3. Summary of Common Metals Waste Treatment System Operation of Option 1

When operated properly, the Option 1 system is a highly reliable method for removing dissolved heavy metals from waste water, although proper system monitoring, control, and preliminary treatment to remove interfering substances are required. Effective operation depends upon attention to proper chemical addition, raw waste load variations, routine maintenance, and solids removal. The control of chemical addition is required to maintain the appropriate pH for precipitation of the metals present and to promote coagulation of the metals precipitated. When fluctuating levels of raw waste loading occur, constant monitoring of the system flow and pH is needed to provide chemical addition at the proper rate. Other raw waste types, such as hexavalent chromium or cyanide, must be appropriately treated before entering the option 1 system. Specifically, hexavalent chromium will not be removed by that system, and cyanide will interfere with the system's ability to remove dissolved metals. (The necessary preliminary treatment for hexavalent chromium and cyanide is discussed in detail later in this chapter.)

An important factor in successful Option 1 system operation is the handling of changes in raw waste load. That is equally true for small batch systems and for large continuous systems. Most system failures, i.e., excessive discharges of pollutants, are the result of inadequate response to raw waste loading changes. Both hydraulic overloading and pollutant shock loads can be avoided by the segregation and bleed-in of concentrated batch dumps. When those practices are not employed, successful operation requires careful monitoring and quick response by the system operator. Appropriate action on his part in the event of an upset usually involves adjusting chemical feed rate, changing residence time, recycling treated waste water, or shutting the system down for maintenance.

The major maintenance requirements involve the periodic inspection and adjustment of monitoring devices, chemical mixing and feeding equipment, feed and sludge pumps, and clarifier mixing and drive components. Removal of accumulated sludge is necessary for efficient operation of precipitation/sedimentation systems. Solids that precipitate must be continually removed and disposed of properly. (Proper disposal practices are discussed later in this chapter under Treatment of Sludges.)

Although the performance of many Option 1 treatment systems (as shown in Fig. 13-7 with sources of wastes) is excellent, those of others are inferior. The major causes of poor performance are low pH, resulting in incomplete metals precipitation, and poor sedimentation, evidenced by high suspended solids in the effluent.

b. Treatment of Common Metals Wastes, Option 2

The Option 2 treatment system for common metals wastes is pictured

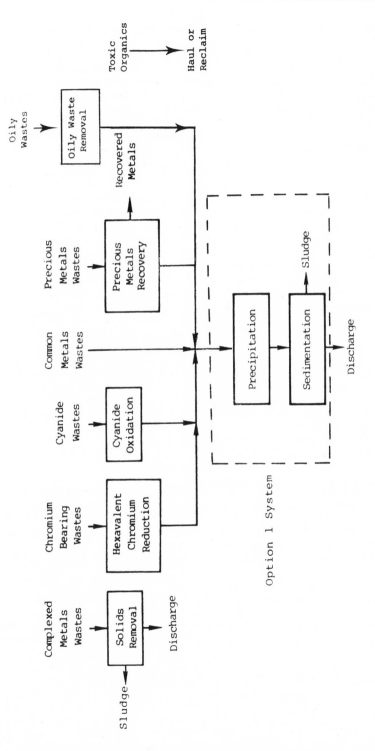

Fig. 13-7 The option 1 treatment methodology of precipitation/sedimentation. Source: Effluent Guidelines Division. EPA 440/1-83/091. June 1983.

schematically in Fig. 13-8. As shown in the figure, the system is identical to the option 1 common metals treatment system with the addition of a filtration phase after the primary solids removal step. The purpose of this filtration unit is to "polish" the effluent, that is, to remove suspended solids, such as metal hydroxides, that did not settle out in the clarifier. The filter also acts as a safeguard against pollutant discharge if an upset should occur in the sedimentation device. Filtration techniques that are applicable for option 2 systems include granular bed filtration and diatomaceous earth filtration.

Filtration is basic to water treatment technology, and experience with the process dates back tot he 1800s. Filtration occurs in nature as the surface ground waters are purified by sand. Silica sand, anthracite coal, and garnet are common filter media used in water treatment plants. The media are usually supported by gravel and may be used singly or in combination. The multi-media filters may be arranged to maintain relatively distinct layers by virtue of balancing the force of gravity, flow, and bouyancy on the individual particles. That is accomplished by selecting appropriate filter flow rates, i.e., gallons per minute per square foot of surface area, media grain size, and density.

Granular bed filters may be classified in terms of filtration rate, filter media, flow pattern, or method of pressurization. Traditional rate classifications are slow sand, rapid sand, and high-rate mixed media. In the slow-sand filter, flux or hydraulic loading is relatively low; therefore removal of collected solids to clean the filter is relatively infrequent. The filter is often cleaned by scraping off the inlet face or top of the sand bed. In the higher rate filters, cleaning is frequent and is accomplished by a periodic backwash, opposite to the direction of normal flow.

A filter may use a single medium, such as sand or diatomaceous earth, but dual and mixed, i.e., multiple, media filters allow higher flow rates and efficiencies. The dual-media filter usually consists of a bed of fine sand under a coarser bed of anthracite coal. The coal removes most of the influent solids, while the sand performs a polishing function. At the end of the backwash, the sand settles to the bottom, because it is denser than the coal, and the filter is ready for normal operation. The mixed-media filter operates on the same principle, with the finer, denser media at the bottom and the coarser, less dense media at the top. The usual arrangement is garnet at the bottom, or outlet end of the bed, sand in the middle, and anthracite coal at 'the top. Some mixing of the layers occurs and is, in fact, desirable.

The flow pattern is usually top-to-bottom, but other patterns are sometimes used. Upflow filters are sometimes used, and in a horizontal filter the flow is horizontal. In a biflow filter the influent enters both the top and the bottom and exits laterally. The advantage of an upflow filter is that with an upflow backwash the particles of a single filter medium are distributed and maintained in the desired coarse-to-fine—i.e., bottom-to-top—arrangement. The disadvantage is that the bed tends to become fluidized, which ruins filtration efficiency. The biflow design is an attempt to overcome that problem.

The usual granular bed filter operates by gravity flow; however, pressure filters are also used. Pressure filters permit higher solids loadings before cleaning and are advantageous when the filter effluent must be pressurized for further downstream treatment. In addition, pressure filter systems are often less costly for low to moderate flow rates.

Fig. 13-9 depicts a granular bed filter. It is a high-rate, dual-media, gravity downflow filter, with self-stored backwash. Both filtrate and backwash are piped

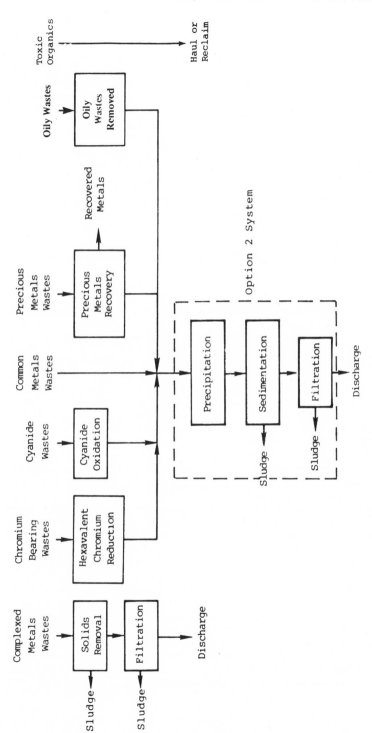

Fig. 13-8 Treatment of common metals wastes showing precipitation/sedimentation followed by filtration as option 2. Source: Effluent Guidelines Division, EPA 440/1-83/091, June 1983.

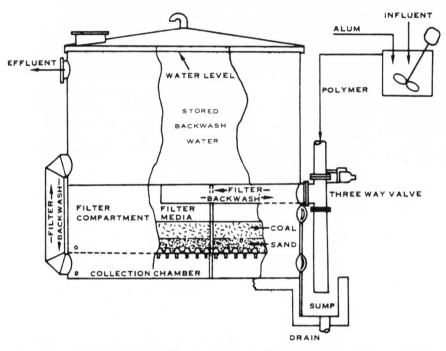

Fig. 13-9   An illustration of granular bed filtration. Source: Effluent Guidelines Division, EPA 440/1-83/091, June 1983.

around the bed in an arrangement that permits upflow of the backwash, with the stored filtrate serving as backwash. Addition of the indicated coagulant and polyelectrolyte usually results in a substantial improvement in filter performance.

Auxiliary filter cleaning is sometimes employed in the upper few inches of filter beds. That is conventionally referred to as surface wash and is accomplished by water jets just below the surface of the expanded bed during the backwash cycle. Those jets enhance the scouring action in the bed by increasing the agitation.

An important feature for successful filtration and backwashing is the underdrain. It is the support structure for the bed, and it provides an area for collection of the filtered water without clogging from either the filtered solids or the media grains. In addition, the underdrain prevents loss of the media with the water, and during the backwash cycle it furnishes even flow distribution over the bed. Failure to dissipate the velocity head during the filter or backwash cycle will result in bed upset and the need for major repairs.

Several standard approaches are employed for filter underdrains. The simplest one consists of a parallel porous pipe embedded under a layer of coarse gravel and manifolded to a header pipe for effluent removal. Other approaches to the underdrain system are known as the Leopold and Wheeler filter bottoms. Both incorporate false concrete bottoms with specific porosity configurations to supply drainage and velocity head dissipation.

Filter system operation may be manual or automatic. The filter backwash cycle may be on a timed principle, a pressure drop basis with a terminal value that triggers backwash or a solids carryover basis from turbidity monitoring of the outlet stream. All of these types have been successfully used.

1. Granular Bed Filters

Granular bed filters are used in metal finishing to remove residual solids from clarifier effluent. Filters in waste water treatment plants are often employed for polishing after sedimentation or other similar operations. Granular bed filtration thus has potential application to nearly all industrial plants. Chemical additives that enhance the upstream treatment equipment may or may not be compatible with or enhance the filtration process. It should be borne in mind that in the overall treatment system, effectiveness and efficiency, not the performance of any single unit, are the objectives. The volumetric fluxes for various types of filters are as follows:

Slow Sand	2.04—5.30	l/min/sq m
Rapid Sand	40.74—51.48	l/min/sq m
High-Rate Mixed Media	81.48—122.22	l/min/sq m

The principal advantages of granular bed filtration are its low initial and operating costs and reduced land requirements over other methods to achieve the same level of solids removal. The filter may require pretreatment, however, if the solids level is high, i.e., from 100 mg/l to 150 mg/l. Operator training requirements demand superior skill levels owing to controls and periodic backwashing and to the fact that backwash must be stored and dewatered to be disposed of economically.

Nevertheless, the recent improvements in filter technology have significantly improved filtration reliability. Control systems, improved designs, and good operating procedures have made filtration a highly reliable method of water treatment. Deep bed filters may be operated with either manual or automatic backwash. In either case they must be periodically inspected for media attrition, partial plugging, and leakage. Where backwashing is not used, collected solids must be removed by shoveling, and filter media must be at least partially replaced. Filter backwash is generally recycled within the waste water treatment system, so that the solids ultimately appear in the clarifier sludge stream for subsequent dewatering. Alternatively, the backwash stream may be dewatered directly, or if there is no backwash, the collected solids may be suitable disposed of; in either of those situations there is a solids disposal problem similar to that of clarifiers.

Suspended solids are commonly removed from waste water streams by filtering through a deep 0.3m-to-0.9-m (1 foot-to-3 feet) granular filter bed. The porous bed formed by the granular media can be designed to remove practically all suspended particles. Even colloidal suspensions (roughly 1 micron to 100 microns) are adsorbed on the surface of the media grains as they pass in close proximity in the narrow bed passages.

Properly operating filters following some pretreatment to reduce suspended solids should produce water averaging 12.8 mg/l, total suspended solids (TSS). Pretreatment with inorganic or polymeric coagulants can improve poor performance.

Deep bed filters are in common use in municipal treatment plants. Their use in polishing industrial clarifier effluent is increasing, and the technology is proven and conventional.

2. Diatomaceous Earth Filtration

Diatomaceous earth filtration, combined with precipitation and sedimentation, is a solids separation device that can further enhance suspended solids removal. The diatomaceous earth filter is used to remove metal hydroxides and other solids from the waste water and provides an effluent of high quality.

A diatomaceous earth filter consists of a filter element, a filter housing, and associated pumping equipment. The filter element consists of multiple leaf screens, which are coated with diatomaceous earth. The size of the filter is a function of flow rate and desired operating time between filter cleanings.

Normal operation of the system involves pumping a mixture of diatomaceous earth and water through the screen leaves. That deposits the diatomaceous earth filter media on the screens and prepares them for treatment of the waste water. Once the screens are completely coated, the pH-adjusted waste water can be pumped through the filter. The metal hydroxides and other suspended solids are removed from the effluent in the diatomaceous earth filter. The buildup of solids in the filter increases the pressure drop across the filter. At a certain pressure the wastewater is stopped, the filter is cleaned, and the cycle is repeated.

The principal advantage of using a diatomaceous earth filter is its increased removal of suspended solids and precipitates. Moreover, sludge removed from the filter is much drier than that removed from a clarifier (approximately 50 percent solids). The high solids content can significantly reduce the cost of hauling and landfill. The major disadvantage in the use of a filter system is an increase in operation and maintenance costs.

3. Summary of Common Metals Wastes Treatment System Operation of Option 2

The entire option 1 system operation summary discussion above, applies equally to Option 2. In addition, the use of a polishing filter necessitates further precautions. Close monitoring is needed to prevent both hydraulic overloading and solids overloading. Either form of overloading may result in pollutant bypassing in a barrier filter, through element breakage or pressure relief, or pollutant reentrainment in a depth filter. Many types of filters must be shut down for solids removal. Waste water flow must not be bypassed during that period. Bypassing can be obviated by use of a holding tank or by installation of dual filters in parallel arrangement. A further consideration concerns disposable elements for filters that use them. Because of the contained toxic metals, those elements must be treated as hazardous waste and should not be placed in the plant trash.

Performance of a properly operating option 2 treatment system, as shown in Fig. 13-8 with its sources of waste, is demonstrated by low effluent levels of total suspended solids (TSS). Generally speaking, a pH range of 8.5 to 9.5 is considered most effective for settling and filtration of precipitated hydroxides in mixed metal-finishing wastes.

c. Treatment of Common Metals Wastes, Option 3

The Option 3 treatment system for metal wastes consists of the Option 1 end-of-pipe treatment system plus the addition of in-plant controls for cadmium. In-plant controls could include evaporative recovery, ion exchange, and recovery rinses. The purpose of those in-plant controls is to eliminate, as nearly as possible, cadmium from the raw waste stream. The additional controls have the added advantage of minimizing the possibility of discharging this highly toxic metal in the event of treatment system failure.

The following sections illustrate two common metals treatment techniques that are applicable to Option 3: evaporation and ion exchange.

1. Evaporation

Evaporation is a concentration process. Water is evaporated from a solution, increasing the concentration of solute in the remaining solution. If the resulting water vapor is condensed back to a liquid, the evaporation-condensation process is called distillation. To be consistent with industry terminology, however, evaporation is used in this chapter to describe both processes. Both atmospheric and vacuum evaporation are commonly used in industry today. Specific evaporation techniques are shown in Fig. 13-10 and discussed below.

Atmospheric evaporation could be accomplished simply by boiling the liquid. To aid evaporation, however, heated liquid is sprayed on an evaporation surface, and air is blown over the surface and subsequently released into the atmosphere. Thus, evaporation occurs by humidification of the air stream, similar to a drying process. Equipment for carrying out atmospheric evaporation is quite similar for most applications. The major element is generally a packed column with an accumulator bottom. Accumulated waste water is pumped from the base of the column, through a heat exchanger, and back into the top of the column, where it is sprayed into the packing. At the same time air drawn upward through the packing by a fan is heated as it contacts the hot liquid. The liquid partially vaporizes and humidifies the air stream. The fan then blows the hot, humid air to the outside atmosphere. A scrubber is often unnecessary because the packed column itself acts as a scrubber.

Another form of atmospheric evaporation combines evaporative recovery of plating chemicals with plating tank fume control. A third form of atmospheric evaporation also works on the air humidification principle, but the evaporated rinse water is recovered for reuse by condensation. Those air humidification techniques operate well below the boiling point of water and can utilize waste process heat to supply the energy required.

In vacuum evaporation the evaporation pressure is lowered to cause the liquid to boil at reduced temperature. All of the water vapor is condensed, and to maintain the vacuum condition, noncondensable gases (air in particular) are removed by a vacuum pump. Vacuum evaporation may be either single or double effect. In double-effect evaporation, 2 evaporators are used, and the water vapor from the first evaporator, which may be heated by steam, is used to supply heat to the second evaporator. As that happens, the water vapor from the first evaporator condenses. Approximately equal quantities of waste water are evaporated in each unit; thus the double-effect system evaporates twice the amount of water that a single-effect system does, at nearly the same cost in energy, but with added capital cost and complexity. The double-effect technique is thermodynamically possible because the second evaporator is maintained at lower pressure, i.e., higher vacuum and therefore lower evaporation temperature. Another means of increasing energy efficiency is vapor recompression, either thermal or mechanical, which enables heat to be transferred from the

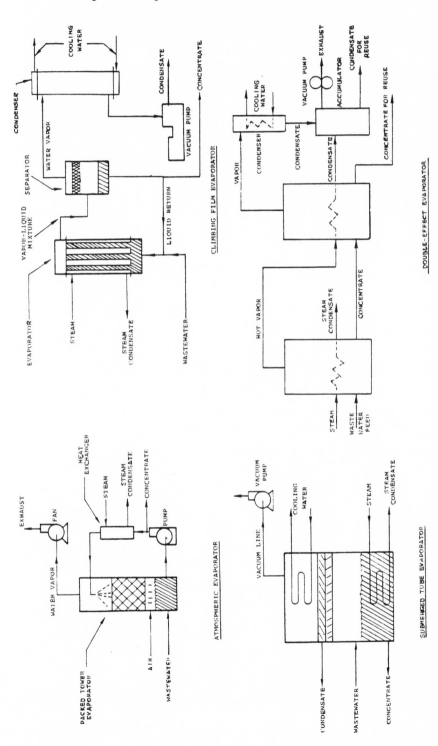

Fig. 13-10   Representative types of evaporation equipment and systems. Source: Effluent Guidelines Division, EPA 440/1-83/091, June 1983. EPA 440/1-83/091, June 1983.

condensing water vapor to the evaporating waste water. Vacuum evaporation equipment may be classified as submerged tube or climbing film evaporation units.

In the most commonly used submerged tube evaporator, the heating and condensing coil are contained in a single vessel to reduce capital cost. The vacuum in the vessel is maintained by an educator-type pump, which creates the required vacuum by the flow of the condenser cooling water through a venturi. Waste water accumulates in the bottom of the vessel, and it is evaporated by means of submerged steam coils. The resulting water vapor condenses as it contacts the condensing coils in the top of the vessel. The condensate then drips off the condensing coils into a collection trough, which carries it out of the vessel. Concentrate is removed from the bottom of the vessel. The major elements of the climbing film evaporator are the evaporator, separator, condenser, and vacuum pump. Waste water is "drawn" into the system by the vacuum so that a constant liquid level is maintained in the separator. Liquid enters the steam-jacketed evaporator tubes, and part of it evaporates so that a mixture of vapor and liquid enters the separator. The design of the separator is such that the liquid is continuously circulated from it to the evaporator. Entering vapor flows out through a mesh entrainment separator to the condenser tubes. The condensate, along with any entrained air, is pumped out of the bottom of the condenser by a liquid ring vacuum pump. The liquid seal provided by the condensate keeps the vacuum in the system from being broken.

Evaporation is used in the metal-finishing industry for recovery of a variety of metals, bath concentrates, and rinse waters. Both atmospheric and vacuum evaporation are used in metal-finishing plants, mainly for the concentration and recovery of plating solutions. Many of those evaporators also recover water for rinsing. In addition, evaporation has been applied to the recovery of phosphate metal cleaning solutions. There is no fundamental limitation on the applicability of evaporation. Recent changes in construction materials used for climbing film evaporators enable them to process a wide variety of waste waters, including cyanide-bearing solutions, as do the other types of evaporators described in this section.

Advantages of the evaporation process are that it permits recovery of a wide variety of process chemicals, and it is often applicable to removal and concentration of compounds that cannot be accomplished by any other means. The major disadvantage is that the evaporation process consumes relatively large amounts of energy for the evaporation of water. The recovery of waste heat from many industrial processes, however, e.g., diesel generators, incinerators, boilers and furnaces, should be considered as a source of this heat for a totally integrated evaporation system. For some applications, pretreatment may be required to remove solids and bacteria that tend to cause fouling in the condenser or evaporator. Sometimes the buildup of scale on the evaporator surfaces reduces the heat transfer efficiency and may present a maintenance problem or increased operating cost; however, it has been demonstrated that fouling of the heat transfer surfaces can be avoided or minimized for certain dissolved solids by maintaining a seed slurry, which provides preferential sites for precipitate deposition. In addition, low temperature differences in the evaporator will eliminate nucleate boiling and supersaturation effects. It must be said also that steam-distillable impurities in the process stream are carried over with the product water and must be handled by pre- or posttreatment.

In theory, evaporation should yield a concentrate and a deionized condensate.

Actually, carryover has resulted in condensate metal concentrations as high as 10 mg/l, although the usual level is less than 3 mg/l, pure enough for most final rinses. The condensates may also contain organic brighteners and antifoaming agents. They can be removed with an activated carbon bed, if necessary. Also, evaporators are available in a range of capacities, typically from 15 gph to 75 gph and may be used in parallel arrangements for processing of higher flow rates. Evaporation is a fully developed, commercially available waste water treatment system. It is used extensively to recover plating chemicals, and a pilot scale unit has been used in connection with phosphate washing of aluminum coil.

2. Ion Exchange

Ion exchange is a process in which ions, held by electrostatic forces to charge functional groups on the surface of the ion exchange resin, are exchanged for ions of similar charge from the solution in which the resin is immersed. That is classified as a sorption process because the exchange occurs on the surface of the resin, and the exchanging ion must undergo a phase transfer from solution phase to solid phase. Thus, ionic contaminants in a waste stream can be exchanged for the harmless ions of the resin.

Although the precise technique may vary slightly according to the application involved, a generalized process description is as follows: The waste water stream being treated passes through a filter to remove any suspended solids and then flows through a cation exchanger, which contains the ion exchange resin. Here metallic impurities, such as copper, iron, and trivalent chromium, are retained. The stream then passes through the anion exchanger and its associated resin. Hexavalent chromium, for example, is retained at that stage. If 1 pass does not reduce the contaminant levels sufficiently, the stream may then enter another series of exchangers. Many ion exchange systems are equipped with more than 1 set of exchangers for that reason.

The other major portion of the ion exchange process concerns the regeneration of the resin, which now holds those impurities retained from the waste stream. An ion exchange unit with in-place regeneration is shown in Fig. 13-11. Metal ions, such as nickel, are removed by an acidic cation exchange resin, which is regenerated with hydrochloric or sulfuric acid, replacing the metal ion with 1 or more hydrogen ions. Anions, such as dichromate, are removed by a basic anion exchange resin, which is regenerated with sodium hydroxide, replacing the anion with 1 or more hydroxyl ions. The 3 principal methods employed by industry for regenerating the spent resin are as follows:

(a) *Replacement Service*—A replacement service replaces spent resin with regenerated resin and regenerates the spent resin at its own facility. The service then has the problem of treating and disposing of the spent regenerant.

(b) *In-Place Regeneration*—Some establishments may find it less expensive to do their own regeneration. The spent resin column is shut down for perhaps an hour, and the spent resin is regenerated. That results in 1 or more waste streams, which must be treated in an appropriate manner. Regeneration is performed only as the resins require it.

(c) *Cyclic Regeneration*—In this process the regeneration of the spent resins takes place in alternating cycles with the ion removal process. A regeneration frequency of twice an hour is typical. That very short cycle time permits operation with a very small quantity of resin and with fairly concentrated solutions, resulting in a very compact system. Again, the process varies according to

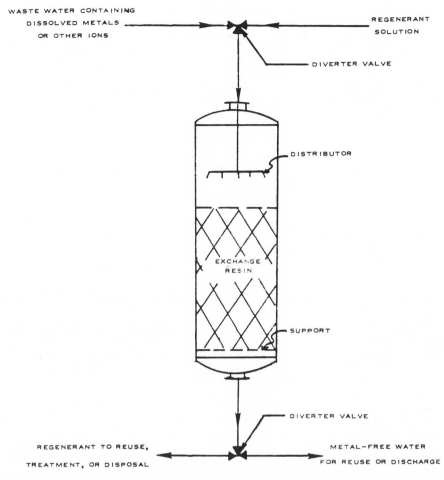

Fig. 13-11   Representative ion exchange unit with regeneration. Source: Effluent Guidelines Division, EPA 440/1-83/091, June 1983.

application, but the regeneration cycle generally begins with caustic being pumped through the anion exchanger, carrying out hexavalent chromium, for example, as sodium dichromate. The sodium dichromate stream then passes through a cation exchanger, converting the sodium dichromate to chromic acid. After concentration by evaporation or other means, the chromic acid can be returned to the process line. Meanwhile, the cation exchanger is regenerated with sulfuric acid, resulting in a waste acid stream containing the metallic impurities removed earlier. Flushing the exchangers with water completes the cycle. Thus, the waste water is purified, and in this example, chromic acid is recovered. The ion exchangers, with newly regenerated resin, then enter the ion removal cycle again.

Many metal-finishing facilities utilize ion exchange to concentrate and purify their plating baths. The list of pollutants for which the system has proved effective includes aluminum, arsenic, cadmium, chromium (hexavalent and trivalent), copper, cyanide, gold, iron, lead, manganese, nickel, selenium, silver, tin, zinc, and more. Thus ion exchange can be applied to a wide variety of industrial processes. Because of the heavy concentrations of metals in the waste water of the metal-finishing industries, they utilize ion exchange in several ways. As an end-of-pipe treatment, ion exchange is certainly feasible, but its greatest value is in recovery applications. It is commonly used, however, as an integrated treatment to recover rinse water and process chemicals. In addition to metal finishing, ion exchange is finding applications in the photography industry for bath purification, in battery manufacturing for heavy metal removal, in the chemical industry, the food industry, the nuclear industry, the pharmaceutical industry, the textile industry, and others. It could also be used in the copper and copper alloys industry for recovery of copper from pickle rinses. Also, many industrial and nonindustrial companies utilize ion exchange for reducing the salt concentrations in their incoming water.

Ion exchange is a versatile technology applicable to a great many situations. That flexibility, along with the compact nature and performance of ion exchange, makes it a very effective method of waste water treatment. The resins in these systems, however, can prove to be a limiting factor. The thermal limits of the anion resins, generally placed in the vicinity of 60° C, could prevent its use in certain situations. Similarly, nitric acid, chromic acid, and hydrogen peroxide can all damage the resins, as will iron, manganese, and copper when present with sufficient concentrations of dissolved oxygen.

Removal of a particular trace contaminant may be uneconomical, because of the presence of other ionic species that are preferentially removed. Also, the regeneration of the resins presents its own problems, since the cost of the regenerative chemicals can be expensive. In addition, the waste streams originating from the regeneration process are extremely high in pollutant concentrations, although low in volume. They must be further processed for proper disposal.

Nevertheless, ion exchange is highly efficient at recovering metal-finishing chemicals. Recovery of chromium, nickel, phosphate solution, and sulfuric acid from anodizing is in commercial use. A chromic acid recovery efficiency of 99.5 percent has been demonstrated. While it may be said that all of the applications mentioned in this section are available for commercial use, the research and development in ion exchange is focusing upon improving the quality and efficiency of the resins, rather than on new applications. At this writing, experimental work is being performed on developing a continuous regenerative process, whereby the resins are contained on a fluid-transfusible belt. It passes through a compartmented tank with ion exchange, washing, and regeneration sections. The resins are, therefore, continually used and regenerated in a continuous rather than a batch-process type operation. The system is still in the pilot stage but promises potential for future industrial processing.

d. Alternative Treatment Methods for Common Metals Removal

In addition to the treatment methods described under Options 1, 2, and 3, there are several other alternative treatment technologies applicable for the treatment of common metals wastes. They may be used in conjunction with or in place of the Option 1, 2, or 3. The following paragraphs, therefore, describe the technologies: peat adsorption, insoluble starch xanthate, sulfide precipitation, flotation, and membrane filtration.

1. Peat Adsorption

Peat moss is a rather complex material, whose major constituents are lignin and cellulose. The constituents, particularly lignin, bear polar functional groups, such as alcohols, aldehydes, ketones, acids, phenolic hydroxides, and ethers, which can be involved in chemical bonding. Because of the polar nature of this material, its adsorption of dissolved solids, such as transition metals and polar organic molecules, is quite high. Those properties have led to the use of peat as an agent for the purification of industrial waste water.

Peat adsorption is a "polishing" process that can achieve very low effluent concentrations for several pollutants. If the concentrations of pollutants are above 10 mg/l, then peat adsorption must be preceded by pH adjustment and settling. The waste water is then pumped into a large metal chamber (a kier), which contains a layer of peat through which the waste stream passes. The water flows to a second kier for further adsorption. The waste water is then ready for discharge. The system may be automated or manually operated.

Peat adsorption can be used in metal-finishing plants for removal of residual dissolved metals from clarifier effluent. Also, peat moss may be used to treat waste waters containing heavy metals, such as mercury, cadmium, zinc, copper, iron, nickel, chromium, and lead as well as organic matter, such as oil, detergents, and dyes. Peat adsorption could be used in metal-finishing industries, coil coating plants, porcelain enameling, battery manufacturing plants, copper products manufacturing facilities, photographic plants, textile manufacturing, newsprint production facilities, and other industries. Peat adsorption is currently used commercially at a textile plant, a newsprint facility, and a metal reclamation operation. The following tabulation contains performance figures obtained from pilot plant studies. Peat adsorption was preceded by pH adjustment for precipitation and by clarification:

Pollutant	Before Treatment (mg/l)	After Treatment (mg/l)
Pb	20.0	0.025
Sb	2.5	0.9
Cu	250.0	0.24
Zn	1.5	0.25
Ni	2.5	0.07
Cr^{+6}	35,000.0	<0.04
CN	36.0	0.7
Hg	>1.0	0.02
Ag	>1.0	0.05

In addition, pilot plant studies have shown that complexed metal wastes as well as the complexing agents themselves are removed by contact with peat moss. Therefore, peat adsorption could be applied to printed circuit board manufacturing, which uses complexing agents extensively.

Only 3 commercial adsorption systems are currently in use in the United States: at a textile manufacturer, a newsprint facility, and a metal reclamation firm. No data have been reported showing the use of peat adsorption in any metal-finishing plants. Its only commercial applications are as stated above.

2. Insoluble Starch Xanthate (ISX)

Insoluble starch xanthate (ISX) is essentially an ion exchange medium used to remove dissolved heavy metals from waste water. ISX is formed by reacting commercial cross-linked starch with sodium hydroxide and carbon disulfide. Magnesium sulfate is also added as a stabilizer and to improve sludge settling.

ISX acts as a cationic ion exchange material removing the heavy metal ions and replacing them with sodium and magnesium. The starch has good settling characteristics as well as good filtering characteristics, and it is well suited for use as a filter precoat. ISX can be added as a slurry for continuous treatment operations, in solid form for batch treatments, and as a precoat to a filter. The ISX process is effective for removal of all uncomplexed metals, including hexavalent chromium and also some complexed metals, such as the copper-ammonia complex. The removal of hexavalent chromium is brought about by lowering the pH to below 3 and subsequently raising it above 7. The hexavalent chromium is reduced by the ISX at the acid pH and is removed at the alkaline pH as chromium starch xanthate, or chromic hydroxide.

At the present time ISX is being used in 2 metal-finishing establishments. One of the plants utilizes the ISX process as a polishing filter and claims to reduce levels of metals in the effluent of its clarifier from 1 mg/l to .020 mg/l. The other plant uses the ISX process to recycle rinse waters on its cleaning line and nickel, copper, and solder plating lines. The results of the sampling are listed below:

	Solder Line		Nickel Line		Cleaning Line	
	Input to Filter	Output from Filter	Input to Filter	Output from Filter	Input to Filter	Output from Filter
Cu	.42	.41	.24	.24	.43	.39
Pb	.56	.53	—	—	—	—
Sn	2.0	1.5	—	—	—	—
Zn	.092	.083	.047	.040	.167	.126
Ni	—	—	552.	547.	—	—
Fe	—	—	—	—	.38	.26

The ISX was not removing a high percentage of metal. Its main purpose was to keep contaminants from building up to a point where the water would not be reusable.

3. Sulfide Precipitation

Hydrogen sulfide or soluble sulfide salts, such as sodium sulfide, are used to precipitate many heavy metal sulfides. Since most metal sulfides are even less soluble than metal hydroxides at alkaline pH levels, greater heavy metal removal can be accomplished through the use of sulfide rather than hydroxide as a chemical precipitant before sedimentation. The solubilities of metallic sulfides are pH dependent and are shown in Fig. 13-12.

Of particular interest is the ability, at a pH of 8 to 9, of the ferrous sulfide process to precipitate hexavalent chromium (Cr^{-6}) without prior reduction to the

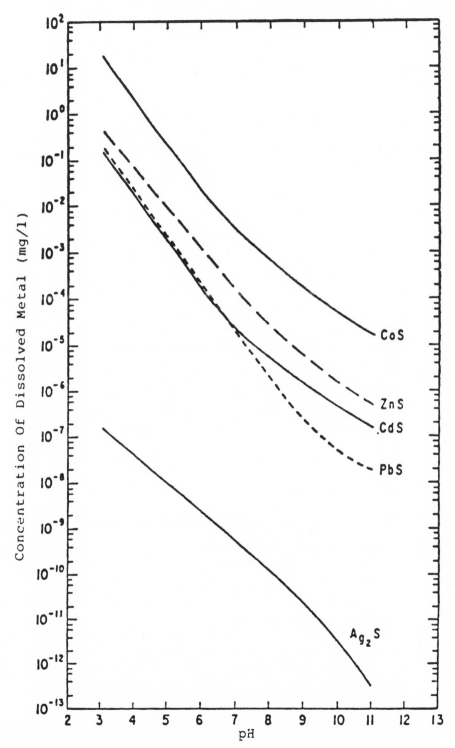

Fig. 13-12 Comparative solubilities of metal sulfides as a function of pH. Source: Effluent Guidelines Division, EPA 440/1-83/091, June 1983.

trivalent state as is required in the hydroxide process, although the chromium is still precipitated as the hydroxide. When ferrous sulfide is used as the precipitant, iron and sulfide act as reducing agents for the hexavalent chromium.

$$Cr_2O^=7 + 2FeS + 7H_2O = 2Fe(OH)_3 + 2Cr(OH)_3 + 2S^\circ + 2OH^-$$

In this case the sludge produced consists mainly of ferric hydroxides and chromic hydroxides. Some excess hydroxyl ions are produced in this process, possibly requiring a downward readjustment of Ph to between 8 to 9 before to discharge of the treated effluent.

In addition to the advantages listed above, the process will precipitate metals complexed with most complexing agents. Care must be taken, however, to maintain the pH of the solution above approximately 8 in order to prevent the generation of toxic hydrogen sulfide gas. For that reason, ventilation of the treatment tanks may be a necessary precaution in some installations. The use of ferrous sulfide, however, virtually eliminates the problem of hydrogen sulfide evolution. As with hydroxide precipitation, excess sulfide must be present to drive the precipitation reaction to completion. Since sulfide itself is toxic, sulfide addition must be carefully controlled to maximize heavy metals precipitation with a minimum of excess sulfide to avoid the necessity of posttreatment.

At very high excess sulfide levels and high pH, soluble mercury-sulfide compounds may also be formed. Where excess sulfide is present, aeration of the effluent stream can aid in oxidizing residual sulfide to the less harmful sodium sulfate (Na_2SO_4). The cost of sulfide precipitants is high in comparison with hydroxide precipitating agents, and disposal of metallic sulfide sludges may pose problems. With improper handling or disposal of sulfide precipitates, hydrogen sulfide may be released into the atmosphere, creating a potential toxic hazard, toxic metals may be leached out into surface waters, and sulfide might oxidize to sulfate and release dilute sulfuric acid into surface waters. An essential element in effective sulfide precipitation is the removal of precipitated solids from the waste water to a site where reoxidation and leaching are not likely to occur. Data from sampling at one plant show the effectiveness of sulfide precipitation on unreduced hexavalent chromium as well as on total chromium. Mean concentrations for the only metals present in the aluminum-anodizing operation were as follows:

Parameter	Influent mg/l	Effluent mg/l
Chromium, hex.	11.5	Undetectable
Chromium, total	18.4	Undetectable
Aluminum	4.18	0.112

One report (Treatment of Metal Finishing Wastes by Sulfide Precipitation, EPA-600/2-75-049, U.S. Environmental Protection Agency, 1977) concluded that (with no complexing agents present) the following effluent quality can be achieved:

Parameter	Effluent mg/l
Cadmium	0.01
Copper	0.01
Zinc	0.01

Nickel	0.05
Chromium, Total	0.05

Total suspended solids (TSS) removal is a reliable indicator of precipitation and sedimentation system performance. Lack of such data makes it difficult to fully evaluate the bench tests, and insufficient solids removal can result in high metals concentrations. Lead is consistently removed to very low levels, less than 0.02 mg/l, in systems using hydroxide precipitation and sedimentation. Therefore, one would expect even lower effluent concentrations of lead resulting from properly operating sulfide precipitation systems owing to the lower solubility of the lead sulfide compound. Because of their effectiveness, full-scale commercial sulfide precipitation units are in operation at numerous installations, including several plants in the metal-finishing industries.

4. Flotation

Flotation is the process of causing particles, such as metal hydroxides or oil, to float to the surface of a tank, where they can be concentrated and removed. That is accomplished by releasing gas bubbles, which attach to the solid particles, increasing their buoyancy and causing them to float. In principle, the process is the opposite of sedimentation. Fig. 13-13 shows one type of flotation system. Flotation processes that are applicable to oil removal are discussed in a later section of this chapter under the heading of the treatment of oily wastes and organics.

Flotation is used primarily in the treatment of waste water containing large quantities of industrial wastes that carry heavy loads of finely divided suspended solids. Solids having specific gravity only slightly greater than 1.0, which would require abnormally long sedimentation times, may be removed in much less time by flotation. The process may be performed in several ways, in which foam,

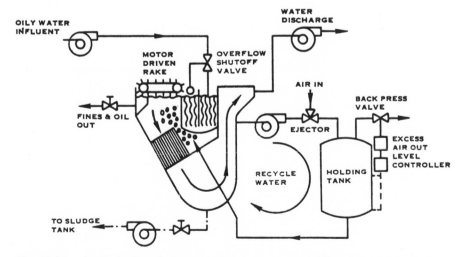

Fig. 13-13   An illustration of dissolved air flotation. Source: Effluent Guidelines Division, EPA 440/1-83/091, June 1983.

dispersed air, dissolved air, gravity, and vacuum flotation are the most commonly used techniques. In practice, chemical additives are often used to enhance the performance of the flotation process.

The principal difference between the various types of flotation techniques is the method of generation of the minute gas bubbles, usually air, in suspension of water and small particles. Addition of chemicals to improve the efficiency may be employed with any of the basic methods. The following paragraphs describe the different flotation techniques and the method of bubble generation for each process.

Foam flotation is based on the utilization of differences in the physiochemical properties of various particles. Wettability and surface properties affect the particles' ability to attach themselves to gas bubbles in an aqueous medium. In froth flotation air is blown through the solution containing flotation reagents. The particles with water-repellent surfaces stick to air bubbles as they rise and are brought to the surface. A mineralized froth layer with mineral particles attached to air bubbles is formed. Particles of other minerals that are readily wetted by water do not stick to air bubbles and remain in suspension.

In dispersed air flotation, gas bubbles are generated by introducing the air by means of mechanical agitation with impellers or by forcing air through porous media. With the dissolved air flotation method, bubbles are produced as a result of the release of air from a supersaturated solution under relatively high pressure. There are 2 types of contact between the gas bubbles and particles. The first type is predominant in the flotation of flocculated materials and involves the entrapment of rising gas bubbles in the flocculated particles as they increase in size. The bond between the bubble and particle is one of physical capture only. The second type of contact is one of adhesion. It results from the intermolecular attraction exerted at the interface between the solid particle and gaseous bubble.

The vacuum flotation process consists of saturating the waste water with air either (a) directly in an aeration tank, or (b) by permitting air to enter on the suction of a waste water pump. A partial vacuum is applied, which causes the dissolved air to come out of solution as minute bubbles. They attach to solid particles and rise to the surface to form a scum blanket, which is normally removed by a skimming mechanism. Grit and other heavy solids that settle to the bottom are generally raked to a contral sludge pump for removal. A typical vacuum flotation unit consists of a covered cylindrical tank in which a partial vacuum is maintained. The tank is equipped with scum and sludge removal mechanisms. The floating material is continuously swept to the tank periphery, automatically discharged into a scum trough, and removed from the unit by a pump, also under partial vacuum. Auxiliary equipment includes and aeration tank for saturating the waste water with air, a tank with a short retention time for removal of large bubbles, vacuum pumps, and sludge and scum pumps.

Flotation applies to most situations requiring separation of suspended materials. It is most advantageous for oils and for suspended solids of low specific gravity, or small particle size. Some advantages of the process are the high levels of solids separation achieved in many applications, the relatively low energy requirements, and the air flow adjustment capability to meet the requirements of treating different waste types. Limitations of flotation are that it often requires addition of chemicals to enhance process performance, and it generates large quantities of solid waste.

5. Membrane Filtration

Membrane filtration is a technique for removing precipitated heavy metals from a waste water stream. It must, therefore, be preceded by those treatment techniques that will properly prepare the waste water for solids removal. Typically, a membrane filtration unit is preceded by cyanide and chromium pretreatment as well as by pH adjustment for precipitation of the metals. Those steps are followed by the addition of a proprietary chemical reagent, which causes the metal precipitates to be nongelatinous, easily dewatered, and highly stable. The resulting mixture of pretreated waste water is continuously recirculated through a filter module and back into a recirculation tank. The filter module contains tubular membranes. While the reagent-metal precipitates mixture flows through the inside of the tubes, the water and any dissolved salts permeate the membrane. The permeate, essentially free of precipitate, is alkaline, noncorrosive, and safely dischargable to sewer or stream. When the recirculating slurry reaches a concentration of 10 percent to 15 percent solids, it is pumped out of the system as sludge.

Membrane filtration can be used in metal finishing in addition to sedimentation to remove precipitated metals and phosphates. Membrane filtration systems are being used in a number of industrial applications, particularly in the metal-finishing industry, and they have also been used for heavy metals removal in the paper industry. They have potential application in coil coating, porcelain enameling, battery, and copper and copper alloy plants.

A major advantage of the membrane filtration system is that installation can utilize most of the conventional end-of-pipe system that may already be in place. Also, the sludge is highly stable in an alkaline state. Removal efficiencies are excellent, even with sudden variation of pollutant input rates. The effectiveness of the membrane filtration system, however, can be limited by clogging of the filters. Because a change in the pH of the waste stream greatly intensifies the clogging problem, the pH must be carefully monitored and controlled. Clogging can force the shutdown of the system and may interfere with production.

The membrane filters must be regularly monitored and cleaned or replaced as necessary. Depending on the composition of the waste stream and its flow rate, cleaning of the filters may be required quite often. Flushing with hydrochloric acid for 6 hours to 24 hours will usually suffice. In addition, the routine maintenance of pumps, valves, and other plumbing is required.

When the recirculating, reagent-precipitate slurry reaches 10 percent to 15 percent solids, it is pumped out of the system. It can then be disposed of directly, or it can undergo a dewatering process. The sludge's leaching characteristics are such that the state of South Carolina has approved the sludge for landfill, provided that an alkaline condition be maintained. Tests carried out by the State indicate that even at the slightly acidic pH of 6.5, leachate from a sludge containing 2,600 mg/l of copper and 250 mg/l of zinc contained only 0.9 mg/l of copper and 0.1 mg/l of zinc.

The permeate is guaranteed by 1 manufacturer to contain less than the effluent concentrations shown in the following table, regardless of the influent concentrations. Those claims have been largely substantiated by the analysis of water samples at various plants, including those shown for comparison in the tabulation, on next page:

MEMBRANE FILTER PERFORMANCE (mg/l)

Metal, or TSS*	Manufacturer's guarantee	Pilot plant		Pilot plant	
		Raw	Treated	Raw	Treated
Aluminum	0.5	—	—	—	—
Chromium, hexavalent	0.03	0.46	0.01	5.25	<.005
Chromium, total	0.02	4.13	0.018	98.4	.057
Copper	0.1	18.8	0.043	8.00	.222
Iron	0.1	288	0.3	21.1	.263
Lead	0.05	.652	0.01	0.288	0.01
Cyanide	0.02	<.005	<.005	<.005	<.005
Nickel	0.1	9.56	.017	194.	.352
Zinc	0.1	2.09	.046	5.00	.051
*Total suspended solids	—	632	0.1	13.0	8.0

There are approximately 20 membrane filtration systems now in use by the metal-finishing and other industries. Bench scale and pilot studies are being run by the EPA in an attempt to expand the list of pollutants for which this system is known to be effective.

H. Treatment of Complexed Metal Wastes

1. Introduction

This section describes the treatment techniques applicable for the removal of complexed metal wastes. Complexed metal wastes within the metal-finishing industry are a product of electroless plating, immersion plating, etching, and printed circuit board manufacture. The metals in those waste streams are tied up or complexed by particular chemicals, complexing agents, whose function is to prevent metals from falling out of solution. That counteracts the precipitation techniques employed by most conventional metals removal methods, so these treatment methods are not always successful when used on complexed metal waste streams.

In order for the EPA's Effluent Guidelines Division to establish the performance of waste treatment systems in which complexed metal wastes were being treated, it was necessary to establish which plants were employing complexing agents. A list of complexing agents was compiled using information contained in plant portfolios and information obtained from a literature search. Table 13-6 presents a listing of the most commonly employed complexing agents.

I. Treatment of Hexavalent Chromium Wastes

1. Introduction

Hexavalent chromium-bearing waste waters are produced by the metal-finishing industry in the following processes:

a. in chromium electroplating;
b. in chromate conversion coatings;
c. in etching with chromic acid; and,

Table 13-6.   Common Complexing Agents

Ammonia	O-phenanthroline
Ammonium Chloride	Oxine, 8-Hydroxyquinoline (Q)
Ammonium Hydroxide	Oxinesulphonic Acid
Ammonium Bifluoride	Phthalocyanine
Acetylacetone	Potassium Ethyl Xanthate
Citric Acid	Phosphoric Acid
Chromotropic Acid (DNS)	Polyethyleneimine (PEI)
Cyanide*	Polymethacryloylacetone
DTPA	Poly (p-vinylbenzyliminodiacetic Acid)
Dipyridyl	Rochelle Salts
Disulfopyrocatechol (PDS)	Sodium Gluconate
Dimethylglyoxime	Sodium Pyrophosphate
Disalicylaldehyde 1, 2-propylenediimine	Succinic Acid
Dimercaptopropanol (BAL)	Sodium Tripolyphosphate
Dithizone	Sulphosalicylic Acid (SSA)
Diethyl Dithiophosphoric Acid	Salicylaldehyde
Ethylenediaminetetraacetic Acid (EDTA)	Salicylaldoxime
Ethylenebis (hydroxypnenylglycine) (EHPG)	Sodium Hydroxyacetate
Ethylenediamine	Sodium Citrate
Ethylenediaminetetra (methylenephosphoric Acid) (EDTPO)	Sodium Fluoride
Glyceric Acid	Sodium Malate
Glycolic Acid	Sodium Amino Acetate
Gluconic Acid	Tartaric Acid
Hydroxyethylethylenediaminetriacetic Acid (HEDTA)	Trisodium Phosphate (TSP)
Hydroxyethylidenediphosphonic Acid (HEDP)	Trifluoroacetylacetone
HEDDA	Thenoyltrifluoroacetone (TTA)
Lactic Acid	Triethylenetetramine
Malic Acid	Triaminotriethylamine
Monosodium Phosphate	Triethanolamine (TEA)
Nitrilotriacetic Acid (NTA)	Tetraphenylporphin
N-Dihydroxyethylglycine	Toluene Dithiol
Nitrilotrimethylenephosphonic Acid (NTPO, ATMP)	Thioglycolic Acid
N-Hydroxyethylethylenediamine	Thiourea

d. in metal-finishing operations carried out on chromium as a basic material.

(1) *Chemical Chromium Reductions*

Reduction is a chemical reaction in which electrons are transferred to the chemical being reduced from the chemical initiating the transfer, the reducing agent. Sulfur dioxide, sodium bisulfite, sodium metabisulfite, and ferrous sulfate form strong reducing agents in aqueous solution and are, therefore, useful in industrial waste treatment facilities for the reduction of hexavalent chromium to the trivalent form. The reduction enables the trivalent chromium to be separated from solution in conjunction with other metallic salts by alkaline precipitation. Gaseous sulfur dioxide is a widely used reducing agent and provides a good example of the chemical reduction process. Reduction using other reagents is chemically similar. The reactions involved may be illustrated as follows:

$$3\ SO_2 + H_2O = 3\ H_2SO_3$$
$$3\ H_2SO_3 + H_2CrO_4 = Cr_2\ (SO_4)_3 + 5\ H_2O$$

The above reaction is favored by low pH. A pH of 2 to 3 is normal for situations requiring complete reduction. At pH levels above 5, the reduction rate is slow. Each reaction tank has an electronic recorder-controller device to control process conditions with respect to pH and oxidation reduction potential, ORP. Gaseous sulfur dioxide is metered to the reaction tanks to maintain the ORP within the range of 250 millivolts to 300 millivolts. Sulfuric acid is added to maintain a pH level of from 1.8 to 2.0. Each of the reaction tanks is equipped with a propeller agitator designed to provide approximately 1 turnover per minute. Following reduction of the hexavalent chromium, the waste is combined with other waste streams for final adjustment to an appropriate alkaline pH to remove chromium and other metals by precipitation and sedimentation. Fig. 13-14 shows a continuous chromium reduction system.

Chromium reduction is used in metal finishing for treating chromium bearing waste streams, including chromium plating baths, chromating baths and rinses. The main application of chemical reduction to the treatment of waste water is in the reduction of hexavalent chromium to trivalent chromium. Rinse waters and cooling tower blowdown are 2 major sources of chromium in waste streams. A study of an operational waste treatment facility chemically reducing hexavalent chromium has shown that a 99.7 percent reduction efficiency is easily achieved. Final concentrations of 0.05 mg/l are readily attained, and those down to 0.01 mg/l are documented in the literature.

The major advantage of chemical reduction of hexavalent chromium is that it is a fully proved technology based on years of experience. Operation at ambient conditions results in minimal energy consumption, and the process, especially when one uses sulfur dioxide, is well suited to automatic control. Furthermore, the equipment is readily obtainable from many suppliers, and operation is straightforward.

One limitation of chemical reduction of hexavalent chromium is that for high concentrations of chromium, the cost of treatment chemicals may be correspondingly high. When that situation occurs, other treatment techniques are likely to be more economical. Chemical interference by oxidizing agents is possible in the treatment of mixed wastes, and the treatment itself may introduce pollutants if not properly controlled. In addition, the storage and handling of sulfur dioxide is somewhat hazardous. The reduction of chromium waste by sulfur dioxide or sodium bisulfite, however, is a well-founded process and is used by numerous plants where chromium compounds are employed in metal finishing and non-contact cooling operations.

(2) Electrochemical Chromium Reduction

That process has been developed to aid in the removal of chromium from metal finishing and cooling tower blowdown waste waters. It involves an electrochemical reaction in which consumable iron electrodes in the presence of an electrical current generate ferrous ions, which react with chromate ions in solution. The reaction produces chromic hydroxides and ferric hydroxides that can be removed in a settling pond or clarifier without the need for further chemical addition. The process has also been shown effective in removing zinc and other heavy metals. The metallic hydroxides formed are gelatinous and highly adsorptive. They can, therefore, coprecipitate other species that might be present in a waste water solution.

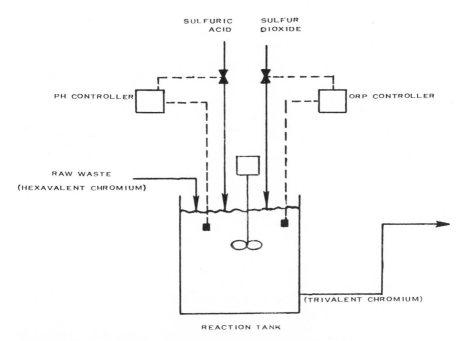

Fig. 13-14   A continuous hexavalent chromium reduction system with sulfur dioxide. Source: Effluent Guidelines Division, EPA 440/1-83/091, June 1983.

In addition to the electrochemical unit, the only equipment required is a pump and a clarifier or pond for settling. As long as the pH of the entering waste stream is between 7.0 and 8.0, no pH adjustment is necessary.

Although the process was developed for removal of chromium and zinc from cooling tower discharge, electrochemical chromium reduction can also be applied to the treatment of metal-finishing waste waters such as chromating baths and rinses. Coil coating and porcelain enameling plants are other potential applications. According to manufacturers, the electrochemical reduction process performs best on low concentration, high-volume waste water streams. Conventional chemical reduction is probably more economical in treating more concentrated effluents.

An advantage of the electrochemical chromium reduction process is that no pH adjustment chemicals are required with incoming pH values between 7 and 8. Retention time is unimportant when the pH is held within that range and the process is continuous and automatic. It is not efficient for effluents with high chromium concentrations, however, and species that consume hydroxide ions interfere with the precipitation of the ferric and ferrous hydroxides.

Since the iron electrodes are consumable, they need to be replaced periodically. Sedimentation is part of the process, and there is consequently a requirement for sludge processing and removal. The precipitation of ferric and chromic hydroxides generates waste sludge, which must eventually be dewatered and properly disposed of. No appreciable amounts of sludge are allowed to settle in the actual electrochemical process tank, which would interfere with the process.

That process is capable of removing hexavalent chromium from waste water to less than 0.05 mg/l with input chromium concentrations up to at least 20 mg/l. Performance that was noted for 1 plant follows:

Pollutant	Influent	Effluent
Hexavalent Chromium	10 mg/l	0.05 mg/l
Zinc	3	0.1

There are more than 50 electrochemical reduction systems in operation in a variety of industries, mostly in organic and inorganic chemicals plants. Five are now in service at plants in the metal-finishing industry. The process has potential for applications in the photographic industry since it has been shown to successfully remove silver from waste waters.

(3) *Electrochemical Chromium Regeneration*

Chromic acid baths must be continually discarded and replenished to prevent buildup of trivalent chromium. An electrochemical system employing a lead anode and nickel cathode has been developed to recover chromium by converting the trivalent form to the hexavalent form. In this process, trivalent chromium is electro-oxidized to hexavalent chromium at the lead anode while hydrogen is released at the cathode. The process is similar to the electrodialytic chromium oxidation process, but no membrane is used to separate the concentrate from the dilute solution. The reaction is carried out at 68° C, a cell voltage of 4.5 volts, and an anode-to-cathode area ratio of 30:1. The same process can also be used to recover chromium from chromic oxide sludges precipitated by conventional chemical chromium waste treatment. The sludges are dissolved in 200 g/l chromic acid and electro-oxidized under slightly different operating conditions than those previously described.

Electrochemical chromium regeneration can be used in metal finishing to prolong the life of chromium plating and chromating baths. Chromic acid baths are used for electroplating, anodizing, etching, chromating, and sealing. The electro-oxidation process has been commercially applied to regeneration of a plastic etchant. In this particular installation, chromic acid dragged out of the etching bath into the first stage of a countercurrent rinse is concentrated by evaporation and returned to the etching bath. This closed loop system tends to cause a rapid buildup of trivalent chromium. When the etchant is recirculated through an electrochemical regeneration unit, however, the trivalent chromium is oxidized to the hexavalent form. The process has, also, been applied to regeneration of a chromic acid sealing bath in the coil-coating industry.

Some advantages of the electrochemical chromium regeneration process are its relatively low energy consumption; its operation at normal bath temperatures, eliminating the need for heating or cooling; its ability for recovering and reusing valuable process chemicals; and its elimination of sludges generated by conventional chromium treatment processes. Some limitations of chromium electrooxidation are low current efficiencies for baths with less than 5.0 g/l trivalent chromium, need for control of impurities that can interfere with the process, and dependence on electrical energy for oxidation to take place. The current efficiency for the process is 80 percent at concentrations above 5 g/l. If a trivalent chromium concentration of less than 5 g/l were treated, research has shown that the current efficiency would drop.

(4) *Evaporation*

Evaporation, which is explained in detail in the "Treatment of Common Metal

Wastes" has found applicability in the treatment of chromium-bearing wastes, especially the rinse waters after chromium plating. The rinse waters following the finishing operation (normally a countercurrent rinse of at least 3 stages) are sent to an evaporator. Here the chromium-bearing solution is broken down into water and process solution (predominantly chromic acid). The water is returned to the last (cleanest) stage of the countercurrent rinse, and the process solution may be returned to the process tank or put aside for sale to a scavenger.

(5) *Ion Exchange*

Ion exchange is another possible method for recovering and regenerating chromic acid solution. As explained under the section on the "Treatment of Common Metal Wastes," anions such as chromates or dichromates can be removed from rinse waters with an anion exchange resin. In order to regenerate the resin, caustic is pumped through the anion exchanger, carrying out sodium dichromate. The sodium dichromate stream is passed through a cation exchanger, converting the sodium dichromate to chromic acid. After some means of concentration, such as evaporation, the chromic acid can be returned to the process bath.

J. Treatment of Cyanide Wastes

1. Introduction

Cyanides are introduced as metal salts for plating and conversion coating or are active components in plating and cleaning baths. Cyanide is generally destroyed by oxidation.

(a) *Oxidation by Chlorination*

Chlorine is used primarily as an oxidizing agent in industrial waste treatment to destroy cyanide. Chlorine can be used in the elemental or hypochlorite form. That classic procedure can be illustrated by the following 2-step chemical reaction:

1. $Cl_2 + NaCN + 2NaOH = NaCNO + 2NaCl + H_2O$
2. $3Cl_2 + 6NaOH + 2NaCNO = 2NaHCO_3 + N_2 + 6NaCl + 2H_2O$

The reaction presented as equation (2) for the oxidation of cyanate is the final step in the oxidation of cyanide. A complete system for the alkaline chlorination of cyanide is shown in Fig. 13-15.

The cyanide waste flow is treated by the alkaline chlorination process for oxidation of cyanides to carbon dioxide and nitrogen. The equipment often consists of an equalization tank followed by two reaction tanks, although the reaction can be carried out in a single tank. Each has an electronic recorder-controller to maintain required conditions with respect to pH and oxidation-reduction potential (ORP). In the first reaction tank, conditions are adjusted to oxidize cyanides to cyanates. To effect the reaction, chlorine is metered to the reaction tank as required to maintain the ORP in the range of 350 millivolts to 400 millivolts, and 50 percent, aqueous caustic soda is added to maintain a pH range of 9.5 to 10.

In the second reaction tank, conditions are maintained to oxidize cyanate to carbon dioxide and nitrogen. The desirable ORP and pH for that reaction are 600 millivolts and a pH of 8.0. Each of the reaction tanks is equipped with a

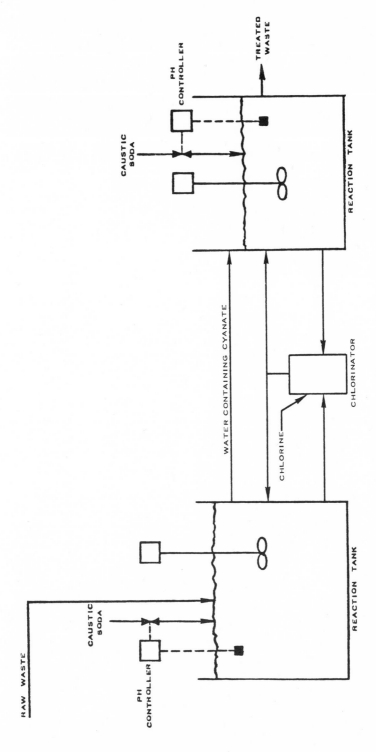

Fig. 13-15   The treatment of cyanide waste by alkaline chlorination. Source: Effluent Guidelines Division, EPA 440/1-83/091, June 1983.

propeller agitator designed to provide approximately 1 turnover per minute. Treatment by the batch process is accomplished by using 2 tanks, one for collection of waste over a specified time period, and 1 tank for the treatment of an accumulated batch. If dumps of concentrated wastes are frequent, another tank may be required to equalize the flow to the treatment tank. When the holding tank is full, the liquid is transferred to the reaction tank for treatment. After treatment, the supernatant is discharged and the sludges are collected for removal and ultimate disposal.

The oxidation of cyanide waste by chlorine is a classic process and is found in most plants using cyanide. The process is capable of achieving efficiencies of 99 percent or greater and effluent levels that are nondetectable. Chlorine has also been used to oxidize phenols, but use of chlorine dioxide for that purpose is much preferred because formation of toxic chlorophenols is avoided.

Some advantages of chlorine oxidation for handling process effluents are operation at ambient temperature, suitability for automatic control, and low cost. Some disadvantages of chlorine oxidation for treatment of process effluents are that toxic, volatile intermediate reaction products must be controlled by careful pH adjustment; chemical interference is possible in the treatment of mixed wastes; and a potentially hazardous situation exists when chlorine gas is stored and handled.

The oxidation of cyanide wastes by chlorine is a widely used process in plants using cyanide in cleaning and plating baths. There has been recent attention to developing chlorine dioxide generators and bromine chloride generators. A problem that has been encountered is that the generators simultaneously produce not only the bromine chloride and chlorine dioxide gas but also chlorine gas. Both of those gases are extremely unstable and corrosive, and both have low vapor pressure, which results in handling difficulties. These generators are in the development stages and as advances are made in their design, they may become competitive with chlorine.

(b) *Oxidation By Ozonation*

Ozone may be produced by several methods, but the silent electrical discharge method is predominant in the field. That process produces ozone by passing oxygen or air between electrodes separated by an insulating material. The electrodes are usually stainless steel or aluminum. The dielectric or insulating material is usually glass. The gap or air space between electrodes or dielectrics must be uniform and is usually on the order of 0.100 to 0.125 inches. The voltage applied is 20,000 volts or more, and a single phase current is applied to the high tension electrode.

Ozone is approximately 10 times more soluble than oxygen on a weight basis in water, although the amount that can be efficiently dissolved is still slight. Ozone's solubility is proportional to its partial pressure and also depends on the total pressure on the system. It should be noted, however, that it is the oxidizable contaminant in the water that determines the quantity of ozone needed to oxidize the contaminants present. A complete ozonation system is represented in Fig. 13-16.

Thorough distribution of ozone in the water under treatment is extremely important for high efficiency of the process. There are 4 methods of mixing ozone with water: (1) diffusers, (2) negative or positive pressure injection, (3) packed columns whereby ozone-containing air or oxygen is distributed throughout the water, and (4) atomizing the aqueous solution into a gaseous atmosphere containing ozone.

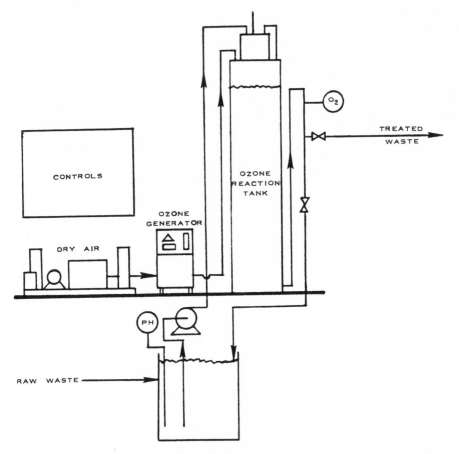

Fig. 13-16 A typical ozonation plant for waste treatment. Source: Effluent Guidelines Division, EPA 440/1-83/091, June 1983.

Ozonation has been applied commercially for oxidation of cyanides, phenolic chemicals, and organo-metal complexes. It is used commercially with good results to treat photo-processing waste waters. Divalent iron hexacyanate complexes—spent bleach—are oxidized to the trivalent form with ozone and reused for bleaching purposes. Ozone is used to oxidize cyanides in other industrial waste waters and to oxidize phenols and dyes to a variety of colorless, nontoxic products.

Oxidation of cyanide to cyanate is illustrated below:

$$CN^{-1} + O_3 = CNO^{-1} + O_2$$

Continued exposure to ozone will convert the cyanate formed to carbon dioxide and ammonia if the reaction is allowed to proceed; however, that is not economically practical, and cyanate can be economically decomposed by biological oxidation to neutral pH.

Ozone oxidation of cyanide to cyanate requires 1.8 pounds to 2.0 pounds of ozone per pound of CN^-, and complete oxidation requires 4.6 pounds to 5.0 pounds of ozone per pound of CN^-. Zinc, copper, and nickel cyanides are easily destroyed to a nondetectable level, but cobalt cyanide is resistant to ozone treatment.

The first commercial plant using ozone in the treatment of cyanide waste was installed by a manufacturer of aircraft. The plant is capable of generating 54.4 Kg (120 pounds) of ozone per day. The concentration of ozone used in the treatment is approximately 20 mg/l. In the process the cyanate is hydrolyzed to CO_2 and NH_3. The final effluent from this process passes into a lagoon. Because of an increase in waste flow the original installation has been expanded to produce 162.3 Kg (360 pounds) of ozone per day.

Some advantages of ozone oxidation for handling process effluents are that it is well suited to automatic control, on-site generation eliminates treatment chemical procurement and storage problems, reaction products are not chlorinated organics, and no dissolved solids are added in the treatment step. Ozone in the presence of ultraviolet radiation or other promoters such as hydrogen peroxide and ultrasound shows promise of reducing reaction time and improving ozone utilization. Some limitations of the process are high capital expense, possible chemical interference in the treatment of mixed wastes, and an energy requirement of 15 kwh to 22 kwh per kilogram of ozone generated. Cyanide is not economically oxidized beyond the cyanate form.

Ozone is useful for application to cyanide destruction. There are at least 2 units now in operation in the country, and additional units are planned. Also, there are numerous orders for industrial ozonation cyanide treatment systems pending.

Ozone is useful in the destruction of waste waters containing phenolic materials, and there are several installations in operation in the United States. Research and development activities within the photographic industry have established that ozone is capable of treating some compounds that are produced as waste products. Solutions of key ingredients in photographic products were composed and treated with ozone under laboratory conditions to determine the treatment of those solutions. It was found that some of them were oxidized almost completely by ozonation, and some were oxidized that were difficult to treat by conventional methods. Ozone breaks down certain developer agents that biodegrade slowly, including color-developing agents, pheniodone, and hydroxylamine sulfate. Developing agents, thiocyanate ions, and formate ions degrade ore completely with ozone than when they are exposed to biological degradation. Thiosulfate, sulfite, formalin, benzyl alcohol, hydroquinone, maleic acid, and ethylene glycol can be degraded to a more or less equal degree with either biological treatment or ozone. Silver thiosulfate complexes were also treated with ozone, resulting in significant recovery of the silver present in solution. Ozone for regeneration of iron cyanide photo-processing bleach and treatment of thiosulfate, hydroquinone, and other chemicals is currently being utilized by the photo-processing industry. There are 40 to 50 installations of this type in use at the present time.

(c) *Oxidation by Ozonation with UV Radiation*

One of the modifications of the ozonation process is the simultaneous application of ultraviolet light and ozone for the treatment of waste water, including treatment of halogenated organics. The combined action of the 2 forms produces reactions by photolysis, photosensitization, hydroxylation, oxygenation, and oxidation. The process is unique because several reactions and reaction species are active simultaneously.

Ozonation is facilitated by ultraviolet absorption because both the ozone and the reactant molecules are raised to a higher energy state so that they react more rapidly. The energy and reaction intermediates created by the introduction of both ultraviolet radiation and ozone greatly reduce the amount of ozone required as compared with a system that utilizes ozone alone to achieve the same level of treatment. Fig. 13-17 shows a three-stage UV/ozone system.

A typical process configuration employs 3 single stage reactors. Each is a closed system that is illuminated with ultraviolet lamps placed in the reactors, and the ozone gas is sparged into the solution from the bottom of the tank. The ozone dosage rate requires 2.6 pounds of ozone per pound of chlorinated aromatic. The ultraviolet power is on the order of 5 watts of useful ultraviolet light per gallon of reactor volume. Operation of the system is at ambient temperature, and the residence time per reaction stage is about 24 minutes. Thorough mixing is necessary, and the requirement for that particular system is 20 horsepower per 1,000 gallons of reactor volume in quadrant baffled reaction stages. A system to treat mixed cyanides requires pretreatment that involves chemical coagulation, sedimentation, clarification, equalization, and pH adjustment. Pretreatment is

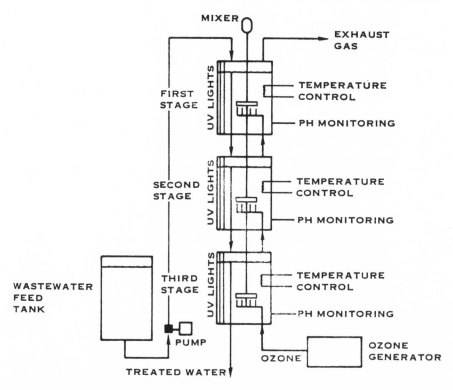

Fig. 13-17 A 3-stage ultraviolet/ozonation system. Source: Effluent Guidelines Division, EPA 440/1-83/091, June 1983.

followed by a single stage reactor, where constituents with low refractory indices are oxidized. That may be followed by a second, multi-stage reactor, which handles constituents with higher refractory indices. Staging in that manner reduces the ultimate reactor volume required for efficient treatment.

The ozonation/UV radiation process was developed primarily for cyanide treatment in the metal-finishing and color photo-processing areas, and it has been successfully applied to mixed cyanides and organics from organic chemicals manufacturing methods. The procedure is particularly useful for treatment of complexed cyanides, such as ferricyanide, copper cyanide, and nickel cyanide, which are resistant to ozone alone but readily oxidized by ozone with UV radiation.

For mixed metal cyanide wastes, consistent reduction in total cyanide concentration to less than 0.1 mg/l is claimed. Metals are converted to oxides, and halogenated organics are destroyed. Total organic carbon (TOC) and chemical oxygen demand (COD) concentrations are reduced to less than 1 mg/l.

A full-scale unit for treating metal complexed cyanides has been installed in Oklahoma, and a large American chemical company in France has installed an on-line unit for the treatment of cyanides and organics. A similar design is scheduled for installation by the same company in the United States. There are also 2 other units known to be in service, 1 for treating mixed cyanides and the other for treatment of copper cyanide.

(d) *Oxidation by Hydrogen Peroxide*

The hydrogen peroxide oxidation treatment process treats both the cyanide and metals in cyanide waste waters containing zinc or cadmium. In this process, cyanide rinse waters are heated to 49° to 54° C (120° to 130° F) to break the cyanide complex, and the pH is adjusted to 10.5 to 11.8. Formalin (37 percent formaldehyde) is added, while the tank is vigorously agitated. After 2 to 5 minutes, a proprietary formulation (41 percent hydrogen peroxide with a catalyst and additives) is likewise added. After an hour of mixing, the reaction is complete. The cyanide is converted to cyanate, and the metals are precipitated as oxides or hydroxides. The metals are then removed from solution by either settling or filtration.

The chemical reactions that take place are as follows:

$$CN + HCHO + H_2O = HOCH_2CN + OH^-$$

The hydrogen peroxide converts cyanide to cyanate in a single step:

$$CN + H_2O_2 = NCO + H_2O$$

The formaldehyde also acts as a reducer, combining with the cyanide ions:

$$Zn(CN)_4^{-2} + 4\ HCHO + 4H_2O = 4\ HOCH_2CN + 4\ OH^- + Zn^{-2}$$

The metals subsequently react with the hydroxyl ions formed and precipitate as hydroxides or oxides:

$$Zn^{+2} + 2\ OH^- = ZnO + H_2O$$

The main pieces of equipment required for this process are 2 holding tanks. They must be equipped with heaters and air spargers or mechanical stirrers. The tanks may be used in a batch or continuous fashion with 1 tank being used for treatment while the other is being filled. A settling tank or a filter is needed to concentrate the precipitate.

The hydrogen peroxide oxidation process is applicable to cyanide-bearing waste waters, especially those from cyanide zinc and cyanide cadmium electroplating. The process has been used on photographic wastes to recover silver and oxidize toxic compounds, such as cyanides, phenols and "hypo" (sodium thiosulfate pentahydrate). Additions of hydrogen peroxide are made regularly at a large waste water treatment plant to control odors and minimize pipe corrosion by oxidizing hydrogen sulfide.

Chemical costs are similar to those for alkaline chlorination and lower than those for treatment with hypochlorite, and all free cyanide reacts and is completely oxidized to the less toxic cyanate state. In addition, metals precipitate and settle quickly, and they are recoverable in many instances. The process requires energy expenditures to heat the waste water before treatment. Furthermore, the addition of formaldehyde results in treated waste water having relatively high biochemical oxygen demand (BOD) values. Although cyanates are much less toxic than cyanide, there is not complete acceptance of the harmlessness of cyanates.

In terms of waste reduction performance, this process is capable of reducing the cyanide level to less than 0.1 mg/l and the zinc or cadmium to less than 1.0 mg/l. That treatment process was introduced in 1971 and is being used in several facilities.

(e) *Electrochemical Cyanide Oxidation*

Electrochemical cyanide oxidation is used to reduce free cyanide and cyanate levels in industrial waste waters. In the process waste water is accumulated in a storage tank and then pumped to a reactor, where an applied DC potential oxidizes the cyanide to nitrogen, carbon dioxide, and trace amounts of ammonia. The gases generated are vented to the atmosphere. The oxidation reaction is accomplished if concentrations are not greater than 1,000 mg/l. If reaction time is critical, the process can be accelerated by augmenting the system with a chemical (hypochlorite) treatment as long as the cyanide concentration level is less than 200 mg/l. The process equipment consists of a reactor, a power supply, a storage tank, and a pump.

Another electrochemical oxidation system employs a low voltage anode with a metallic oxide coating. Upon application of an electrical potential, several oxidation reactions occur at the anode. They include the oxidation of chloride from common salt to chlorine or hypochlorite and the formation of ozone as well as direct oxidation at the anode. Although untested on cyanide-bearing waste waters, this system shows good potential in that area.

The electrochemical cyanide oxidation system has been used commercially only for heat treating applications; however, it should be equally appropriate for other cyanide-bearing wastes. Its application for plating and photographic process waste waters is still in the development stage. The process can also be applied to the electrochemical oxidation of nitrite to nitrate.

Electrochemical cyanide oxidation has the advantage of low operating costs with moderate capital investment, relative to alternative processes. There is no requirement for chemicals, thereby eliminating both their storage and control, and there is no need to dilute or pretreat the waste water, as the process is most efficient at high cyanide concentration levels. It is less efficient, however, than chemical destruction at cyanide concentrations less than 100 mg/l, and it is relatively slow when not accelerated by addition of treatment chemicals. Moreover, it will not work well in the presence of sulfates.

Performance has been demonstrated on a commercial scale and shown to result in a reduction in the cyanide concentration level from 3,500 mg/l to less than

1.0 mg/l in 160 hours. The process emits no noticeable odor with adequate ventilation. Currently in operation is a unit that is handling the cyanide-bearing waste water generated by a heat-treating operation. The manufacturer claims there is a potential for future use of the process in both the electroplating and photographic industries. Despite a variety of experimental programs, however, industry has not been enthusiastic about the electrolytic approach to cyanide oxidation.

(f) *Chemical Precipitation*

Chemical precipitation is a classic waste treatment process for metals removal as described under the "Treatment of Common Metal Wastes" heading. The precipitation of cyanide can be accomplished by treatment with ferrous sulfate. That process precipitates the cyanide as a ferrocyanide, which can be removed in a subsequent sedimentation step. Waste streams with a total cyanide content of 2 mg/l or above have an expected waste reduction of 1.5 to 2 orders of magnitude.

(g) *Evaporation*

Evaporation is another recovery alternative applicable to cyanide process baths, such as copper cyanide, zinc cyanide, and cadmium cyanide and was described in detail for common metals removal.

K. Treatment of Oily Wastes

Oily wastes include process coolants and lubricants, wastes from cleaning operations directly following many other unit operations, wastes from painting processes, and machinery lubricants. Oily wastes generally are of 3 types: free oils, emulsified or water soluble oils, and greases. Techniques commonly employed in the metal-finishing industry to remove oil include skimming, coalescing, emulsion breaking, flotation, centrifugation, ultrafiltration, reverse osmosis, and removal by contractor hauling. Oil removal techniques may also afford additional removal of toxic organics, and the applicability and performance of these techniques for toxic organics is discussed under "Treatment of Toxic Organics."

L. The Treatment of Toxic Organics

Toxic organics may be removed from spent solvents, from total plant process waste waters, and from oily waste streams. The major sources of those toxic organics in metal-finishing processes are waste solvents from degreasing operations. Thus, the primary control technology for toxic organics is not to dump them in concentrated form directly into the waste streams or to combine them with any waste that will, subsequently, enter the waste treatment system. The method for controlling toxic organics in the waste water, therefore, is to segregate concentrated toxic organic wastes for contract haulage, i.e., bodily remove them from the plant or segregate them for reclamation. It is also possible to use alternative methodologies for solvent degreasing in order to reduce the dependence on solvents and to eliminate the quantity of waste solvents that are generated.

14

Acid Rain

A. Introduction

No text on hazardous materials and waste would be complete without a discussion of such hazardous wastes as sulfuric acid and nitrogen oxide. During the past several decades the composition of moisture in the form of rain, snow, sleet, fog, dew, and frost that falls on the earth has changed drastically. We now have acid rain, acid snow, acid dew, and so on to complete the roster. If there is any one problem that has surfaced recently to affect our quality of life on this planet it is acid rain.

B. Weather Systems

Weather systems approaching our western coast bring moisture that is slightly acidic, with a chemical composition much like that of the ocean. As those weather systems pass over industrial areas, air pollutants react with the moisture, making it increasingly acidic. Studies show that rain falling in some parts of western Washington is 10 times more acidic than that in the coastal areas. Although parts of Washington do receive acidic rain, at this time there is little definite evidence that lakes are becoming acidic, or that other environmental damage is occurring. In fact, very little is known about the nature and extent of acid rain in the Northwest.

High-altitude lakes in the Cascades and Olympics are potentially very vulnerable to acid rain, as has been discovered. In lowland drainage areas, soils and vegetation are able to absorb and neutralize large amounts of acid rain, because moisture is filtered through sand, soil, and rock formations. The rocks, soils, and vegetation in high-elevation drainages are often unable to perform that filtration and neutralizing function as well. If enough acid water runoff from a surrounding area reaches a lake, its natural balance may be thrown off, resulting in a more acid condition. As the process continues, the lake gradually becomes less able to support fish, because most fish cannot exist in water that is more acid than alkaline.

Despite the widespread occurrence of acid rain in the United States and the world, scientists still do not completely understand the complex chemical reactions that take place in the atmosphere. Apparently, emissions of sulfur and nitrogen oxides from coal-burning power plants, smelters, factories, and automobiles cause acid rain. Rising high into the air and transported long distances by winds, the gases are transformed to sulfuric and nitric acid. The acids are then absorbed by rain and snow and fall to earth as acid rain.

Acid deposition may also occur as dry "fallout" from polluted air. It usually happens close to the pollution sources but may occur hundreds of miles away during a long dry spell with no precipitation. The dry particles may react with the rain to increase the acidity of precipitation and make the acid rain problem worse.

C. Natural and Human Causes of Acid Rain

There are both natural and human contributions to this fallout problem. As an example, in the state of Washington, the ASARCO copper smelter in Tacoma emitted up to 90,000 tons of sulfur dioxide a year, which accounted for about 60 percent of the sulfur dioxide in western Washington and about 40 percent of the statewide total. Since the plant stopped smelting copper in early 1985, the Washington State Department of Ecology has been assessing the reduction of emissions on acid rain levels in the state. Pacific Power and Light Company's coal-fired power plant in Centralia produces 55,000 tons of sulfur dioxide and 35,000 tons of nitrogen oxides per year. Other major pollution sources in the state are pulp mills, aluminum plants, petroleum refineries and motor vehicles. Mt. St. helens, the active volcano that erupted violently a few years ago, is a significant natural contributor to Washington's total sulfur dioxide emission problem.

D. How Acidic is Rain?

The acidity of any solution or liquid, including rain or snow, is measured on a scale known as a pH scale. This scale is graduated from 0 to 14, with 7 considered neutral. Liquids with a pH of less than 7 are acidic. The closer the pH is to zero, the greater the acidity. Readings more than 7 indicate alkalinity, and the closer to 14, the more alkaline the liquid. The pH scale is a logarithmic measure. That means each change of 1 pH unit, say from pH to 6 to pH 5, represents a tenfold increase and a drop from pH 6 to pH 4 represents a hundredfold increase in acidity.

Scientists generally consider unpolluted rainwater, which is slightly acidic, to have a pH around 5.6. That pH value takes into consideration the amount of acidity created by reaction of rainwater with normal levels of carbon dioxide in the atmosphere. See Fig. 14-1, in which the pH scale is illustrated.

The average precipitation in most states east of the Mississippi River has a pH value between 4 and 5, with individual storms well below those averages. Hardly any of the states have been spared from acid rain. Storm systems approaching the coast of western Washington have a pH of about 5.5 to 5.7. As the moist air passes over industrial areas, the acidity is increased (and pH lowered) by reaction with acid-forming air pollutants. Rainfall collected and analyzed in Olympia, Seattle, Bellingham, and the Cascade Mountains has an average pH of 4.5 to 4.7. A low pH of 3.6 was measured in Seattle when a storm carrying acid-forming pollutants from the Tacoma area passed directly over the rainfall sampler.

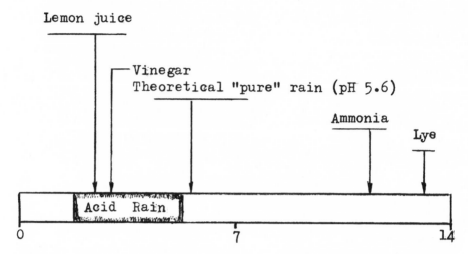

Fig. 14-1   The logarithmic pH scale. Note that it registers numbers from 0 to 14. The closer to 0, the greater the acidity, the closer to 14 the greater the alkalinity of a liquid.

E. The Acid Rain Problem is Widespread

The damage caused by acid rain has been felt not only in the Western States but also in eastern United States, Canada, Europe, and Scandinavia. Sensitive lakes in those areas have become so acidic that they no longer support fish life. Large forest areas are affected by a problem thought to be associated with acid rain. It is also responsible for the increased corrosion of monuments and buildings, including the Statue of Liberty and the U.S. Capitol as well as other public buildings.

The extent of change in the acidity of a lake or stream resulting from acid rain is determined mainly by the neutralizing or buffering capacity of nearby soils and other characteristics of the surrounding drainage area. If the soil of the drainage area, or "watershed," is alkaline or has the capacity to neutralize incoming acids, most of the water entering the lakes and streams will be less acidic than the incoming rain. Therefore, aquatic life in those lakes and streams will be far less vulnerable to harm. Lakes in watersheds with large amounts of exposed, resistant bedrock, such as granite, generally have very little acid-neutralizing capacity and are more likely to become acidic. (See Fig. 14-2, for an illustration of a typical, watershed area.)

The alkalinity of a lake is a measure of its acid-neutralizing or buffering capacity. It is the natural protective buffering action that consumes acidic materials entering the lake. If the watershed cannot produce enough alkaline material through erosion of rocks and soil leaching to keep up with the incoming acid, the alkalinity will gradually decrease. The pH will not significantly decrease, however, until almost all the alkalinity in the lake is used up. In some states, many sensitive lakes have naturally low alkaline levels, but no observable decrease in lake alkalinity has been measured so far.

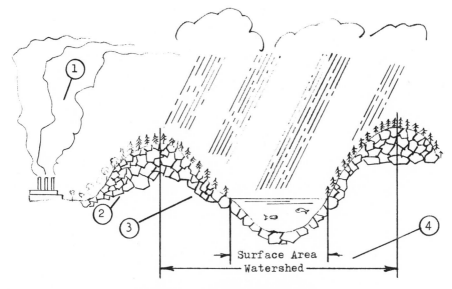

Fig. 14-2   A typical watershed area.

F. The Effect on Fish Life

Acid rain can have a devastating impact on fish in lakes that have already become acidic, although the process resulting in fish losses is not well understood. The scientific community is divided on that question, because some scientists believe it is related to pH, while others believe it results from a reduction of food sources or an excess of aluminum found in low pH waters.

The following facts indicate that whatever the cause, the effect of acid rain is taking its toll:

(1) Five thousand lakes in southwest Sweden are not capable of supporting fish;

(2) Fish in 1,500 lakes and 7 Atlantic Salmon rivers in southern Norway have been lost;

(3) Ontario, Canada, has lost fish in an estimated 1,200 lakes, and Canadian authorities calculate that the province stands to lose the fish in 48,000 more lakes within the next 20 years if acid rain continues at the present rate;

(4) In New York State acid rain has rendered 212 lakes in the Adirondacks unfit for fish.

Paradoxically, in some lakes, as young fish fail to develop and cannot compete for food, the older fish grow larger, providing a bonanza for anglers who report catching the "big ones" just before acidification reaches lethal levels. In addition to suffering fish losses, an acidified body of water loses hundreds of other organisms, including certain types of algae, aquatic crustaceans and mollusks, and insects, thus indicating tremendous damage to the ecological balance.

G. The Effect on Forests and Other Vegetation

The forest products industry has noted unquestionable damage to tree growth in many regions of the Northern Hemisphere, particularly in Germany, where up to 60 percent of some forest areas are dying. Although acid rain cannot be specifically blamed for such damage, it is widely thought that the combination of acid rain and other environmental and climatogical factors may be the cause of the problems.

The effect of acid rain on growing crops is difficult to determine scientifically under natural conditions. Simulated acid rain, however, has caused reduced growth, and reduced yield for such crops as lettuce, soybeans, tomatoes, beets, carrots, and broccoli. Other crops that may be damaged include alfalfa, rye, and oats.

H. The Effect on Human Life

It is thought that acid rain can also affect people's health by introducing heavy metals into our drinking water supplies. That happens by increased leaching of metals in groundwater and the corrosion of metal pipes. Shallow wells, where rainfall rapidly penetrates into the groundwater, may have higher concentrations of aluminum, mercury, and cadmium as well as other contaminants. Since the rain entering the shallow wellground system has not had time to be neutralized, it dissolves those contaminants from the soils and rocks, carrying them into the drinking water. Acid waters also dissolve metal piping, which may add iron, copper, lead, or zinc to the drinking water supply.

Some recent scientific theory has advanced the proposition that dissolved aluminum may be a contributing factor in Alzheimer's disease, a disease usually of old age, which brings on senility.

I. Controlling Acid Rain

Extensive research regarding acid rain is being performed by government, industry, and independent scientists both in the United States and overseas. Also, many of our states have established rainfall-monitoring programs to assess the effects of acid rain in its various forms. In addition, many studies are under way to investigate a number of ecological factors, including trends in air and water quality. Thus, there is no doubt that the problems of acid rain have captured the imagination and creative genius of the scientific brotherhood worldwide.

It is not only to the scientists and academicians that the gauntlet has been flung, but it remains for the world's legislative bodies to take the necessary actions to reduce acid rain so that the quality of life may be maintained. The only long-term solution to the problem of acid rain is to reduce and then eliminate the sources of acid rain pollution. It is well known that the public utilities with their belching smokestacks pump thousands of tons of sulfur and nitrogen oxides into the air every day of the year. The so-called smokestack industries and automobiles contribute their share of pollutants, but thanks to increased vigilance of some consumer groups scrubbers and automated emission controls are reducing that overall effect. We know also that because of public awareness emission levels of pollutants is on the decline, but is it declining fast enough?

There are several ways that emission levels may be reduced to further low levels:

(1) By the treatment of fuels and smokestack gases in existing oil and coal-burning utilities;

(2) By using nuclear energy and solar power instead of fossil fuels;

(3) By conserving energy through more efficient fuel use; and,

(4) By the use of fuels that are low in sulfur content.

15

Oil Spills

A. Introduction

Petroleum meets almost half of our total energy needs, making it the largest direct source of energy in the United States. This large demand coupled with a limited domestic supply has required us to import nearly one-half of our oil at an annual cost of more than $30 billion. Demand has grown to the extent that at 1 point during the last decade we were importing an average of more than 240 million gallons of oil each day, predominantly by ship. See Chart 15-1.

As the world has learned in the past few years, the transportation of petroleum involves significant elements of risk. The 60 million-gallon *Amoco Cadiz* tanker spill off the northern coast of France provided a graphic example of the severe impact of oil spills on the environment.

Some oil spills frequently result in extensive damage to plant and animal life, unsightly fouling of beaches and waterways, and great economic hardship for individuals and communities. In addition to the more familiar damage, petroleum products have been shown to affect plant and animal growth patterns by decreasing the rates of photosynthesis, disrupting feeding and reproductive behavior, and seriously impairing other vital biological processes.

The need for a comprehensive Federal oil spills research program became increasingly apparent in the aftermath of such major spills as the Santa Barbara, California, offshore well blowout of 1969 and the December 1976 grounding of the *Argo Merchant* near Nantucket, Massachusetts. (See Figs. 15-1 and 15-2.) Numerous agencies including the EPA, National Oceanic and Atmospheric Administration, the Coast Guard, and the Departments of Energy and Interior, are engaged in research projects directed toward preventing and minimizing the effects of oil spills. One major group coordinating a portion of this research is the Interagency Energy/Environment R&D Program, sponsored by the Office of Research and Development (ORD) within the EPA. In the Interagency Program, nearly $4 million is being spent annually on oil-spills-related research.

Oil spills research is also 1 of the aspects of ocean pollution research being considered by the Interagency Committee on Ocean Pollution Research and Development and Monitoring. The committee, chaired by the National Oceanic and Atmospheric Administration and composed of several Federal agencies including the EPA, is developing a 5-year plan for a comprehensive marine-pollution-related Federal program. The planning activity is mandated by the Ocean Pollution Research and Development and Monitoring Planning Act of 1978.

Federal responsibilities for regulating oil spills are divided between the EPA and the Coast Guard under 2 broad areas: response and prevention. The EPA is charged with setting regulations for responses to inland spills, while the Coast Guard is concerned with spills occurring in coastal waters and the Great Lakes.

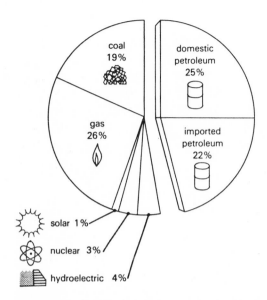

Chart 15-1   Energy sources of the U.S., relative percentages. Source: EPA-600/8-79-007, February 1979.

With regard to prevention, the EPA and the Coast Guard are responsible for nontransportation-related and transportation-related spills, respectively. To enable the EPA to fulfill its regulatory functions, the Office of Research and Development has developed a comprehensive oil-spills research program. Major initiatives are under way or have been completed in the following areas:

(1) Development of oil spill prevention, control, cleanup, and monitoring methods and technologies to minimize adverse effects on the environment.

(2) Improvement of understanding of the impacts of spills on ecosystems and our capability to predict and assess the effects of petroleum on plants and animals.

(3) Implementation of efficient methods of transferring research information through technical reports, manuals, workshops, and other avenues of information dissemination.

Every year more than 10 million gallons of oil escape into U.S. waters as a result of more than 10,000 spills. (See Chart 15-2.) The National Oil and Hazardous Substances Pollution Contingency Plan, developed as a result of section 311 of the Federal Water Pollution Control Act, designates which local, state, or Federal organization is responsible for the cleanup of a given spill.

Upon request, ORD's Oil and Hazardous Materials Spills Team in Edison, New Jersey, provides technical support to personnel responsible for spill cleanup by assisting in the control, removal, and recovery of petroleum and other hazardous substances. The team also obtains and analyzes samples of spilled material and compiles initial environmental impact data. A fully equipped mobile chemical

Fig. 15-1   The breakup of the *Argo Merchant,* a 7.5 million-gallon tanker, off Nantucket in 1976. Source: U.S. EPA Environmental Monitoring and Support Laboratory, Las Vegas, Nevada.

laboratory was recently developed to facilitate rapid on-site analyses during spill decontamination and removal operations.

B. Spill Prevention, Control, and Countermeasure (SPCC)

In December 1973 the EPA announced oil pollution regulations requiring the preparation of a Spill Prevention, Control, and Countermeasure (SPCC) plan by the owner or operator of any facility that could reasonable be expected to spill oil into U.S. waters. To assist in the preparation of SPCC plans, ORD's Edison, New Jersey, branch of the Cincinnati, Ohio, Industrial Environmental Research Laboratory, is funding a study to determine the feasibility of technologies that could be used by operators of oil storage facilities to comply with those regulations.

The EPA Oil and Hazardous Materials Simulated Environmental Test Tank (OHMSETT) facility is also operated by the Edison, New Jersey, team and permits environmentally safe testing of spill cleanup methods and technologies.

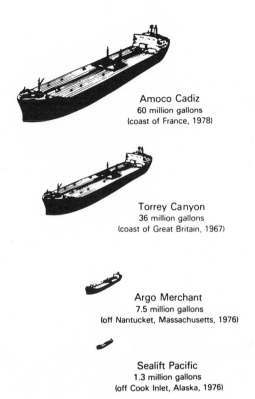

Amoco Cadiz
60 million gallons
(coast of France, 1978)

Torrey Canyon
36 million gallons
(coast of Great Britain, 1967)

Argo Merchant
7.5 million gallons
(off Nantucket, Massachusetts, 1976)

Sealift Pacific
1.3 million gallons
(off Cook Inlet, Alaska, 1976)

Fig. 15-2 Comparative sizes of oil tankers that have caused major spills.

(See Fig. 15-3.) The facility consists of a 2.6 million-gallon concrete tank, with mobile bridges overhead capable of carrying monitoring instrumentation, including closed-circuit television. The facility has a wave generator and simulated beach available to help duplicate actual environmental conditions. The efficiency of oil spill containment booms, skimmers, and gelling agents can be evaluated with repeated duplication of tests to ensure statistically significant results. This method of environmentally safe testing permits the evaluation of equipment under varying marine or freshwater conditions without the cost, time, and uncertainties involved in natural environmental tests. The OHMSETT facility is routinely made available to other government agencies, such as the Coast Guard, Navy, Department of Energy, U.S. Geological Survey, and National Oceanic and Atmospheric Administration, for their oil spill experiments.

Pipeline leaks, ruptures, and related accidents account for about 14 percent, by volume, of all oil spilled in the nation. (See Fig. 15-4.) With the development of additional oil and gas fields on the outer continental shelf, pipelines will be the principal means of moving products to shore installations. For this reason, the EPA's Office of Research & Development is evaluating existing techniques that facilitate the rapid detection of leaks and that allow a pipeline to be shut down before significant quantities of petroleum are spilled.

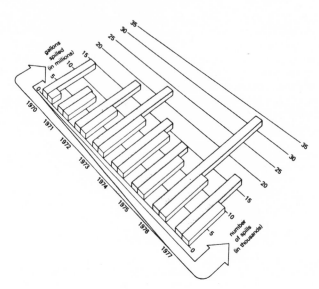

Chart 15-2   Oil spills in U.S. waters during 1970-1977.
Source: U.S. Coast Guard.

The Edison team recognizes the difficulty of oil containment and removal operations in rough seas, and it has centered its efforts upon the evaluation of oil-dispersing chemicals and their delivery systems for coping with large-scale spills. In addition, the capabilities for applying dispersants from surface vessels and aircraft at environmentally acceptable concentrations and flow rates are being developed.

Even with the most rapid response and the best available control technology, many oil spill incidents result in significant damage and environmental impact to shorelines. Those shorelines include estuarine, ocean, and inland areas that often have considerable recreational, aesthetic, or commercial value. The state-of-the-art for cleanup of oil from shorelines currently requires extensive use of manpower and equipment that, in some cases, can be more environmentally damaging to the shoreline than the oil itself. A promising research area currently being addressed by the Edison team is the application of chemical agents to protect the shoreline area by forming a thin film that prevents the oil from adhering to the beach. The agents are intended to be applied just before the oil slick arrives. As the tide recedes, the oil can then be washed by wave action off the beach and back into the water, where it may be collected and removed.

The research supported by ORD's Environmental Research Laboratory (ERL) in Athens, Georgia, is improving current petroleum identification techniques that are frequently used to locate sources of spills. Using high resolution gas chromatography and computer analysis, the laboratory can assign an oil sample a unique "fingerprint," or pattern of data that will distinguish it from other samples. Those data can then be compared with fingerprinted samples taken from oil tankers, pipelines, or storage facilities, and a determination can be made as to the sources of the spill. It is because of the difficulty of determining whether oil samples are indigenous to a given area or whether they are present as a result of a spill, that this project has been designed to improve methods of distinguishing between naturally and artificially occurring petroleum compounds.

In order to enforce regulatory standards, the early detection of violations is essential. While it is not practical to monitor the entire United States on a continuous basis, the use of aerial photography and remote sensing technologies offers an effective approach to monitoring environmental stresses. Techniques developed by ORD's Environmental Monitoring and Support Laboratory (EMSL) in Las Vegas, Nevada, are frequently used to rapidly detect spills, assist in locating violators, and routinely monitor pipeline and storage facilities. EMSL-Las Vegas annually responds to numerous spills across the nation, providing aerial photographs and other data needed for EPA and Coast Guard spill analysis.

ORD's Environmental Monitoring and Support Laboratory (EMSL) in Cincinnati, Ohio, is developing a device for detecting low concentrations of oil in waste waters from treatment processes involved in coal liquefaction, shale oil recovery, and petroleum refining. Through the use of liquid chromatography and optical fiber technologies, this device will be capable of identifying petroleum in water at very small concentrations. Such monitoring methods are being developed for new and emerging fossil fuel extraction technologies to help avoid adverse environmental impacts in the future.

A major ORD program is under way at several labs throughout the country to assess the ecological impacts of petroleum and petroleum products on the environment. The Environmental Research Laboratory (ERL) at Narragansett, Rhode Island, is researching the problem of oil in the marine environment in

Fig. 15-3   The EPA Oil and Hazardous Materials Simulated Environmental Test Tank (OHMSETT) facility at Leonardo, N.J.

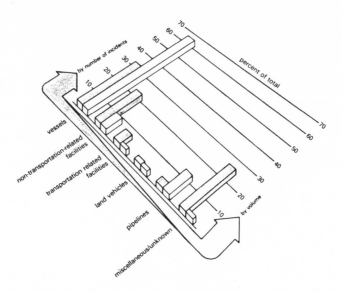

Fig. 15-4 Oil spills in the United States by number and percent of total volume spilled. Source: U.S. Coast Guard.

several areas. First, with regard to damage assessment, expertise was provided in the early stages of determining the biological effects of the massive *Amoco Cadiz* spill in France and again at the subsequent *Ocean 250* gasoline spill in local waters. The capability for improved response to oil spills is currently being developed in order to further the state of the art of the measurement of biological change as a result of a spill.

Other major efforts at Narragansett concerning oil spill research include a new approach to the determination of permissible levels of various petroleum components, as required by the Toxic Substances Control Act (TOSCA). That is being done through the evaluation of a variety of sublethal effects. Those effects or responses, i.e., the disruption of such biological processes as normal feeding behavior and reproductive patterns, can be elicited in particular marine organisms under certain conditions by very small amounts of oil. The mechanisms by which sublethal effects are brought about are very complex and poorly understood, and ERL-Narragansett is striving to develop biological data and protocols for chemical analysis that will help explain some of the processes.

In addition, the Narragansett lab is conducting research to determine the possible histopathological and mutagenic effects of oil. Carcinogenesis of the reproductive tract and breakdown of connective tissue were found in soft-shell clams taken in the vicinity of an oil spill in Maine. Carefully controlled laboratory experiments are being conducted to determine whether those effects are the direct result of the spill. Precisely how petroleum and petroleum compounds affect organisms is not clear at this time. Experiments will continue, and others will be undertaken, particularly in the field of genetic toxicology, to attempt to define and clarify the mechanisms.

Sublethal effects, while perhaps not immediately fatal to a given individual, have a direct bearing on the survival of the species as a whole and, consequently,

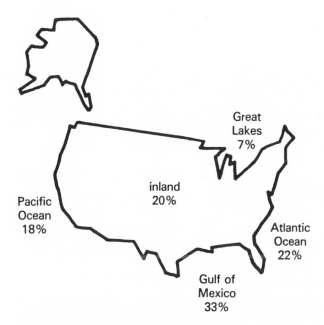

Fig. 15-5 Percentage of occurrence of U.S. oil pollution incidents by area. Source: U.S. Coast Guard.

on the balance of the ecosystem of which it is a member. ERL-Narragansett is developing the use of advanced simulated marine environments or microcosms to study the effects of oil and other organic chemicals on marine ecosystems. The concept of the "persistence limit" of the ecosystem—its ability to return to its former state of equilibrium following a perturbation such as an oil spill—is being evaluated as a means of establishing permissible levels of toxic substances in the marine environment.

Shellfish, such as mussels and oysters, when exposed to toxic compounds often store and concentrate those substances in their body tissues, allowing their possible use for pollutant surveillance. ERL-Narragansett is using that principle to develop data on several categories of marine pollutants, including petroleum hydrocarbons. The Mussel Watch project, coordinated by the Scripps Institution of Oceanography in La Jolla, California, under contract with ERL-Narragansett, is designed to monitor pollutant levels in U.S. coastal waters by systematically analyzing samples of mussels and oysters taken from 100 stations along the Pacific and Atlantic coasts and the Gulf of Mexico. The Mussel Watch program has been recognized internationally as one of the most promising ongoing marine monitoring efforts. The project will play an increasingly important role in assisting the EPA in setting congressionally mandated standards and regulations as techniques for quantifying pollutant concentrations are improved.

A series of projects supported by ORD's Environmental Research Laboratory in Corvallis, Oregon, is being undertaken to document the effects of oil spills on high-latitude arctic lakes, subarctic coastal intertidal environments, and Alaska's salt marsh communities. Alaska's permafrost areas also provide a prime location for studying the biodegradation, cleanup, and remedying of oil spills complicated by temperature extremes. The information derived from those experiments will be especially useful in protecting areas overrun by the development of Alaksan oil resources. Since tanker movement through the Northwest Passage has been attempted and is considered impractical at this time, it means that oil from Alaska's North Slope oil fields is moved to southern markets by pipeline to Valdez. Thus, documenting the potential effects of oil spilled in the terrestrial or freshwater environments is important in the event of a pipeline failure.

Other research projects being performed at or through ERL-Corvallis are aimed at demonstrating the effects of oil pollution on marine organisms. One project is aimed at assessing the sublethal effects of petroleum constituents on the fatty acid metabolism of fish. Sometime earlier, polychlorinated biphenyls (PCBs) had been shown to adversely affect fish fatty acid metabolism, causing abnormal biochemical structure of cell membranes; therefore, it is suspected that petroleum constituents may have similar effects. A fish's ability to adapt to temperature changes depends to a large extent on the chemical makeup of those complex membranes. A second project is directed at determining the effects of selected petroleum refinery effluent components on the growth and population structure of phytoplankton. A related project is investigating changes in the enzyme activities of fish exposed to sublethal concentrations of these refinery effluents.

ORD's Environmental Research Laboratory in Gulf Breeze, Florida, is coordinating several research projects related to offshore oil drilling and the ecological effects of petroleum. Since offshore petroleum exploration, extraction, and transportation activities may adversely affect marine ecosystems in the Gulf of Mexico and other coastal areas, toxicity tests are being performed on various organisms to determine the effects of drilling muds and other extraction-related emissions on the environment. An offshore laboratory, established on a U.S. naval research platform in the Gulf, is being used by scientists from several universities and the Gulf Breeze lab to study the developmental and behavioral effects of drilling fluids and their components on marine life indigenous to oil and gas drilling sites. In 1977 more than 3,500 oil spills were reported in the Gulf area. (See Fig. 15-5.) Such research will be instrumental in setting standards and regulations designed to lessen the impact of pollution associated with offshore oil development.

C. Office of Research and Development

The EPA's Office of Research and Development publishes numerous documents related to oil spill research that are distributed to Federal, state, and local governments, academia, industry, and the interested public.

A manual of practice for the environmentally acceptable use of oil spill dispersants has been published by the Edison, New Jersey, branch of the Industrial Environmental Research Laboratory in Cincinnati. The manual includes an inventory of available equipment and a description of the factors that must be considered when one uses dispersants for spill control and cleanup. The Edison lab also has prepared a "Manual for Oil Spill Cleanup Priorities" to assist cleanup teams as they face decisions regarding aesthetic, commercial, recreational, and other environmental aspects of spill cleanup.

In addition, Edison is preparing or has prepared a manual for the "Protection,

Cleanup, and Restoration of Salt Marshes Endangered by Oil Spills" and a Manual of Practice for the "Protection and Cleanup of Ocean, Estuarine, and Inland Shorelines." Both emphasize the decision-making processes through which field personnel can arrive at effective cleanup recommendations.

Oil spills in cold climates, such as those in Alaska, are more difficult to deal with than those occurring in warmer climates. Aside from the problems of working in sub-zero temperatures, natural ocean cleansing actions are slowed, and the oil tends to congeal and even solidify. Accordingly, the Edison lab is preparing or, has prepared, a manual on "Cold Climate Oil Spills—Shoreline Restoration and Protection." It will assemble the field methods and techniques proven most efficient in cold climate spill cleanup.

The EPA annually sponsors several conferences and workshops throughout the country to promote information transfer both among scientists performing oil-spills-related research and the "user community"—design engineers and state and local officials. For example, a recent workshop in Hartford, Connecticut, established response strategies applicable to oil spill emergencies in the New England region, while a similar workshop in Anchorage, Alaska, identified scientific needs and capabilities later incorporated into a regional response plan for assessing ecological damage resulting from major Alaskan spills. Similar workshops have been held for the Gulf of Mexico, Mid-Atlantic, and South Atlantic regions.

Presentations at national and regional conferences for governmental and industrial representatives who deal with spill response and control are prepared by the Environmental Monitoring and Support Laboratory in Las Vegas to disseminate information on aerial monitoring applications. Topics include types of film for different conditions, data collection and processing procedures, and photography analysis techniques. They address oil recognition and differentiation, environmental damage assessment, cleanup analysis, and potential spill situation recognition.

The research projects that have been listed below are being performed, or have been completed, by the EPA laboratories and associated organizations that head each tabulation.

Industrial Environmental Research Laboratory Cincinnati, Ohio (including the Oil and Hazardous Materials Spills Branch, Edison, New Jersey)

1. Evaluation of Oil Spill Prevention
2. Field Verification of Pollution Control Rationale for Offshore Platforms
3. Pollution Assessment of Advanced Oil and Gas Recovery Programs
4. Environmental Guidelines for Onshore Impact of Offshore Petroleum Development
5. Petroleum Pipeline Leak Detection
6. Methods Manual for Oil Spill Source Identification
7. Methods of Quantification for Petroleum Oils in Water
8. Oil Slick Dispersal Mechanics
9. Performance Testing—Inland/Harbor Oil Spill Equipment
10. Performance Testing—Offshore Oil Spill Equipment
11. Multi-agency Project for Oil Spill Equipment Evaluation
12. Manual of Practice for Oil Spill Dispersants
13. Manual of Practice for Surface Collecting Agents

14. Development and Demonstration of Effective Dispersant Application Techniques
15. Manual of Practice for Cold-Climate Shoreline Protection and Restoration
16. Surface Treatment Agents for Shoreline and Marsh Area Protection
17. Manuals of Practice for Ocean, Estuarine, and Inland Shorelines
18. Amine Carbamate Gelation for Oil Spill Recovery
19. Users Manual for Oil Spill Damage Assessment

Office of Energy, Minerals, and Industry Washington, D.C. (via Interagency Energy/Environment R&D Program)

1. Environmental Assessment of Northern Puget Sound and the Strait of Juan de Fuca (U.S. Department of Commerce—NOAA)
2. Environmental Assessment of an Active Oil Field in the Northwestern Gulf of Mexico (U.S. Department of Commerce—NOAA)
3. Major Coastal Ecosystem Characterization and Methodology with Emphasis on Fish and Wildlife as Related to Oil and Gas Development (U.S. Department of Interior)
4. Fate and Effects of Petroleum Hydrocarbons and Selected Toxic Metals in Selected Marine Ecosystems and Organisms (U.S. Department of Commerce—NOAA)
5. Energy-Related Water Pollutant Analyses Instrumentation (U.S. Department of Commerce—NOAA)
6. Ecological and Physiological/Toxicological Effects of Oil on Birds (U.S. Department of the Interior)
7. Ocean Oil Spill Concentration and Trajectory Forecast (U.S. Department of Commerce—NOAA)

Environmental Monitoring and Support Laboratory— Las Vegas, Nevada

1. Aerial Remote Sensor Data Collection Processing and Analysis for Environmental Monitoring
2. Development of Deployable Oil Sensors Package (through interagency agreement with NOAA)

Environmental Monitoring and Support Laboratory Cincinnati, Ohio

1. Production of Water Quality Control Supplies for the Quality Assurance Program
2. Development of Oil in Water Monitor

Environmental Research Laboratory—Corvallis, Oregon

1. Consequences of Crude Oil Contamination on Cold Climate Salt Marshes and Inshore Ecosystems
2. Ecological Effects of Oil and Derived Hydrocarbons and Guidelines for Damage Assessment and Methods for Predicting Impact

3. Alaskan Oil Seeps—Their Chemical and Biological Effects on the Environment
4. Oil Spills—Effects on Arctic Lake Systems
5. Effect of Petroleum Hydrocarbons on Fatty Acid Metabolism in Marine Fishes and Possible Sublethal Effect on the Physiology of Temperature Acclimation
6. Effects of Petroleum Refinery Discharges on West Coast Marine Organisms
7. Effects of Crude Oil Spills on Benthic/Intertidal Organisms
8. Evaluation of the Effect of Crude Oil on Permafrost Underlain Ecosystems
9. Survey of Chemical, Physical, and Biological Conditions Existing in Major Streams Before and After Oilfield Development in the Alaskan Arctic
10. Response of Microorganisms to Hot Crude Oil Spills on a Subarctic Taiga Soil

Environmental Research Laboratory— Narragansett, Rhode Island

1. Onshore Survey of Macrobenthos Along the Brittany Coast of France, Following the Amoco Cadiz Oil Spill
2. Onshore Biological Survey of the Effects of the Ocean 250 Gasoline Spill, Fishers Island, New York
3. Oil Spill Response Research, North Atlantic Coast (Norfolk, Virginia, to Eastport Maine)
4. Culture of Marine Algae for Experimental Use for the Detection of Toxic Substances and for the Effects of Oil on Reproductive Stages of Marine Macroalgae
5. Effects of No. 2 Fuel Oil on the Chemically Evoked Feeding Behavior of the Mud Snail, Illynassa obsoleta
6. Effects of No. 2 Fuel Oil on Filter Feeding in Blue Mussels, Mytilus edulis
7. Effects of No. 2 Fuel Oil on the Reproduction of Winter Flounder, Pseudopleuronectes americanus
8. Sublethal Feeding Response of Three Commercially Important Fish Species to Oil-Tainted Prey
9. Biological Consequences of Exposure of Blue Crabs to No. 2 Fuel Oil
10. Recovery of Natural Benthic Marine Communities Following Experimental Oiling of Sediments
11. Chemical Studies Directed Toward Ecological Damage Assessment of Petroleum Discharges into the Marine Environment
12. Relation Between Hydrocarbon Contamination and Tumors in the Soft Shell Clam, Mya arenaria
13. The Use of Microcosms as a Method for Determining the Persistence Limits of Marine Ecosystems
14. Monitoring Levels of Several Classes of Pollutants, Including Petroleum Hydrocarbons, in Mussels and Oysters at Over 100 Stations Nationwide in the Mussel Watch Program.

Environmental Research Laboratory— Gulf Breeze, Florida

1. Determine Toxicity to Marine Organisms of Petrochemicals and Energy Related Organic Solvents Derived from Offshore Activities and Ocean Dumping

2. Toxic, Sublethal and Latent Effects of Selected Petroleum Hydrocarbons and Barium Sulfate on Marine Organisms
3. Effects of Petroleum Compounds on Estuarine Fishes
4. Environmental Effects of Offshore Drilling and Oil on the Marine Environment (with U.S. Navy)
5. Effects of Drilling Fluids and Oil on Corals

Environmental Research Laboratory— Athens, Georgia

1. High-Resolution Separation of Organics in Water (fingerprinting)

Environmental Research Laboratory— Ada, Oklahoma

1. Treatment of Oil Refinery Waste Waters for Reuse Using a Sand Filter-Activated Carbon System

16

Hospital Handling of Hazardous Materials and Hazardous Wastes

A. Introduction

In most hospitals the handling of materials of all types has been somewhat dignified during the last 2 decades by the creation of materials management departments. A compelling impetus to perform the functions of distribution and logistics within the hospital environs as effectively as possible has been the skyrocketing costs of hospital care and the reduction of local, state and Federal subsidy in some areas.

Since a great deal of money is involved in hospital operations, it has attracted the attention of a segment of the business and financial world to the extent that certain hospital operations have been syndicated and there are hospital chains that are the core of profit-making entities, much in the manner of hotel chains.

While hospital managements have been traditionally conservative in their views on materials handling mechanization, there appears to be a growing awareness on the part of some hospital administrations that some of the areas of costs where it is possible to hold the line are, indeed, in the materials handling and housekeeping activities, and that there are cost benefits to be realized there. On the other hand, there is the realization that the economics of scale apply primarily to the larger, several hundred-bed hospitals, and that the small and medium-sized hospitals must cope with their logistics problems much as they have in the past with their dependence solely upon how well they can train their staff personnel. In addition to the increases in cost owing to inflation, there is the added burden of government regulations that have been imposed on all hospitals by County Boards of Health, by the state, and of course by the U.S. Environmental Protection Administration.

Since the logistics of hospital materials handling is not within the purview of this text, that is to say, the overall distribution system itself, the reader will find a discussion in this chapter concerning the way hospital wastes and hazardous materials are generally handled in a large number of hospitals.

B. Hospital Hazardous Materials

By and large, there are about a dozen different gases used in hospitals that, because they are under pressure in cylinders or steel bottles, are kept on the hospital grounds in a special cage in the service yard area. The quantities of the gases, such as nitrogen and nitrous oxide, used by hospitals are not very large, and they are handled only by qualified personnel. The largest single gas used is oxygen, and since it is consumed in fairly large amounts in most hospitals, it is received by bulk tanker and stored in its own specially fenced-off storage point on the grounds, fairly well isolated from other installations or facilities. From this point it is piped underground into the hospital operating rooms and special units.

Other hazardous substances, such as those used in radiation therapy, are carefully handled and controlled in all hospitals as far as the new supplies are concerned; therefore, it is the postuse handling that sometimes constitutes a problem, especially since there have been new governmental regulations, primarily the ones promulgated by the EPA, that must be addressed by each hospital.

C. Hospital Waste Management Program

The basis for each hospital's policy on the subject of waste management is to collect, transport, and dispose of all types of waste in a manner consistent with the local, state and Federal regulations. The purpose of the policy is to protect the patients, visitors, employees, visitors of the hospital, and the general public as well as the environment from any harm or contamination from improperly handled hospital waste.

Usually, the staff personnel manages the waste program. They are responsible for the hospital's housekeeping and custodial operations. It is the responsibility of the head of housekeeping to keep abreast of current regulations, and to communicate those changes to all of the concerned departments and personnel.

This text discusses 4 major types of waste as having the largest impact on hospital waste management operations. They are as follows: (a) infectious waste, (b) injurious waste, (c) non-infectious waste, and (d) hazardous waste.

1. Infectious Waste

That means waste from medical and intermediate care facilities, research centers, and other similar facilities that may contain sufficient concentration of pathogenic material that is capable of transmitting disease from any personal contact with the waste material either directly or indirectly. It includes, but is not limited to, the following substances:

1. Cultures and stocks of etiologic agents;
2. Blood and blood products, pathological wastes, other wastes from surgery, and contaminated laboratory wastes;
3. Sharps (hypodermic needles, cutting instruments, etc.) and dialysis unit wastes;
4. Body parts;

5. Discarded biologicals, contaminated equipment, contaminated food, and other products.

2. Injurious Waste

That refers to waste from medical and intermediate-care facilities capable of inflicting wounds; it includes, but is not limited to, needles, scalpel blades, lancets, broken glass, and pipettes.

3. Noninfectious Waste

All waste originated from medical and intermediate care facilities that is not infectious or injurious and includes among other things packaging materials, paper towels, and other trash.

4. Hazardous Waste

Hazardous waste comprises all wastes regulated as hazardous by the EPA.

The typical hospital breakdown by departments as waste is classified is illustrated in Chart 16-1.

D. Handling Procedures

All well-managed hospitals have written procedures and training manuals for personnel. While most of the training done is largely the vestibule type, or on-the-job training, every employee in the housekeeping department should have his or her own written, preferably printed copy of handling and disposal procedures.

A typical set of printed instructions would follow this general outline:

WASTE MANAGEMENT DISPOSAL PROCEDURES

I. *NONINFECTIOUS WASTE*
   A. Collected in plastic-lined trash containers located in all departments.
   B. Plastic liners removed daily, or as necessary by housekeeping personnel, tied, and transported to refuse chute rooms located on all floors.
   C. Sent down refuse chute, transported to trash compactor and compacted as necessary by housekeeping personnel.
   D. XYZ Disposal Corporation transports compactor weekly to an approved landfill sight, empties contents, and returns compactor.

II. *INFECTIOUS WASTE*
   A. *Isolation Trash:*
      1. Nursing Units
         • double bagged in red plastic liner labeled infectious waste by nursing personnel.
         • transported to incinerator by housekeeping personnel.
         • incinerated daily by engineering personnel.

Chart 16-1. Typical Departmental Classifications of Wastes

| Department | Classification of Wastes | | | |
	Noninfectious	Infectious	Injurious	Hazardous
Emergency Services	General waste; paper, plastic, metal, etc.	Known isolation trash	Needles, scalpel blades, broken glass.	None.
Nursing Units	General waste, I.V. tubing, chux, blood tubing	Waste from isolation rooms.	Needles, razor blades, broken glass.	None.
Labor and Delivery, Nursery, and Post-Partum	General waste, peri pads, disp. diapers.	Waste from isolation cases, placentas, fetuses.	Needles, etc.	None.
Radiology	General waste.	Waste from isolation cases.	Needles, etc.	Radioactive waste.
Pharmacy, Medical Oncology, and Chemotherapy	General waste.	None.	Needles, outdated syringes	Chemotherapy drugs and tubing, outdated narcotics, and other dangerous drugs. Dangerous chemicals.
Laboratory and Pathology	General waste, urine specimens, any autoclaved material.	Pathological wastes, cultures, blood, etiological agents, waste from isolation cases.	Needles, test tubes, Petri dishes, pipettes, glass slides, broken glass.	
Surgery	General waste.	Waste from contaminated cases. Body parts.	All sharps.	None.
Other	General waste, incinerator ashes.	Any known isolation or contaminated waste.	Any broken glass, needles, or sharp objects.	

2. Surgery and Labor and Delivery
   - double bagged by department personnel in red bags.
   - transported to incinerator by housekeeping personnel.
   - incinerated daily by engineering personnel.

B. *Placentas:*
   - double bagged by housekeeping personnel and transported to incinerator.
   - incinerated as necessary by engineering personnel.

C. *Fetuses and Recognizable Body Parts:*
   - transported via dumbwaiter to pathology lab.
   - after lab work completed, transported to tissue bank and/or funeral director by pathologist.

D. *Pathological Wastes, Cultures, Blood, Etiological Wastes:*
   - placed in rigid waste containers lined with red bags.
   - double bagged by housekeeping at least once daily and transported to incinerator.
   - incinerated by engineering personnel.

III. *INJURIOUS WASTE*

A. *Needles, Broken Thermometers, Razor Blades, Disposable Sharps:*
   - placed in red needle boxes in all nursing units and clinical areas. (Needles should be left intact with syringe.)
   - when needle boxes full, housekeeping personnel tape all openings closed and double bag in regular see-through plastic liners.
   - Housekeeping transports to refuse chute room for later hand transport to incinerator. *NOT TO BE THROWN DOWN REFUSE CHUTE.*
   - incinerated daily by engineering personnel.

B. *Outdated Syringes:*
   - user departments contact housekeeping to pick up outdated syringes in original packaging.
   - Housekeeping personnel hand-transports to incinerator.
   - incinerated, as necessary, by engineering personnel.

C. *Pipettes, Glass Slides, Petrie Dishes:*
   - collected in plastic lined, rigid, disposable containers (usually a cardboard box).
   - picked-up daily by housekeeping personnel, double-bagged, and transported to: incinerator if infectious; compactor if noninfectious.

D. *All Disposable Glass (unbroken):*
   - place in glass receptacles lined with 2 mil plastic found in all soiled utility rooms. *Do not dispose of in regular trash containers.*
   - Housekeeping personnel empties receptacles into 32-gallon rubbermaid containers located in refuse chute rooms. *Must not be thrown down refuse chute.*
   - Rubbermaid containers transported daily by housekeeping personnel and emptied into trash compactor.

E. *Broken Glass:*
   - when glass accidentally breaks, *do not touch* and call housekeeping personnel immediately.
   - Housekeeping personnel will remove glass and thoroughly sweep or vacuum to rid surface of glass fragments.

- broken glass is placed in glass receptacles found in all soiled utility rooms.
- disposed of in same manner as unbroken glass.

IV. *HAZARDOUS WASTE*
  A. Radioactive Waste:
    - kept in original lead-lined containers in radiology storage.
    - picked up regularly by provider for appropriate disposition.
    - not currently disposed of on-site.
  B. *Chemotherapy Chemicals* (waste and contaminated disposables):
    - placed in metal, double red plastic lined container by user departments (pharmacy and nursing unit).
    - picked-up by gloved housekeeping personnel and double-bagged again in red plastic liners.
    - hand transported to incinerator by housekeeping personnel.
    - incinerated, as necessary, by engineering personnel.
  C. *Outdated Narcotics and/or Dangerous Drugs:*
    - Housekeeping notified to pickup drugs in original containers.
    - transported to incinerator by housekeeping personnel.
    - incinerated, as necessary, by engineering personnel.
  D. *Hazardous Chemicals:*
    - Engineering regularly neutralizes acidic chemicals for discharge into system or disposal in compacter.

17

Handling Radioactive Waste

A. Introduction

In several chapters throughout this text we have referred to the labeling, placarding, hospital handling, etc., of radioactive materials. (Chapter 6, B, 10; Chapter 7, C, 4-h, and 5-g; Chapter 9, B, G-9; and Chapter 16.) The handling of radioactive materials, such as spent fuel rods, or neutron thermal shields of nuclear power reactors, however, presents somewhat different requirements than are customarily encountered in the general run of materials-handling situations.

While some of the equipment components required for handling radioactive waste are off-the-shelf, e.g., straddle carriers or lifting cranes, they must be adapted for remote operations in order to avoid the exposure of the operators to radioactive dosages that in some instances could be lethal. For example, the removal of an embrittled neutron thermal shield from a power reactor could not be accomplished except by the use of mechanisms that are guided and operated in a remote environment. The reason is quite simple. Human exposure of approximately 30 seconds would result in a lethal dose of radioactive poisoning, because the shield is so hot radioactively, 163,000 Curies with 5,000 R/hr. radiation peaks, in 1 shield for which data were available. The shield itself, the Millstone 2NTS, was composed of stainless steel approximately 7.6 cm (3 in.) thick, 3.5 m (11.5 ft.) long, and having an outside diameter of 4.1 m (13.6 ft.). It weighed approximately 27.2 t (30 tons).

The design of the equipment used to handle this type of radioactive load requires the coordination of many engineering disciplines. For example: structural, dynamics, engineering mechanics, heat transfer, etc., augmented by state-of-the-art computer science methodologies and electronic and electrical engineering. Overall, there is required analytical skill, in addition to the knowledge of the current regulatory requirements imposed by Federal, state, and local authorities.

B. Shielded Transportation Casks and Systems

The design and fabrication of shielded transportation casks used in the nuclear industry is a highly specialized business, as we have indicated above. Not very many companies have the necessary experience, expertise, and skills as well as the nuclear regulatory agency compliance credibility to perform satisfactorily in that field. One of the largest and most respected is the Nuclear Packaging company, a wholly owned subsidiary of Pacific Nuclear Systems Inc., of Federal Way, Washington.

Mr. David F. Snedeker, Vice President of Nuclear Packaging, has indicated that his company has produced more than 1,000 casks and shipping containers, which are fabricated in accordance with the company's Nuclear Regulatory Commission's (NRC) quality assurance program. Each container is designed to minimize exposure to operating personnel. Some of the features that are designed into each container system are:

a. high-strength, quick-acting binders to secure the primary lid and to reduce turnaround time and operator exposure by as much as 50 percent;
b. NRC approval for low-temperature materials;
c. high-quality epoxy-phenolic paint and internal stainless steel liners to facilitate decontamination; and,
d. large payload capacities.

Some of the features that are designed into these specially engineered containers may be seen in Fig. 17-1. The illustration shows the new overpack for transporting uranium hexafluoride (UF6) from enrichment plants to fuel-processing facilities around the world. The container is designed to hold a cylinder 30 inches in diameter weighing two and a half tons.

1. High Integrity Containers (HICs)

Reinforced concrete has been used by Nuclear Packaging for the fabrication of HIC containers for the burial of EPICOR 2 prefilter wastes from the Three Mile Island reactor Unit 2. That container possesses special design characteristics in excess of 10 CFR 61 for corrosion resistance, deep burial, i.e., external pressure resistance, and gas generation owing to radiolytic decomposition of waste materials.

Several HIC containers are shown in Fig. 17-2. They are fabricated from a space age alloy, Ferralium 255, which is a registered trademark of Bonar Langley Alloys, Limited. The containers are highly resistant to chemicals in the radioactive waste streams, including oils, organics, acids, and caustics. The duplex alloy is extremely resistant to corrosion, including intergranular stress corrosion (IGSC). These NuPac HIC containers resist pitting, and, in addition, are impervious to ultraviolet radiation.

2. On-site Transfer System for Remote Handling of Low-Level Radioactive Waste

Increased uncertainties regarding the future of low-level radioactive waste disposal sites have caused many commercial nuclear-power utilities to investigate and implement alternatives to radioactive (radwaste) storage and disposal. One current methodology being explored by the nuclear power industry centers on the capability of storing large quantities of low-level radwaste on the plant site itself. Implementation of on-site radwaste storage requires well-defined engineering parameters, carefully planned project cost analysis, and the design of remotely operated equipment that will ensure safe and reliable handling of radwaste containers.

Mr. Duane S. Schmoker, Director of Engineering Operations for Nuclear Packaging, Inc., has described a remotely operated waste-handling and transfer system

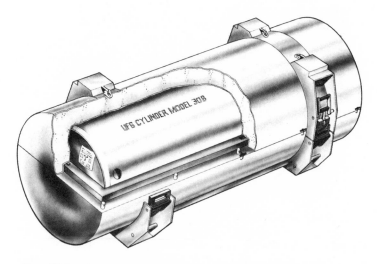

Fig. 17-1 Overpack for transporting uranium hexafluoride (UF6).
Source: Nuclear Packaging Inc., Federal Way, Washington.

that contains several unique features and was designed, fabricated, and installed at Southern California Edison's San Onofre Nuclear Generating Station. This NuPac system incorporates modular components, such as a waste-processing shield, a bottom and top loading shielded cask, and a transportation system with remote grappling equipment that makes it adaptable to multi-task waste-handling operations. The system has proved to be operationally flexible and has contributed significantly to reducing waste-processing personnel exposure.

The system (see Fig. 17-3) comprises the following modular components:

(1) *Transfer Cask:* This unit provides shielding for radwaste containers during waste processing and remote transfer operations;

(2) *Cask Transport Trailer:* This modular unit provides the transportation mode for the Transfer Cask between the radwaste-processing facilities within the plant site confines;

(3) *Mobile Straddle Crane:* This component provides for remote handling and positioning of the Transfer Cask during waste container transfer operations.

The waste handling system, Fig. 17-3, design requirements included the capability of:

(a) Processing variable-size radwaste liners in a shielded environment;
(b) Transfer liners within the confines of the site boundaries;
(c) Remotely transfer waste liners from the processing-transfer cask into storage modules or shipping casks;
(d) Keep personnel radiation exposure as low as is reasonably achievable (ALARA) during transfer operations.

In addition, the on-site transfer system (OTS) was designed to operate in an outdoor, inclement-weather, salt-air environment with temperature ranges from

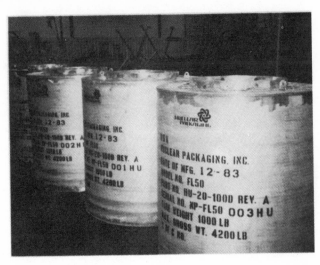

Fig. 17-2   Nuclear Packaging company's patented HIC containers.

0° to 49°C. (32° to 120°F.) Furthermore, design considerations were incorporated into the OTS for equipment repair in the event of mechanical or electrical system malfunctions during "hot" transfer operations.

a. Transfer Cask.   The top and bottom loading transfer cask (see Fig. 17-4) is the primary component of the OTS. It provides shielding from radwaste containers during waste processing and subsequent transfer operations. It has an internal cavity sized to handle containers up to 5.9m³ (210 ft³) and is designed with a shielding equivalent of 8.8cm, (3.5 in.) of lead. Externally, it is 3.2m (10.6 ft.) high, 2.2m (7.1 ft.) in diameter and weighs 29.3 tonne (32.5 tons) empty.

The cask has an integrally attached, remotely operated, 15.2cm (6 in.) thick steel closure door assembly located at its base. The door assembly enables containers to be transferred through the cask bottom and into shipping or storage vessels. The door assembly is electrically operated either from a locally mounted control panel or from the OTS remote control station. In the event of an electrical system failure the door is designed to be manually operated in order to complete or abort a container transfer operation.

Both interior and exterior surfaces of the transfer cask are designed for decontamination. The interior surface of the cask, including the doors, is clad in stainless steel. The cask exterior finish is a heavy coat of epoxy paint.

The transfer cask is equipped with 2 interchangeable shielding lids. The processing lid is utilized during container processing including radwaste solidification or resin drying operations. The 15.2cm (6 in.) thick lid is clad in stainless steel. It has a 71.1cm (28 in.) diameter hole in its center that enables the processing equipment to be remotely connected or disconnected from the container inside the transfer cask. The lid is adjustable in height for processing various size containers.

The grappling lid provides for remote grappling and handling of the radwaste container. it is equipped with a 13.5 tonne, (15 ton) capacity chain hoist, grapple,

Fig. 17-3 A remotely operated radioactive waste-handling and transfer system designed and fabricated by Nuclear Packaging Inc., Federal Way, Washington, for the Southern California Edison's San Onofre plant.

load cell, grapple depth encoder and a programmable logic controller. The grapple lid is designed for quick installation and removal from the transfer cask. Keyed guide ways, quick disconnect electrical connectors, and ball-lock pins enable the

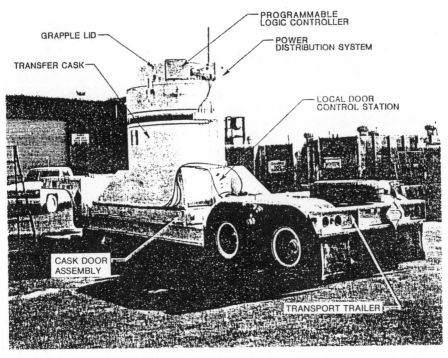

Fig. 17-4   The Nuclear Packaging Inc., transfer cask.

grapple lid to be installed or removed without personnel exposure during waste container handling operations.

The container grapple is suspended under the grapple lid (see Fig. 17-5). It contains 2 grappling fingers, which engage specially designed pockets on the waste container. The grapple is equipped with guides to ensure alignment with the container during remote grappling operations. The guides enable positive grapple engagement with as much as 44mm (1¾ in.) misalignment between the vertical axes of the container and grapple.

The grapple fingers are engaged and disengaged by a linear actuator positioned in the grapple body. Limit switches are used to indicate when the grapple is seated on the container and provide position status on the grapple fingers (i.e., extended or retracted).

A grapple depth encoder is mounted on the grapple lid and connected to the grapple by a retractable cable. The depth encoder enables the operator to determine the grapple's vertical position during container transfer. Adjustable limit switches on the depth encoder's remote console readout allows operational set points to be established and controlled during operation.

A load cell is used on the grapple hoist for accurately weighing the waste container. The load cell also provides the operator with an accurate indication as to whether the container transfer operation is proceeding smoothly. Sudden fluctuations in the load indicator during container transferring would signify an operational anomaly that may require further operator attention.

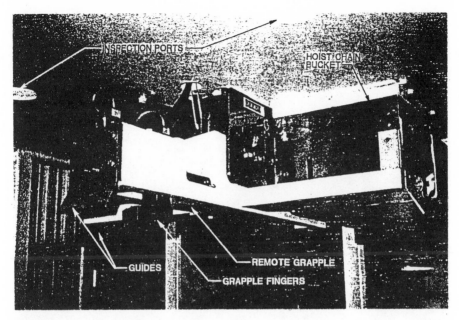

Fig. 17-5   Remote grapple under the transfer container lid. Note the grapple fingers. Courtesy of the Nuclear Packaging Inc.

Three inspection ports are provided on the grapple lid. They can be used for insertion of sampling equipment or remote viewing equipment. They can also be used for releasing the grapple with long-reach manual tools if the actuator should fail.

The modularity of the grapple lid enables it to be operated as a remotely controlled component, completely separate from the transfer cask. In that mode it can be used for placing empty containers in the transfer cask or as a remote handling device in hazardous environments within the plant's radwaste facilities.

b. Cask Transport Trailer.   The transport trailer (see Fig. 17-4) is a highway legal trailer utilized for tractor transportation of the transfer cask between in-plant facilities. It also provides a working base for the transfer cask during radwaste processing. It is 10.7m (35 ft.) long and has a 45 tonne (50 ton) load capacity. A specially designed suspension system provides a maximum tire ground pressure of 655/kPa (95 psi) under full load.

The trailer is equipped with a standard fifth wheel tractor attachment and has a hydraulically actuated landing gear system for leveling of the cask during radwaste processing and long-term storage. Specially designed guides are incorporated into the trailer deck to enable expedient transfer cask positioning. Tie-down cables and adjustable ratchet binders secure the transfer cask to the trailer during transport.

System modularity enables the transport trailer to be used with the transfer system or as a separate trailer for highway transportation of other plant-related cargo. That helps ensure maximum system cost effectiveness during nonrad-waste processing activities.

c. Mobile Straddle Crane.   The mobile straddle (Fig. 17-6) crane provides a means for transporting and positioning the transfer cask during container-transferring operations. The crane is designed to "straddle" the transport trailer, lift the transfer cask, and transport it to the container transfer area. The crane has a 45 tonne (50 ton) capacity and is equipped with a hydrostatic drive system that enables precision positioning and alignment of the cask over the container transfer area (see Fig. 17-7).

The crane is equipped with 2, 22.7 tonnes (25 tons) each, variable speed, dual

Fig. 17-6   The mobile straddle crane adapted by Nuclear Packaging Inc.

Fig. 17-7 The container grapple equipment designed and fabricated by Nuclear Packaging Inc.

hook, lifting systems to enable forward and aft leveling of the cask during transport on unlevel ground. The 2 individual hoists can also be used for cask leveling before container transfer operations.

A lift fixture is provided for interfacing the transfer cask to the straddle crane hoists and trolley. The lift fixture is designed for quick disconnect from the cask by removal of locking pins. Ratchet binders are utilized to connect the transfer cask and lift fixture to the straddle crane's structural frame during cask transport. That prevents unwanted sway motion between the crane and cask, which could cause structural damage during transit.

The crane's design includes the following additional features:

Maximum height—6.7m (22 ft.)
Maximum outside width—5.4m (17ft.-9 in.)
Minimum internal clear width—3m (10 ft.)
Maximum ground pressure at full load 655k/Pa (95 psi)
Capable of negotiating a 5 percent grade for 305m (1,000 ft.)
    under a full load of 45 tonnes (50 ton)
Less than a 7.6m (25 ft.) outside turning radius
Diesel-powered engine with an auxiliary 10 KW A.C. generator system
Emergency and safety interlocks for remote operation

In addition to the crane's standard cab controls, it has been modified to enable remote control of its engine, hoist and trolley travel functions. A 10 kw electric a.c. generator, belt driven by the diesel engine, provides self-contained electrical power to the cask and control station during container transferring operations.

SYSTEM CONTROL STATION

The OTS control station provides remote operation of the transfer cask and straddle crane during waste container handling operations. It is designed to operate all of the modular subcomponents of the on-site transfer system. During container transfers involving the straddle crane, the control station remains mounted to a storage and operating location on the side of the straddle crane (see Fig. 17-8). When the cask is trailer mounted (such as during radwaste processing or autonomous grapple lid operation), the control station is removed from its crane storage location and placed on a portable pedestal up to 6.1m (20 ft.) from the transfer cask. The control station portability enables the operator to minimize his radiation exposure while simultaneously maximizing the system's operational flexibility.

The control station includes the following OTS controls and features:

Master power control switch
Controls for raising and lowering the remote grapple
Grapple extend-retract control and status indicators
Grapple seated indicator
Cask door open-closed control and status indicators
Grapple depth readout with adjustable set points
Grapple load readout with maximum-minimum load indicators
Isolated; manual interlock override switch
Emergency system shut-down switch
Crane, forward trolley-hoist control

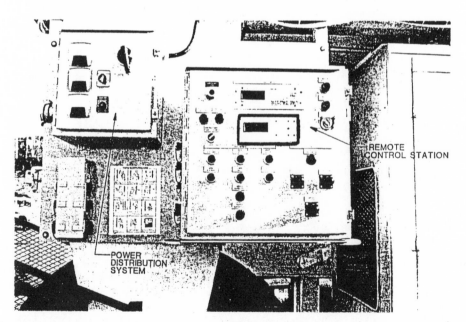

Fig. 17-8   The system control station designed by Nuclear Packaging Inc. becomes portable enabling the operator to minimize radiation exposure.

Crane, aft trolley-hoist control
Master crane trolley-hoist control

A programmable logic controller (PLC) located on the grapple lid of the transfer cask provides the interlock logic for the OTS operation. It aids in ensuring that waste container grappling, hoisting, and transfer functions are properly sequenced so that operator and equipment safety is not compromised.

The following provides a representative example of the interlock functions controlled by the PLC:

Cask doors will not open or close if grapple is not in the full "up" position
Automatic hoist shut down if preset maximum load on grapple is achieved or exceeded
Automatic hoist shut down if preset negative load on grapple is achieved
Grapple fingers will not operate if grapple is not fully seated on waste container
Grapple fingers will not retract if grapple is not fully seated on the waste container and the load cell weight is not at or below preset limits

All control system wiring is interconnected with quick disconnect electrical connectors. That enables continued system modularity and ease in startup and maintenance under adverse field conditions.

Following the on-site installation of the OTS, Nuclear Packaging company's field engineering staff conducted an operator-training program. Part of the instruction consisted in having the San Onofre's plant operators verify system design parameters. They gained operating proficiency by positioning and transferring mockup radwaste containers into a shipping cask.

The system was certified as operational when the first 4 cubic meters (142 c. ft.) High Integrity Container (HIC) filled with radioactive bead resins was processed and transferred to a shielded storage module utilizing the OTS. Since that time continuous operation of the OTS for shielded processing and transferring of radwaste has resulted in no major system anomalies; also, there have been significant reductions in personnel exposure.

3. Shielded and Spent Fuel Transportation Casks and Other Radwaste Equipment

It is fairly obvious from a perusal of the preceding material that the design and fabrication of radwaste containers requires highly specialized engineering expertise. On the following pages we would like to familiarize the reader with a number of those containers and system components that have been licensed by the Nuclear Regulatory Commission.

a. Shielded Transportation Casks. The Type A cask illustrated in Fig. 17-9 is shown prepared for shipment. The restraining cables shown in the picture are for mounting on a rail car of flat-bed trailer. The flat-bed trailer shown in the property of the Tri-State Motor Transit Co. of Joplin, Missouri. It has a proven track record of nuclear transportation dating back to the 1950s, with more than 64 million road miles without a single release of radioactive material. Fig. 17-10 shows a TSMT tractor and trailer with a shielded transportation cask on board the trailer.

Fig. 17-9   A Nuclear Packaging model NuPa 14-210 H, Type "A" shielded transportation cask. Note the cable restraints.

A Type "B" transportation cask is illustrated in Fig. 17-11. Shielding thickness for type "A" or "B" containers can vary from 1.5 in. to 4.4 in. of lead. Type A or B containers must be licensed by the NRC and must also be designed and fabricated to an NRC-approved quality assurance program.

In Fig. 17-12, a technician is seen hoisting a spent fuel transportation cask.

Fig. 17-10   A Tri-State Motor Transit Co. tractor-trailer combination with a shielded transportation cask on board.

Fig. 17-11   A Nuclear Packaging company Type "B" shielded transportation cask.

The illustration shows a Nuclear Packaging T-3 cask. It is the first spent fuel transportation cask to be licensed by the NRC since 1976. The T-3 was designed and fabricated for The U.S. Department of Energy's Fast Flux Test Facility (FFTF). Nuclear Packaging is also designing the NuPac— 125 B rail cask to transport canisters of damaged fuel from the Unit 2 reactor at Three Mile Island to the Idaho Falls National Engineering Laboratory. The NuPac rail cask employs a unique double-containment system that safely encapsulates the damaged fuel canisters. The NuPac—125 B represents a prototype of a new generation of

Fig. 17-12   A Nuclear Packaging's T-3, spent fuel cask.

irradiated fuel shipping casks that will see service throughout this decade and probably the next. In the 1990s casks such as the NuPac—125 B will be used at monitored retrievable storage facilities or in repository operations, since those casks are designed for "well-aged" fuel. In addition to their licensability and convenient handling, they are compatible with robotic systems; therefore, they lend themselves to robotic systems and remote handling operations.

Another container designed for the transfer of radioactive reactor filter elements is shown in Fig. 17-13. This unit has been designed by Nuclear Packaging to enable direct observation of the filter change-out operation when one is using a remotely controlled grapple system. It is currently in use in more than 12 nuclear reactor units around the world. Fig. 17-14 shows a 55-ton cask and transportation system designed to remotely handle a 10,000-pound cannister of radioactive material. The shielded cask has been designed with top and bottom loading capability, remote computer-controlled doors, and a grapple system. A Large Equipment Transfer System (LETS) was also designed by Nuclear Packaging to interface with a reactor's hot cell examination ports and to remotely handle contaminated material. It is mounted on an air-bearing transporter.

Fig. 17-13 A container designed and fabricated by Nuclear Packaging for the transfer of radioactive reactor filter elements.

Fig. 17-14   A 55-ton cask and transportation system designed to handle a 10,000-pound cannister of radioactive material; designed and fabricated by Nuclear Packaging Inc.

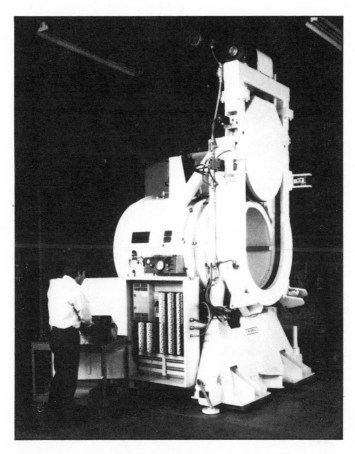

Fig. 17-15   A Large Equipment Transfer System (LETS) mounted on an air bearing; designed and fabricated by Nuclear Packaging Inc.

18

Superfund, A Remedial Response Program

A. Introduction

In the first chapter of this text, a grim picture was painted that involved hazardous wastes in what appeared to be several isolated instances. In order to place the subject in its proper perspective, it is necessary to retrace our path and tell the reader what really happened at Love Canal.

For more than 20 years, hundreds of tons of toxic wastes were dumped into an unfinished canal built by William T. Love in Niagara Falls, New York. The canal was covered when it was full; houses and a school were later built near and above the canal. In the late 1970s, alarmed by unusual health symptoms, residents of the Love Canal area informed government officials of the fact that hazardous substances were rising to the surface, seeping into basements, and migrating from the site. The rest is history.

In another bizarre circumstance, on April 21, 1980, a fire of unknown origin broke out at an abandoned waste dump in Elizabeth, New Jersey. The site was littered with some 20,000 leaking and corroded drums containing pesticides, explosives, radioactive wastes, acid, and other hazardous substances. A cloud of toxic gases drifted around heavily populated areas one-quarter mile from the site. Significant quantities of contaminated water from the fire-fighting efforts ran off into the Elizabeth River.

Those are, unfortunately, not isolated examples of the careless disposal of hazardous wastes over the past decades of our country's history. They are current problems, and our government has taken steps to prevent the continuation of such situations that are a constant threat to the quality of life and the environment. In the 1970s the United States made dramatic improvements in the quality of its air and water. In the 1980s large-scale environmental improvements are now being made by controlling the hazardous wastes being generated by the vast industrial complex of the country and by cleaning the land of hazardous wastes that were so carelessly tossed aside in years gone by.

In response to increased public awareness, Congress passed the Resource Conservation and Recovery Act of 1976 (RCRA), which established a regulatory system permitting the states to track hazardous wastes from the time of generation to disposal. RCRA requires safe and secure procedures to be used in treating, storing, and disposing of hazardous wastes. It was designed to prevent the creation of new Love Canals, but it did not give general authority to the

Federal or state governments to respond directly to problems caused by the uncontrolled hazardous waste disposal sites already in existence. The Federal Government had been able to respond to some of those problems under the Clean Water Act (CWA), which authorized the EPA and the U.S. Coast Guard to take action when spills and accidents involving oil or some 300 hazardous substances threatened navigable waters. But the CWA did not permit the government to act when hazardous substances were released elsewhere into the environment.

The Comprehensive Environmental Response, Compensation, and Liability Act of 1980 (CERCLA)—commonly known as the "Superfund" law—was passed to provide the needed general authority and to establish a Trust Fund for Federal and State governments to respond directly to any problems at uncontrolled hazardous waste disposal sites, not only in emergency situations but also at sites where longer-term permanent remedies are required.

CERCLA filled the gap in the national system to protect public health and the environment from hazardous substances by authorizing Federal action to respond to the release (or threatened release) from any source, including abandoned hazardous waste sites, into any part of the environment.

Costs of that response are to be financed by the Trust Fund, which is supported largely by taxes on producers and importers of petroleum and 42 basic chemicals. Over the 5-year expected lifespan of this "Superfund," $1.6 billion is to be collected, 86 percent from industry and the remainder from Federal appropriations.

Under the Superfund program, U.S. land can thus be freed of the most threatening of its abandoned wastes, the leaking drums and the contaminated soil. At the same time, the quality of the nation's air and water resources will be significantly improved.

B. The Need for a Superfund

The EPA is responsible for managing this cleanup program under the Superfund law but may call on 130 other Federal agencies, such as the Federal Emergency Management Agency, the Department of Health and Human Services, and the Army Corps of Engineers, for assistance in their respective areas of expertise.

Until CERCLA was passed, the Federal Government lacked the general authority to clean up hazardous waste sites or to respond to spills of hazardous substances onto land or into the air or nonnavigable waters. Congress had addressed hazardous waste problems before, but Federal responsibilities were mostly regulatory.

The Resource Conservation and Recovery Act (RCRA), passed in 1976 and reauthorized in 1984, has established a regulatory system to track hazardous wastes from the time they are generated to their final disposal. The act also requires safe and secure procedures to be used in treating, storing, and transporting hazardous wastes. RCRA is designed to prevent the creation of new Love Canals; it does not, however, authorize the Federal Government to respond directly to the problems at uncontrolled hazardous waste disposal sites already in existence, such as at the Love Canal.

The Clean Water Act and its predecessors enable the Federal Government to take action when oil or certain hazardous substances are discharged into navigable waterways. But those laws do not authorize the government to act when hazardous substances are released elsewhere in the environment, threatening to contaminate groundwater or to emit dangerous fumes.

Those and other environmental laws, such as the Clean Air Act, authorize the Federal Government to take legal action to compel individuals or companies—generators, transporters, or disposers of hazardous substances—to clean up problems for which they are responsible. When a dump site is old and abandoned, however, it may be impossible to find anyone responsible for the problem or anyone able to afford the cost of a cleanup. Moreover, many releases of hazardous substances demand prompt attention if serious damage is to be averted. There often is not enough time for legal proceedings before action must be taken.

Some states had established their own programs for responding to spills or cleaning up uncontrolled waste disposal sites. Like the Federal Government, however, state governments often lacked the funds and the legal authority needed to deal fully with the problem.

That is the reason, therefore, that Congress enacted CERCLA, to establish a program to spearhead both Federal and state efforts to respond to releases of hazardous substances into the environment.

C. The CERCLA, Superfund

CERCLA authorizes the Federal Government to respond directly to releases (or threatened releases) of hazardous substances and pollutants or contaminants that may endanger public health or environment. The costs are to be covered by a Trust Fund, approximately 87 percent of which is financed by taxes on the manufacture or import of certain chemicals and petroleum; the remainder comes from the general revenues. As indicated, above, the fund will have brought in about $1.6 billion over the first 5 years of the program. Sometimes the Federal Government may be able to take action to recover its cleanup costs from those subsequently identified as responsible for the release. Anyone liable for a release who fails to take ordered actions is (under specified conditions) liable for punitive damages equal to 3 times the government's response costs.

D. The National Contingency Plan (NCP)

The guidelines and procedures that the Federal Government must follow in implementing the Superfund law are spelled out in a regulation known as the National Contingency Plan (NCP).

Responses and cleanups under the Superfund program must be tailored to the specific needs of each site or to each release of hazardous substances. The EPA's strong enforcement effort seeks to ensure that private, responsible parties finance cleanup actions when possible. Direct government action, when called for, may take the following forms:

(1) *Removal actions,* when a prompt response is needed to prevent harm to public health or welfare or the environment: Removals may be ordered, for example, to avert fires or explosions, to prevent exposure to acutely toxic substances, or to protect a drinking water supply from contamination. Actions may include the

installation of security fencing, the construction of physical barriers to control a discharge, or the removal of hazardous substances from the site. Removal actions may also be taken when an expedited, but not necessarily immediate, response is needed. They are intended to minimize increases in danger or exposure that would otherwise occur if response were delayed. Ordinarily, each removal is limited by law to 6 months and a total cost of $1 million. Removals can be taken at sites on EPA's National Priorities List (NPL) sites as well as non-NPL sites.

(2) *Remedial actions,* which are longer-term and usually more expensive, are aimed at permanent remedies. They may be taken only at sites on the NPL. Identifying the nation's uncontrolled hazardous waste sites is an ongoing process. After publishing 2 preliminary lists, the EPA proposed the first NPL, consisting of 406 sites, in September 1983. By March 1985, the list had grown to 540 sites, with an additional 272 in the proposed category awaiting a decision on listing, for a total of 812 sites. The NPL may eventually include as many as 2,200 sites.

A remedial action may include the removal of drums containing wastes from the site; the installation of a clay "cap" over the site; the construction of ditches and dikes to control surface water or drains, liners, and grout "curtains" to control ground water; the provision of an alternate water supply; or the temporary or permanent relocation of residents.

The primary responsibility for carrying out the Superfund program has been assigned to the EPA. The Coast Guard, however, responds to spills that occur in coastal areas. Other Federal agencies provide assistance during a response. States are encouraged to take responsibility for an increasing number of Superfund-financed remedial actions. Under the law, state governments may plan and manage responses under agreement with the Federal Government. In remedial actions for which the Federal Government has lead responsibility, the Army Corps of Engineers will manage the design and construction stages for the EPA. Private contractors will perform the work at a site under Federal or state government supervision.

An important part of the Superfund program is to encourage voluntary cleanup by private industries and individuals when they are responsible for releases. In fact, since the full extent of the problem has become understood, industry has spent millions of dollars for cleanups as well as for the upgrading of existing facilities. In addition, industrial research and development have resulted in significant advances in technologies to control hazardous wastes.

Working with the local community is a key aspect of every Superfund response. At each site officials responsible for technical work ensure that the concerns of local citizens and officials are taken into account in the development of solutions and that information about the site is widely distributed.

E. Federal, State, and Private Contributions

The Trust Fund is large, but the cost of responding to a hazardous substance release is usually large, too, and there are many sites and spills in need of attention. Consequently, while CERCLA authorizes the Federal Government to respond to releases of hazardous substances, it does not require the government to respond to every release. At present, private parties handle about 90 percent of all releases that would otherwise require a removal action. In addition,

CERCLA specifies that Superfund money can be spent only under carefully prescribed conditions.

For example, a Superfund-financed response may not be taken if the EPA determines that the owner, operator, or other responsible party is undertaking an appropriate cleanup. Moreover, removals are taken only to bring a release of hazardous substances under control; they are not intended to eliminate every long-term problem.

Response under Superfund is not authorized in specified situations that may be covered by other laws (e.g., for certain releases of source, by-product, or special nuclear material from a nuclear incident). Also, remedial-action costs must be shared by a state in the design and construction phases. States must contribute 10 percent on sites that were privately owned at the time of disposal of hazardous substances and at least 50 percent on sites that were publicly owned. The state must also finance operation and maintenance costs, except for an initial period, when the EPA will share in the cost to certify that the remedy actually functions as planned.

Because remedial actions may be confronted by complex problems that are expensive to resolve, they are subject to further conditions. Technical measures can be selected only after evaluation of all feasible alternatives on the basis of economic, engineering, and environmental factors. The NCP explains how to determine the extent of cleanup that is appropriate and most cost-effective for a particular site. In addition:

The law requires that wherever possible, the remedy selected should avoid the costly step of excavating hazardous wastes and transporting them off the site for disposal elsewhere.

The benefits to be derived from continued work on a remedial action at a site must be weighed against the benefits of working at other sites in the nation. A project could be delayed or terminated to allow funds to be shifted where they are most needed.

In summation, the intent of these conditions is to derive the maximum benefit from Superfund for the nation as a whole. Therefore, the Superfund program is a coordinated effort of the Federal, state, and local governments, private industry, and citizens. The problems are widespread and often will require time to resolve, but the Superfund program is a significant part of our national response to 1 of the major environmental challenges of the decade.

F. EPA's National Priorities List (NPL)

The National Priorities List (NPL) identifies the targets for long-term action under the "Superfund" law, the Comprehensive Environmental Response, Compensation, and Liability Action of 1980 (CERCLA). The law sets up a trust fund (expected to total $1.6 billion over the 5-year life of the act) to help pay for cleaning up abandoned or uncontrolled hazardous waste sites that threaten public health, welfare, or the environment. The EPA has the primary responsibility for managing cleanup activities under Superfund. See Chart 18-1 for the EPA regional organization.

To date, the EPA has inventoried almost 19,000 uncontrolled hazardous waste sites. Some require emergency action because they represent an immediate threat. The agency cleans up such sites promptly through its removal program. To be eligible for a long-term "remedial action" under Superfund, however, a

United States Environmental Protection Agency
Regional Organization

EPA Superfund Offices

REGION 1
Director
Waste Management Division
John F. Kennedy Building
Boston, MA 02203
CML: (617)223-5186
FTS: 223-5186

REGION 2
Director
Office of Emergency & Remedial
Response
26 Federal Plaza
New York, NY 10278
CML: (212) 264-3082
FTS: 264-2647

REGION 3
Director
Hazardous Waste Management
Division
6th & Walnut Streets
Philadelphia, PA 19106
CML: (215) 597-8131
FTS: 597-8131

REGION 4
Director
Air & Waste Management Division
345 Courtland Street, N.E.
Atlanta, GA 30365
CML: (404) 881-3454
FTS: 257-3454

REGION 5
Director
Waste Management Division
111 West Jackson Boulevard
Chicago, IL 60604
CML: (312) 886-7579
FTS: 886-7579

REGION 6
Director
Air & Waste Management Division
1201 Elm Street
Dallas, TX 75270
CML: (214) 767-2730
FTS: 729-2730

REGION 7
Director
Air & Waste Management Division
324 E. 11th Street
Kansas City, MO 64106
CML: (816) 374-6529
FTS: 758-6529

REGION 8
Director
Air & Waste Management Division
1860 Lincoln Street
Denver, CO 80295
CML: (303) 844-2407
FTS: 564-2407

REGION 9
Director
Toxics & Waste Management
Division
215 Fremont Street
San Francisco, CA 94105
CML: (415) 974-7460
TS: 454-7460

REGION 10
Director
Air & Waste Division
1200 6th Avenue
Seattle, WA 98101
CML: (206) 442-1918
FTS: 399-1918

Chart 18-1   EPA regional organization and Superfund offices.

site must be listed on the NPL. As of October 1984, 538 sites have met the criteria for inclusion in the NPL. Another 248 sites have been proposed for addition. Those totals result from the following notices published in the Federal Register.

Date	Number of Sites
October 23, 1981	115 ("Interim Priority List")
July 23, 1982	45 ("Expanded Eligibility List")
December 30, 1982	418 (proposal of first NPL, including 153 of 160 sites published previously)
March 4, 1983	1 (proposal of Times Beach, Missouri)
September 8, 1983	406 (promulgation of first NPL; seven sites continue to be proposed)
September 8, 1983	133 (proposal of Update #1)
May 8, 1984	4 (promulgation of 4 San Gabriel Valley Sites in California)
September 21, 1984	128 (promulgation of Update #1; 4 sites continue to be proposed)
October 15, 1984	244 (proposal of Update #2)

The EPA published the first official NPL of hazardous waste sites in September 1983. It was the result of EPA's consideration of comments received during the 60-day public comment period on the 419 sites proposed in December 1982 and March 1983. Of the 419, 7 were not promulgated in the September 1983 NPL because their scores on EPA's Hazard Ranking System (HRS) were below the cutoff for listing; another 7 sites were held over in proposed status because the EPA needed more time for analyzing the technical comments and resolve other issues; 1 was split into 2 sites. The September 1983 NPL thus consisted fo 406 sites.

In September 1983, the EPA also proposed 133 new sites for listing (Proposed NPL Update #1). Four of the proposed update sites were added to the NPL in a separate action in May 1984 because the EPA recognized that a serious problem existed at the sites that required taking immediate remedial action. The sites, which involve the drinking water supplies for 500,000 people, are the San Gabriel Valley, Areas 1-4, in Los Angeles County, California. This action raised the NPL to 410 sites.

After completing its evaluation of most sites on Proposed NPL Update #1 as well as the 7 in the continuing-to-be-proposed status the EPA added 128 sites to the NPL in September 1984—123 of the Proposed Update #1 sites and 5 of the continuing-to-be-proposed sites. The details of that action are discussed below. It brought the total sites on the NPL to 538.

The EPA received 128 comments on proposed Update #1, of which 112 pertained to 50 sites; the remainder involved locations that were not proposed or generic or technical issues that did not pertain to specific sites.

Of the 133 sites proposed in Update #1, the four San Gabriel sites were added to the final NPL earlier, 2 sites were dropped, and 4 continued to be proposed. The EPA thus added 123 Update #1 sites to the NPL in September 1984.

Two sites were not listed because of changes in HRS scores resulting from the EPA's review of public comments and other information. A total of 14 scores changed, but in only 2 cases did the final scores drop below the cut-off point.

The two sites are:

(1) Old Brine Sludge Landfill, Delaware City, Delaware. The ground water contaminated by the site is not used as a source of drinking water. New information indicated that the contaminated ground water is not connected to deep ground water (which is a source of drinking water), as had been originally reported.

(2) Rosch Property, Roy, Washington. The EPA determined that drums on the site were not leaking, as had been originally reported. In addition, a reduction in the estimate of hazardous wastes present lowered the score.

Four sites continue to be proposed:

(1) Olin Corp. (Areas 1, 2 & 4), Augusta, Georgia. The EPA determined that certain sections of the appropriate scoring documents for the site were not in the public docket and so were not available during the comment period. The agency provided an additional 60-day comment period for this site following the September 21, 1984, promulgation of Update #1 in the Federal Register.

(2) Quail Run Mobile Manor, Gray Summit, Missouri. This site does not meet the current criteria for placing a site on the NPL. Those criteria are specified in the National Contingency Plan (NCP), the Federal regulation by which CERCLA is implemented. The EPA is considering modifying the NCP in such a way that Quail Run and similar sites will qualify for the NPL on the basis of additional criteria. Those criteria for listing are explained in the Preamble to Proposed Update #1 (48 Federal Register 40675, September 8, 1983).

(3) Sand Springs Petrochemical Complex, Sand Springs, Oklahoma. The EPA determined that certain sections of the appropriate scoring documents for the site were not in the public docket and so were not available during the comment period. The agency provided an additional 60-day comment period for this site following the September 21, 1984, promulgation of Update #1 in the Federal Register.

(4) Pig Road, New Waverly, Texas. The agency determined that further sampling and analysis are necessary for verifying the site's score.

The EPA completed its analysis of technical comments and resolved technical issues for the 7 sites proposed in December 1982 and continued to be proposed in the September 1983 rulemaking. The following 2 sites were dropped from consideration:

(1) Kingman Airport Industrial Area, Kingman, Arizona. The EPA determined that the elevated levels of hexavalent chromium occur naturally in ground water in the area and are probably not related to any industrial or other human activity. The agency has decided not to include sites in the NPL if the contamination is determined to occur naturally.

(2) Littlefield Township Dump, Oden, Michigan. The EPA found that an error had been made in the population affected by the site. The correct population drops the HRS score to below the cutoff point.

Based on the completion of additional analysis, EPA added these 5 sites to the NPL:

1. Airco, Calvert, Kentucky
2. Bayou Sorrel, Bayou Sorrel, Louisiana
3. Clare Water Supply, Clare, Michigan
4. Electrovoice, Buchanan, Michigan
5. Whitehall Municipal Wells, Whitehall, Michigan

Each entry on the list contains the name of the site, the state and city or county in which it is located and the corresponding EPA Region. The entries are arranged according to their scores on the HRS. HRS scores are designed to take into account a standard set of factors related to risks from potential or actual migration of substances through ground water, surface water, and air. Listing by rank serves both as an information and management tool, allowing the EPA and the states to decide which sites warrant detailed investigation to determine what, if any, response is needed.

The 538 sites on the NPL are in 11 groups, 50 to a group (except for the last group of 38). The scores on the NPL (except for some sites designated by states as their top priority) range from 75.60 to 28.62. An HRS score of 28.50 was selected as the cutoff point for the first proposed NPL to identify at least 400 sites, the minimum specified by CERCLA. The range within most of the groups of 50 is only 2 to 4 points. For convenience, the sites are numbered. However, the EPA considers sites within a group to represent approximately the same threat to public health, welfare, or the environment.

CERCLA allows a state or territory to designate a top-priority site, and 36 did so. Those sites must be listed in the first 100, regardless of score. Of the 36 top priority sites, 9 are included in the first 100 on the basis of their scores. The remaining 27 have scores that would not place them in the top 100. They are included at the bottom of the second group of 50. Of the 27, 8 have scores below 28.50.

Each entry on the NPL is also accompanied by 1 or more notations referencing the status of response and cleanup activities at the site at the time the list was prepared (September 21, 1984). In the past the EPA categorized NPL sites based on the type of response at each site (fund-financed, enforcement, or voluntary action). The agency is now expanding the prior categorization system in 2 ways: First, Federal enforcement actions are separated from state enforcement actions. Second, the status of site cleanup activities is designated by 3 new cleanup status codes. The codes identify sites where significant cleanup activities are under way or completed. Five response categories are used to designate the type of response under way. One or more categories may apply to each site. The 5 are as follows:

V   The Voluntary or Negotiated Response category includes sites where private parties have started or completed response actions pursuant to settlement agreements or consent decrees to which the EPA or the state is a party.

R   The Federal and/or State Response category includes sites at which EPA or State agencies have started or completed response actions.

F   The Federal Enforcement category includes sites where the United States has filed a civil complaint (including cost recovery actions) or issued an administrative order. It also includes sites at which a Federal court has mandated some form of response action following a judicial proceeding.

S   The State Enforcement category includes sites where a state has filed a civil complaint or issued an administrative order. It also includes sites at which a state court has mandated some form of response action following a judicial proceeding.

D   Category to Be Determined. The category includes all sites not listed in any other category. A wide range of activities may be in progress at sites in this category. The EPA or a State may be evaluating the type of response action to undertake, or an enforcement case may be under consideration. Responsible parties may be undertaking cleanup actions that are not covered by a consent decree or an administrative order.

The EPA has decided to indicate for informational purposes the status of fund-financed field activities under way or completed, as well as the status where responsible parties are conducting cleanup activities under a consent decree, court order, or an administrative order. Remedial planning or engineering studies do not receive a cleanup status code.

Many sites are cleaned up in stages or "operable units"—that is, a discrete action taken as part of the entire site cleanup that significantly decreases or eliminates contamination, threat of contamination, or pathway of exposure. One or more operable units may be necessary before the EPA will consider the cleanup of a hazardous waste site completed. A simple action such as constructing a fence is not considered an operable unit for coding purposes.

Three cleanup status codes are used. (Only 1 code can be used at a site because the codes are mutually exclusive.) The 3 codes are as follows:

I   Implementation activities are under way for 1 or more operable units. Field work is in progress at the site, but no operable units are completed.

O   Implementation activities for 1 or more (but not all) operable units are completed. Some field work has been completed, but additional work is necessary.

C   Implementation activities for all operable units are completed. All actions agreed upon for remedial action at the site have been completed, and work has started to monitor the effectiveness of the cleanup. Further site activities could occur if EPA considers them necessary.

Of the 50 States, the District of Columbia, and 6 territories, 6 have no sites on the current NPL. They are Alaska, the District of Columbia, Hawaii, Nebraska, Nevada, and the Virgin Islands. New Jersey has the largest number of sites (85), followed by Michigan (47), Pennsylvania (39), and Florida and New York (29 each).

National Priorities List (by Rank)

EPA RANK	REG	ST	SITE NAME *	CITY/COUNTY	RESPONSE CATEGORY#	CLEANUP STATUS @
Group 1			(HRS SCORES 75.60 - 58.41)			
1	02	NJ	Lipari Landfill	Pitman	R F	O
2	03	DE	Tybouts Corner Landfill *	New Castle County	V R F	
3	03	PA	Bruin Lagoon	Bruin Borough	R	I
4	02	NJ	Helen Kramer Landfill	Mantua Township	R	
5	01	MA	Industri-Plex	Woburn	V R	I
6	02	NJ	Price Landfill *	Pleasantville	R F	O
7	02	NY	Pollution Abatement Services *	Oswego	R F	O
8	07	IA	LaBounty Site	Charles City	V_ F S	O
9	03	DE	Army Creek Landfill	New Castle County	V F	
10	02	NJ	CPS/Madison Industries	Old Bridge Township	S	

RANK	EPA REG	ST	SITE NAME *	CITY/COUNTY	RESPONSE CATEGORY #	CLEANUP STATUS @

Group 1 (continued)

RANK	EPA REG	ST	SITE NAME *	CITY/COUNTY	RESPONSE CATEGORY #	CLEANUP STATUS @
11	01	MA	Nyanza Chemical Waste Dump	Ashland	R	
12	02	NJ	Gems Landfill	Gloucester Township	R	I
13	05	MI	Berlin & Farro	Swartz Creek	V R F S	O
14	01	MA	Baird & McGuire	Holbrook	R F	O
15	02	NJ	Lone Pine Landfill	Freehold Township	R	
16	01	NH	Somersworth Sanitary Landfill	Somersworth	R	
17	05	MN	FMC Corp. (Fridley Plant)	Fridley	V F S	O
18	06	AR	Vertac, Inc.	Jacksonville	V F	I
19	01	NH	Keefe Environmental Services	Epping	V R S	O
20	08	SD	Whitewood Creek *	Whitewood	V	
21	08	MT	Silver Bow Creek	Sil Bow/Deer Lodge	R	
22	06	TX	French, Ltd.	Crosby	R F	O
23	01	NH	Sylvester *	Nashua	R S	O
24	05	MI	Liquid Disposal, Inc.	Utica	R F	O
25	03	PA	Tysons Dump	Upper Merion Twp	R	O
26	03	PA	McAdoo Associates *	McAdoo Borough	R	
27	06	TX	Motco Inc. *	La Marque	R	O
28	05	OH	Arcanum Iron & Metal	Darke County	R F	
29	08	MT	East Helena Site	East Helena	F	
30	06	TX	Sikes Disposal Pits	Crosby	R F	O
31	04	AL	Triana/Tennessee River	Limestone/Morgan	V R F	
32	09	CA	Stringfellow *	Glen Avon Heights	R F	O
33	01	ME	McKin Co.	Gray	R S	O
34	06	TX	Crystal Chemical Co.	Houston	R F	O
35	02	NJ	Bridgeport Rental & Oil Services	Bridgeport	R	O
36	08	CO	Sand Creek Industrial	Commerce City	F	I
37	06	TX	Geneva Industries/Fuhrmann Energy	Houston	R F	O
38	01	MA	W. R. Grace & Co. (Acton Plant)	Acton	V F	I
39	05	MN	Reilly Tar (St. Louis Park Plant)	St. Louis Park	R F S	I
40	02	NJ	Burnt Fly Bog	Marlboro Township	R S	O
41	02	NJ	Vineland Chemical Co., Inc.	Vineland	D	
42	04	FL	Schuylkill Metals Corp.	Plant City	D	O
43	05	MN	New Brighton/Arden Hills	New Brighton	R	O
44	02	NY	Old Bethpage Landfill	Oyster Bay	V S	
45	02	NJ	Shieldalloy Corp.	Newfield Borough	D	
46	04	FL	Reeves SE Galvanizing Corp.	Tampa	D	O
47	08	MT	Anaconda Co. Smelter	Anaconda	V R	
48	10	WA	Western Processing Co., Inc.	Kent	V R F	O
49	05	WI	Omega Hills North Landfill	Germantown	D	
50	04	FL	American Creosote Works	Pensacola	R F	O

Group 2

(HRS SCORES 58.30 - 55.79, EXCEPT FOR STATE PRIORITY SITES)

RANK	EPA REG	ST	SITE NAME *	CITY/COUNTY	RESPONSE CATEGORY #	CLEANUP STATUS @
51	02	NJ	Caldwell Trucking Co.	Fairfield	R S	
52	02	NY	GE Moreau	South Glen Falls	V	
53	05	IN	Seymour Recycling Corp. *	Seymour	V R F	O
54	05	OH	United Scrap Lead Co., Inc.	Troy	D	
55	06	OK	Tar Creek (Ottawa County)	Ottawa County	R	I
56	07	KS	Cherokee County	Cherokee County	R	
57	02	NJ	Brick Township Landfill	Brick Township	D	
58	05	MI	Northernaire Plating	Cadillac	R	O
59	05	WI	Janesville Old Landfill	Janesville	D	
60	10	WA	Frontier Hard Chrome, Inc.	Vancouver	R	
61	04	SC	Independent Nail Co.	Beaufort	D	
62	04	SC	Kalama Specialty Chemicals	Beaufort	S	
63	05	WI	Janesville Ash Beds	Janesville	D	
64	04	FL	Davie Landfill	Davie	D	
65	05	OH	Miami County Incinerator	Troy	F	
66	04	FL	Gold Coast Oil Corp.	Miami	D	
67	05	WI	Wheeler Pit	La Prairie Township	D	
68	09	AZ	Tucson Intl Airport Area	Tucson	R	
69	02	NY	Wide Beach Development	Brant	R	
70	09	CA	Iron Mountain Mine	Redding	R	
71	02	NJ	Scientific Chemical Processing	Carlstadt	S	
72	08	CO	California Gulch	Leadville	D	
73	02	NJ	D'Imperio Property	Hamilton Township	R	
74	05	MI	Gratiot County Landfill *	St. Louis	V R F S	
75	01	RI	Picillo Farm *	Coventry	R S	O
76	01	MA	New Bedford Site *	New Bedford	V R F S	I
77	06	LA	Old Inger Oil Refinery *	Darrow	R	O
78	05	OH	Chem-Dyne *	Hamilton	V R F S	O
79	04	SC	SCRDI Bluff Road *	Columbia	V R F	O
80	01	CT	Laurel Park, Inc. *	Naugatuck Borough	V S	

* = STATES' DESIGNATED TOP PRIORITY SITES.
#: V = VOLUNTARY OR NEGOTIATED RESPONSE; R = FEDERAL AND STATE RESPONSE;
F = FEDERAL ENFORCEMENT; S = STATE ENFORCEMENT;
D = ACTIONS TO BE DETERMINED.
@: I = IMPLEMENTATION ACTIVITY UNDERWAY, ONE OR MORE OPERABLE UNITS;
O = ONE OR MORE OPERABLE UNITS COMPLETED, OTHERS MAY BE UNDERWAY;
C = IMPLEMENTATION ACTIVITY COMPLETED FOR ALL OPERABLE UNITS.

	EPA				RESPONSE	CLEANUP
RANK	REG	ST	SITE NAME *	CITY/COUNTY	CATEGORY#	STATUS @

Group 2 (continued)

RANK	REG	ST	SITE NAME *	CITY/COUNTY	RESPONSE CATEGORY#	CLEANUP STATUS @
81	08	CO	Marshall Landfill *	Boulder County	V	I
82	05	IL	Outboard Marine Corp. *	Waukegan	R F S	
83	06	NM	South Valley *	Albuquerque	R F	
84	01	VT	Pine Street Canal *	Burlington	V	I
85	03	WV	West Virginia Ordnance *	Point Pleasant	V	O
86	07	MO	Ellisville Site *	Ellisville	R	O
87	08	ND	Arsenic Trioxide Site *	Southeastern N.D.	R	
88	09	TT	PCB Wastes *	Pacific Trust Terr	R	C
89	03	VA	Matthews Electroplating *	Roanoke County	R	
90	07	IA	Aidex Corp. *	Council Bluffs	R F	O
91	09	AZ	Mountain View Mobile Homes *	Globe	R F	I
92	09	AS	Taputimu Farm *	American Samoa	R	C
93	04	TN	North Hollywood Dump *	Memphis	R S	
94	04	KY	A.L. Taylor (Valley of Drums) *	Brooks	R F	O
95	04	NC	PCB Spills *	210 Miles of Roads	R F	C
96	09	GU	Ordot Landfill *	Guam	R	
97	04	MS	Flowood Site *	Flowood	D	
98	08	UT	Rose Park Sludge Pit *	Salt Lake City	V	
99	07	KS	Arkansas City Dump *	Arkansas City	R	
100	09	CM	PCB Warehouse *	Marianas	R	C

Group 3   (HRS SCORES 55.71 - 51.35)

RANK	REG	ST	SITE NAME *	CITY/COUNTY	RESPONSE CATEGORY#	CLEANUP STATUS @
101	05	MN	Oakdale Dump	Oakdale	F	
102	05	IL	A & F Material Reclaiming, Inc.	Greenup	V R F S	O
103	03	PA	Douglassville Disposal	Douglassville	R	
104	02	NJ	Krysowaty Farm	Hillsborough	R	
105	05	MN	Koppers Coke	St. Paul	D	
106	01	MA	Plymouth Harbor/Cannon Engnrng	Plymouth	V R S	O
107	10	ID	Bunker Hill Mining & Metallurg	Smelterville	D	
108	02	NY	Hudson River PCBs	Hudson River	R	
109	02	NJ	Universal Oil Products(Chem Div)	East Rutherford	S	
110	09	CA	Aerojet General Corp.	Rancho Cordova	S	
111	10	WA	Com Bay, South Tacoma Channel	Tacoma	R F	O
112	03	PA	Osborne Landfill	Grove City	V S	
113	01	CT	Old Southington Landfill	Southington	D	
114	02	NY	Syosset Landfill	Oyster Bay	D	
115	09	AZ	Nineteenth Avenue Landfill	Phoenix	S	
116	10	OR	Teledyne Wah Chang	Albany	D	
117	02	NY	Sinclair Refinery	Wellsville	V R	
118	04	AL	Mowbray Engineering Co.	Greenville	R	O
119	05	MI	Spiegelberg Landfill	Green Oak Township	R	
120	04	FL	Miami Drum Services	Miami	R	O
121	02	NJ	Reich Farms	Pleasant Plains	D	
122	10	ID	Union Pacific Railroad Co.	Pocatello	D	
123	02	NJ	South Brunswick Landfill	South Brunswick	V	I
124	04	AL	Ciba-Geigy Corp. (McIntosh Plant)	McIntosh	D	
125	04	FL	Kassauf-Kimerling Battery	Tampa	R F	
126	05	IL	Wauconda Sand & Gravel	Wauconda	R	
127	01	NH	Ottati & Goss/Kingston Steel Drum	Kingston	V R F S	O
128	05	MI	Ott/Story/Cordova	Dalton Township	R	O
129	02	NJ	NL Industries	Pedricktown	S	O
130	05	MN	St. Regis Paper Co.	Cass Lake	D	
131	02	NJ	Ringwood Mines/Landfill	Ringwood Borough	V	
132	04	FL	Whitehouse Oil Pits	Whitehouse	R	
133	04	GA	Hercules 009 Landfill	Brunswick	D	
134	05	MI	Velsicol Chemical (Michigan)	St. Louis	V F S	O
135	05	OH	Summit National	Deerfield Township	R	
136	02	NY	Love Canal	Niagara Falls	R F S	O
137	05	IN	Fisher-Calo	LaPorte	F	
138	04	FL	Pioneer Sand Co.	Warrington	R S	
139	05	MI	Springfield Township Dump	Davisburg	R	
140	03	PA	Hranica Landfill	Buffalo Township	D	
141	04	NC	Martin Marietta, Sodyeco, Inc.	Charlotte	D	
142	04	FL	Zellwood Ground Water Contam	Zellwood	F	
143	05	MI	Packaging Corp. of America	Filer City	F	
144	05	WI	Muskego Sanitary Landfill	Muskego	D	
145	02	NY	Hooker (S Area)	Niagara Falls	F S	
146	03	PA	Lindane Dump	Harrison Township	D	
147	08	CO	Central City-Clear Creek	Idaho Springs	R	
148	02	NJ	Ventron/Velsicol	Wood Ridge Borough	S	
149	04	FL	Taylor Road Landfill	Seffner	V F	O
150	01	RI	Western Sand & Gravel	Burrillville	R S	O

Group 4   (HRS SCORES 51.27 - 47.10)

RANK	REG	ST	SITE NAME *	CITY/COUNTY	RESPONSE CATEGORY#	CLEANUP STATUS @
151	04	SC	Koppers Co., Inc (Florence Plant)	Florence	S	
152	02	NJ	Maywood Chemical Co.	Maywood/Rochelle Pk		I
153	02	NJ	Nascolite Corp.	Millville	V R	
154	06	OK	Hardage/Criner	Criner	F	
155	05	MI	Rose Township Dump	Rose Township	R	
156	05	MN	Waste Disposal Engineering	Andover	V R F	
157	02	NJ	Kin-Buc Landfill	Edison Township	V R F	O

RANK	EPA REG	ST	SITE NAME *	CITY/COUNTY	RESPONSE CATEGORY #			CLEANUP STATUS @

Group 4 (continued)

RANK	REG	ST	SITE NAME	CITY/COUNTY	V/R	F	S	CLEANUP
158	05	OH	Bowers Landfill	Circleville	V			
159	02	NJ	Ciba-Geigy Corp.	Toms River	R			
160	05	MI	Butterworth #2 Landfill	Grand Rapids		F		
161	02	NJ	American Cyanamid Co.	Bound Brook			S	
162	03	PA	Heleva Landfill	North Whitehall Twp	R			
163	02	NJ	Ewan Property	Shamong Township				D
164	02	NY	Batavia Landfill	Batavia	V			
165	05	MN	Boise Cascade/Onan/Medtronics	Fridley			S	I
166	01	RI	L&RB, Inc.	North Smithfield	V		S	
167	04	FL	NW 58th Street Landfill	Hialeah	R			
168	02	NJ	Delilah Road	Egg Harbor Township	R			
169	03	PA	Mill Creek Dump	Erie	R			O
170	04	FL	Sixty-Second Street Dump	Tampa	R			
171	05	MI	G&H Landfill	Utica	R			
172	02	NJ	Metaltec/Aerosystems	Franklin Borough	R			
173	05	WI	Schmalz Dump	Harrison				D
174	02	NJ	Lang Property	Pemberton Township				D
175	02	NJ	Sharkey Landfill	Parsippany Troy His	R			
176	09	CA	Selma Treating Co.	Selma			S	
177	06	LA	Cleve Reber	Sorrento	V R			O
178	05	IL	Velsicol Chemical (Illinois)	Marshall				D
179	05	MI	Tar Lake	Mancelona Township				D
180	08	CO	Lowry Landfill	Arapahoe County	V R			
181	05	MN	MacGillis & Gibbs/Bell Lumber	New Brighton			S	
182	02	NJ	Combe Fill North Landfill	Mount Olive Twp	R			
183	01	MA	Re-Solve, Inc.	Dartmouth	R	F		I
184	02	NJ	Goose Farm	Plumstead Township	R	F		O
185	04	TN	Velsicol Chem (Hardeman County)	Toone				D
186	02	NY	York Oil Co.	Moira	R	F		O
187	04	FL	Sapp Battery Salvage	Cottondale	R			I
188	04	SC	Wamchem, Inc.	Burton				D
189	02	NJ	Chemical Leaman Tank Lines, Inc.	Bridgeport				D
190	05	WI	Master Disposal Service Landfill	Brookfield				D
191	07	KS	Doepke Disposal Site (Holliday)	Johnson County				D
192	02	NJ	Florence Land Recontouring LF	Florence Township	R			
193	01	RI	Davis Liquid Waste	Smithfield	R		S	
194	01	MA	Charles-George Reclamation Lf	Tyngsborough	R	F		O
195	02	NJ	King of Prussia	Winslow Township				D
196	03	VA	Chisman Creek	York County	R			
197	05	OH	Nease Chemical	Salem				D
198	02	NJ	W. R. Grace & Co. (Wayne Plant)	Wayne Township	R			O
199	02	NJ	Chemical Control	Elizabeth	R		S	O
200	04	SC	Leonard Chemical Co., Inc.	Rock Hill			S	O

Group 5 (HRS SCORES 47.05 - 43.23)

RANK	REG	ST	SITE NAME	CITY/COUNTY	V/R	F	S	CLEANUP
201	05	OH	Allied Chemical & Ironton Coke	Ironton	R	F		
202	05	MI	Verona Well Field	Battle Creek	R	F	S	O
203	01	CT	Beacon Heights Landfill	Beacon Falls	R			
204	04	AL	Stauffer Chem (Cold Creek Plant)	Bucks				D
205	05	MN	Burlington Northern (Brainerd)	Brainerd/Baxter		F	S	
206	03	PA	Malvern TCE	Malvern				D
207	02	NY	Facet Enterprises, Inc.	Elmira	V			
208	03	DE	Delaware Sand & Gravel Landfill	New Castle County	R			O
209	04	TN	Murray-Ohio Dump	Lawrenceburg			S	
210	05	IN	Envirochem Corp.	Zionsville	V R	F	S	O
211	05	IN	MIDCO I	Gary	R	F		O
212	05	OH	South Point Plant	South Point				D
213	04	FL	Coleman-Evans Wood Preserving Co.	Whitehouse			S	
214	03	PA	Dorney Road Landfill	Upper Macungie Twp	R			
215	05	IN	Northside Sanitary Landfill, Inc	Zionsville		F		
216	04	FL	Florida Steel Corp.	Indiantown				D
217	09	AZ	Litchfield Airport Area	Goodyear/Avondale		F		
218	02	NJ	Spence Farm	Plumstead Township	R			
219	06	AR	Mid-South Wood Products	Mena		F		
220	09	CA	Atlas Asbestos Mine	Fresno County				D
221	09	CA	Coalinga Asbestos Mine	Coalinga				D
222	04	FL	Brown Wood Preserving	Live Oak		F		
223	02	NY	Port Washington Landfill	Port Washington				D
224	02	NJ	Combe Fill South Landfill	Chester Township	R			
225	02	NJ	JIS Landfill	Jamesburg/S. Brnswck			S	
226	03	PA	Centre County Kepone	State College Boro			S	O
227	05	OH	Fields Brook	Ashtabula				D
228	01	CT	Solvents Recovery Service	Southington	V			
229	08	CO	Woodbury Chemical Co.	Commerce City	R			
230	01	MA	Hocomonco Pond	Westborough	R			
231	04	KY	Distler Brickyard	West Point	R	F		O
232	02	NY	Ramapo Landfill	Ramapo	V			
233	09	CA	Coast Wood Preserving	Ukiah			S	
234	02	NY	Mercury Refining, Inc.	Colonie				D
235	04	FL	Hollingsworth Solderless Terminal	Fort Lauderdale				D
236	02	NY	Olean Well Field	Olean	V R			O
237	04	FL	Varsol Spill	Miami	R			
238	05	MN	Joslyn Manufacturing & Supply Co.	Brooklyn Center		F	S	
239	08	CO	Denver Radium Site	Denver	R			

RANK	EPA REG	ST	SITE NAME *	CITY/COUNTY	RESPONSE CATEGORY/		CLEANUP STATUS @

Group 5 (continued)

RANK	EPA REG	ST	SITE NAME *	CITY/COUNTY	R	S	D	Status
240	04	FL	Tower Chemical Co.	Clermont	R F			O
241	07	MO	Syntex Facility	Verona	V F			I
242	08	MT	Milltown Reservoir Sediments	Milltown	R			
243	05	MN	Arrowhead Refinery Co.	Hermantown	R			
244	02	NJ	Pijak Farm	Plumstead Township	R			
245	02	NJ	Syncon Resins	South Kearny	R			O
246	09	CA	Liquid Gold Oil Corp.	Richmond		S		
247	09	CA	Purity Oil Sales, Inc.	Malaga	R			
248	01	NH	Tinkham Garage	Londonderry	R	S		O
249	04	FL	Alpha Chemical Corp.	Galloway			D	
250	02	NJ	Bog Creek Farm	Howell Township	R			

Group 6   (HRS SCORES 43.19 - 40.74)

RANK	EPA REG	ST	SITE NAME *	CITY/COUNTY	R	F	S	D	Status
251	01	ME	Saco Tannery Waste Pits	Saco	R				O
252	04	FL	Pickettville Road Landfill	Jacksonville				D	
253	01	MA	Iron Horse Park	Billerica	R				
254	03	PA	Palmerton Zinc Pile	Palmerton		F			
255	05	IN	Neal's Landfill (Bloomington)	Bloomington	V	F	S		
256	05	WI	Kohler Co. Landfill	Kohler				D	
257	01	MA	Silresim Chemical Corp.	Lowell	R		S		O
258	01	MA	Wells G&H	Woburn	V	F			
259	02	NJ	Chemsol, Inc.	Piscataway			S		
260	05	WI	Lauer I Sanitary Landfill	Menomonee Falls				D	
261	05	MI	Petoskey Municipal Well Field	Petoskey		F			
262	05	MN	Union Scrap	Minneapolis			S		
263	02	NJ	Radiation Technology, Inc.	Rockaway Township	V				
264	02	NJ	Fair Lawn Well Field	Fair Lawn			S		I
265	05	IN	Main Street Well Field	Elkhart				D	
266	05	MN	Lehillier/Mankato Site	Lehillier	R				O
267	10	WA	Lakewood Site	Lakewood	R		S		I
268	03	PA	Industrial Lane	Williams Township		F			
269	05	WI	Onalaska Municipal Landfill	Onalaska				D	
270	02	NJ	Monroe Township Landfill	Monroe Township			S		O
271	02	NJ	Rockaway Borough Well Field	Rockaway Township	R				
272	05	IN	Wayne Waste Oil	Columbia City	R		S		
273	10	ID	Pacific Hide & Fur Recycling Co.	Pocatello	R	F			O
274	07	IA	Des Moines TCE	Des Moines		F			
275	02	NJ	Beachwood/Berkley Wells	Berkley Township	R				
276	02	NY	Vestal Water Supply Well 4-2	Vestal			S		
277	02	PR	Vega Alta Public Supply Wells	Vega Alta	R				
278	05	MI	Sturgis Municipal Wells	Sturgis				D	
279	05	MN	Washington County Landfill	Lake Elmo			S		
280	09	AZ	Indian Bend Wash Area	Scottsdale/Tempe		F			
281	09	CA	San Gabriel Valley (Area 1)	El Monte	R				I
282	09	CA	San Gabriel Valley (Area 2)	Baldwin Park Area	R				
283	10	WA	Com Bay, Near Shore/Tide Flats	Pierce County	R				
284	05	IL	LaSalle Electric Utilities	LaSalle	R				
285	05	IL	Cross Brothers Pail (Pembroke)	Pembroke Township	R				
286	02	PR	Upjohn Facility	Barceloneta				D	
287	09	CA	McColl	Fullerton	R	F			I
288	03	PA	Henderson Road	Upper Merion Twp				D	
289	10	WA	Colbert Landfill	Colbert	R				O
290	06	LA	Petro-Processors	Scotlandville	V	F			
291	02	PR	Frontera Creek	Rio Abajo				D	
292	02	PR	Barceloneta Landfill	Florida Afuera				D	
293	03	MD	Sand, Gravel & Stone	Elkton	R				I
294	05	MI	Spartan Chemical Co.	Wyoming				D	
295	02	NJ	Roebling Steel Co.	Florence	R				
296	03	PA	East Mount Zion	Springettsbury Twp	R				
297	04	TN	Amnicola Dump	Chattanooga				D	
298	02	NJ	Vineland State School	Vineland				D	
299	03	PA	Enterprise Avenue	Philadelphia	R		S		O
300	01	MA	Groveland Wells	Groveland	V R		S		

Group 7   (HRS SCORES 40.71 - 37.93)

RANK	EPA REG	ST	SITE NAME *	CITY/COUNTY	R	F	S	D	Status
301	02	NY	General Motors (Cent Foundry Div)	Massena		F			
302	04	SC	SCRDI Dixiana	Cayce	R	F	S		O
303	07	MO	Fulbright Landfill	Springfield				D	
304	03	PA	Presque Isle	Erie				D	
305	02	NJ	Williams Property	Swainton	R				
306	02	NJ	Renora, Inc.	Edison Township				D	
307	02	NJ	Denzer & Schafer X-Ray Co.	Bayville				D	
308	02	NJ	Hercules, Inc. (Gibbstown Plant)	Gibbstown				D	
309	05	IN	Ninth Avenue Dump	Gary	V				
310	06	AR	Gurley Pit	Edmondson	V R	F			O
311	01	RI	Peterson/Puritan, Inc.	Lincoln/Cumberland				D	
312	07	MO	Times Beach Site	Times Beach	R				O
313	05	MI	Wash King Laundry	Pleasant Plains Twp			S		
314	05	MN	Whittaker Corp.	Minneapolis			S		
315	05	MN	NL Industries/Taracorp/Golden	St. Louis Park				D	
316	01	CT	Kellogg-Deering Well Field	Norwalk	R				
317	01	MA	Cannon Engineering Corp. (CEC)	Bridgewater	R		S		

	EPA				RESPONSE CATEGORY#					CLEANUP STATUS @
RANK	REG	ST	SITE NAME *	CITY/COUNTY	V	R	F	S	D	

Group 7   (continued)

RANK	REG	ST	SITE NAME *	CITY/COUNTY	V	R	F	S	D	CLEANUP
318	02	NY	Niagara County Refuse	Wheatfield					D	
319	04	FL	Sherwood Medical Industries	Deland					D	
320	04	AL	Olin Corp. (McIntosh Plant)	McIntosh					D	
321	05	MI	Southwest Ottawa County Landfill	Park Township				S		
322	02	NY	Kentucky Avenue Well Field	Horseheads		R				
323	02	NJ	Asbestos Dump	Millington			F			
324	04	KY	Lee's Lane Landfill	Louisville			F			
325	06	AR	Frit Industries	Walnut Ridge	V		F			I
326	05	OH	Fultz Landfill	Jackson Township		R				
327	04	FL	Tri-City Oil Conservationist, Inc	Tampa		R	F			O
328	05	OH	Coshocton Landfill	Franklin Township			F			
329	03	PA	Lord-Shope Landfill	Girard Township	V			S		I
330	10	WA	FMC Corp. (Yakima Pit)	Yakima					D	
331	05	WI	Northern Engraving Co.	Sparta					D	
332	01	MA	PSC Resources	Palmer				S		I
333	05	MI	Forest Waste Products	Otisville		R	F			I
334	03	PA	Drake Chemical	Lock Haven		R	F			O
335	01	NH	Kearsarge Metallurgical Corp.	Conway				S		
336	04	SC	Palmetto Wood Preserving	Dixianna					D	
337	05	MI	Clare Water Supply	Clare					D	
338	03	PA	Havertown PCP	Haverford		R				
339	03	DE	New Castle Spill	New Castle County					D	
340	05	MN	Morris Arsenic Dump	Morris		R				
341	05	IN	Lake Sandy Jo (M&M Landfill)	Gary		R				
342	05	IL	Johns-Manville Corp.	Waukegan	V		F			
343	05	MI	Chem Central	Wyoming Township				S		
344	05	MI	Novaco Industries	Temperance			F			
345	02	NJ	Jackson Township Landfill	Jackson Township					D	
346	05	MI	K&L Avenue Landfill	Oshtemo Township					D	
347	10	WA	Kaiser Aluminum Mead Works	Mead	V					O
348	05	MN	Perham Arsenic Site	Perham		R				
349	05	MI	Charlevoix Municipal Well	Charlevoix		R				I
350	02	NJ	Montgomery Township Housing Dev	Montgomery Township		R				

Group 8   (HRS SCORES 37.93 - 35.51)

RANK	REG	ST	SITE NAME *	CITY/COUNTY	V	R	F	S	D	CLEANUP
351	02	NJ	Rocky Hill Municipal Well	Rocky Hill Borough		R				
352	02	NY	Brewster Well Field	Putnam County		R				
353	02	NY	Vestal Water Supply Well 1-1	Vestal		R				
354	05	MN	Nutting Truck & Caster Co.	Faribault				S		
355	02	NJ	U.S. Radium Corp.	Orange		R				
356	06	TX	Highlands Acid Pit	Highlands		R				
357	03	PA	Resin Disposal	Jefferson Borough					D	
358	08	MT	Libby Ground Water Contamination	Libby					D	
359	04	KY	Newport Dump	Newport					D	
360	03	PA	Moyers Landfill	Eagleville		R	F			
361	04	FL	Parramore Surplus	Mount Pleasant					D	
362	01	NH	Savage Municipal Water Supply	Milford		R		S		O
363	05	IN	Poer Farm	Hancock County		R				O
364	05	MI	Hedblum Industries	Oscoda			F			
365	06	TX	United Creosoting Co.	Conroe	V	R				C
366	08	WY	Baxter/Union Pacific Tie Treating	Laramie					D	
367	02	NJ	Sayreville Landfill	Sayreville					D	
368	01	NH	Dover Municipal Landfill	Dover		R				
369	02	NY	Ludlow Sand & Gravel	Clayville					D	
370	05	WI	City Disposal Corp. Landfill	Dunn					D	
371	02	NJ	Tabernacle Drum Dump	Tabernacle Twp	V		F			
372	02	NJ	Cooper Road	Voorhees Township					D	
373	07	MO	Minker/Stout/Romaine Creek	Imperial		R				O
374	01	CT	Yaworski Waste Lagoon	Canterbury		R		S		
375	03	WV	Leetown Pesticide	Leetown		R				O
376	04	FL	Cabot/Koppers	Gainesville		R		S		
377	02	NJ	Evor Phillips Leasing	Old Bridge Township					D	
378	03	PA	Wade (ABM)	Chester		R	F	S		O
379	03	PA	Lackawanna Refuse	Old Forge Borough		R	F			O
380	06	OK	Compass Industries (Avery Drive)	Tulsa		R				
381	02	NJ	Mannheim Avenue Dump	Galloway Township					D	
382	02	NY	Fulton Terminals	Fulton					D	O
383	01	NH	Auburn Road Landfill	Londonderry		R		S		
384	03	WV	Fike Chemical, Inc.	Nitro	V		F			I
385	05	MN	General Mills/Henkel Corp.	Minneapolis				S		
386	05	OH	Laskin/Poplar Oil Co.	Jefferson Township		R	F			O
387	05	OH	Old Mill	Rock Creek		R	F			O
388	07	KS	Johns' Sludge Pond	Wichita	V		F			I
389	09	CA	Del Norte Pesticide Storage	Crescent City		R				
390	02	NJ	De Rewal Chemical Co.	Kingwood Township					D	
391	02	NJ	Swope Oil & Chemical Co.	Pennsauken	V	R				I
392	04	GA	Monsanto Corp. (Augusta Plant)	Augusta	V					
393	01	NH	South Municipal Water Supply Well	Peterborough				S		
394	01	ME	Winthrop Landfill	Winthrop	V		F			I
395	06	AR	Cecil Lindsey	Newport		R				
396	05	OH	Zanesville Well Field	Zanesville	V					
397	05	WI	Eau Claire Municipal Well Field	Eau Claire					D	
398	04	GA	Powersville Site	Peach County					D	
399	05	MI	Grand Traverse Overall Supply Co.	Greilickville					D	
400	05	MI	Metamora Landfill	Metamora					D	

EPA RANK	REG	ST	SITE NAME *	CITY/COUNTY	RESPONSE CATEGORY#	CLEANUP STATUS@

Group 9   (HRS SCORES 35.45 – 33.73)

EPA RANK	REG	ST	SITE NAME	CITY/COUNTY	RESPONSE CATEGORY	CLEANUP STATUS
401	05	MI	Whitehall Municipal Wells	Whitehall	R	
402	05	MN	South Andover Site	Andover		D
403	02	NJ	Diamond Alkali Co.	Newark	V	I
404	05	MI	Kentwood Landfill	Kentwood		D
405	05	MI	Electrovoice	Buchanan		D
406	02	PR	Fibers Public Supply Wells	Jobos		D
407	05	IN	Marion (Bragg) Dump	Marion		D
408	05	OH	Pristine, Inc.	Reading	F	I
409	05	WI	Mid-State Disposal, Inc. Landfill	Cleveland Township	R	
410	08	CO	Broderick Wood Products	Denver	R	
411	05	OH	Buckeye Reclamation	St. Clairsville		D
412	06	TX	Bio-Ecology Systems, Inc.	Grand Prairie	R	D
413	02	NJ	Woodland Route 532 Dump	Woodland Township		D
414	05	IN	American Chemical Service, Inc.	Griffith		D
415	01	VT	Old Springfield Landfill	Springfield	V R F	O
416	02	NY	Solvent Savers	Lincklaen		D
417	03	VA	U.S. Titanium	Piney River	F S	
418	05	IL	Galesburg/Koppers Co.	Galesburg		D
419	02	NY	Hooker (Hyde Park)	Niagara Falls	V F S	O
420	05	MI	SCA Independent Landfill	Muskegon Heights		D
421	09	CA	MGM Brakes	Cloverdale	S	
422	06	LA	Bayou Sorrel	Bayou Sorrel	F	
423	05	MI	Duell & Gardner Landfill	Dalton Township		D
424	02	NJ	Ellis Property	Evesham Township	R	O
425	04	KY	Distler Farm	Jefferson County	R F	O
426	10	WA	Harbor Island (Lead)	Seattle		D
427	05	WI	Lemberger Transport & Recycling	Franklin Township		D
428	05	OH	E.H. Schilling Landfill	Hamilton Township		D
429	05	MI	Cliff/Dow Dump	Marquette	F	
430	10	WA	Queen City Farms	Maple Valley	F	
431	05	WI	Scrap Processing Co., Inc.	Medford	S	
432	06	NM	Homestake Mining Co.	Milan	V F	I
433	05	MI	Mason County Landfill	Pere Marquette Twp	R	D
434	05	MI	Cemetery Dump	Rose Center	R	
435	02	NJ	Hopkins Farm	Plumstead Township		D
436	01	RI	Stamina Mills, Inc.	North Smithfield	R	
437	05	IN	Reilly Tar (Indianapolis Plant)	Indianapolis		
438	01	ME	Pinette's Salvage Yard	Washburn	R	O
439	06	TX	Harris (Farley Street)	Houston	V	
440	02	NJ	Wilson Farm	Plumstead Township		D
441	03	PA	Old City of York Landfill	Seven Valleys		D
442	05	IL	Byron Salvage Yard	Byron	R	I
443	03	PA	Stanley Kessler	King of Prussia	F	
444	02	NJ	Friedman Property	Upper Freehold Twp	R	
445	02	NJ	Imperial Oil/Champion Chemicals	Morganville		D
446	02	NJ	Myers Property	Franklin Township	R	I
447	02	NJ	Pepe Field	Boonton	R	
448	05	MI	Ossineke Ground Water Contam	Ossineke		D
449	03	WV	Follansbee Site	Follansbee	F	
450	09	CA	Koppers Co.,Inc. (Oroville Plant)	Oroville	S	

Group 10   (HRS SCORES 33.66 – 30.77)

EPA RANK	REG	ST	SITE NAME	CITY/COUNTY	RESPONSE CATEGORY	CLEANUP STATUS
451	05	MI	U.S. Aviex	Howard Township	S	
452	03	PA	Walsh Landfill	Honeybrook Township	R	
453	02	NJ	Landfill & Development Co.	Mount Holly	S	I
454	02	NJ	Upper Deerfield Township Slf	Upper Deerfield Twp		D
455	06	NM	AT & SF (Clovis)	Clovis	V F	
456	02	NY	American Thermostat Co.	South Cairo	V	
457	04	TN	Lewisburg Dump	Lewisburg		D
458	05	MI	McGraw Edison Corp.	Albion	V	
459	04	KY	Airco	Calvert City		D
460	03	PA	Metal Banks	Philadelphia	V F	
461	04	KY	B.F. Goodrich	Calvert City		D
462	05	MI	Organic Chemicals, Inc.	Grandville		D
463	01	MA	Sullivan's Ledge	New Bedford	R	
464	02	PR	Juncos Landfill	Juncos	V F	O
465	05	IN	Bennett Stone Quarry	Bloomington	S	O
466	04	FL	Munisport Landfill	North Miami		D
467	04	AL	Stauffer Chem (Le Moyne Plant)	Axis		D
468	02	NJ	M&T Delisa Landfill	Asbury Park	V R	
469	04	SC	Geiger (C & M Oil)	Rantowles		D
470	05	WI	Moss-American(Kerr-McGee Oil Co.)	Milwaukee		D
471	05	WI	Waste Research & Reclamation Co.	Eau Claire	.	D
472	10	OR	Gould, Inc.	Portland	V	I
473	05	MN	St. Louis River Site	St. Louis County	V	
474	05	MI	Auto Ion Chemicals, Inc.	Kalamazoo	V	
475	04	SC	Carolawn, Inc.	Fort Lawn	R F	O
476	03	PA	Berks Sand Pit	Longswamp Township	R	O
477	05	MI	Sparta Landfill	Sparta Township	S	
478	05	IL	ACME Solvent (Morristown Plant)	Morristown	R	
479	04	FL	Hipps Road Landfill	Duval County		D
480	04	FL	Pepper Steel & Alloys, Inc.	Medley	R F	O

RANK	EPA REG	ST	SITE NAME *	CITY/COUNTY	RESPONSE CATEGORY#	CLEANUP STATUS @

Group 10 (continued)

481	01	ME	O'Connor Co.	Augusta		D	
482	05	WI	Oconomowoc Electroplating Co. Inc	Ashippin		D	
483	05	MI	Rasmussen's Dump	Green Oak Township	R		
484	03	PA	Westline Site	Westline	R		O
485	05	OH	Powell Road Landfill	Dayton		D	
486	05	MI	Ionia City Landfill	Ionia	F		I
487	08	CO	Lincoln Park	Canon City		D	
488	05	IN	Wedzeb Enterprises, Inc.	Lebanon		D	I
489	02	PR	GE Wiring Devices	Juana Diaz	V F		
490	05	OH	New Lyme Landfill	New Lyme	V		
491	02	NJ	Woodland Route 72 Dump	Woodland Township		D	
492	02	PR	RCA Del Caribe	Barceloneta		D	C
493	03	PA	Brodhead Creek	Stroudsburg	R F		O
494	10	OR	United Chrome Products, Inc.	Corvallis	R		
495	05	MI	Anderson Development Co.	Adrian		D	
496	05	MI	Shiawassee River	Howell		D	
497	03	PA	Taylor Borough Dump	Taylor Borough	R		O
498	03	DE	Harvey & Knott Drum, Inc.	Kirkwood	R F		O
499	04	TN	Gallaway Pits	Gallaway	R F		O
500	05	OH	Big D Campground	Kingsville		D	

Group 11

(HRS SCORES 30.61 - 28.62)

501	03	DE	Wildcat Landfill	Dover		D	
502	05	MI	Burrows Sanitation	Hartford		D	
503	03	PA	Blosenski Landfill	West Cain Township	F		
504	03	DE	Delaware City PVC Plant	Delaware City	V F		
505	03	MD	Limestone Road	Cumberland	R		
506	02	NY	Hooker (102nd Street)	Niagara Falls	V F S		
507	03	DE	New Castle Steel	New Castle County		D	
508	06	NM	United Nuclear Corp.	Church Rock	F		
509	06	AR	Industrial Waste Control	Fort Smith	F		
510	09	CA	Celtor Chemical Works	Hoopa	R		O
511	04	AL	Perdido Ground Water Contam	Perdido			O
512	02	NY	Marathon Battery Corp.	Cold Springs	R	D	
513	03	PA	Lehigh Electric & Engineering Co.	Old Forge Borough	R F		O
514	05	OH	Skinner Landfill	West Chester		D	
515	04	NC	Chemtronics, Inc.	Swannanoa		D	
516	07	MO	Shenandoah Stables	Moscow Mills	V F		O
517	06	LA	Bayou Bonfouca	Slidell	R		
518	03	VA	Saltville Waste Disposal Ponds	Saltville	R		
519	03	PA	Kimberton Site	Kimberton Borough		D	
520	03	MD	Middletown Road Dump	Annapolis	R		I
521	10	WA	Pesticide Lab (Yakima)	Yakima		D	
522	05	IN	Lemon Lane Landfill	Bloomington	R S		I
523	10	ID	Arrcom (Drexler Enterprises)	Rathdrum	R		O
524	03	PA	Fischer & Porter Co.	Warminster	V F		
525	09	CA	Jibboom Junkyard	Sacramento	R		
526	02	NJ	A. O. Polymer	Sparta Township		D	O
527	02	NJ	Dover Municipal Well 4	Dover Township		D	
528	02	NJ	Rockaway Township Wells	Rockaway	V		I
529	05	WI	Delavan Municipal Well #4	Delavan		D	
530	09	CA	San Gabriel Valley (Area 3)	Alhambra	R		
531	09	CA	San Gabriel Valley (Area 4)	La Puente	R		
532	10	WA	American Lake Gardens	Tacoma	V		O
533	10	WA	Greenacres Landfill	Spokane County		D	
534	06	TX	Triangle Chemical Co.	Bridge City	R F		O
535	02	NJ	PJP Landfill	Jersey City	S		I
536	03	PA	Craig Farm Drum	Parker		D	
537	03	PA	Voortman Farm	Upper Saucon Twp	R		
538	05	IL	Belvidere Municipal Landfill	Belvidere		D	

TOTAL SITES LISTED: 538

19

The Resource Conservation and Recovery Act (RCRA)

A. Introduction

The EPA administers 2 major waste management programs. They are the Comprehensive Environmental Response, Compensation and Liability Act (CERCLA), or "Superfund," as it is called, which is designed to clean up the nation's worst, abandoned hazardous waste depositories, and the Resource Conservation and Recovery Act (RCRA), which is empowered to regulate current and future hazardous waste disposal activities. RCRA was greatly expanded by Congress in 1984, and both the EPA and the states face a large challenge in implementing the new provisions of the law. It has been said that the success of RCRA will require the close cooperation of Federal, state, local governments, industry, public-interest groups, and private citizens. The onus of responsibility, however, falls directly on those who generate, transport, treat, recycle, and dispose of regulated wastes of both hazardous and nonhazardous types.

B. Historical Perspectives

Looking backward, we find that when population centers were relatively compact and produced manageable volumes of conventional waste, the disposal of such material was not a major issue in urban or environmental affairs. In recent decades, however, the tonnage and chemical complexity of the nation's waste has grown dramatically, posing a threat to air, water, and land resources, to the balance of nature and even to human health. Congress recognized the problem in 1965 and passed the Solid Waste Disposal Act to fund research and technical assistance for state and local planners.

In 1970 the original legislation was enlarged and restructured in the form of the Resource Recovery Act, which promoted the adoption of sanitary landfills and encouraged a shift from mere disposal toward conservation, recycling and advanced control technology. A cabinet-level interagency resource conservation committee was set up, and the EPA funded 6 major resource-recovery projects at a cost of some $25 million.

During the mid to late '70s the EPA invested about $10 million in direct technical assistance to a host of cities for experiments like separate collection of newsprint, computerized routing and scheduling, and new management systems.

Between 1979 and 1981, the agency allocated $28 million to 63 communities to help them plan the development of large-scale resource-recovery facilities.

Up to this point legislation had focused mainly on the traditional kinds of municipal trash—paper, glass, cans, garbage. However, mounting scientific evidence indicated that wastes generated by chemical and other industrial processes could be hazardous. That persuaded Congress first to strengthen existing regulations and then, in 1976, to pass the Resource Conservation and Recovery Act (RCRA), which amended the Solid Waste Disposal Act.

Under RCRA, the EPA set standards for generators and transporters of hazardous waste and for owners and operators of hazardous waste treatment, storage, and disposal facilities. This cradle-to-grave system has identified 52,864 waste generators, 12,000 transporters, and about 5,000 treatment, storage, and disposal facilities, and it has brought a greater degree of order to the management of large-scale wastes. The EPA estimates, based on 1981 data, that some 264 million metric tons of hazardous waste are being generated annually in this country. The combined total of all forms of solid waste, hazardous and otherwise, amounts to almost 6 billion tons per year.

C. The RCRA Law is Strengthened

When Congress reauthorized RCRA late in 1984, it imposed new and far-reaching requirements on a vastly larger regulated community, notably the 175,000 enterprises that generate small amounts (between 220 and 2,200 pounds) of waste per month and those that own or operate a million underground storage tanks. Controls for land disposal will be tightened, while certain wastes will be banned from landfills altogether. In addition, burners and blenders of fuels derived from hazardous waste will be subject to EPA regulation. The new RCRA represents a clear shift in national policy away from land disposal and toward waste reduction, recycling and new treatments for flammable, reactive, corrosive, and toxic wastes that now threaten air quality and vital surface and ground water resources. The amended RCRA embraces more than 70 new provisions and 58 action deadlines. For example, the EPA is now required to establish a program to control underground tanks containing petroleum and other designated hazardous substances.

The agency must issue regulations by February 1987 for petroleum tanks, by August 1987 for new tanks containing chemical products listed as hazardous under the Superfund, and by August 1988 for existing tanks containing such Superfund chemicals. Installation of certain underground tanks is prohibited. The Underground Storage Tank (UST) program may require the EPA to inspect and regulate a million tanks nationwide. New statutory controls may be imposed on as many as 100,000 new tanks installed each year.

The new RCRA bans the land disposal of hazardous wastes unless the EPA finds they will not endanger human health and the environment. Landfilling of bulk or noncontainerized liquids is now prohibited. By February 1986 the EPA must promulgate regulations to minimize the landfilling of containerized liquid hazardous waste.

No bulk liquids may be disposed of in salt domes. Using oil contaminated with hazardous waste as a dust suppressant and infection of hazardous wastes into or above an underground source of drinking water are both outlawed.

The new act further requires those who produce, burn, distribute, or market fuel derived from hazardous wastes to notify the EPA of their operations. The agency must then issue record-keeping requirements and technical standards.

In addition, anyone who wants to operate a waste management facility must meet minimum technological requirements, including double liners, leachate-collection systems, and extensive ground-water monitoring. Facility owners and operators are required by the new law to take corrective action if any part of a RCRA facility not on a permanent control plan suffers an uncontrolled release. Such action can now be accomplished through new permit requirements or legal remedies.

The amendments also strengthen Federal controls over the disposal of non-hazardous municipal wastes; Federal enforcement authority can be applied in cases where states do not mandate a permit program for municipal landfills. Finally, RCRA strengthens Federal enforcement by expanding criminal offenses and raising maximum penalties. Any citizen can file an "imminent hazard" lawsuit, and the EPA is authorized to issue an administrative order to correct any release of hazardous waste from a facility that is or was subject to temporary permit requirements.

Another of the several purposes of the 1984 RCRA amendments is protection of precious ground water supplies from contamination by seepage from the land surface. Major parts of regulations governing small quantity generators (SQGs) and underground storage tanks are designed to prevent such damage to aquifers. The law is also intended to control air pollution resulting from combustion of hazardous waste mixed with various fuels and the evaporation of volatile organic materials from landfills and storage depots.

D. Restriction on Land Disposal of Hazardous Wastes

It was the intent of Congress to discourage land disposal of hazardous waste, because of long-term uncertainties about its persistence, toxicity, mobility, and accumulation in plants, animals, and human tissue. Certain materials will be banned unless they receive specific EPA approval. Land disposal can be permitted if the waste meets pretreatment levels or standards.

The land disposal program features tight deadlines and "hammers"—automatic bans if the EPA fails to meet them. Some of these deadlines are as follows:

(1) Dioxin-containing waste and spent or discarded solvents are banned as of November 16, 1986.

(2) Wastes listed as hazardous by the state of California, including liquid hazardous wastes containing certain metals, free cyanides up to 1,000 milligrams per liter, PCBs up to 50 milligrams per liter and acids with a pH rating lower than 2.0 are banned after July 8, 1987.

(3) Liquid hazardous wastes containing arsenic up to 500 milligrams per liter, cadmium up to 100 milligrams per liter, chromium VI up to 500 milligrams per liter, lead up to 500 milligrams per liter, mercury up to 20 milligrams per liter, nickel up to 134 milligrams per liter, selenium up to 100 milligrams per liter, and thallium up to 130 milligrams per liter are banned as of July 8, 1987.

(4) Liquid or solid hazardous waste containing halogenated organic compounds up to 1,000 milligrams per liter are banned as of July 8, 1987.

(5) Contaminated soil and debris from CERCLA response or corrective action

under RCRA are exempted until November 1988 and waste injected into deep wells until August 8, 1988.

(6) The deadline for promulgation of the EPA waste review schedule is November 8, 1986.

(7) The EPA must review at least one-third of the wastes by August 8, 1988, at least two-thirds of them by June 8, 1989, and all ranked waste and all "characteristic" waste by May 8, 1990.

Those schedules are based on the toxicity and volume of waste disposed on land. The most hazardous are to be examined first. A new leaching test will determine if hazardous waste constituents exceed allowable health thresholds.

If no available treatment methods can safeguard public health, performance standards will be imposed where possible, based on best demonstrable achievable technology. Effective dates of land-disposal bans can be extended by petition for 2 1-year periods on a case-by-case basis if alternative disposal capacity is not available. Petitions must demonstrate a reasonable certainty, however, that there will be no migration of constituents as long as wastes remain hazardous.

E. The Effects on Small Quantity Generators (SQGs)

Before the reauthorized Act, the EPA regulated only establishments generating more than 1,000 kilograms (2,200 pounds) of hazardous waste per month. Under the new law those that generate 100 kilograms (220 pounds, or roughly half a 55-gallon drum) but less than 1,000 kilograms per month will have to comply with rules covering transportation and disposal of wastes hazardous to human health and the environment.

The agency estimates that the new RCRA will increase the number of Federally regulated waste generators from about 15,000 to about 175,000. An EPA survey released in March 1985 suggested that 85 percent of SQGs are in vehicle maintenance, equipment repair, construction, printing, photography, laboratories, schools, laundries, dry cleaners and pesticide applicators. Most of the remainder are in manufacturing or finishing of metals. (See Fig. 19-1.)

The new requirements will have their greatest impact on firms in the 28 states that do not currently impose some regulation on SQGs.

Starting August 5, 1985, SQGs shipping hazardous waste off premises must, like large-volume generators, attach a manifest (see Chapter IX) required by the EPA and the DOT. It must include the generator's name, address and signature; DOT waste nomenclature and classification; number and type of containers; weights and quantities being transported and name and address of cosignee. The manifest will help prevent confusion and illegal dumping by permitting the EPA and the states to track shipments from origin to final disposal.

By March 31, 1986, the agency must issue final regulations protecting human health and environment from small quantities of hazardous waste. At a minimum, the regulations, now in process, will:

(1) require that hazardous waste from generators of more than 100 kilograms per month be treated, stored or disposed of at an approved hazardous waste facility;

(2) allow small quantity generators to store waste on premises up to 180 days

States Regulating Small Quantity Generators As Of February 1984

*No exemption for small quantity generators regardless of quantity

Who They Are...

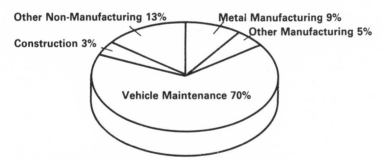

The Wastes They Produce...

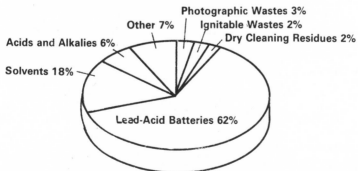

Fig. 19-1   States that regulate small quantity generators, types of generators, and wastes they produce. Source: EPA Office of Solid Waste and Emergency Response, October 1985.

without a storage permit, or 270 days for waste to be transported more than 200 miles, provided that no more than 6,000 kilograms are stored.

If the EPA fails to issue final regulations by March 31, 1986, hazardous waste from SQGs automatically becomes subject to those minimum requirements. In addition, for waste shipped off-site, SQGs will be required to include the name of the transporter on the manifest, retain manifests signed and returned by the hazardous waste facility for at least 3 years, and notify the EPA at least twice per year of any manifests not returned, so the agency can follow up for possible violations.

Because the new provisions regulate a large number of generators for the first time, the EPA is conducting a comprehensive educational assistance program to alert SQGs to their responsibilities under the law. For provisions that must be implemented by August, 1985, the EPA will:

(1) identify potential SQGs;

(2) provide information through EPA regional offices, state and local governments, trade associations, and other groups to help SQGs determine if they are subject to the regulations. EPA will identify wastes by trade, chemical, and colloquial names, and will correlate the waste with DOT identification numbers wherever possible;

(3) inform SQGs of the need to prepare a Uniform Hazardous Waste Manifest to accompany any materials they ship and explain how and where to obtain forms.

For the final regulations, to be issued or take effect automatically by April 1, 1986, the EPA will:

(1) alert SQGs to the new regulations plus additional requirements;

(2) provide them with complete instructions and industry-specific information to help them comply.

F. The Underground Storage Tank Program (UST)

There are approximately a million underground storage tanks (USTs) in the United States that contain hazardous materials and petroleum products. The new RCRA provisions apply, for the first time, not only to wastes but also to the storage of useful materials.

Under a new Subtitle I, RCRA now regulates underground tank storage of all petroleum products (including gasoline and crude oil) and any substance defined as hazardous under the Comprehensive Environmental Response, Compensation, and Liability Act of 1980 (the Superfund law), which authorizes cleanup of abandoned or uncontrolled hazardous waste sites. An underground storage tank is defined as any tank with at least 10 percent of its volume buried below ground, including any pipes which are attached. Thus, above-ground tanks with extensive underground piping may now be regulated.

The EPA's UST program does *not* apply to the following:

1. tanks holding a hazardous waste regulated under the RCRA hazardous waste program (Subtitle C);
2. farm and residential tanks holding less than 1,100 gallons of motor fuel;

3. on-site heating oil tanks;
4. septic tanks;
5. pipelines regulated under other laws;
6. systems for collecting storm and waste water;
7. flow-through process tanks; and,
8. liquid traps or gathering lines related to oil and natural-gas operations.

Among other things, the UST program bans the installation of corrodible tanks. It also initiates a tank notification program and sets technical standards for all tanks; it coordinates Federal and state efforts; and it provides Federal inspection and enforcement.

In addition, a provision banning underground installation of unprotected new tanks went into effect on May 7, 1985. After that date no person may install an underground storage tank unless the following provisions are adhered to:

(1) the tank will prevent release of the stored substance resulting from corrosion of structural failure for the life of the tank;

(2) the tank is protected against corrosion, constructed of noncorrosive material, or designed to prevent release of the stored substance; and,

(3) the tank construction or lining materials are electrolytically compatible with the substance to be stored.

Also, the new law requires state and local notification that applies to several million tank owners and to distributors of regulated substances and owners of tanks that have been taken out of operation within the past 10 years but which are still in the ground. That notification program applies equally to owners of operational tanks.

Another provision of the law is that by May 1985, state governors must have designated the state or local agency that will receive the notifications.

There are several other provisions of the notification program:

(1) by November, 1985, the EPA must prescribe the form of the notice;

(2) by May 1986, owners of existing underground storage tanks must notify the state or local agency of each tank's age, size, type, location, and uses;

(3) by May 1986, owners of underground storage tanks taken out of operation after January 1, 1974, but still in the ground, must notify the state or local agency of each tank's age, decommission date, size, type, location and type and quantity of substance left in the tank;

(4) after May 1986, owners of newly installed underground storage tanks must notify the state or local agency of certain operational data within 30 days of startup;

(5) within 30 days of the date on which EPA prescribes the notification form (and for 18 months thereafter), any person who deposits regulated substances in an underground storage tank must inform its owner of the requirement to notify the state or local agency; and,

(6) within 30 days of the date on which the EPA prescribes the notification form, sellers of tanks must notify purchasers of the need to notify the state or local agency.

Under the new RCRA provisions, the EPA must develop and promulgate performance standards for new tanks, as well as standards covering leak detection, leak prevention, and corrective action for both new and existing underground storage tanks on the following schedule:

Item	Petroleum	Hazardous chemicals
Standards for new tanks	February 1987	August 1987
Regulation of leak detection/prevention and corrective action	February 1987	August 1988
Study/report to Congress	November 1985	November 1987

The law, also, specifies that leak detection/prevention and corrective action regulations must require owners and operators of underground storage tanks to:

1. be able to detect releases;
2. keep records of release-detection methods;
3. take corrective action when leaks occur;
4. report leaks and corrective action;
5. provide for proper tank closure; and,
6. provide evidence, as the EPA deems necessary, of financial capability of taking corrective action and compensating third parties for injury or damages from instant or continuous releases. In addition, states may finance corrective action and compensation programs by a fee levied on owners and operators.

Several states already have or are developing regulatory programs for underground storage tanks. The new law is designed to avoid interfering with those programs and to encourage other states to press ahead on their own. By May 1987, states may apply to the EPA for authorization to operate an Underground Storage Tank program. It may cover petroleum tanks or hazardous substance tanks or both. State programs must include all the regulatory elements of the Federal program and provide for adequate enforcement. After a one-to-three-year grace period, state requirements must be no less stringent than the Federal.

Federal and State personnel are authorized to request pertinent information from tank owners, inspect and sample tanks, and monitor and test tanks and surrounding soils, air, surface water, and ground water. The EPA may issue compliance orders for any violation of this statute or regulations. Offenders are subject to civil penalties of up to $10,000 per tank for each day of violation. However, criminal penalties are not authorized.

G. Permits Required for Hazardous Waste Management

The new Amendments apply immediately to facilities in all states, whether or not the state is authorized to administer its own hazardous waste program. If a facility is located in an authorized state, the latter will continue to be responsible for that portion of the RCRA program for which it is already authorized. The EPA will be responsible for aspects of the program initiated by 1984 Amendments. So for the time being, RCRA permits will need to be issued jointly by the states and the EPA.

To receive a final permit for approved operation all facilities will need to take corrective action for releases of hazardous waste or hazardous waste constituents from any solid-waste management unit on the property, regardless of when the waste was placed in the unit or whether the unit is closed. If corrective action cannot be completed before a permit is received, the EPA may put a compliance schedule into the permit to allow development of data to determine corrective action or to complete it. Owners and operators must provide financial assurance that they can complete corrective action.

New authority allows the EPA to add to permits whatever conditions are necessary for protecting human health and the environment. After September 1, 1985, a generator with an on-site facility must certify annually a reduction in volume and toxicity of waste to the maximum degree economically practicable, and that management methods minimize risk to the extent technically practicable.

A waiver from ground water monitoring requirements is no longer permissible for units located above the seasonal high-water table or for units where owners and operators have installed 2 liners and a leachate collection system or inspect liners.

An application for interim status operation and certification of compliance with ground water monitoring and financial assurance requirements must be submitted by November 8, 1985, to avoid loss of such status.

Land-disposal permits must be reviewed every 5 years and modified as needed. Any such modification must consider improvements in the state of control and measurement technology and regulations that then apply to the facility.

After August 8, 1985, each application for interim-status operation must be accompanied by exposure information, which must address potential hazardous waste releases in the course of transportation to or from the waste disposal unit, normal operations and accidents, and potential pathways, magnitude and nature of human exposure to such releases. If an application for interim-status operation has already been submitted, exposure information must be transmitted by August 8, 1985.

The EPA will determine whether a facility poses a substantial health risk because of releases of hazardous constituents. If so, the EPA will make information available to the Agency for Toxic Substances and Disease Registry (ATSDR) in the Centers for Disease Control, which will undertake a health assessment.

Permits will require ground water monitoring and installation of 2 or more liners with leachate collection above or between liners, as appropriate. Surface impoundments and landfills outside of Alabama can obtain a waiver from the double-liner requirement if alternative design and operating practices, along with local characteristics, are shown to be at least as effective in preventing migration of hazardous constituents to aquifers.

Facility owner-operators can also obtain a waiver from the double-liner and leachate collection requirements for certain monofills and foundry wastes. Landfills or surface impoundments that have received approval for interim-status operation and that receive waste into new units and/or lateral expansions or replacements of existing units after May 8, 1984, must meet the double-liner and leachate-collection requirements and waiver conditions.

If liner and leachate collection systems are installed in good faith compliance with the EPA regulations or guidance documents, a different system may not be required when a facility receives its first permit, unless the EPA has reason to believe the liner is performed.

If an owner or operator intends to receive waste into new interim-status units on or after May 8, 1985, he must notify the agency at least 60 days before receiving waste. He must submit his Part B permit application within 6 months of this notification.

As of May 8, 1985, a facility owner or operator will not be able to dispose of bulk or noncontainerized liquid hazardous waste or free liquids contained in hazardous waste (regardless of whether absorbents have been added) in a landfill. After November 8, 1985, he will not be able to dispose of any nonhazardous liquid wastes in his landfill. A waiver of this prohibition may be obtained under certain conditions.

If an interim-status surface impoundment was in existence on November 8, 1984, 2 or more liners with leachate collection between the liners must be installed and ground water must be monitored by November 8, 1988. Permits issued to surface impoundments will require retrofitting within four years, unless the owner-operator applies for and receives a waiver, which is available for units that meet certain criteria. Interim-status waste piles that receive waste into new units or lateral expansions or replacements of existing units on or after May 8, 1985, must meet current Part 264 standards for liners and leachate collection systems.

Additional provisions that will impact selected facilities (e.g., prohibitions on placing some hazardous waste in salt domes, underground mines, or caves until the facility receives a permit) are not outlined in this section because they are germane only to small segments of the regulated community. Moreover, this section does not treat provisions that become effective more than 1 year after enactment of the RCRA amendments.

H. RCRA Enforcement Provisions

The EPA already had considerable compliance authority before the enactment of the Hazardous and Solid Waste Amendments of 1984. The agency could require information, obtain access for inspections, and conduct monitoring. In addition, violators were subject to compliance orders, penalties, and criminal fines and imprisonment. Finally, the administration could bring an action whenever the handling, treatment, transportation, or disposal of a solid or hazardous waste threatened an imminent and substantial endangerment of public health or the environment. The Amendments enlarged most of those authorities, including corrective action.

The Amendments require periodic inspection of facilities. Beginning 12 months after the date of enactment, and no less than every 2 years thereafter, the EPA or authorized states must inspect all treatment, storage, and disposal facilities for which a permit is required. The EPA must annually inspect every treatment, storage, or disposal facility operated by a state or municipality. The EPA must, and a state may, inspect all treatment, storage, or disposal facilities owned or operated by a Federal agency annually, starting 12 months from the date of enactment. No waivers or variances are permitted. Within 6 months of the date of enactment, the EPA had to submit a report to Congress on the potential for private inspectors to support government inspection efforts.

As discussed elsewhere in this chapter, the amendments established numerous new regulatory requirements. These are subject to conventional enforcement to compel compliance and impose penalties for violations. To eliminate certain substandard operations, temporary operational permits terminate for land disposal facilities unless they certify compliance with all applicable ground water monitoring and financial responsibility requirements, and operators submit applications for a final determination regarding issuance of a permit within 1 year of the date of enactment. Existing interim-status incinerators have until November 8, 1986, to submit Part-B applications to maintain interim status. All other existing interim-status facilities have 4 years to submit such applications.

Corrective action authorities are a major component of the Amendments. The EPA can issue orders requiring corrective action or other response measures at interim-status facilities when the agency determines that a release of hazardous waste is now or has been taking place. Such orders can include a "suspend or revoke" authorization to operate under interim status. Facilities that fail to comply with the terms and schedules specified in these orders are subject to civil penalties of up to $25,000 per day of noncompliance. The EPA can also initiate a civil action for appropriate relief, including a temporary or permanent injunction.

In addition, the agency must promulgate regulations as promptly as practicable requiring owners or operators of all permitted facilities and regulated units to take corrective action beyond the property boundary where necessary to protect human health and the environment, unless they can demonstrate to the agency's satisfaction an inability to obtain permission despite their best efforts. Pending such regulation, the EPA can issue corrective action orders for that purpose on a case-by-case basis.

In addition, criminal sanctions were expanded by the new amendments. Any treatment, storage, and disposal facility or generator and transporter of hazardous waste is subject to criminal penalties for knowing violations of interim-status standards, material omissions, and failure to file required reports, transportation without a manifest and causing such waste to be transported to an unpermitted facility. The maximum penalty for those violations is boosted to $50,000 for each day of noncompliance. The maximum prison sentence for transporting to an unpermitted facility and treating or disposing of wastes without a permit or in violation thereof or of interim-status standards is raised to 5 years.

The class of "knowing endangerment" crimes is also expanded. The requirement that those responsible for a "knowing endangerment" display an unjustifiable and inexcusable disregard for, or extreme indifference to, human life before they can be prosecuted is deleted. The maximum prison sentence for any person convicted of "knowing endangerment" is extended to 15 years.

Citizens are authorized to bring actions in cases where past or present management of hazardous waste presents an imminent and substantial endangerment. This right is circumscribed where the EPA or a state is diligently prosecuting an action under the Comprehensive Response, Compensation and Liability Act (CERCLA), is engaged in a removal action, or has issued an administrative order to compel a cleanup. Citizens are expressly authorized to sue open-dump operations.

Actions or issues for which owners or operators have previously had opportunity to obtain review (the permit process, state authorization) cannot be subject to judicial review in civil or criminal enforcement proceedings. The administrator's action in issuing, denying, modifying or revoking any permit, or granting, denying, or withdrawing authorization is not subject to review in an enforcement action if review could have been obtained elsewhere under the law.

The U.S. Attorney General is authorized to deputize EPA employees to act as marshals in RCRA criminal investigations. EPA is also authorized to conduct criminal investigations and to refer the results to the Attorney General for prosecution.

Under the Amendments, the endangerment provision was clarified. Those contributing to the endangerment, including past and present generators, transporters, and owners or operators of treatment, storage, and disposal facilities are liable for conditions resulting from past as well as present activities. When the EPA proposes to settle such actions, the agency is to provide public notice, an opportunity to comment in writing and a public meeting to discuss proposed settlement terms.

I. State Programs and Agencies for Hazardous Waste Management

RCRA required the EPA to institute a national program to control hazardous waste. It was the intent of Congress, however, that where possible states assume responsibility for controlling such waste within their borders, and Federal financial and technical assistance is available for program development. Section 3006 of the act specifically authorizes states to operate their own hazardous waste programs after approval by the EPA.

To receive final approval, a state must show that it has the resources to administer and enforce a hazardous waste program equivalent to and consistent with the Federal program. Equivalent means "equal in effect." States may set more stringent standards, but they may not impose any requirement that might interfere with the free movement of hazardous wastes across state boundaries to treatment, storage, or disposal facilities holding a RCRA permit.

The deadline for a state to obtain final authorization to administer the national hazardous waste program is January 31, 1986. Eligibility for final authorization will be determined on the basis of standards in effect before application, or January 26, 1983, whichever is later. Requirements imposed under the 1984 amendments apply immediately in all states and will be administered by the EPA until the state is authorized to do so. The EPA and a state may enter into a cooperative agreement regarding the administration of the program, and joint permits may be issued for those requirements not yet incorporated into the state program. As of August, 1985, 26 states have been granted final authorization to administer the program. Eight others have submitted applications, and 2 have been granted tentative approval for final authorization. For a list of State solid and hazardous waste agencies, see Appendix F.

J. What RCRA May Encompass

The RCRA Amendments of 1984 have provided sweeping changes to existing legislation and have definitely put teeth into the law. Therefore, in order that the reader may obtain a more complete grasp of the total complexity of this important piece of legislation, the following discussion has been included in this Chapter.

1. Small Quantity Generators (SQGs)

After August, 1985, SQG waste not managed at a permitted Subtitle C installation may be disposed of only at a state-approved municipal or industrial facility.
By March 31, 1986, the EPA must promulgate standards for waste generated in
quantities between 100–1000 kg/month; the rules may vary from conventional
Subtitle C regulations but must protect human health and the environment.

At a minimum, the standards must not allow on-site storage more than 180
days without a permit, and all other management of SQG waste must occur at
a permitted Subtitle C facility; however, the on-site storage period may be
extended to 270 days for waste transported more than 200 miles if it does not
exceed 6,000 kgs.

If the EPA fails to promulgate standards on time. SQG waste generated above
100 kg/month becomes subject to the minimum requirements described above
plus exception reports and retention of manifests for three years. By August,
1985 waste generated in quantities between 100 and 1000 kg/month must be
accompanied by a Uniform Manifest.

By April 1, 1985, the EPA must submit a study characterizing the generators,
wastes, practices, and risks posed by wastes in quantities less than 1000
kg/month. By April 1, 1987, the agency must submit studies on the feasibility
of establishing a licensing system whereby transporters assume the responsibilities of SQGs, the merits of retaining the existing manifest system for SQG
waste, and the problems associated with the disposal of hazardous waste
generated by educational institutions.

Within 30 months of enactment, the EPA must inform SQGs of their responsibilities under the 1984 amendments, for which $500,000 a year is authorized
for FY 1985 through 1987.

2. Banned Wastes

The EPA must determine whether to ban the land disposal of a third of wastes
listed as hazardous in 45 months, two-thirds of listed wastes in 55 months, all
listed and characteristic wastes in 66 months and for wastes listed after enactment, 6 months after listing.

The land disposal of a hazardous waste must be banned unless the EPA
determines that the prohibition is not required in order to protect human health
and the environment. A petitioner must demonstrate there will be no migration
from the disposal unit/injection zone for as long as the waste remains hazardous.
The EPA must promulgate regulations specifying levels or methods of treatment,
if any, which substantially reduce the likelihood of migration of hazardous
constituents such that threats to human health and environment are minimized.
"Otherwise banned" wastes so treated are exempt.

Other than for disposal in injection wells, the EPA must decide whether to
ban the land disposal of dioxins and solvents within 24 months of enactment
and, 8 months later, the "California wastes." The decision whether to ban those
wastes from injection wells must be made within 45 months of enactment.
Within 24 months the agency must publish a schedule for determining whether
to ban the land disposal of *listed* hazardous wastes. High-hazard and high-volume
wastes must be scheduled first. The schedule is not subject to paperwork
reduction or judicial review.

Land disposal prohibitions are effective immediately unless:

(1) another date is selected because, on a national scale, alternative capacity is unavailable; postponement beyond 2 years is not permissible;

(2) a variance is granted to an individual facility. Variances can be granted for 1 year and renewed for an additional year upon a showing of a binding contractual commitment to provide alternative capacity, but disposal in a landfill or surface impoundment must be at a facility in compliance with the minimum technology described.

The prohibitions for dioxins, solvents, and the California wastes do not apply for 45 months to contaminated soil debris from cleanup and removal actions. If the EPA fails to make a determination during the allotted time for the California wastes and the "last third" of the listed wastes, land disposal of such waste is prohibited.

If the EPA fails to make a determination during the allotted time for the "first and second thirds" of the listed wastes, disposal in a landfill or surface impoundment is permissible only if the generator certifies no alternative capacity and disposal is at a facility in compliance with minimum technology requirements. However, if the agency fails to make a determination within 66 months of enactment, land disposal is prohibited.

3. Other Land-Disposal Restrictions

Within 6 months of enactment, the landfilling of bulk or noncontainerized liquids is prohibited. Within 12 months, the disposal of nonhazardous liquids is prohibited in Subtitle C facilities unless the only reasonable alternative is disposal in a non-Subtitle C landfill or unlined impoundment that contains or may contain hazardous waste, and such disposal will endanger a potable-water aquifer.

Within 15 months the EPA must promulgate regulations to minimize landfilling of containerized hazardous liquids and prohibit the landfilling of liquids absorbed in materials that biodegrade or release liquids when compressed. Also, the placement of bulk liquids in salt domes, salt beds, underground mines, or caves is prohibited until the EPA promulgates placement rules and the facility receives the permit. Containerized hazardous wastes cannot be dumped therein until the facility receives a permit. The Waste Isolation Pilot Project in New Mexico is not subject to those restrictions. Oil contaminated with hazardous, except ignitable, wastes cannot be used as a dust suppressant.

Within 6 months of enactment (or sooner if a state has primacy), hazardous waste cannot be injected into or above any formation that contains, within one-quarter mile of the well, a potable-water aquifer unless it is part of certain actions under CERCLA or RCRA.

4. Retrofitting Surface Impoundments

Interim status impoundments must either comply with the double-liner, leachate collection, and ground water monitoring requirements for new impoundments or stop receiving, storing, or treating hazardous waste within four years of enactment for current impoundments or within 4 years of the date such an impoundment becomes subject to Subtitle C. Exempted impoundments other than waste water impoundments that no longer qualify for the exemption must comply within 2 years of discovery of the disqualifying condition. Subsequently disqualified waste water impoundments have 3 years to retrofit.

Impoundments not located within one-quarter mile of a drinking-water aquifer have at least 1 liner that complies with the current Part 264 standards for new impoundments and for which there is no evidence the liner is leaking, and are in compliance with the Part 264 ground water monitoring requirements, are exempt from the above regulations. Waste water impoundments conducting "aggressive" biological treatment and various downstream impoundments subject to a Sec. 402 CWA permit, in compliance with the Part 264 ground water monitoring requirements, and part of a facility in compliance with Best Available Technology (BAT) effluent guidelines are also exempt.

The same applies where no BAT guideline is applicable, the facility is not implementing BAT based on a BPJ permit, and the impoundment is part of a facility with a Sec. 402 CWA permit achieving a significant degradation of hazardous constituents in the untreated wastestream, and where impoundments are designed, located, and operated to prevent the migration of any hazardous constituent into ground water or surface water at any future time for which the EPA has modified retrofitting requirements. Also exempt: impoundments for which, before the date of enactment, the EPA or an authorized state had entered into a consent decree, order, or agreement mandating corrective action equivalent to double-liner and leachate collection.

If the EPA determines an exempted impoundment is likely to leak hazardous constituents to ground water, it may impose any requirement necessary to protect health and environment including retrofitting. An exempted impoundment, other than a waste water impoundment found to be leaking or otherwise no longer qualifying for the exemption, must be retrofitted. An exempted waste water impoundment found to be leaking must be retrofitted unless the EPA determines within 3 years of enactment that it is not necessary to protect health and environment.

To obtain an exemption, owners or operators must apply within 24 months of enactment, submit a Part B application and leakage-to-groundwater monitoring data, and provide certification by a registered professional engineer that the impoundment meets applicable criteria. The EPA must comment and process the application within 12 months of receipt. It must submit a report to Congress on the environmental impoundment exemption and on the feasibility and cost of deleting it.

5. Storage of Banned Wastes

Surface impoundments operators that store or treat hazardous wastes banned from land disposal units must remove hazardous residues within 1 year and comply with requirements for new impoundments unless the impoundment meets the conditions for a retrofitting waiver and be solely for the purpose of accumulating sufficient quantities for proper subsequent management.

6. Minimum Technology Standards

A landfill or impoundment for which a Part B application has not been received by the date of enactment must have a double-liner with leachate collection above and between the liners respectively and monitor ground water. With 2 years of enactment, the EPA must promulgate implementing regulations or issue guidance documents. Meanwhile, a synthetic or clay liner system may be installed. Where the owner or operator can demonstrate that an alternative design, when one considers location characteristics, is as effective in preventing migration of hazardous constituents to ground water, a double-liner will not be required except in Alabama. Certain monofills containing foundry wastes are also exempt.

Expansions and replacements of interim status landfills, impoundments, and piles that receive waste 6 months after enactment are subject to the same requirements as above. Owners and operators of such landfills and impoundments must notify the EPA 60 days before the unit receives waste and submit a Part B application 6 months afterward. The EPA may not require those who installed double-liner systems in good-faith compliance with regulations or guidance documents to alter these systems in order to receive a permit. However, if a liner is leaking, the EPA may require replacement.

Interim status landfills, impoundments, land treatment facilities, and piles that received waste after August 20, 1983, are subject to the requirements for ground water monitoring, unsaturated zone monitoring, and corrective action applicable to new facilities.

Incinerators receiving permits after the date of enactment must achieve a 99.99 percent DRE.

Within 18 months of enactment, the EPA must publish guidance criteria identifying areas of vulnerable hydrogeology and write regulations for the acceptable location of new and existing hazardous waste facilities. Within 30 months, the EPA must promulgate standards for leak detection and air emissions.

7. Ground Water Monitoring

Part 264/5 variance from ground water monitoring standards for certain double-lined facilities is eliminated. Thus, the EPA is authorized to exempt from ground water-monitoring requirements land-disposal units designed to prevent liquids from entering the unit and equipped with multiple-leak detection systems.

8. Corrective Action

The EPA must promulgate regulations that require evidence of financial capacity for cleanups, and as soon as practicable, amend hazardous waste regulations to require corrective action beyond the facility boundary. The regulations will take effect immediately upon promulgation and apply to all permitted facilities and interim status landfills, impoundments, and piles that received waste after August 26, 1982. Until then the EPA must issue corrective action orders on a case-by-case basis as necessary to protect health and environs.

All permits issued after the date of enactment must address releases of hazardous waste or constituents regardless of type of unit, when waste was placed in it, or whether the unit is closed. Owners and operators must prove their financial ability to clean up. The EPA is authorized to issue administrative orders requiring corrective action for releases of hazardous waste from interim status facilities, and to sue those responsible.

9. Permits

Permits must be renewed every 10 years, and land-disposal permits must be reviewed every 5 years. Renewals are subject to regulations applicable to new permits and must reflect improvements in control and measurement technology. Interim status terminates unless a Part B application is submitted according to the following schedule:

Facility	Interim status terminates	Unless Part B Submitted by
Land Disposal	Oct 1985	Oct 1985
Incinerators	Oct 1989	Oct 1986
Others	Oct 1992	Oct 1988

Land disposal owner-operators must also certify compliance with ground water-monitoring and financial responsibility requirements to retain interim status. The EPA or the states must process permit applications within 4, 5, and 8 years of the date of enactment for land disposal units, incinerators, and other facilities, respectively. The EPA is authorized to issue 1-year permits, renewable each year up to 4 years, for experimental facilities without first issuing permitting standards under Sec. 3004.

Interim status is granted to facilities that become subject to Subtitle C as a result of the 1984 amendments or implementing regulations. A permit is required before construction of a hazardous waste facility can begin, except for PCB incinerators approved under TSCA.

10. Exposure Assessments

Within 9 months of enactment, permit applications for landfills and surface impoundments must be accompanied by an assessment of the potential public exposure to hazardous substances. Facilities whose applications have already been submitted have 9 months to submit the assessment. The Agency for Toxic Substance and Disease Registration (ATSDR) is to conduct health assessments of communities where evidence indicates substantial risk.

11. Waste Minimization

After September 1, 1985, manifests must contain generator certification that volume and/or quantity and toxicity of waste has been reduced to the maximum degree economically practicable. Thereafter, generators must annually certify their efforts to reduce waste volume and the reduction actually achieved. By October 1, 1986, the EPA must submit a report to Congress on the feasibility and desirability of establishing waste minimization rules.

12. Listing and Other Measures to Add New Wastes

The EPA must determine whether to list the following wastes within the following dates of enactment: 6 months: chlorinated dioxins and dibenzofurans; 12 months: other halogenated dioxins and dibenzofurans; 15 months: coal slurry pipeline effluent, coke by-products, chlorinated aliphatics, dioxin, dimethyl hydrazine, TDI, carbamates, bromacil, linuron, organobromines, solvents, refining wastes, chlorinated aromatics, dyes and pigments, inorganic wastes, lithium batteries, and paint-production wastes.

In addition, the EPA and ATSDR must identify wastes hazardous solely because they contain, for example, recognized carcinogens at levels beyond which human health is endangered. Within 2 years of enactment the EPA must identify additional characteristics, including measures of toxicity. Within 28 months the agency must select media that accurately predict leaching potential of wastes that threaten health and environs when mismanaged.

13. Delistings of Hazardous Wastes

The EPA must consider factors in addition to those for which the waste was listed when processing delisting petitions and must provide notice and comment beforehand. Temporary delistings not finalized within 24 months of enactment lapse summarily. To the extent practical, new petitions must be processed within 24 months of the date the EPA receives a complete application. Temporary exclusions are prohibited without prior notice and comment.

14. Burning and Blending Hazardous Wastes

Within 15 months of enactment, individuals who produce, burn, and distribute or market hazardous waste-derived fuel must notify the EPA. Within 15 months the agency must promulgate record-keeping requirements for such activity, and within 2 years it must promulgate technical standards for them and those who transport such fuel. Within 90 days of enactment, invoices for hazardous waste-derived fuel must bear a warning label, except for fuels from petroleum-refining operations where oil-containing hazardous wastes are reintroduced to the refining process.

Until regulations are promulgated, certain cement kilns cannot burn hazardous waste-derived fuel unless they comply with incinerator standards. Hazardous-waste-derived coke is exempt from labeling and record keeping requirements and regulations for producers, burners, and distributors, provided the coke is derived from on-site refinery wastes and does not meet any of the listing characteristics. The EPA must exempt from these requirements facilities burning "de minimis" quantities of hazardous waste, provided they meet certain requirements.

15. Used Oil and Automotive Oil

Within 12 months of enactment, the EPA must propose whether to list used automotive oil as hazardous waste and, within 24 months, make a final determination regarding automotive and other used oil. The performance standard for used oil regulations under Sec. 3014 is to protect human health and the environment (as well as to promote recycling). Recycled used oil is exempt from the Sections 3002 and 3003 standards for generators and transporters. Instead, the EPA must promulgate within 24 months of enactment special standards and subject recyclers to conventional Sec. 3004 standards.

Generators who enter into an agreement to deliver used oil to a permitted recycling facility are exempt from the manifest requirements, provided they do not mix hazardous waste with the oil and keep records as the administrator deems necessary. The EPA is authorized to abjure issuance of class permits for certain generators and transporters who treat or recycle used oil but may tailor permits to individual cases.

16. The Burning of Municipal Solid Waste

The combustion of municipal solid waste at a resource recovery facility is exempt from the Subtitle C requirements, provided that the owner or operator takes precautions to ensure that hazardous wastes are not burned. As soon as practicable, the EPA must submit a report to Congress on the risk of dioxin emissions from resource recovery facilities that burn municipal solid wastes and on the means to control them.

17. Domestic Sewage

Within 15 months of enactment, the EPA must report to Congress on hazardous wastes exempt from Subtitle C because they are mixed with domestic sewage or other wastes that pass through sewers to POTWs. Then, within 18 months, the EPA must promulgate rules to ensure that those wastes are adequately controlled to protect human health and the environment. Within 36 months of enactment, the agency must submit a report to Congress on waste water lagoons at POTWs and their effect on ground water. RCRA inspection and notification requirements apply as much to solid or dissolved materials in domestic sewage as to other hazardous wastes.

18. Hazardous Wastes That are Exported

Within 24 months of enactment, no one may export hazardous waste unless he or she has filed a notification, the receiving country has agreed in writing to accept the waste and a copy of the consent is attached to the manifest, and the shipment conforms to the terms of the consent agreement.

Within 12 months of enactment the EPA must promulgate implementing regulations. Bilateral agreements between the United States and the receiving country establishing hazardous waste export procedures supersede, as above, but exporters must file annual reports to the agency.

19. Mining Wastes, Utility Company Wastes, and Cement Kiln Dust

According to the reauthorized RECRA regulations concerning mining wastes, cement kiln dust, and wastes from the production of power, the EPA is authorized to modify certain requirements for those wastes to take into account their special characteristics and sites, provided that health and environs are protected. That discretion is restricted to 3 aspects of Section 3004 requirements for landfills and surface impoundments: double-liners (including retrofitting for surface impoundments), prior releases, and land-disposal restrictions.

20. State Implementation of RCRA Requirements

For final authorization, states must meet standards in effect before state application, or January 23, 1983, whichever is later. To obtain or maintain authorization, states must make available to the public information they have obtained from TSDFs to the extent such information would be available if the EPA were running the program. The deadline to obtain final authorization is extended by one year.

Any requirement imposed under the 1984 amendments applies immediately in authorized states until their programs are revised to incorporate the requirements. The EPA administers the requirements until the states receive authorization; states with provisions substantially equivalent to the new requirements may apply for interim authorization to administer them. The agency is authorized to enter into cooperative agreement with states to assist in the administration of the 1984 amendments. The EPA is also authorized to jointly issue permits with the states for those requirements not yet incorporated into state programs.

States are authorized to require that copies of manifests for intrastate shipments be sent to them.

21. RCRA Subtitle D Criteria

Within 36 months of enactment, the EPA must submit a report to Congress, determining whether the Sections 1008(a) and 4004 criteria are adequate to protect health and environs from ground water contamination and recommending whether additional authority is needed to enforce them. By March 21, 1988, the agency must revise the criteria for facilities that may receive hazardous household or SQG waste. The criteria must protect health and environs. At a minimum, the EPA will require ground water monitoring, establish location criteria, and provide for corrective action as appropriate.

Within 36 months of enactment, each state must develop a program to ensure that municipal facilities comply with existing criteria. States must develop a plan to ensure compliance; if they fail to do so, the EPA may enforce them. Fifteen million dollars of the 1985 appropriation for state grants and $20 million per year of the 1986–88 appropriation can be used to implement the criteria.

State solid-waste plans for waste-to-energy facilities must consider present and future needs of recycling and resource recovery interests, including those created by the implementation of Sec. 6002. By October 1, 1986, the EPA must submit a report to Congress on methods for extending the useful life of sanitary landfills and for putting closed landfills to more efficient use.

22. Government Agency Procurement

Each procuring agency is required to encourage the preferential purchase of items containing recovered materials. The EPA must promulgate guidelines for paper within 180 days of enactment and for 3 additional products (including tires) by October 1, 1985. The Office of Procurement Policy must submit biennial reports to Congress on Federal progress in promoting the use of recovered materials.

23. Inventory of Wells Injected with Hazardous Wastes

The EPA has been chartered to submit to Congress an inventory of all wells that have been injected with hazardous wastes as a disposal methodology. It is feared that that method of disposing of hazardous waste may contaminate ground water; therefore, the problem becomes one of sampling, monitoring, and taking countermeasures.

24. Inventory of Federal Treatment, Disposal, and Storage (TSD) Facilities

Each federal Agency must submit to the EPA biennially an inventory of each treatment, storage, or disposal facility (TSD) it owns or operates. Agencies need not resubmit information already submitted under Sec. 103 of CERCLA or Sections 3005 and 3010 of RCRA. The EPA must conduct that inventory where Federal agencies decline to do so.

25. Annual EPA Inspection Requirements and Enforcement

The EPA must inspect annually each hazardous-waste facility operated by a state or municipality. It must, and authorized states may, inspect each federally owned or operated TSD facility annually. At least every 2 years, the EPA (or authorized states) must inspect privately operated facilities; the EPA will promulgate regulations governing the minimum frequency and manner of such inspections. Within 6 months the EPA is required to submit a report to Congress on the merits of using private inspectors to supplement government inspections.

The EPA is authorized to assess civil penalties administratively for past as well as present violations of RCRA. The agency was provided a new corrective-action order authority regarding releases. In addition, the broad imminent and substantial endangerment authority is clarified. Section 7003 applies to past generators and to situations or sites where past acts or failures to act may have contributed to present endangerment of health and environs. The EPA must notify local officials and post a sign at sites posing in imminent and substantial threat to health and environs. The agency must provide for public notice and comment before entering into a settlement or covenant not to sue under Sec. 7003.

The maximum criminal penalties are raised and criminal actions are expanded to include: violations of interim-status standards; failure to file required reports; and transportation of hazardous waste without a manifest. The category of actions subject to the "knowing endangerment" provision of Sec. 3008(e) is also expanded.

The Attorney General is authorized to deputize EPA employees to act as special marshals in RCRA criminal investigations. The EPA is authorized to conduct investigations and refer the results to the Attorney General for prosecution.

Defendants in enforcement proceedings cannot challenge permit terms and conditions of state program provisions that could have been challenged at the time of permit issuance.

26. Rights of Citizens

Citizens are authorized to bring actions under Sec. 7003 in cases where past or present management of hazardous waste permits presents an imminent hazard. That right is circumscribed in several ways (e.g., where the EPA or the state is diligently bringing and prosecuting an action under Sec. 7003 of RCRA or Sec. 106 of CERCLA or has settled the action by entering into a consent decree).

Common carriers are immunized from citizen suits for imminent hazards arising after shipments are delivered to the cosignee. Section 7003 applies to past generators and to situations or sites where past acts or failures to act may have contributed to a present endangerment to human health and the environment. The administrator is prevented from bringing an action against common carriers for imminent hazards arising after delivery of the shipment to the cosignee.

Citizens are also authorized to bring action against persons engaged in open-dump activities.

There are other citizens rights in the Amendment also, and they are as follows:

(1) *Immediate Notice:* The EPA must notify local officials and post a sign at sites posing an imminent and substantial threat to human health and the environment.

(2) *Public Participation:* The EPA must provide for public notice and comment before entering into a settlement or covenant not to sue under Sec. 7003.

(3) *Ombudsman:* The EPA must establish an Office of Ombudsman to provide information, receive complaints, and assist in their resolution. The office terminates 4 years after enactment.

27. Groundwater Commission Established

Among other regulatory provisions and effects of the reauthorized RCRA Amendments of 1984 is the establishment until January 1, 1987, of a Groundwater Commission to assess ground water issues and to submit several reports to Congress. The budgetary allotment authorized for this work has been $7 million for the years 1985 through 1987.

20

The Toxic Substances Control Act

A. Introduction

The chemical industry plays a vital part in the economic life of the United States with a contribution of approximately 5 percent of the Gross National Product. Not only does it affect each and every one of us in our daily lives, but also the industry employs more than 1 million of our population. More than 60,000 chemical substances are produced or processed for commercial use in the United States, and almost 1,000 more are introduced every year.

Although most chemicals present little or no danger to the environment or human health when used properly, in the past few decades some chemicals commonly used and idly dispersed have been found to be significantly harmful. An example is the family of chemicals called polychlorinated biphenyls, or PCBs. It was not until after tens of millions of pounds of PCBs were produced and released into the environment, however, that scientists realized how persistent and potentially toxic they were. Over the past few years, we have found those chemicals in our bodies and even in the milk of nursing mothers, as we have indicated in Chapter X of this text.

In 1971 the President's Council on Environmental Quality developed a legislative proposal for coping with the increasing problems of toxic substances. After 5 years of public hearings and debate, Congress enacted the Toxic Substances Control Act (TSCA) in the fall of 1976. When TSCA was enacted by Congress, it established a number of new requirements and authorities for identifying and controlling toxic chemical hazards to human health and the environment. Programs now exist under TSCA for gathering information about the toxicity of particular chemicals and the extent to which people and the environment are exposed to them, to assess whether they cause unreasonable risks to humans and the environment, and to institute appropriate control actions after weighing their potential risks against their benefits to the nation's economic and social well-being.

The following paragraphs briefly touch upon the major provisions of the law in order to familiarize the reader with its content. It is not intended to be an authoritative, legal statement concerning the legislation.

B. A TSCA Overview

To ensure informed decision making by the government, TSCA gives the EPA authority to gather certain kinds of basic information on chemical risks from those who manufacture and process chemicals. The law also enables the EPA to

383

require companies to test selected existing chemicals for toxic effects and requires the agency to review most new chemicals before they are manufactured.
To prevent unreasonable risks, the EPA may select from a broad range of control
actions under TSCA, from requiring hazard-warning labels to outright bans on
the manufacture or use of especially hazardous chemicals. The EPA may regulate
a chemical's unreasonable risks at any stage in its life cycle, the manufacturing,
processing, distribution in commerce, use, or disposal.

It is important to note that 8 product categories are exempt from TSCA's
regulatory authorities: pesticides, tobacco, nuclear material, firearms and ammunition, food, food additives, drugs, and cosmetics. Many of those product
categories are regulated under other Federal laws and are discussed in other
sections of this text.

C. Testing Requirements

TSCA gives the EPA authority to require manufacturers or processors of certain
existing chemicals, i.e., those already being distributed in commerce, to test their
health and environmental effects. The EPA exercises that authority only when
it can make certain statutory findings about the substance involved and when
industry fails to develop the needed data on its own. Those required findings
are (a) that there are insufficient data already available with which to perform
a reasonable risk assessment; (b) that testing is necessary to provide such data;
(c) that a chemical may present an unreasonable risk of injury to human health
or the environment; or (d) that the chemical is produced in substantial quantities
resulting in significant human exposure or environmental release.

Testing requirements are imposed only after a rule-making proceeding that
includes opportunities for both public comments and an oral presentation at a
hearing. An Interagency Testing Committee of government experts on chemical
substances advises the EPA on those that should be tested; however, actions are
not limited to those recommended by the committee. The eight committee
members represent the Departments of Labor, Commerce, Health and Human
Services, including the National Cancer Institute, the National Institute for
Occupational Safety and Health, and the National Institute of Environmental
Health Sciences, and the National Science Foundation, the Council on Environmental Quality, and the EPA. The committee designates priority chemicals for
testing. Then the EPA either initiates rule making for testing requirements for
designated chemicals or publishes the reasons why testing is not required.

D. Introduction of New Chemical Substances

TSCA recognizes that health and environmental considerations are more easily
addressed before, rather than after, a chemical is produced and introduced into
commerce. Therefore, manufacturers or importers of new chemicals must give
the EPA a 90-day advance notification of their intent to manufacture or import
a new chemical, except for those chemical categories specifically excluded by
TSCA. Any chemical that is not listed on the inventory of existing chemicals
published by the agency is considered "new" for purposes of this premanufacture
notice requirement. In addition, the agency may designate a use of a chemical

as a significant new use, based on consideration of several factors, including the anticipated extent and type of exposure. Anyone who intends to manufacture, import, or process a chemical for such a significant new use, even if the chemical is on the inventory or went through premanufacture notification review, must notify the agency 90 days before manufacturing, importing, or processing the chemical for that use.

The 90-day review period for new chemicals and significant new uses can be extended by the EPA for an additional 90 days for a good cause. Notices submitted for new chemicals or significant new uses of chemicals are to include the identity of the chemical; its molecular structure; proposed categories of use; an estimate of the amount to be manufactured, imported, or processed; the by-products resulting from the manufacture, processing, use and disposal of the chemical; estimates of exposure and any test data related to the health and environmental effects of the chemical. In addition, if a rule requiring testing of the chemical class has been issued, the notice must include test data developed from that testing along with the other information.

Chemicals produced in small quantities, solely for experimental or research and development purposes, are automatically exempt from the premanufacture and significant new-use notification requirements. In addition, any person may apply for an exemption for chemicals used solely for test marketing purposes or those determined by the EPA not to present an unreasonable risk of injury to human health or the environment.

If the EPA determines that insufficient information is in a notification to evaluate potential risk, the agency may order that the manufacture or importation of the chemical be prohibited until adequate data are developed. The company is under no time limit to submit the information, but until it does, EPA's ban remains in effect. After reviewing a premanufacture notification, containing sufficient data, if the EPA determines that the new chemical presents or will present an unreasonable risk of injury to health or the environment, the agency can, during the review period, prohibit the manufacturing, processing, or distribution of that chemical.

E. The Control of Hazardous Chemical Substances

TSCA has the authority to regulate the disposal stage of a chemical's life cycle on a chemical-by-chemical basis, that is, once a particular chemical is determined to be an unreasonable risk to human health and the environment (e.g., PCBs). The Resource Conservation and Recovery Act (RCRA) has the authority to establish regulations and programs to ensure safe waste treatment and disposal of any number of chemicals and generally deals with waste streams rather than individual chemicals.

Under TSCA, the EPA has the authority to prohibit or limit the manufacture, importation, processing, distribution in commerce, use, or disposal of a chemical when those activities are found to pose an unreasonable risk of injury to human health or the environment. A number of possible control options are available, ranging from total prohibition to labeling.

A manufacturer or processor may be required to make and keep records of the processes used in manufacturing a chemical and to conduct tests to assure compliance with any regulatory requirements. Furthermore, the agency may require a manufacturer or processor to give notice of any unreasonable risk of injury presented by his chemical to those who purchase or may be exposed to that substance. A manufacturer or processor may also be required to recall a substance that presents an unreasonable risk. Also, in proposing regulatory actions, the EPA must provide an opportunity for comments by all interested parties.

A rule limiting, but not banning, a chemical may be made immediately effective when initially proposed in the *Federal Register* if the agency determines that the chemical is likely to present an unreasonable risk of serious or widespread injury to health or the environment before normal rule-making procedures could be completed. In the case of a rule prohibiting the manufacture of the chemical, the EPA must first obtain a court injunction before the rule can be made immediately effective.

For those chemicals that present an imminent and unreasonable risk of serious or widespread injury to health or the environment, the EPA must ask a court to require whatever action may be necessary to protect against such risk. The EPA also cooperates with other agencies in toxic chemical control.

In addition to using various control options, the EPA warns the public about chemical hazards through its chemical advisory program. Chemical advisories encourage individuals or organizations to voluntarily reduce the risks associated with a chemical, such as the danger of continual contact with used motor oil. Chemical advisories discuss toxic effects of chemicals of concern, routes of exposure, and alternative methods of reducing risks.

1. Reporting and Records

A major objective of the TSCA program was to develop a mechanism for identifying those chemicals likely to damage human health or the environment so that appropriate action could be taken by industry or the EPA. In fact, Congress recognized during its discussion on TSCA that very little was known about chemicals in the environment.

When TSCA was passed, it was not even known how many chemicals there were, in what quantities they were produced and where, what their by-products were, who was exposed to them, and under what conditions. That information was available only for a handful of existing chemicals. Therefore, Congress gave the EPA the authority to compile an inventory of existing chemical substances and to develop additional information on those basic questions.

The first inventory was published in 1979, based on information reported to the EPA by chemical manufacturers, importers, and processors. The inventory, to which new chemicals are added when they go into production, shows that nearly 58,000 commercial chemical substances are, or have been, manufactured or imported into the United States since January 1, 1975. As a matter of note, there are well over 5 million known chemical compounds, but most are research and development chemicals that are not used commercially.

It is important to remember also that the chemical inventory is not a list of toxic or hazardous chemicals; rather, it lists existing chemicals by their specific chemical name (e.g., acetonitrile, bromobenzene, chlormethane, etc.), giving, for the first time, an overall picture of the chemicals used for commercial purposes in the United States. In addition to being unprecedented, this list is of major importance because chemicals not on the inventory must be reviewed by the EPA under the premanufacture notification program before they are allowed into U.S. commerce.

In addition to compiling the inventory, the EPA has used its TSCA reporting authority to obtain production, use, release, and exposure data on a number of chemicals, including asbestos and chemicals recommended by the Interagency Testing Committee for test rule consideration. TSCA also requires any person who manufacturers, processes, or distributes in commerce any chemical substance or mixture to keep records of significant adverse reactions to health or the environment that allegedly were caused by the chemical. Records concerning health effects on employees must be kept for 30 years; other records must be retained for 5 years.

The agency may require the submission of health and safety studies that are known or available to those who manufacture, process, or distribute in commerce specified chemicals. In addition, if the chemical industry has information indicating that a chemical presents a substantial risk of injury to health or the environment, the EPA must be notified. Substantial-risk reporting has heightened industry's awareness of potential chemical risks, often resulting in manufacturers, processors, and distributors taking action on their own to minimize exposure to hazardous substances.

2. Importing and Exporting

Any intention to import a chemical substance, mixture, or article containing a chemical substance or mixture will not be allowed into the U.S. customs territory if it fails to comply with any TSCA regulation or otherwise violates the provisions of TSCA.

If anyone intends to export a chemical that is subject to certain requirements under TSCA they must first notify the EPA. The agency is responsible for notifying the importing country's government of the export and of EPA's regulatory action or the availability of information.

TSCA requirements generally related to substances that are produced and distributed in U.S. commerce. However, chemical substances produced in the United States but for export only can be covered by TSCA if the EPA finds that the chemical substance will present an unreasonable risk within the United States. The EPA may require that the chemical substance be tested, and may further regulate it.

3. Date Disclosure

Confidential data, such as trade secrets and privileged financial data, will be protected from disclosure by the EPA. All health and safety information, submitted under TSCA, on chemicals in commerce is subject to disclosure. A person submitting other types of data to the agency may designate any part of them as confidential until it determines that the information is not entitled to such protection. If the release of confidential business information is essential for the protection of health or the environment, the EPA may disclose it after notifying the person who submitted the data in advance of any contemplated release.

4. The Impact of TSCA on the States

With certain exceptions, TSCA will not affect the authority of any state or political subdivision to establish regulations concerning chemicals. If the EPA issues a testing requirement for a chemical, a state may not establish a similar one for the same purposes. If the EPA restricts the manufacture or otherwise regulates a chemical under TSCA, a state may only issue requirements that are identical, are mandated by other Federal laws, or prohibit the use of the chemical. In response to a request by a state, the EPA may grant an exemption to allow the state to regulate differently from TSCA regulatory actions under certain conditions. Specifically, the EPA can grant exemptions if the state requirements: (a) would not cause a person or activity to be in violation of a requirement under TSCA and (b) would provide a greater degree of protection and not unduly burden interstate commerce. The EPA is actively committed to a policy of delegating more authority and decision-making power to the states to implement TSCA.

TSCA authorizes grants to be awarded to states to help establish programs to prevent or eliminate unreasonable risks associated with toxic substances. Seven States and Puerto Rico were recipients of such grants in 1979, 1980, and 1981. The projects funded, which were to be completed and evaluated in 1983 and 1984, ranged in scope from developing toxic substances data bases to establishing state strategies to investigate and, if necessary, control unreasonable risks.

In addition, a regional and state relations program to promote cooperative approaches to dealing with toxic substances problems was begun. The plan's design and initial efforts came from recommendations made in a 1981 report submitted by the National Governors' Association that studied, with the help of an EPA grant, how states manage toxic substances problems. Subsequently, grants to the NGA were awarded to coordinate the states' access to EPA's Chemical Substances Information Network (CSIN) to act as a clearinghouse to publicize state toxic substances management practices and to allow experts from one state to advise their counterparts in another. In addition, a conference attended by 26 state representatives was held.

5. Judicial Review

Not later than 60 days after a rule is promulgated under certain TSCA provisions, a person may file a petition for judicial review of such rule with the U.S. Court of Appeals for the District of Columbia Circuit or with the U.S. Court of Appeals for the circuit of his or her residence or business.

6. Actions by Citizens and Employees

Any person may bring a civil suit to restrain a violation of TSCA by any party or to compel the agency to perform any nondiscretionary duty required by this law.

In addition, any person may petition the EPA to issue, amend, or repeal a rule under the testing, reporting, or restriction sections of TSCA. The agency has 90 days to respond to a petition. If no action is taken or a petition is denied, the party has the opportunity for judicial review in a U.S. district court. In both civil suits and citizens' petitions, the court may award reasonable legal costs and attorneys' fees, if appropriate.

If an employee believes that his or her employer has discriminated against him because of the employee's participation in carrying out TSCA, he may file a complaint with the secretary of labor. The secretary shall investigate the alleged discrimination and, if warranted, may order the employer to remedy the effects of any such discriminatory action. Employees and employers alike may obtain judicial review in the U.S. courts of appeals.

The EPA will evaluate the potential effects on employment of regulatory actions under TSCA. In response to a petition by an employee, the EPA may investigate and hold public hearings concerning job losses or other adverse effects allegedly resulting from a requirement under TSCA. The agency will make public those findings and recommendations.

7. Violations and Enforcement

Any person who fails or refuses to comply with any requirement made under TSCA may be subject to a civil penalty of up to $25,000 for each day of violation. Persons who knowingly or willfully violate the law, in addition to any civil penalties, may be fined up to $25,000 for each day of violation, imprisoned up to a year, or both.

The law has also given the EPA the authority to inspect any establishment in which chemicals are manufactured, processed, imported to, stored, or held before or after their distribution in commerce. No inspection shall include financial, sales, pricing, personnel, or research data, unless specified in an inspection notice.

The agency can subpoena witnesses, documents, and other information as necessary to carry out TSCA. Also, civil actions concerning violations of or lack of compliance with TSCA may be brought to a U.S. district court to restrain or compel the taking of an action. Any chemical substance or mixture that was manufactured, processed, or distributed in commerce in violation of TSCA may be subject to seizure.

The agency has developed specific enforcement strategies for implementing TSCA regulations. Those strategies identify and rank possible violations of a particular regulation, identify the tools available for compliance monitoring and how they will be used, provide a formula for determining application of inspection resources, and establish policy for determining civil penalties under the regulation.

F. Research and Development and Information Systems

The EPA Research Committee system has been established to serve as a foundation for program planning in the Office of Research and Development (ORD) and to effect a formal liaison between ORD and the other program offices within the EPA. The Chemical Testing and Assessment Research Committee (CTARC) plans research activities relevant to TSCA.

In addition to establishing a data system within itself for information submitted under TSCA, the EPA is designing and establishing a system for toxicological and other scientific data accessible to all Federal agencies. The agency has developed the Chemicals-in-Commerce Information System (CICIS) to store and retrieve TSCA data. That system contains TSCA confidential business information and state-of-the-art computer security techniques. The computerized TSCA Inventory became operational in late 1979, and several information services have derived from it, including subsystems for Freedom of Information Act requests, inventory profiles for EPA Regional Offices, support for the TSCA premanufacture review process, and health and safety study submissions.

The Interagency Toxic Substances Data Committee (ITSDC), formed in February 1978 by the EPA and the Council on Environmental Quality, is continuing its work to construct a comprehensive Chemical Substances Information Network (CSIN). CSIN enables toxic substances information users to have access to a number of independent and autonomous data banks in the public and private sectors. Users can make use of 1 computer to manage the logging in, accessing, and processing of their queries for relevant records in and among many data and information systems, i.e., 1 simple access point to a "library of systems." Through the ITSDC, the EPA is responsible for the day-to-day administration of CSIN, which became operational in November 1981. CSIN significantly advances the availability of chemical data to both governmental and private-sector organizations to efficiently resolve and manage issues concerning chemical substances.

As required by TSCA, the EPA has established an office to provide technical and other nonfinancial assistance to chemical manufacturers, processors, and others who are interested in requirements and activities under this law. To help people understand TSCA's requirements, the TSCA Assistance Office (TAO) provides a toll-free telephone information service, a bimonthly bulletin, field consultants, and other technical assistance upon request.

To obtain up-to-date information on TSCA requirements, call the TSCA Assistance Office's toll-free line: 800-424-9065 (554-1404 in Washington, D.C.) Outside of the United States call 202-554-1404.

Glossary

abatement. The reduction in degree or intensity of pollution.

abrasive belt grinding. Roughing and/or finishing a workpiece by means of a power-driven belt coated with an abrasive, usually in particle form, which removes material by scratching the surface.

abrasive belt polishing. Finishing a workpiece with a power-driven abrasive-coated belt in order to develop a very good finish.

abrasive blasting. (Surface treatment and cleaning.). Using dry or wet abrasive particles under air pressure for short durations of time to clean a metal surface.

abrasive cutoff. Severing a workpiece by means of a thin abrasive wheel.

abrasive jet machining. Removal of material from a workpiece by a high-speed stream of abrasive particles carried by gas from a nozzle.

abrasive machining. Used to accomplish heavy stock removal at high rates by use of a free-cutting grinding wheel.

adsorption. The penetration of 1 substance into or through another.

acceleration. See activation.

accelerator. In radiation science, a device that speeds up charged particles, such as electrons or protons.

acceptance testing. A test, or series of tests, and inspections that confirms product functioning in accordance with specified requirements.

acclimation. The adaptation over several generations of a species to a marked change in the environment.

acetic acid. (Ethanoic acid, vinegar acid, methanecarboxylic acid) CH3 COOH. Glacial acetic acid is the pure compound (99.8 percent min.), as distinguished from the usual water solutions known as acetic acid. Vinegar is a dilute acetic acid.

acid cleaning. Using any acid for the purpose of cleaning any material. Some methods of acid cleaning are pickling and oxidizing.

acid dip. An acidic solution for activating the workpiece surface before electroplating in an acidic solution, especially after the workpiece has been processed in an alkaline solution.

acidity. The quantitative capacity of aqueous solutions to react with hydroxyl ions. It is measured by titration with a standard solution of a base to a specified end point. Usually expressed as milligrams per liter of calcium carbonate.

act. Federal Water Pollution Control Act Amendments of 1972.

activated carbon. A highly adsorbent form of carbon used to remove odors and toxic substances from gaseous emissions. In advanced waste treatment, it is used to remove dissolved organic matter from waste water.

activated sludge. Sludge that has been aerated and subjected to bacterial adtion; used to speed breakdown of organic matter in raw sewage during secondary waste treatment.

activated sludge process. Removes organic matter from sewage by saturating it with air and biological active sludge.

activation. The process of treating a substance by heat, radiation or the presence of another substance so that the first-mentioned substance will undergo chemical or physical change more rapidly or completely.

acute toxicity. Any poisonous effect produced by a single short-term exposure, that results in severe biological harm or death.

adaptation. A change in structure or habit of an organism that produces better adjustment to its surroundings.

additive circuitry. 1. Full-circuitry produced by the buildup of an electroless copper pattern upon an unclad board. 2. Semi-Circuitry produced by the selective "quick" etch of an electroless layer; this cipper layer was previously deposited on an unclad board.

adhesion. Molecular attraction that holds the surfaces of 2 substances in contact, such as water and rock particles.

administrator.   Means the Administrator of the United States Environmental Protection Agency.

adsorption.   The attachment of the molecules of a liquid or gaseous substance to the surface of a solid.

adulterants.   Chemical impurities or substances that by law do not belong in a good, plant, animal, or pesticide formulation.

advanced waste water treatment.   The tertiary stage of sewage treatment.

aeration.   To circulate oxygen through a substance, as in waste water treatment where it aids in purification.

aerobic.   Life or processes that depend on the presence of oxygen.

aerobic biological oxidation.   Any waste treatment process utilizing organisms in the presence of air or oxygen to reduce the pollution load or oxygen demand of organic substance in water.

aerobic digestion.   (sludge processing) The biochemical decomposition of organic matter, by organisms living or active only in the presence of oxygen, which results in the formation of mineral and simpler organic compounds.

aerosol.   A suspension of liquid or solid particles in a gas.

afterburner.   As air pollution control device that removes undesirable organic gases by incineration.

aging.   The change in properties (eg. increase in tensile strength and hardness) that occurs in certain metals at atmospheric temperature after heat treatment.

agitation of parts.   The irregular movement given to parts when they have been submerged in a plating or rinse solution.

agricultural pollution.   The liquid and solid wastes from farming, including: runoff from pesticides, fertilizers, and feedlots; erosion and dust from plowing; animal manure carcasses, crip residues, and debris.

air agitation.   The agitation of a liquid medium through the use of air pressure injected into the liquid.

air curtain.   A method of containing oil spills, air bubbling through a perforated pipe causes an upward water flow that slows the spread of oil. It can also be used to stop fish from entering polluted water.

air flotation.   See flotation.

air-liquid interface.   The boundary layer between the air and the liquid in which mass transfer is diffusion controlled.

air mass.   A widespread body of air that gains certain characteristics while set in 1 location. The characteristics change as it moves away.

air monitoring.   See monitoring.

air pollution.   The presence of contaminant substances in the air that do not disperse properly and interfere with human health.

air pollution episode.   A period of abnormally high concentration of air pollutants, often caused by low winds and temperature inversion that can cause illness and death.

air quality control region.   An area designated by the Federal Government in which communities share a common air pollution problem, sometimes involving several states.

air quality criteria.   The levels of pollution and lengths of exposure above which adverse effects may occur on health and welfare.

air quality standards.   The level of pollutants prescribed by law that cannot be exceeded during a specified time in a defined area.

aldehydes group.   A group of various highly reactive compounds typified by actaldehyde and characterized by the group CHO.

algae.   Simple rootless plants that grow in bodies of water in relative proportion to the amounts of nutrients available. Algal blooms or sudden growth spurts can affect water quality adversely.

algicides.   Chemicals for preventing the growth of algae.

alkaline cleaning.   A process for cleaning basis material where mineral and animal fats and oils must be removed from the surface. Solutions at high temperatures containing caustic soda, soda ash, alkaline silicates and alkaline phosphates are commonly used.

alkalinity.   The capacity of water to neutralize acids, a property imparted by the water's content of carbonates, bicarbonates, hydroxides, and occasionally borates, silicates, and phosphates.

alloy steels. Steels with carbon content between 0.1 percent to 1.1 percent and containing elements such as nickel, chromium, molybdenum, and vanadium. (The total of all such alloying elements in these type steels is usually less than 5 percent.)

alpha particles. The least penetrating type of radiation, usually not harmful to life.

aluminizing. Forming an aluminum or aluminum alloy coating on a metal by hot dipping, hot spraying or diffusion.

ambient air. Any unconfined portion of the atmosphere: open air.

amines. A class of organic compounds of nitrogen that may be considered as derived from ammonia (NH3) by replacing 1 or more of the hydrogen atoms by organic radicals, such as CH3 or C6H5, as in methylamine and analine. The former is a gas at ordinary temperature and pressure, but other amines are liquids or solids. All amines are basic in nature and usually combine readily with hydrochloric or other strong acids to form salts.

anadromous. Fish that swim upriver to spawn like salmon.

anaerobic. Life or processed that can occur without free oxygen.

anaerobic biological treatment. Any waste treatment process utilizing anderobic or facultative organisms in the absence of air to reduce the organic matter in water.

anaerobic digestion. The process of allowing sludges to decompose naturally in heated tanks without a supply of oxygen.

anaerobic waste treatment. (sludge processing) Waste stabilization brought about through the action of microorganisms in the absence of air or elemental oxygen.

anhydrous. Containing no water.

anions. The negatively charged ions in solution, e.g., hydroxyl.

annealing. A process for preventing brittleness in a metal part. The process consists of raising the temperature of the metal to a preestablished level and slowly cooling the steel at a prescribed rate.

annual capital recovery cost. Allocates the initial investment and the interest to the total operating cost. The capital recovery cost is equal to the initial investment multiplied by the capital recovery factor.

anode. The positively charged electrode in an electrochemical process.

anodizing. The production of a protective oxide film on aluminum or other light metal by passing a high voltage electric current through a bath in which the metal is suspended.

anticoagulant. A chemical that interferes with blood clotting.

antidegradation clause. Part of air quality and water quality laws that prohibits deterioration where pollution levels are within the legal limit.

aquifer. An underground bed or layer of earth, gravel, or porous stone that contains water.

area source. In air pollution, any small individual fuel combustion source, including vehicles. A more precise legal definition is available in Federal regulations.

asbestos. A mineral fiber that can pollute air or water and cause cancer if inhaled or ingested.

a-scale sound level. A measurement of sound approximating the sensitivity of the human ear, used to note the intensity or annoyance of sounds.

ash. The solid residue left after complete combustion.

assembly. The fitting together of manufactured parts into a complete machine, structure, or unit of a machine.

assimilation. The ability of a body of water to purify itself of pollutants.

atmosphere. The body of air surrounding the Earth.

atmospheric evaporation. Evaporation at ambient pressure utilizing a tower filled with packing material. Air is drawn in from the bottom of the tower and evaporates feed material entering from the top. There is no recovery of the vapors.

atomic absorption. Quantitative chemical instrumentation used for the analysis of elemental constituents.

atomic pile. A nuclear reactor.

attractant. A chemical or agent that lures insects or other pests by stimulating their sense of smell.

attrition.   Wearing or grinding down a substance by friction. A contributing factor in air pollution, as with dust.

audiometer.   An instrument that measures hearing sensitivity.

automatic plating.   1. Full—Plating in which the workpieces are automatically conveyed through successive cleaning and platine tanks. 2. Semi—Plating in which the workpieces are conveyed automatically through only 1 plating tank.

autotrophic.   An organism that produces food from inorganic substances.

austempering.   Heat-treating process to obtain greater toughness and ductility in certain high-carbon steels. The process is characterized by interrupted quenching and results in the formation of bainite grain structure.

austenitizing.   Heating a steel to a temperature at which the structure transforms to a solution of 1 or more elements in face-centered cubic iron. Usually performed as the essential preliminary of heat treatment, in order to get the various alloying elements into solid solution.

backfill.   The material used to refill an excavation, or the process of doing so.

background level.   In air pollution, the level of pollutants present in ambient air from natural sources.

bacteria.   Single-celled microorganisms that lack chlorophyll. Some cause diseases; other aid in pollution control by breaking down organic matter in air and water.

baffle.   A deflector that changes the direction of flow or velocity of water, sewage, or particulate matter. Also used to deaden sound.

baghouse.   An air pollution abatement device used to trap particulates by filtering gas streams through large fabric bags usually made of glass fibers.

baling.   Compacting solid waste into blocks to reduce volume.

ballistic separator.   A machine that sorts organic from inorganic matter for composting.

band application.   In pesticides, the spreading of chemicals over or next to each row of plants in a field.

barrel finishing.   The process of polishing a workpiece using a rotating or vibrating container and abrasive grains or other polishing materials to achieve the desired surface appearance.

barrel plating.   Electroplating of workpieces in barrels (bulk).

bar screen.   In waste water treatment, a device that removes large solids.

basal application.   In pesticides, the spreading of a chemical on stems or trunks just above the soil line.

basis metal or material.   That substance of which the workpieces are made and that receives the electroplate and the treatments in preparation for plating.

batch treatment.   A waste treatment method where waste water is collected over a period of time and then treated before discharge.

bending.   Turning or forcing by brake press or other device from a straight or even to a curved or angular condition.

benthic region.   The bottom layer of a body of water.

benthos.   The plants and animals that inhabit the bottom of a water body.

beryllium.   A metal that can be hazardous to human health when inhaled. It is discharged by machine shops, ceramic and propellant plants, and foundries.

best available technology economically achievable (BAT).   Level of technology applicable to effluent limitations to be achieved for industrial discharges to surface waters as defined by Section 310 (b) (1) (A) of the act.

beta particle.   An elementary particle emitted by radioactive decay that may cause skin burns. It is halted by a think sheet of metal.

bioassay.   Using living organisms to measure the effect of a substance, factor, or condition.

biochemical oxygen demand (BOD).   The dissolved oxygen required to decompose organic matter in water. It is a measure of pollution since heavy waste loads have a high demand for oxygen.

biodegradable.   Any substance that decomposes quickly through the action of microorganisms.

biological control. Using means other than chemicals to control pests, such as predatory organisms, sterilization, or inhibiting hormones.

biological magnification. The concentration of certain substances up a food chain. A very important mechanism in concentrating pesticides and heavy metals in organisms such as fish.

biological oxidation. The way that bacteria and microorganisms feed on and decompose complex materials. Used in self-purification of water bodies and activated sludge waste water treatment.

biomass. The amount of living matter in a given unit of the environment.

biomonitoring. The use of living organisms to test water quality at a discharge site or downstream.

biosphere. The portion of Earth and its atmosphere that can support life.

biostabilizer. A machine that converts solid waste into compost by grinding and aeration.

biota. All living organisms that exist in an area.

blanking. Cutting desired shapes out of sheet metal by means of dies.

bleve. Boiling liquid-expanding vapor explosion.

bloom. A proliferation of algae and/or high aquatic plants in a body of water, often related to pollution.

blowdown. The minimum discharge of recirculating water for the purpose of discharging materials contained in the water, the further buildup of which would cause concentration in amounts exceeding limits established by best engineering practice.

BOD$_5$. The amount of dissolved oxygen consumed in 5 days by biological processes breaking down organic matter in an effluent.

bog. Wet, spongy land usually poorly drained, highly acid and rich in plant residue, the result of lake eutrophication.

bonding. The process of uniting using an adhesive of fusible ingredient.

boom. A floating device used to contain oil on a body of water.

boring. Enlarging a hole by removing metal with a single or occasionally a multiple point cutting tool moving parallel to the axis of rotation of the work or tool. 1. Single-point boring—Cutting with a single-point tool. 2. Precision boring—cutting to tolerances held within narrow limits. 3. Gun boring—cutting of deep holes. 4. Jug boring—cutting of high-precision and accurate location holes. 5. Groove boring—cutting accurate recesses in hole walls.

botanical pesticide. A plant-produced chemical used to control pests; for example micotine or strychine.

brackish water. A mixture of fresh and salt water.

brazing. Joining metals by flowing a thin layer, sapillary thickness, of nonferrous filler metal into the space between them. Bonding results from the intimate contact produced by the dissolution of a small amount of base metal in the **molten filler metal.** Without fusion of the base metal. The term brazing is used where the temperature exceeds 425°C (800°F).

breeder. A nuclear reactor that produces more fuel than it consumes.

bright dipping. The immersion of all or part of a workpiece in a media designed to clean or brighten the surface and leave a protective surface coating on the workpiece.

brine. An aqueous salt solution.

broaching. Cutting with a tool that consists of a bar having a single edge or a series of cutting edges (i.e., teeth) on its surface. The cutting edges of multiple-tooth, or successive single-tooth, broaches increase in size and/or change in shape. The broach cuts in a straight line or axial direction when relative motion is produced in relation to the workpiece, which may also be rotating. The entire cut is made in single or multiple passes over the workpiece to shape the required surface contour. 1. Pull broaching—tool pulled through or over workpiece. 2. Push broaching—tool pushed over or through workpiece. 3. Chain broaching—a continuous high production broach. 4. Tunnel broaching—work travels through an enclosed area containing broach inserts.

broadcast application. In pesticides, to spread a chemical over an entire area.

bromine water. A nonmetallic halogen liquid, normally deep red, corrosive and toxic, which is used as an oxidizing agent.

buffer strips. Strips of grass or other erosion-resisting vegetation between or below cultivated strips or fields.

buffing. An operation to provide a high luster to a surface. The operation, which is not intended to remove much material, usually follows polishing.

buffing compounds. Abrasive contained by a liquid or solid binder composed of fatty acids, grease, or tallow. The binder serves as lubricant, coolant, and an adhesive of the abrasive to the buffing wheel.

burial ground (graveyard). A disposal site for unwanted radioactive materials that uses earth or water for a shield.

burnishing. Finish sizing and smooth finishing of a workpiece (previously machined or ground) by displacement, rather than removal, of minute surface irregularities with smooth point or line-contact, fixed or rotating tools.

cadmium. A heavy metal element that accumulates in the environment.

calendering. Process of forming a continuous sheet by squeezing the material between 2 or more parallel rolls to impart the desired finish or to ensure uniform thickness.

calibration. The application of thermal, electrical, or mechanical energy to set or establish reference points for a part, assembly or complete unit.

calibration equipment. Equipment used for calibration of instruments.

capital recovery costs. Allocates the initial investment and the interest to the total operating cost. The capital recovery cost is equal to the initial investment multiplied by the capital recovery factor.

capital recovery factor. Capital recovery factor is defined as: $i + l/(a - 1)$ where $i =$ interest rate, $a = (1 + i)$ to the power n, n = interest period in years.

captive facility. A facility that owns more than 50 percent (annual area basis) of the materials undergoing metal finishing.

captive operation. A manufacturing operation carried out in a facility to support subsequent manufacturing, fabrication, or assembly operations.

carbides. Usually refers to the general class of pressed and sintered tungsten carbide cutting tools that contain tungsten carbide plus smaller amounts of titanium and tantalum carbides along with cobalt, which acts as a binder. (It is also used to describe hard compounds in steels and cast irons.)

carbon adsorption. Activated carbon contained in a vessel and installed in either a gas or liquid stream to remove organic contaminates. Carbon is regenerable when subject to steam, which forces contaminant to desorb from media.

carbon bed catalytic destruction. A nonelectrolytic process for the catalytic oxidation of cyanide wastes using filters filled with low-temperature coke.

carbon dioxide (CO_2). A colorless, odorless nonpoisonous gas normally part of ambient air, a result of fossil fuel combustion.

carbon monoxide (CO). A colorless, odorless, poisonous gas produced by incomplete fossil fuel combustion.

carbon steels. Steel that owes its properties chiefly to various percentages of carbon without substantial amounts of other alloying elements.

carbonate. A compound containing the acid radical of carbonic acid (CO_3 group).

carbonitriding. (Physical property modification) Increasing the carbon content of a metal by heating with a carburizing medium (which may be solid, liquid, or gas) usually for the purpose of producing a hardened surface by subsequent quenching.

carcenogenic. Cancer-producing.

carrying capacity. 1. In recreation, the amount of use a recreation area can sustain without deterioration of its quality. 2. In wildlife, the maximum number of animals an area can support during a given period of the year.

case hardening. A heat-treating method by which the surface layer of alloys is made substantially harder than the interiors. (Carburizing and nitriding are common ways of case-hardening steels.)

cast. A state of the substance after solidification of the molten substance.

casthouse. The facility that melts metal, holds it in furnaces for degassing (fluxing) and alloying and then casts the metal into pigs, ingots, billets, rod, etc.

casting. The operation of pouring molten metal into a mold.

catalytic bath. A bath containing a substance used to accelerate the rate of chemical reaction.

catalytic converter. An air pollution abatement device that removes organic contaminants by oxidizing them into carbon dioxide and water.

category. Also point source category. A segment of industry for which a set of effluent limitations has been established.

cathode. The negatively charged electrode in an electrochemical process.

cation. The positively charged ions in a solution.

caustic. Capable of destroying or eating away by chemical action. Applies to strong bases and characterized by the presence of hydroxyl ions in solution.

caustic soda. Sodium hydroxide (NaOH), a strong alkaline substance used as the cleaning agent in some detergents.

cells. In solid waste disposal, holes where waste is dumped, compacted, and covered with layers of dirt daily.

cementation. The electrochemical reduction of metal ions by contact with a metal of higher oxidation potential. It is usually used for the simultaneous recovery of copper and reduction of hexavalent chromium with the aid of scrap iron.

centerless grinding. Grinding the outside or inside of a workpiece mounted on rollers rather than on centers. The workpiece may be in the form of a cylinder or the frustrum of a cone.

central treatment facility. Treatment plant that co-treats process waste waters from more than 1 manufacturing operation or co-treats process waste waters with noncontact cooling water or with nonprocess waste waters (e.g., utility blowdown, miscellaneous runoff, etc.).

centrifugal collector. A mechanical system using centrifugal force to remove aerosols from a gas stream or to dewater sludge.

centrifugation. An oil recovery step employing a centrifuge to remove water from waste oil.

centrifuge. A device having a rotating container in which centrifugal force separates substances of differing densities.

cfs. Cubic feet per second, a measure of the amount of water passing a given point.

channelization. To straighten and deepen streams so water will move faster, a flood reduction or marsh drainage tactic that can interfere with waste assimilation capacity and disturb fish habitat.

chelated compound. A compound in which the metal is contained as an integral part of a ring structure and is not reading ionized.

chelating agent. A coordinate compound in which a central atom (usually a metal) is joined by covalent bonds to 2 or more other molecules or ions (called ligands) so that heterocyclic rings are formed with the central (metal) atom as part of each ring. Thus, the compound is suspending the metal in solution.

chemical brightening. Process utilizing an addition agent that leads to the formation of a bright plate or that improves the brightness of the deposit.

chemical deposition. Process used to deposit a metal oxide on a substrate. The film is formed by hydrolysis of a mixture of chlorides at the hot surface of the substrate. Careful control of the water mixture ensures that the oxide is formed on the substrate surface.

chemical etching. To dissolve a part of the surface of a metal or all of the metal laminated to a base.

chemical machining. Production of derived shapes and dimensions through selective or overall removal of metal by controlled chemical attack or etching.

chemical metal coloring. The production of desired colors on metal surfaces by appropriate chemical or electrochemical action.

chemical milling. Removing large amounts of stock by etching selected areas of complex workpieces. This process entails cleaning, masking, etching, and demasking.

chemical oxidation. (including cyanide) The addition of chemical agents to waste water for the purpose of oxidizing pollutant material.

chemical oxygen demand (COD). A measure of the oxygen required to oxidize all compounds in water, organic and inorganic.

chemical precipitation. A chemical process in which a chemical in solution reacts with another chemical introduced to that solution to form a third substance which is partially or mainly insoluble and, therefore, appears as a solid.

chemical recovery systems. Chemical treatment to remove metal or other materials from waste water for later reuse.

chemical reduction. A chemical reaction in which one or more electrons are transferred to the chemical being reduced from the chemical initiating the transfer (reducing agent).

chemical treatment. Treating contaminated water by chemical means.

chip dragout. Cutting fluid or oil adhering to metal chips from a machining operation.

chemosterilant. A chemical that controls pests by preventing reproduction.

chilling effect. The lowering of the Earth's temperature because of increased particles in the air blocking the Sun's rays.

chlorinator. A device that adds chlorine to water in gas or liquid form.

chlorine-contact chamber. That part of a waste treatment plant where effluent is disinfected by chlorine before being discharged.

chlorosis. Discoloration of normally green plant parts that can be caused by disease, lack of nutrients, or various air pollutants.

chromate conversion coating. Protective coating formed by immersing metal in a aqueous acidified solution consisting substantially of chromic acid or water soluble salts of chromic acid together with various catalysts or activators.

chromatizing. To treat or impregnate with a chromate (salt of ester of chromic acid) or dichromate, especially with potassium dichromate.

chrome-pickle process. Forming a corrosion-resistant oxide film on the surface of magnesium base metals by immersion in a bath of an alkaline bichromate.

chromium. See heavy metals.

chronic. Long-lasting or frequently recurring, as a disease.

clarification. Clearing action that occurs during waste water treatment when solids settle out, often aided by centrifugal action and chemically induced coagulation.

clarifier. A settling tank where solids are mechanically removed from waste water.

cleaning. The removal of soil and dirt (including grit and grease) from a workpiece using water with or without a detergent or other dispersing agent.
See vapor degreasing
  solvent cleaning
  contaminant factor
  acid cleaning
  emulsion cleaning
  alkaline cleaning
  salt bath descaling
  pickling
  passivate
  abrasive blast cleaning
  sonic and ultrasonic cleaning

closed-loop evaporation system. A system used for the recovery of chemicals and water from a chemical-finishing process. An evaporator concentrates flow from the rinse water holding tank. The concentrated rinse solution is returned to the bath, and distilled water is returned to the final rinse tank. The system is designed for recovering 100 percent of chemicals normally lost in dragout for reuse in the process.

closed loop rinsing. The recirculation of rinse water without the introduction of additional makeup water.

coagulation. A clumping of particles in waste water to settle out impurities, often induced by chemicals such as lime or alum.

coastal zone. Ocean waters and adjacent lands that exert an influence on the uses of the sea and its ecology.

coating.   See aluminum coating
                   hot dip coating
                   ceramic coating
                   phosphate coating
                   chromate conversion coating
                   rust-preventive compounds
                   porcelain enameling

cod.   See chemical oxygen demand

coefficient of the haze (COH).   A measurement of visibility interference in the atmosphere.

coffin.   A thick-walled container (usually lead) used for tansporting radioactive materials.

cold drawing.   A process of forcing material through dies or other mandrels to produce wire, rod, tubular, and some bars.

cold heading.   A method of forcing metal to flow cold into enlarged sections by endwise squeezing. Typical cold-headed parts are standard screws, bolts under 1 in. diameter and a large variety of machine parts such as small gears with stems.

cold rolling.   A process of forcing material through rollers to produce bars and sheet rock.

coliform index.   A rating of the purity of water based on a count of fecal bacteria.

coliform organism.   Organisms found in the intestinal tract of humans and animals; their presence in water indicates pollution and potentially dangerous bacterial contamination.

colorimetric.   A procedure for establishing the concentration of impurities in water by comparing its color to a set of known color impurity standards.

combined sewers.   A system that carries both sewage and storm water runoff. In dry weather all flow goes to the waste treatment plant. During a storm only part of the flow is intercepted owing to overloading. The remaining mixture of sewage and storm water overflows untreated into the receiving stream.

combustion.   Burning, or a rapid oxidation accompanied by release of energy in the form of heat and light, a basic cause of air pollution.

comminution.   Mechanical shredding or pulverizing of waste, used in solid waste management and waste water treatment.

comminutor.   A machine that grinds solids to make waste treatment easier.

common metals.   Copper, nickel, chromium, zinc, tin, lead, cadmium, iron, aluminum, or any combination thereof.

compaction.   Reduction of the bulk of solid waste by rolling and tamping.

compatible pollutants.   Those pollutants that can be adequately treated in publicly owned treatment works without upsetting the treatment process.

complexing agent.   A compound that will join with a metal to form an ion that has a molecular structure consisting of a central atom (the metal) bonded to other atoms by coordinate covalent bonds.

composite wastewater sample.   A combination of individual samples of water or waste water taken at selected intervals, generally hourly for some specified period, to minimize the effect of the variability of the individual sample. Individual samples may have equal volume or may be proportioned to the flow at time of sampling.

compost.   Relatively stable decomposed organic material.

composting.   A controlled process of organic breakdown of matter. In mechanical composting the materials are constantly mixed and aerated by a machine. The ventilated cell method mixes and aerates materials by dropping them through a vertical series of serated chambers. Using windows, compost is placed in piles out in the open air and mixed or turned periodically.

conductance.   See electrical conductivity.

conductivity meter.   An instrument that displays a quantitative indication of conductance.

conductivity surface.   A surface that can transfer heat or electricity.

contact pesticide. A chemical that kills pests when it touches them, rather than by being eaten (stomach poison).

contact water. See process waste water.

contamination. Intrusion of undesirable elements.

continuous treatment. Treatment of waste streams operating without interruption as opposed to batch treatment; sometimes referred to as flow-through treatment.

contractor removal. Disposal of oils, spent solutions, or sludge by a scavenger service.

contrails. Long narrow clouds caused when high-flying jets disturb the atmosphere.

contour plowing. Farming methods that break ground following the shape of the land in a way that discourages erosion.

conversion coating. A coating produced by chemical or electrochemical treatment of a metallic surface that gives a superficial layer containing a compound of the metal. For example, chromate coating on zinc and cadmium, oxide coatings on steel.

coolant. A liquid or gas used to reduce the heat generated by power production in nuclear reactors or electric generators.

cooling tower. A device that aids in heat removal from water used as a coolant in electric power generating plants.

cooling water. Water that is used to absorb and transport heat generated in a process or machinery.

copper flash. Quick preliminary deposition of copper for making surface acceptable for subsequent plating.

coprecipitation of metals. Precipitation of a metal with another metal.

core. The uranium-containing heart of a nuclear reactor, where energy is released.

corrosion resistant steels. A term often used to describe the stainless steels with high nickel and chromium alloy content.

cost of capital. Capital recovery costs minus the depreciation.

counterboring. Removal of material to enlarge a hole for part of its depth with a rotary, pilot guided, end cutting tool having 2 or more cutting lips and usually having straight or helical flutes for the passage of chips and the admission of a cutting fluid.

countercurrent rinsing. Rinsing of parts in such a manner that the rinse water is removed from tank to tank counter to the flow of parts being rinsed.

countersinking. Beveling or tapering the work material around the periphery of a hole creating a concentric surface at an angle less than 90 degrees with the center line of the hole for the purpose of chamfering holes or recessing screw and rivet heads.

cover material. Soil used to cover compacted solid waste in a sanitary landfill.

cover. Vegetation or other material providing protection.

criteria. The standards the EPA has established for certain pollutants, which not only limit the concentration but also set a limit to the number of violations per year.

crystalline solid. A substance with an ordered structure, such as a crystal.

crystallization. 1. Process used to manufacture semiconductors in the electronics industry. 2. A means of concentrating pollutants in waste waters by crystallizing out pure water.

cultural eutrophication. Increasing the rate at which water bodies "die" by pollution from human activities.

curcumine or carmine method. A standard method of measuring the concentration of boron (B) within a solution.

curie. A measure of radioactivity.

cutie-pie. An instrument used to measure radiation levels.

cutting fluids. Lubricants employed to ease metal and machining operations, produce surface smoothness and extend tool life by providing lubricity and cooling. Fluids can be emulsified oils in water, straight mineral oils when better smoothness and accuracy are required, or blends of both.

cyaniding. A process of case hardening an iron-base alloy by the simultaneous absorption of carbon and nitrogen by heating in a cyanide salt. Cyaniding is usually followed by quenching to produce a hard case.

cyclone collector. A device that uses centrifugal force to pull large particles from polluted air.

cyclone separator. A device that removes entrained solids from gas streams.

DDT. The first chlorinated hydrocarbon insecticide (chemical name: 1, 1, 1-trichlorous-2, 2-bis [p-chloriphenyl]-ethane.) It has a shelf-life of 15 years and can collect in fatty tissues of certain animals. The EPA banned registration and interstate sale of DDT for virtually all but emergency uses in the U.S. in 1972 because of its persistence in the environment and accumulation in the food chain.

dead rinse. A rinse step in which water is not replenished or discharged.

deburring. Removal of burrs or sharp edges from parts by filing, grinding, or rolling the work in a barrel with abrasives suspended in a suitable medium.

decibel (dB). A unit of sound measurement.

decomposition. The breakdown of matter by bacteria. It changes the chemical makeup and physical appearance of materials.

deep bed filtration. The common removal of suspended solids from waste water streams by filtering through a relatively deep (0.3–0.9 m) granular bed. The porous bed formed by the granular media can be designed to remove practically all suspended particles by physical-chemical effects.

degassing. (Fluxing) The removal of hydrogen and other impurities from molten primary aluminum in a casthouse holding furnace by injecting chlorine gas (often with nitrogen and carbon).

degradable. That which can be reduced, broken down, or chemically separated.

demineralization. The removal from water of mineral contaminants usually present in ionized form. The methods used include ion-exchange techniques, flash distillation of electrolysis.

denitrification (biological). The reduction of nitrates to nitrogen gas by bacteria.

deoxidizing. The removal of an oxide film from an alloy such as aluminum oxide.

depletion curve (hydraulics). A graphical representation of water depletion from storage-stream channels, surface soil, and ground water. A depletion curve can be drawn for base flow, direct runoff, or total flow.

depreciation. Decline in value of a capital asset caused either by use or by obsolescence.

DES (diethylstilbestrol). A synthetic estrogen used as a growth stimulant in food animals. Residues in meat are thought to be carcinogenic.

desalinization. Removing salt from ocean or brackish water.

descaling. The removal of scale and metallic oxides from the surface of a metal by mechanical or chemical means. The former includes the use of steam, scale-breakers, and chipping tools; the latter method includes pickling in acid solutions.

desiccant. A chemical agent that dries out plants or insects, causing death.

desmutting. The removal of smut (matter that soils or blackens) generally by chemical action.

desulfurization. Removal of sulfur from fossil fuels to cut pollution.

detergent. Synthetic washing agent that helps water to remove dirt and oil. Most contain large amounts of phosphorous compounds, which may kill useful bacteria and encourage algae growth in the receiving water.

dewatering. (sludge processing) Removing water from sludge.

diaminabenzidene. A chemical used in a standard method of measuring the concentrations of selenium in a solution.

diatomaceous earth (diatomite). A chalklike material used to filter out solid wastes in waste water treatment plants, also found in powdered pesticides.

dibasic acid. An acid capable of donating 2 protons (hydrogen ions).

dichromate reflux. A standard method of measuring the chemical oxygen demand of a solution.

die casting. (hot chamber, vacuum, pressure) Casting is produced by forcing molten metal under pressure into metal molds called dies. In hot chamber machines, the pressure cylinder is submerged in the molten metal, resulting in a minimum of time and metal cooling during casting. Vacuum-feed machines use a vacuum to draw a measured amount of melt from the molten bath into the feed chamber. Pressure-feed systems use a hydraulic or pneumatic cylinder to feed molten metal to the dies.

diffused air. A type of aeration that forces oxygen into sewage by pumping air through perforated pipes inside a holding tank.

digester. In waste water treatment a closed tank, sometimes heated to 95° F. where sludge is subjected to intensified bacterial action.

digestion. The biochemical decomposition of organic matter. Digestion of sewage sludge occurs in tanks where it breaks down into gas, liquid, and mineral matter.

dilution ratio. The relationship between the volume of water in a stream and the volume of incoming waste. It can affect the ability of the stream to assimilate waste.

dipping. Material coating by briefly immersing parts in a molten bath, solution or suspension.

direct labor costs. Salaries, wages and other direct compensation earned by the employee.

discharge of pollutant(s). 1. The addition of any pollutant to navigable waters from any point source. 2. Any addition of any pollutant to the waters of the continguous zone or the ocean from any point source, other than from a vessel or other floating craft. The term "discharge" includes either the discharge of a single pollutant or the discharge of multiple pollutants.

disinfection. A chemical or physical process that kills organisms that cause infectious disease. Chlorine is often used to disinfect sewage treatment effluent.

dispersant. A chemical agent used to break up concentrations of organic material, such as spilled oil.

dispersed-air flotation. Separation of low density contaminants from water, using minute air bubbles attached to individual particles to provide or increase the buoyancy of the particle. The bubbles are generated by introducing air through a revolving impeller or porous media.

dissolved-air flotation. Separation of low-density contaminants from water using minute air bubbles attached to individual particles to provide or increase the buoyancy of the particle. The air is put into solution under elevated pressure and later released under atmospheric pressure or put into solution by aeration at atmospheric pressure and then released under a vacuum.

dissolved oxygen (DO). A measure of the amount of oxygen available for biochemical activity in a given amount of water. Adequate levels of DO are needed to support aquatic life. Low dissolved oxygen concentrations can result from inadequate waste treatment.

dissolved solids. The total of disintegrated organic and inorganic material contained in water. Excesses can make water unfit to drink or use in industrial processes.

distillation. Purifying liquids through boiling. The steam condenses to pure water and pollutants remain in a concentrated residue.

distillation refining. A metal with an impurity having a higher vapor pressure than the base metal can be refined by heating the metal to the point where the impurity vaporizes.

distillation-silver nitrate titration. A standard method of measuring the concentration of cyanides in a solution.

distillation-SPADNS. A standard method of measuring the concentration of fluoride in a solution.

dollar base. A period in time in which all costs are related. Investment costs are related by the Sewage Treatment Plant Construction Cost Index. Supply costs are related by the "Industrial Commodities" Whole Price Index.

dose. In radiology, the quantity of energy or radiation absorbed.

dosimeter. An instrument that measures exposure to radiation.

drag-in. Water or solution carried into another solution by the work and the associated handling equipment.

dragout. The solution that adheres to the objects removed from a bath, more precisely defined as that solution that is carried past the edge of the tank.

dragout reduction. Minimization of the amount of material (bath or solution) removed from a process tank by adherring to the part or its transfer device.

drainage phase. Period in which the excess plating solution adhering to the part or workpiece is allowed to drain off.

drawing.   Reduction of cross section area and increasing the length by pulling metal through conical taper dies.

drawing compounds.   See wire-forming lubricants.

dredging.   To remove earth from the bottom of water bodies using a scooping machine. This disturbs the ecosystem and causes silting that can kill aquatic life.

drilling.   Hole making with a rotary, end-cutting tool having 1 or more cutting lips and 1 or more helical or straight flutes or tubes for the ejection of chips and the passage of a cutting fluid. 1. Center drilling—drilling a conical hole in the end of a workpiece. 2. Core drilling—enlarging a hole with a chamer-edged, multiple-flute drill. 3. Spade drilling—drilling with a flate blade drill tip. 4. Step drilling—using a multiple diameter drill. 5. Gun drilling—using special straight flute drills with a single lip and cutting fluid at high pressures for deep hole drilling. 6. Oil hole or pressurized coolant drilling—using a drill with 1 or more continuous holes through its body and shank to permit the passage of a high pressure cutting fluid, which emerges at the drill point and ejects chips.

drip station.   Empty tank over which parts are allowed to drain freely to decrease end dragout.

drip time.   The period during which a past is suspended over baths in order to allow the excessive dragout to drain off.

dry limestone process.   An air pollution control method that uses limestone to absorb the sulfur oxides in furnaces and stack gases.

drying beds.   Areas for dewatering of sludge by evaporation and seepage.

dump.   A site used to dispose of solid wastes without environmental controls.

dust.   Fine grain particles light enough to be suspended in air.

dustfall jar.   An open container used to collect large particles from the air for measurement and analysis.

dystrophic lakes.   Shallow bodies of water that contain much humus and organic matter. They contain many plants but few fish and are almost eutrophic.

ecological impact.   The total effect of an environmental change, natural or man-made, on the community of living things.

ecology.   The relationships of living things to one another and to their environment, or the study of such relationships.

economic poisons.   Chemicals used to control pests and to defoliate cash crops, such as cotton.

ecosphere.   See biosphere.

ecosystem.   The interacting system of a biological community and its nonliving surroundings.

EDTA titration.   EDTA—Ethylenediamine tetraacetic acid (or its salts). A standard method of measuring the hardness of a solution.

effluent.   Waste material discharged into the environment. It can be treated or untreated. Generally refers to water pollution.

effluent limitation.   Any restriction (including schedules of compliance) established by a state or the federal EPA on quantities, rates, and concentrations of chemical, physical, biological, and other constituents that are discharged from point sources into navigable waters, the waters of the contiguous zone, or the ocean.

electrical conductivity.   The property that allows an electric current to flow when a potential difference is applied. It is the reciprocal of the resistance in ohms measured between opposite faces of a centimeter cube of an aqueous solution at a specified temperature. It is expressed as microohms per centimeter at temperature degrees Celsius.

electrical discharge machining.   Metal removal by a raped spark discharge between different polarity electrodes, 1 the workpiece and the other the tool separated by a gap distance of 0.0005 in. to 0.035 in. The gap is filled with dielectric fluid and metal particles that are melted, in part vaporized and expelled from the gap.

electrobrightening.   A process of reversed electro-deposition that results in anodic metal taking a high polish.

electrochemical machining (ECM).   A machining process whereby the part to be machined is made the anode and a shaped cathode is maintained in close proximity to the work. Electrolyte is pumped between the electrodes and a potential applied with the result that metal is rapidly dissolved from the workpiece in a selective manner and the shape produced on the workpiece complements that of the cathode.

electrocleaning.   The process of anodic removal of surface oxides and scale from a workpiece.

electrode.   Conducting material for passing electric current into or out of a solution by adding electrons to or taking electrons from ions in the solution.

electrodialysis.   A process that uses electrical current applied to permeable membranes to remove minerals from water. Often used to desalinize salt or brackish water.

electroless plating.   Deposition of a metallic coating by a controlled chemical reduction that is catalyzed by the metal or alloy being deposited.

electrolysis.   The chemical decomposition by an electric current of a substance in a dissolved or molten state.

electrolyte.   A liquid, most often a solution, that will conduct an electric current.

electrolytic cell.   A unit apparatus in which electrochemical reactions are produced by applying electrical energy or which supplies electrical energy as a result of chemical reactions and which includes 2 or more electrodes and 1 or more electrolytes contained in a suitable vessel.

electrolytic decomposition.   An electrochemical treatment used for the oxidation of cyanides. The method is practical and economical when applied to concentrated solutions such as contaminated baths, cyanide dips, stripping solutions, and concentrated rinses. Electrolysis is carried out at a current density of 35 amp/sq. ft. at the anode and 70 amp/sg. ft. at the cathode. Metal is deposited at the cathode and can be reclaimed.

electrolytic oxidation.   A reaction by an electrolyte in which there is an increase in valence resulting from a loss of electrons.

electrolytic reduction.   A reaction in which there is a decrease in valence resulting from a gain in electrons.

electrolytic refining.   The method of producing pure metals by making the impure metal the anode in an electrolytic cell and depositing a pure cathode. The impurities either remain undissolved at the anode or pass into solutions in the electrolyte.

electrometallurgical process.   The application of electric current to a metallurgical process either for electrolytic deposition or as a source of heat.

electrometric titration.   A standard method of measuring the alkalinity of a solution.

electron beam machining.   The process of removing material from a workpiece by a high-velocity focused stream of electrons that melt and vaporize the workpiece at the point of impingement.

electroplating.   The production of a thin coating of 1 metal on a surface by electrodeposition.

electropolishing.   Electrolytic corrosion process that increases the percentage of specular reflectance from a metallic surface.

electrostatic precipitator.   An air pollution control device that imparts an electrical charge to particles in a gas stream, causing them to collect on an electrode.

embossing.   Raising a design in relief against a surface.

emergency episode.   See air pollution episode.

emission.   Like effluent but used in regard to air pollution.

emission factor.   The relationship between the amount of pollution produced and the amount of raw material processed. For example, an emission factor for a blast furnace making iron would be the number of pounds of particulates per ton of raw materials.

emission inventory.   A listing, by source, of the amounts of air pollutants discharged into the atmosphere of a community daily. It is used to establish emission standards.

emission standard.   The maximum amount of discharge legally allowed from a single source, mobile or stationary.

emulsified oil and grease.   An oil or grease dispersed in an immiscible liquid usually in droplets of larger than colloidal size. In general suspension of oil or grease within another liquid (usually water).

emulsifying agent.   A material that increases the stability of a dispersion of one liquid in another.

emulsion breaking.   Decreasing the stability of dispersion of one liquid in another.

emuslion cleaning. A cleaning process using organic solvents dispersed in an aqueous medium with the aid of an emulsifying agent.

end-of-pipe treatment. The reduction and/or removal of pollutants by treatment just before actual discharge.

enrichment. Sewage effluent or agricultural runoff adding nutrients (nitrogen, phosphorus, carbon compounds) to a water body, greatly increasing the growth potential for algae and aquatic plants.

environment. The sum of all external conditions affecting the life, development and survival of an organism.

environmental impact statement. A document required of Federal agencies by the National Environmental Policy Act for major projects or legislative proposals. They are used in making decisions about the positive and negative effects of the undertaking and list alternatives.

Environmental Protection Agency. The United States Environmental Protection Agency.

EPA. See Environmental Protection Agency.

epidemiology. The study of diseases as they affect populations.

episode (pollution). An air pollution incident in a given area caused by a concentration of atmospheric pollution reacting with meteorological conditions that may result in a significant increase in illnesses or deaths.

equalization. (Continuous flow) The balance of flow or pollutant load using a holding tank for a system that has widely varying inflow rates.

equilibrium concentration. A state at which the concentrations of chemicals in a solution remain in a constant proportion to one another.

erosion. The wearing away of land surface by wind or water. Erosion occurs naturally from weather or runoff but can be intensified by land-clearing practices.

ester. An organic compound corresponding in structure to a salt in inorganic chemistry. Esters are considered as derived from the acids by the exchange of the replaceable hydrogen of the latter for an organic alkyl radica. Esters are not ionic compounds, but salts usually are.

estuaries. Areas where fresh water meets salt water (bays, mouths of rivers, salt marshes, lagoons). These brackish water ecosystems shelter and feed marine life, birds, and wildlife.

etchant. The material used in the chemical process of removing glass fibers and epoxy between neighboring conductor layers of a PC board for a given distance.

etching. A process where material is removed by chemical action.

eutrophication. The slow aging process of a lake evolving into a marsh and eventually disappearing. During eutrophication the lake is choked by abundant plant life. Human activities that add nutrients to a water body can speed up this action.

eutrophic lakes. Shallow murky water bodies that have lots of algae and little oxygen.

evaporation ponds. Areas where sewage sludge is dumped and allowed to dry out.

excess capacity factor. A multiplier on process size to account for shutdown for cleaning and maintenance.

extrusion. A material that is forced through a die to form lengths of rod, tube or special section.

4-AAP colorimetric. A standard method of measurement for phenols in aqueous solutions.

fabric filter. A cloth device that catches dust and particles from industrial emissions.

fecal coliform bacteria. A group of organisms found in the intestinal tracts of people and animals. Their presence in water indicates pollution and possible dangerous bacterial contamination.

feedlot. A relatively small, confined area for raising cattle that results in lower costs but may concentrate large amounts of animal wastes. The soil cannot absorb such large amounts of excrement, and runoff from feedlots pollutes nearby waterways with nutrients.

fen. Low-lying land partly covered with water.

fermentation. A chemical change to break down biodegradable waste. The change is induced by a living organism or enzyme, specifically bacteria or microorganisms occurring in unicellular plants such as yeast, molds, or fungi.

ferrite. A solid solution in which alpha iron is present.

ferrous. Relating to or containing iron.

filling. Depositing dirt and mud, often raised by dredging into marshy areas to create more land for real estate development. It can destroy the marcs ecology.

film badge. A piece of masked photographic film worn by nuclear workers to monitor their exposure to radiation. Nuclear radiation darkens the film.

filtrate. Liquid after passing through a filter.

filtration. Removing particles of solid materials from water, usually by passing it through sand.

flameless atomic absorption. A method of measuring low concentration values of certain metals in a solution.

flame hardened. Surface hardened by controlled torch heating followed by quenching with water or air.

flame spraying. The process of applying a metallic coating to a workpiece whereby finely powdered fragments or wire, together with suitable fluxes, are projected through a cone of flame onto the workpiece.

flash evaporation. Evaporation using steam heated tubes with feed material under high vacuum. Feed material "flashes off" when it enters the evaporation chamber.

floc. A clump of solids formed in sewage by biological or chemical action.

flocculation. Separation of suspended solids during waste water treatment by chemical creation of clumps of flocs.

flotation. The process of removing finely divided particles from a liquid suspension by attaching gas bubbles to the particles, increasing their buoyancy, and thus concentrating them at the surface of the liquid medium.

flowmeter. A gauge that shows the speed of waste water moving through a treatment plant.

flue gas. The air coming out of a chimney after combustion. It can include nitrogen oxides, carbon oxides, water vapor, sulfur oxide, particles, and many chemical pollutants.

fluorides. Gaseous, solid, or dissolved compounds containing fluorine that result from industrial processes.

fluorocarbons. A gas used as a propellant in aerosols, thought to be modifying the ozone layer in the stratosphere thereby allowing more harmful solar radiation to reach the Earth's surface.

flume. A natural or man-made channel that diverts water.

fluxing. (Degassing) The removal of oxides and other impurities from molten primary aluminum in a casthouse holding furnace by injecting chlorine gas (often with nitrogen and carbon monoxide).

fly ash. Noncombustible particles carried by flue gas.

fog. Suspended liquid particles formed by condensation of vapor.

fogging. Applying a pesticide by rapidly heating the liquid chemical so that it forms very fine droplets that resemble smoke. It is used to destroy mosquitoes and blackflies.

food waste. Discarded animal and vegetable matter, also called garbage.

forming compounds (sheet). Tightly adhering lubricants composed of fatty oils, fatty acids, soaps, and waxes and designed to resist the high surface temperatures and pressures the metal would otherwise experience in forming.

forming compounds (wire). Tightly adhering lubricants composed of solids (white lead, talc, graphite, or molybdenum disulfide) and soluble oils for cooling and corrosion protection. Lubricants typically contain sulfur, chlorine, or phosphate additives.

fossil fuels. Combustibles derived from the remains of ancient plants and animals, such as coal, oil, and natural gas.

free cyanide. 1. True—the actual concentration of cyanide radical or equivalent alkali cyanide not combined in complex ions with metals in solutions. 2. Calculated—the concentration of cyanide or alkali cyanide present in solution in excess of that calculated as necessary to form a specific complex ion with a metal or metals present in solution. 3.

Analytical—the free cyanide content of a solution as determined by a specified analytical method.

freezing/crystallization. The solidification of a liquid into aggregations of regular geometric forms (crystals) accomplished for removal of solids, oils, greases, and heavy metals from industrial waste water.

fume. Tiny particles trapped by vapor in a gas stream.

fumigant. A pesticide that is vaporized to kill pests; often used in buildings or greenhouses.

fungi. Tiny plants that lack chlorophyll. Some cause disease; others stabilize sewage and break down solid wastes for compost.

galvanizing. The deposition of zinc on the surface of steel for corrosion protection.

game fish. Species like trout, salmon, bass, etc., caught for sport. They show more sensitivity to environmental changes than do "rough" fish.

gamma ray. The most penetrating waves of radiant nuclear energy. They can be stopped by dense materials like lead.

garbage. See food waste.

garbage grinding. Use of a household disposal to crush food waste and wash it into the sewer system.

gas carburizing. The introduction of carbon into the surface layers of mill steel by heating in a current of gas hish in carbon.

gas chromotagrophy. Chemical analytical instrumentation generally used for quantitative organic analysis.

gasification. Conversion of a solid material, such as coal, into a gas for use as fuel.

gas nitriding. Case hardening metal by heating and diffusing nitrogen gas into the surface.

gas phase separation. The process of separating volatile constituents from water by the application of selective gas permeable membranes.

gear forming. Process for making small gears by rolling the gear material as it is pressed between hardened gear shaped dies.

geiger counter. An electrical device that detects the presence of radioactivity.

generator. A device that converts mechanical energy into electrical energy.

germicide. Any compound that kills disease-carrying microorganisms. These must be registered as pesticides with the EPA.

glass fiber filtration. A standard method of measuring total suspended solids.

good housekeeping. (in-plant technology) Good and proper maintenance minimizing spills and upsets.

gpd. Gallons per day.

grab sample. A single sample of waste water taken without regard to time or flow.

grain. A unit of weight equal to 65 milligrams or 2/1,000 of an ounce.

grain loading. The rate at which particles are emitted from a pollution source—measurement is made by the numbers of grains per cubic foot of gas emitted.

gravimetric 103-105C. A standard method of measuring total volatile solids in aqueous solutions.

gravity filtration. Settling of heavier and rising of lighter constituents within a solution.

gravity flotation. The separation of water and low density contaminants such as oil or grease by reduction of the waste water flow velocity and turbulence for a sufficient time to permit separation owing to difference in specific gravity. The floated material is removed by some skimming technique.

gray cast irons. Alloys primarily of iron, carbon, and silicon along with other alloying elements in which the graphite is in flake form. (These irons are characterized by low ductility but have many other properties, such as good castability and good damping capacity.)

grease. In waste water, a group of substances including fats, waxes, free fatty acids, calcium and magnesium soaps, mineral oils, and certain other nonfatty materials. The type of solvent and method used for extraction should be stated for quantification.

grease skimmer. A device for removing floating grease or scum from the surface of waste water in a tank.

green belts. Buffer zones created by restricting development from certain land areas.

greenhouse effect. The warming of our atmosphere caused by buildup of carbon dioxide, which allows light from the Sun's rays to heat the Earth but prevents loss of the heat.

grinding. The removal of stock from a workpiece by use of abrasive grains held by a rigid or semi rigid binder. 1. Surface grinding—producing a flat surface with a rotating grinding wheel as the workpiece passes under the wheel. 2. Cylindrical grinding—grinding the outside diameters of cylindrical workpieces held between centers. 3. Internal grinding—grinding the inside of a rotating workpiece by use of a wheel spindle, which rotates and reciprocates through the length of depth of the hole being ground.

grinding fluids. Water-based, straight oil, or synthetic-based lubricants containing mineral oils, soaps, or fatty materials lubricants serve to cool the part and maintain the abrasiveness of the grinding wheel face.

ground cover. Plants grown to keep soil from eroding.

ground water. The supply fo fresh water under the Earth's surface that forms a natural reservoir.

habitat. The sum of environmental conditions in a specific place that is occupied by an organism, population, or community.

half-life. The time taken by certain materials to lose half their strength. For example the half-life of DDT is 15 years; of radium 1,580 years.

hammer forging. Heating and pounding metal to shape it into the desired form.

hammermill. A high-speed machine that uses hammers and cutters to crush, grind, chip, or shred solid wastes.

hardened. Designates condition produced by various heat treatments such as quench hardening, age hardening, and precipitation hardening.

hardness. A characteristic of water, imparted by salts of calcium, magnesium, and iron such as bicarbonates, carbonates, sulfates, chlorides and nitrates, that cause curdling of soap, deposition of scale, damage in some industrial processes and sometimes objectionable taste. It may be determined by a standard laboratory procedure or computed from the amounts of calcium and magnesium as well as iron, aluminum, manganese, barium, strontium, and zinc and is expressed as equivalent calcium carbonate.

hard water. Alkaline water containing dissolved mineral salts, which interfere with some industrial processes and prevent soap from lathering.

hazardous air pollutant. Substances covered by Air Quality Criteria, which may cause or contribute to illness or death; asbestos, beryllium, mercury, and vinyl chloride.

hazardous waste. Waste materials that by their nature are inherently dangerous to handle or dispose of, such as old explosives, radioactive materials, some chemicals, and some biological wastes; usually produced in industrial operations.

heading. (material forming) Upsetting wire, rod, or bar stock in dies to form parts having some of the cross-sectional area larger than the original. Examples are bolts, rivets, and screws.

heat island effect. A haze dome created in cities by pollutants combining with the heat trapped in the spaces between tall buildings. This haze prevents natural cooling of air and in the absence of strong winds can hold high concentrations of pollutants in one place.

heat resistant steels. Steel with high resistance to oxidation and moderate strength at high temperature above 500 Degrees C.

heat treatment. The modification of the physical properties of a workpiece through the application of controlled heating and cooling cycles. Such operations as heat treating, tempering, carburizing, cyaniding, nitriding, annealing, normalizing, austenizing, quenching, austempering, siliconizing, martempering, and malleabilizing are included in this definition.

heating season. The coldest months of the year when pollution increases in some areas because people burn fossil fuels to keep warm.

heavy metals. Metallic elements like mercury, chromium, cadmium, arsenic, and lead, with high molecular weights. They can damage living things at low concentrations and tend to accumulate in the food chain.

herbicide.   A chemical that controls or destroys undesirable plants.

herbivore.   An animal that feeds on plants.

heterotrophic organism.   Humans and animals that cannot make food from inorganic chemicals.

high density polyethylene.   A material used to make plastic bottles that produces toxic fumes when burned.

high energy forming.   Processes where parts are formed at a rapid rate by using extremely high pressures. Examples: explosive forming, electrohydraulic forming.

high energy rate forging (HERF).   A closed dis process where hot or cold deforming is accomplished by a high velocity ram.

hi-volume sampler.   A device used to measure and analyze suspended particulate pollution.

hobbing.   Gear cutting by use of a tool resembling a worm gear in appearance, having helically spaced cutting teeth. In a single-thread hob, the rows of teeth advance exactly one pitch as the hob makes 1 revolution. With only 1 hob, it is possible to cut interchangeable gears of a given pitch of any number of teeth within the range of the hobbing machine.

holding pond.   A pond or reservoir usually made of earth to store polluted runoff.

honing.   A finishing operation using fine grit abrasive stones to produce accurate dimensions and excellent finish.

hot.   Slang for radioactive material.

hot compression molding.   (plastic processing) A technique of thermoset molding in which preheated molding compound is closed and heat and pressure (in the form of a downward moving ram) are applied until the material has cured.

hot dip coating.   The process of coating a metallic workpiece with another metal by immersion in a molten bath to provide a protective film.

hot rolled.   A term used to describe alloys that are rolled at temperatures above the recrystallization temperature. Many alloys are hot rolled, and machinability of such alloys may vary because of differences in cooling conditions from lot to lot.

hot stamping.   Engraving operation for marking plastics in which roll leaf is stamped with heated metal dies onto the face of the plastics. Ink compounds can also be used.

hot upset forging.   The diameter is locally increased, i.e., to upset the head of a bolt, the end of the barstock is heated and then deformed by an axial blow often into a suitably shaped die.

humus.   Decomposed organic material.

hydrocarbons.   Compounds found in fossil fuels that contain carbon and hydrogen and may be carcinogenic.

hydrofluoric acid.   Hydrogen fluoride in aqueous solution.

hydrogen sulfide (H_2S).   The gas emitted during organic decomposition that smells like rotten eggs. It is also a by-product of oil refining and burning and can cause illness in heavy concentrations.

hydrogen embrittlement.   Embrittlement of a metal or alloy caused by absorption of hydrogen during a pickling, cleaning, or plating process.

hydrology.   The science dealing with the properties, distribution, and circulation of water.

hydrometallurgical process.   The treatment of ores by wet processes such as leaching.

hydrophilic.   A surface having a strong affinity for water or being readily wettable.

hydrophobic.   A surface which is nonwettable or not readily wettable.

hydrostatic pressure.   The force per unit area measured in terms of the height of a column of water under the influence of gravity.

immersed area.   Total area wetted by the solution or plated area plus masked area.

immersion plate.   A metallic deposit produced by a displacement reaction in which 1 metal displaces another from solution. For example: $Fe + Cu(+2) = Cu + Fe(+2)$

impact deformation.   The process of applying impact force to a workpiece such that the workpiece is permanently deformed or shaped. Impact deformation operations such as shot peening, peening, forging, high energy forming, heading, or stamping.

impedance.   The rate at which a substance absorbs and transmits sound.

implementation plan.   An outline of steps needed to meet environmental quality standards by a set time.

impoundment.   A body of water confined by a dam, dike, floodgate, or other barrier.

incineration.   (sludge disposal) The combustion (by burning) of organic matter in wastewater sludge after dewatering by evaporation.

incinerator.   Disposal of solid, liquid, or gaseous wastes by burning.

incompatible pollutants.   Those pollutants that would cause harm to, adversely affect the performance of, or be inadequately treated in publicly owned treatment works.

independent operation.   Job shop or contract shop in which electroplating is done on workpieces owned by the customers.

indicator.   In biology, an organism, species, or community that shows the presence of certain environmental conditions.

indirect labor costs.   Labor-related costs paid by the employer other than salaries, wages and other direct compensation such as social security and insurance.

induction hardened.   Surface hardened using induction heating followed by quenching with water or air.

industrial water.   Any industry that introduces pollutants into public sewer systems and whose wastes are treated by a publicly owned treatment facility.

industrial wastes.   The liquid wastes from industrial processes, as distinct from domestic or sanitary wastes.

inert gas.   A vapor that does not react with other substances under ordinary conditions.

inertial separator.   A device that uses centrifugal force to separate waste particles.

infiltration.   The action of water moving through small openings in the earth as it seeps down into the ground water.

inhibition.   The slowing down or stoppage of chemical or biological reactions by certain compounds or ions.

inoculum.   Bacteria placed in compost to start biological action.

in-process control technology.   The regulation and the conservation of chemicals and the reduction of water usage throughout the operations as opposed to end-of-pipe treatment.

inspection.   A checking or testing of something against standards or specification.

intake water.   Gross water minus reuse water.

integrated chemical treatment.   A waste treatment method in which a chemical rinse tank is inserted in the plating line between the process tank and the water rinse tank. The chemical rinse solution is continuously circulated through the tank and removes the dragout while chemicals react with it.

integrated circuit (IC).   1. A combination of interconnected circuit elements inseparably associated on or within a continuous substrate. 2. Any electronic device in which both active and passive elements are contained in a single package. Methods of making an integrated circuit are by masking process, screening, and chemical deposition.

integrated pest management.   Combining the best of all useful techniques—biological, chemical, cultural, physical, and mechanical—into a custom-made pest control system.

interceptor sewers.   The collection system that connects main and trunk sewers with the waste water treatment plant. In a combined sewer system interceptor sewers allow some untreated wastes to flow directly into the receiving streams so the plant will not be overloaded.

interstate carrier water supply.   A source of water for planes, buses, trains, and ships operating in more than one state. These sources are regulated by the Federal Government.

interstate waters.   Defined by law as: (1) waters that flow across or form a part of state or international boundaries (2) the Great Lakes and (3) coastal waters.

intraforming.   A method of forming by means of squeezing.

inversion.   An atmospheric condition caused by a layer of warm air preventing the rise of cool air trapped beneath it. That holds down pollutants that might otherwise be dispersed and can cause an air pollution episode.

investment costs.   The capital expenditures required to bring the treatment or control technology into operation.

ion exchange.   A reversible chemical reaction between a solid (ion exchanger) and a fluid (usually a water solution) by means of which ions may be interchanged from 1 substance to another. The superficial physical structure of the solid is not affected.

ion exchange resins.   Synthetic resins containing active groups (usually sulfonic, carboxylic, phenol, or substituted amino groups) that give the resin the property of combining with or exchanging ions between the resin and a solution.

ion flotation technique.   Treatment for electroplating rinse waters (containing chromium and cyanide) in which ions are separated from solutions by flotation.

ionization chamber.   A device that detects ionizing radiation.

iridite dip process.   Dipping process for zinc or zinc-coated objects that deposits protective film that is a chromium gel, chromium oxide, or hydrated chromium oxide.

isolation.   Segregation of a waste for separate treatment and/or disposal.

isotope.   A variation of an element that has the same atomic number but a different weight because of its neutrons. Isotopes of an element may have different radioactive behavior.

lagoon.   A shallow pond where sunlight, bacterial action, and oxygen work to purify waste water.

laminate.   1. A composite metal, wood or plastic usually in the form of sheet or bar, composed of 2 or more layers so bonded that the composite forms a structural member. 2. To from a product of two or more bonded layers.

landfill.   Disposal of inert, insoluble waste solids by dumping at an approved site and covering with earth.

lapping.   An abrading process to improve surface quality by reducing roughness, waviness, and defects to produce accurate as well as smooth surfaces.

lateral sewers.   Pipes running underneath city streets that collect sewage.

LC$_{50}$.   Median lethal concentration, a standard measure of toxicity. It tells how much of a substance is needed to kill half of a group of experimental organisms.

leachate.   Materials that pollute water as it seeps through solid waste.

leach field.   An area of ground to which waste water is discharged. Not considered an acceptable treatment method for industrial wastes.

leaching.   The process by which nutrient chemicals or contaminants are dissolved and carried away by water or are moved into a lower layer of soil.

lead.   A heavy metal that may be hazardous to health if breathed or swallowed.

life cycle.   The stages an organism passes through during its existence.

lift.   In a sanitary landfill, a compacted layer of solid waste and the top layer of cover material.

ligands.   The molecules attached to the central atom by coordinate covalent bonds.

limiting factor.   A condition whose absence, or excessive concentration, exerts some restraining influence upon a population through incompatibility with species requirements or tolerance.

limonology.   The study of the physical, chemical, meteorological, and biological aspects of fresh water.

liquefaction.   Changing a solid into a liquid form.

liquid/liquid extraction.   A process of extracting or removing contaminant(s) from a liquid by mixing contaminated liquid with another liquid that is immiscible and has a higher affinity for the contaminating substance(s).

liquid mitriding.   Process of case hardening a metal in a molten cyanide bath.

liquid phase refining.   A metal with an impurity possessing a lower melting point is refined by heating the metal to the point of melting of the low-temperature metal. It is separated by sweating out.

machining.   The process of removing stock from a workpiece by forcing a cutting tool through the workpiece, removing a chip of basis material. Machining operations, such as turning, milling, drilling, boring, tapping, planing, broaching, sawing and filing, and chamfering are included in this definition.

maintenance.   The upkeep of property or equipment.

malleablizing. Process of annealing brittle white cast iron in such a way that the combined carbon is wholly or partly transformed to graphitic or temper carbon nodules in a ferritic or pearlitic microstructure, thus providing a ductile and machinable material.

manual plating. Plating in which the workpieces are conveyed manually through successive cleaning and plating tanks.

maraged. Describes a series of heat treatments used to treat high-strength steels of complex composition (maraging steels) by aging of martensite.

marsh. Wet, soft, low-lying land that provides a niche for many plants and animals. It can be destroyed by dredging and filling.

martensite. An acicular or needlelike microstructure that is formed in quenched steels. (It is very hard and brittle in the quenched form and therefore is usually tempered before being placed into service. The harder forms of tempered martensite have poorer machinability.)

martempering. Quenching an austentized ferrous alloy in a medium at a temperature in the upper part of the martensite range or slightly above that range and holding it in the medium until the temperature throughout the alloy is substantially uniform. The alloy is then allowed to cool in air through martensite range.

masking. Blocking out 1 sight, sound, or smell with another.

material modification. (In-plant technology) Altering the substance from which a part is made.

mechanical agitation. The agitation of a liquid medium through the use of mechanical equipment, such as impellers or paddles.

mechanical finish. Final operations on a product performed by a machine or tool. See polishing, buffing, barrel finishing, shot peening, power brush finishing.

mechanical plating. Providing a coating wherein fine metal powders are peened onto the part by tumbling or other means.

mechanical turbulence. The erratic movement of air caused by local obstructions, such as buildings.

membrane. A thin sheet of synthetic polymer through the apertures of which small molecules can pass, while larger ones are retained.

membrane filtration. Filtration at pressures ranging from 50 to 100 psig with the use of membranes or thin films. The membranes have accurately controlled pore sites and typically low flux rates.

metal ion. An atom or radical that has lost or gained 1 or more electrons and has thus acquired an electric charge. Positively charged ions are cations, and those having a negative charge are anions. An ion often has entirely different properties from the element (atom) from which it was formed.

metal-oxidation refining. A refining technique that removes impurities from the base metal because the impurity oxidizes more readily than the base. The metal is heated and oxygen supplied. The impurity upon oxidizing separates by gravity or volatilizes.

metal paste production. Manufacture of metal pastes for use as pigments by mixing metal powders with mineral spirits, fatty acids, and solvents. Grinding and filtration are steps in the process.

metal powder production. Production of metal particles for such uses as pigments either by milling and grinding or scrap or by atomization of molten metal.

metal spraying. Coating metal objects by spraying molten metal upon the surface with gas pressure.

methane. A colorless, nonpoisonous, flammable gas emitted by marshes and dumps undergoing anaerobic decomposition.

mgd. Millions of gallons per day. Mgd is a measurement of water flow.

microbes. Tiny plants and animals, some that cause disease and are found in sewers.

microstraining. A process for removing solids form water, which consists of passing the water stream through a microscreen with the solids being retained on the screen.

milling. Using a rotary tool with one or more teeth that engage the workpiece and remove material as the workpiece moves past the rotating cutter. 1. Face milling—milling a surface perpendicular cutting edges remove the bulk of the material while the face cutting edges provide the finish of the surface being generated. 2. End milling—milling accomplished with a tool having cutting edges on its cylindrical surfaces as well as on its end. In end milling—peripheral, the peripheral cutting edges on the cylindrical surface are used; while in end milling—slotting, both end and peripheral cutting edges remove metal.3. Slide and slot milling—milling of the side or slot of a workpiece using a peripheral cutter. 4. Slab milling—milling of a surface parallel to the axis of a helical, multiple-toothed cutter mounted on an arbor. 5. Straddle milling—peripheral milling a workpiece on both sides at once using 2 cutters spaced as required.

mist.   Liquid particles measuring 40 to 500 microns, that are found by condensation of vapor. By comparison, fog particles are smaller than 40 microns.

mixed liquor.   Activated sludge and water containing organic matter being treated in an aeration tank.

mobile source.   A moving producer of air pollution, mainly forms of transportation—cars, motorcycles, planes.

molecule.   Chemical units composed of 1 or more atoms.

monitoring.   Periodic or continuous sampling to determine the level of pollution or radioactivity.

muck oils.   Earth made from decaying plant materials.

mulch.   A layer of material (wood chips, straw, leaves) placed around plants to hold moisture, prevent weed growth, and enrich soil.

multi-effect evaporator.   A series of evaporations and condensations with the individual units set up in series and the latent heat of vaporization from 1 unit used to supply energy for the next.

multiple operation machinery.   Two or more tools are used to perform simultaneous or consecutive operations.

multiple subcategory plant.   A plant discharging processed wastewater from more than one manufacturing process subcategory.

multiple use.   Harmonious use of land for more than 1 purpose; i.e., grazing of livestock, wildlife production, recreation, watershed, and timber production. Not necessarily the combination of uses that will yield the highest economic return or greatest unit output.

mutagen.   Any substance that causes changes in the genetic structure in subsequent generations.

National Pollutant Discharge Elimination System (NPDES).   The Federal mechanism for regulating point source discharge by means of permits.

natural gas.   A natural fuel containing cethane and hydrocarbone that occurs in certain geologic formations.

natural selection.   The process of survival of the fittest, by which organisms that adapt to their environment survive and those that do not disappear.

navigable waters.   All navigable waters of the United States, tributaries of navigable waters of the United States, interstate waters, intrastate lakes, rivers, and streams that are utilized for recreational or other purposes.

necrosis.   Death of cells that can discolor areas on a plant or kill the entire plant.

neutralization.   Chemical addition of either acid or base to a solution so that the pH is adjusted to 7.

new source.   Any building, structure, facility, or installation from which there is or may be the discharge of pollutants, the construction of which is commenced after the publication of proposed regulations prescribing a standard of performance under Section 306 of the act which will be applicable to such source if such standard is thereafter promulgated in accordance with Section 306 of the act.

nitric oxide (NO_2).   A gas formed by combustion under high temperature and high pressure in an internal combustion engine. It changes into nitrogen dioxide in the ambient air and contributes to photochemical smog.

nitriding.   A heat treating method in which nitrogen is diffused into the surface of iron-base alloys. (This is done by heating the metal at a temperature of about 950 degrees F in contact with ammonia gas or other suitable nitrogenous materials. The surface, because of formation of nitrides becomes much harder than the interior. Depth of the nitrided surface is a function of the length of time of exposure and can vary from .0005

m. to .032 m. thick. Hardness is generally in the 65 to 70 R*c* range, and, therefore, those structures are almost always ground.)

nitriding steels.   Steels that are selected because they form good case-hardened structures in the nitriding process. (In these steels, elements such as aluminum and chromium are important for producing a good case.)

nitrification (biological).   The oxidation of nitrogenous matter into nitrates by bacteria.

nitrogen dioxide (NO$_2$).   The result of nitric oxide combining with oxygen in the atmosphere, a major component of photochemical smog.

nitrogenous wastes.   Animal or plant residues that contain large amounts of nitrogen.

NO.   A notation meaning oxides of nitrogen. See nitric oxide.

noble metals.   Metals below hydrogen in the electromotive force series; includes antimony, copper, rhodium, silver, gold, bismuth.

noise.   Any undesired noise.

noncontact cooling water.   Water used for cooling that does not come into direct contact with any raw material, intermediate product, waste product, or finished product.

nonferrous.   No iron content.

nonpoint source.   A contributing factor in water pollution that cannot be traced to a specific spot, such as agricultural fertilizer runoff, sediment from construction.

nonwater quality environmental impact.   The ecological impact as a result of solid, air, or thermal pollution owing to the application of various waste water technologies to achieve the effluent guidelines limitations. Associated with the nonwater quality aspect is the energy impact of waste water treatment.

normalizing.   Heat treatment of iron-base alloys above the critical temperature, followed by cooling in still air. (This is often done to refine or homogenize the grain structure of castings, forgings, and wrought steel products.)

notching.   Cutting out various shapes from the edge or side of a sheet, strip, blank, or part.

NPDES.   See National Pollutant Discharge Elimination System.

NTA.   Nitrilotriacetic acid, a compound proposed for use to replace phosphates in detergents.

nuclear power plant.   A device that converts atomic energy into usable power; heat produced by a reactor makes steam to drive electricity-generating turbines.

nutrients.   Elements of compounds essential to growth and development of living things; carbon, oxygen, nitrogen, potassium, and phosphorus.

off-road vehicles.   Forms of motorized transportation that do not require prepared surfaces—they can be used to reach remote areas.

oil cooker.   Open-topped vessel containing a heat source and typically maintained at 68°C (180°F) for the purpose of driving off excess water from waste oil.

oil "fingerprinting."   A method that identifies oil spills so they can be traced back to their sources.

oil spill.   Accidental discharge into bodies of water, can be controlled by chemical dispersion, combustion, mechanical containment, and absorption.

eligotrophic lakes.   Deep, clear lakes with low nutrient supplies. They contain little organic matter and have a high dissolved organic level.

oncogenic.   A substance that causes tumors, whether benign or malignant.

opacity.   The amount of light obscured by an object or substance; a window has zero opacity, a wall 100 percent opacity.

open burning.   Uncontrolled fires in an open dump.

open space.   A relatively undeveloped green or wooded area provided usually within an urban development to minimize feelings of congested living.

operation and maintenance costs.   The cost of running the waste water treatment equipment. That includes labor costs, material and supply costs, and energy and power costs.

organic.   Referring to or derived from living organisms. In chemistry, any compound containing carbon.

organic compound.   Any substance that contains the element carbon, with the exception of carbon dioxide and various carbonates.

organism.   Any living thing.

organophosphates.   Pesticide chemicals that contain phosphorus, used to control insects. They are short-lived, but some can be toxic when first applied.

ORP recorders.   Oxidation-reduction potential recorders.

osmosis.   The tendency of a fluid to pass through a permeable membrane, as the wall of a living cell, into a less concentrated solution, so as to equalize concentrations on both sides of the membrane.

outfall.   The place where an effluent is discarded into receiving waters.

overfire air.   Air forced into the top of an incinerator to fan the flame.

overturn.   The period of mixing (turnover), by top to bottom circulation, of previously stratified water masses. This phenomenon may occur in spring and/or fall. The result is a uniformity of physical and chemical properties of the water at all depths.

oxidant.   A substance containing oxygen that reacts chemically in air to produce a new substance; primary source of photochemical smog.

oxidation.   Oxygen combining with other elements.

oxidation pond.   A holding area where organic wastes are broken down by aerobic bacteria.

oxidizable cyanide.   Cyanide amenable to oxidation.

oxidizing.   Combining the material concerned with oxygen.

ozone (O_3).   A pungent, colorless, toxic gas that contributes to photochemical smog.

packed tower.   A pollution control device that forces dirty air through a tower packed with crushed rock or wood chips while liquid is sprayed over the packing material. The pollutants in the air stream either dissolve or chemically react with the liquid.

paint stripping.   The term "paint stripping" shall mean the process of removing an organic coating from a workpiece or painting fixture. The removal of such coatings using processes such as caustic, acid, solvent, and molten salt stripping are included.

PAN.   (Peroxyacetyl nitrate) A pollutant created by the action of sunlight on hydrocarbons and nitrogen oxides in the air. An ingredient in smog.

pandemic.   Widespread throughout the area.

parameter.   A characteristic element of constant factor.

particulates.   Fine liquid or solid particles such as dust, smoke, mist, fumes, or smog, found in the air or emissions.

particulate loading.   The introduction of particulates into ambient air. Passivation—the changing of the chemically active surface of a metal to a much less reactive state by means of an acid dip.

pathogenic.   Capable of causing disease.

patina.   A blue green oxidation of copped.

PCBs (Polychlorinated biphenyls).   A group of toxic, persistent chemicals used in transformers and capacitors. Further sale or new use is banned in 1979 by law.

pearlite.   A microstituent found in iron-base alloys consisting of a lamellar (patelike) composite of ferrite and iron carbide. (This structure results from the decomposition of austenite and is very common in cast irons and annealed steels.)

peening.   Mechanical working of metal by hammer blows or shot impingement.

percolation.   Downward flow or filtering of water through pores or spaces in rock or soil.

persistent pesticides.   Pesticides that do not break down chemically and remain in the environment after a growing season.

pesticide.   Any substance used to control pests ranging from rats, weeds, and insects to algae and fungi. Pesticides can accumulate in the food chain and can contaminate the environment if misused.

pesticide tolerance.   The amount of pesticide residue allowed by law to remain in or on the harvested crop. By using various safety factors, EPA sets these levels well below the point where the chemicals might be harmful to consumers.

pH.   A measure of the acidity or alkalinity of a material, liquid or solid. pH is represented on a scale of 0 to 14 with 7 being a neutral state. 0 is most acid, and 14 most alkaline.

pH buffer.   A substance used to stabalize the acidity or alkalinity in a solution.

phenols. Organic compounds that are by-products of petroleum refining, tanning, textile, dye, and resin manufacture. Low concentrations can cause taste and odor problems in water, higher concentrations can kill aquatic life.

phosphates. Chemical compounds containing phosphorus.

phosphate coating. Process of forming a conversion coating on iron or steel by immersing in a hot solution of manganese, iron or zinc phosphate. Often used on a metal part before painting or procelainizing.

phosphatizing. Process of forming rust-resistant coating on iron or steel by immersing in a hot solution of acid manganese, iron, or zinc phosphates.

phosphorus. An essential food element that can contribute to the eutrophication of water bodies.

photochemical exidants. Air pollutants formed by the action of sunlight on oxides of nitrogen and hydrocarbons.

photochemical smog. Air pollution caused by not 1 pollutant but by chemical reactions of various pollutants emitted from different sources.

photoresists. Thin coatings produced from organic solutions that when exposed to light of the proper wave length are chemically changed in their solubility to certain solvents (developers). This substance is placed over a surface that is to be protected during processing such as in the etching of printed circuit boards.

photosensitive coating. A chemical layer that is receptive to the action of radiant energy.

photosynthesis. The manufacture by plants of carbohydrates and oxygen from carbon dioxide and water in the presence of chlorophyll, using sunlight as an energy source.

phytotoxic. Something that harms plants.

pickling. The immersion of all or part of a workpiece in a corrosive media such as acid to remove scale and related surface coatings.

pig. A container, usually lead, used to ship or store radioactive materials.

planing. Producing flat surfaces by linear reciprocal motion of the work and the table to which it is attached relative to a stationary single-point cutting tool.

plankton. Tiny plants and animals that live in water.

plant effluent or discharge after treatment. The waste water discharged from the industrial plant. In this definition, any waste treatment device (pond, trickling filter, etc.)is considered part of the industrial plant.

plasma arc machining. The term "plasma arc machining" shall mean the process of material removal or shaping of a workpiece by a high velocity jet of high-temperature ionized gas.

plated area. Surface upon which an adherent layer of metal is deposited.

plastics. Nonmetallic compounds that result from a chemical reaction and are molded or formed into rigid or pliable structural material.

plating. Forming an adherent layer of metal upon an object.

plume. Visible emission from a flue or chimney.

point source. A stationary location where pollutants are discharged, usually from an industry.

point source category. See category.

polishing. The process of removing stock from a workpiece by the action of loose or loosely held abrasive grains carried to the workpiece by a flexible support. Usually, the amount of stock removed in a polishing operation is only incidental to achieving a desired surface finish or appearance.

polishing compounds. Fluid or grease stick lubricants composed of animal tallows, fatty acids, and waxes. Selection depends on surface finish desired.

pollen. A fine dust produced by plants; a natural or background air pollutant.

pollutant. Any introduced substance that adversely affects the usefulness of a resource.

pollutant parameters. Those constituents of waste water determined to be detrimental and, therefore, requiring control.

pollution. The presence of matter or energy whose nature, location, or quantity produces undesired environmental effects.

polychlorinated Biphenyl (PCB). A family of chlorinated biphenyls with unique thermal properties and chemical inertness that have a wide variety of uses as plasticizers, flame retardants, and insulating fluids. They represent a persistent contaminant in waste streams and receiving waters.

polyectrolytes. Synthetic chemicals that help solids to clump during sewage treatment.

polyvinyl chloride. A plastic that releases hydrochloric acid when burned.

post curring. Treatment after changing the physical properties of a material by chemical reaction.

potable water. Appetizing water that is safe for drinking and use in cooking.

pouring. (Casting and molding) Transferring molten metal from a furnace or a ladle to a mold.

power brush finishing. This is accomplished (wet or dry) using a wire or nonmetallic-fiber-filled brush used for deburring, edge blending, and surface finishing of metals.

ppm. Parts per million; a way of expressing tiny concentrations. In air ppm is usually a volume/volume ratio; in water, a weight/volume ratio.

precious metals. Gold, silver, iridium, palladium, platinum, rhodium, ruthenium, indium, osmium, or combination thereof.

precipitate. A solid that separates from a solution because of some chemical or physical change.

precipitation hardening metals. Certain metal compositions that respond to precipitation hardening or aging treatment.

precipitators. Air pollution control devices that collect particles from an emission by mechanical or electrical means.

pressure deformation. The process of applying force (other than impact force) to permanently deform or shape a workpiece. Pressure deformation operations may include operations such as rolling, drawing, bending, embossing, coining, swaging, sizing, extruding, squeezing, spinning, seaming, piercing, necking, reducing, forming, crimping, coiling, twisting, winding, flaring, or weaving.

pressure filtration. The process of solid-liquid phase separation effected by passing the more permeable liquid phase through mesh that is impenetrable to the solid phase.

pretreatment. Processes used to reduce the amount of pollution in water before it enters the sewers or the treatment plant.

primary settling. The first treatment for the removal of settleable solids from waste water that is passed through a treatment works.

primary treatment. The first stage of waste water treatment; removal of floating debris and solids by screening and sedimentation.

printed circuit boards. A circuit in which the interconnecting wires have been replaced by conductive strips prined, etched, etc., onto an insulating board. Methods of fabrication include etched circuit, electroplating, and stamping.

printing. A process whereby a design or pattern in ink or types of pigments are impressed onto the surface of a part.

process modification. (In-plant technology) Reduction of water pollution by basic changes in a manufacturing process.

process waste water. Any water that, during manufacturing or processing, comes into direct contact with or results from the production or use of any raw material, intermediate product, finished product, by-product, or waste product.

process water. Water before its direct contact use in a process or operation. (This water may be any combination of raw water, service water, or either process waste water or treatment facility effluent to by recycled or reused).

process weight. The total weight of all materials, including fuel, used in a manufacturing process. It is used to calculate the allowable rate of emission of pollutant matter from the process.

pulverization. The crushing or grinding of materials into small pieces.

pumping station. A machine installed on sewers to pull the sewage uphill. In most sewer systems waste water flows by gravity to the treatment plant.

putrescible. A substance that can rot quickly enough to cause odors and attract flies.

pyrolysis. Chemical decomposition by extreme heat.

pyrazolone chlorimetric. A standard method of measuring cyanides in aqueous solutions.

quantity GPD. Gallons per day.

quenching. Rapid cooling of alloys by immersion in water, oil, or gasses after heating.

quench tank. A water-filled tank used to cool incinerator residues or hot materials during industrial processes.

racking. The placement of parts on an apparatus for the purpose of plating.

rack plating. Electroplating of workpieces on racks.

rad. A unit of measurement of any kind of radiation absorbed by humans.

radiography. A nondestructive method of internal examination in which metal or other objects are exposed to a beam of X-ray or gamma radiation. Differences in thickness, density, or absorption, caused by internal discontinuities, are apparent in the shadow image either on the fluorescent screen or on photographic film placed behind the object.

radiation. The emission of particles or rays by the nucleus of an atom.

radiation standards. Regulations that govern exposure to permissible concentrations of and transportation of radioactive materials.

radioactive. Substances that emit rays either naturally or as a result of scientific manipulation.

radiobiology. The study of the principles, mechanisms, and effects of radiation on living things.

radioecology. The study of the effects of radiation on plants and animals in natural communities.

radioisotopes. Radioactive forms of chemical compounds, such as cobalt-60, used in the treatment of diseases.

rasp. A machine that grinds waste into a manageable material and helps prevent odor.

raw sewage. Untreated waste water.

raw water. Plant intake water before any treatment or use.

reaming. An operation in which a previously formed hole is sized and contoured accurately by using a rotary cutting tool (reamer) with 1 or more cutting elements (teeth). The principal support of r, the reamer, during the cutting action is obtained from the workpiece. 1. Form reaming—reaming to a contour shape. 2. Taper reaming—using a special reamer for taper pins. 3. Hand reaming—using a long lead reamer that permits reaming by hand. 4. Pressure coolant reaming (or gun reaming)—using a multiple-lip, end-cutting tool through which coolant is forced at high pressure to flush chips ahead of the tool or back through the flutes for finishing of deep holes.

receiving waters. Any body of water where untreated wastes are dumped.

recharge. Process by which water is added to the zone of saturation, as in a recharge of an aquifer.

recirculating spray. A spray rinse in which the drainage is pumped up to the spray and is continually recirculated.

recycled water. Process waste water or treatment facility effluent that is recirculated to the same process.

recycle lagoon. A pond that collects treated waste water, most of which is recycled as process water.

recycling. Converting solid waste into new products by using the resources contained in discarded materials.

red tide. A proliferation of ocean plankton that may kill large numbers of fish. This natural phenomenon may be stimulated by the addition of nutrients.

reduction. A reaction in which there is a decrease in valence resulting from a gain in electrons.

redox. A term used to abbreviate a reduction-oxidation reaction.

refuge, wildlife. An area designated for the protection of wild animals, within which hunting and fishing is either prohibited or strictly controlled.

refuse. See solid waste.

refuse reclamation. Conversion of solid waste into useful products, e.g., composting organic wastes to make a soil conditioner.

rem. A measurement of radiation by biological effect on human tissue. (Acronym for roentgen equivalent man.)

rep. A measurement of radiation by energy development in human tissue. (Acronym for roentgen equivalent physical.)

reservoir. Any holding area, natural or artificial, used to store, regulate, or control water.

residual chlorine. The amount of chlorine left in the treated water that is available to oxidize contaminants.

resource recovery. The process of obtaining matter or energy from materials formerly discarded, e.g., solid waste, wood chips.

reverberation. The echoes of a sound that persist in an enclosed space after the sound source has stopped.

reverse osmosis. An advanced method of waste treatment that uses a semipermeable membrane to separate water from pollutants.

reused water. Process waste water or treatment facility effluent that is further used in a different manufacturing process.

Ringelman chart. A series of shaded illustrations used to measure the opacity of air pollution emissions. The chart ranges from light gray (number 1) through black (number 5) and is used to enforce emission standards.

ring rolling. A metals process in which a doughnut-shaped piece of stock is flattened to the desired ring shape by rolling between variably spaced rollers. The process produces a seamless ring.

rinse. Water for removal of dragout by dipping, spraying, fogging, etc.

riparian rights. Entitlement of a land owner to the water on or bordering his property, including the right to prevent diversion or misuse of it upstream.

river basin. The land area drained by a river and its tributaries.

riveting. Joining of 2 or more members of a structure by means of metal rivets, the unheaded end being upset after the rivet is in place.

rodenticide. A chemical or agent used to destroy rats or other rodent pests or to prevent them from damaging food, crops, etc.

rough fish. Those species not prized for game purposes or for eating, gar, suckers, etc. Most are more tolerant of changing environmental conditions than game species are.

routing. Cutting out and contouring edges of various shapes in a relatively thin material using a small diameter rotating cutter, which is operated at fairly high speeds.

rubbish. Solid waste, excluding food waste and ashes, from homes, institutions, and workplaces.

running rinse. A rinse tank in which water continually flows in and out.

runoff. Water from rain, snow melt, or irrigation that flows over the ground surface and returns to streams. It can collect pollutants from air or land and carry them to the receiving waters.

rust prevention compounds. Coatings used to protect iron and steel surfaces against corrosive environment during fabrication, storage, or use.

salinity. The degree of salt in water.

salt. 1. The compound formed when the hydrogen of an acid is replaced by a metal or its equivalent (e.g., an NH4 radical).
Example: $HCl + NaOH = NaCl + H2O$
This is typical of the general rule that the reaction of an acid and a base yields a salt and water. Most salts ionize in water solution. 2. Common salt, sodium chloride, occurs widely in nature, both as deposits left by ancient seas and in the ocean, where its average concentration is about 3 percent.

salt bath descaling. Removing the layer of oxides formed on some metals at elevated temperatures in a salt solution. See reducing, oxidizing, electrolytic.

salt water intrusion. The invasion of fresh surface or ground water by salt water. If the salt water comes from the ocean, it is called sea water intrusion.

salvage. The utilization of waste materials.

sand bed drying. The process of reducing the water content in a wet substance by transferring that substance to the surface of a sand bed and allowing the processes of drainage through the sand and evaporation to effect the required water separation.

sand blasting. The process of removing stock, including surface films, from a workpiece by the use of abrasive grains pneumatically impinged against the workpiece.

sand filtration. A process of filtering waste water through sand. The waste water trickles over the bed of sand, where air and bacteria decompose the wastes. The clean water flows out through drains in the bottom of the bed. The sludge accumulating at the surface must be removed from the bed periodically.

sanitary landfill, landfilling. Protecting the environment when disposing of solid waste. Waste is spread in thin layers, compacted by heavy machinery and covered with soil daily.

sanitary sewers. Underground pipes that carry only domestic or commercial waste, not stormwater.

sanitation. Control of physical factors in the human environment that can harm development, health, or survival.

sanitary water. The supply of water used for sewage transport and the continuation of such effluents to disposal.

save rinse. See dead rinse.

sawing. Using a toothed blade or disc to sever parts or cut contours. 1. Circular sawing—using a circular saw fed into the work by motion of either the workpiece or the blade. 2. Power band sawing—using a long, multiple-tooth continuous band, resulting in a uniform cutting action as the workpiece is fed into the saw. Power hace sawing—sawing in which a reciprocating saw blade is fed into the workpiece.

scale. Oxide and metallic residues.

scrap. Materials discarded from manufacturing operations that may be suitable for reprocessing.

screening. Use of racks of screens to remove course floating and suspended solids from sewage.

scrubber. An air pollution control device that uses a spray of water to trap pollutants and coal emissions.

secondary settling. Effluent from some prior treatment process flows for the purpose of removing settleable solids.

secondary treatment. Biochemical treatment of waste water after the primary stage, using bacteria to consume the organic wastes. Use of trickling filters or the activated sludge process, removes floating and settleable solids and about 90 percent of oxygen-demanding substances and suspended solids. Disinfection with chlorine is the final stage of secondary treatment.

sedimentation. Letting solids settle out of waste water by gravity during waste water treatment.

sedimentation tanks. Holding areas for waste water where floating wastes are skimmed off and settled solids are pumped out for disposal.

seepage. Water that flows through the soil.

selective pesticide. A chemical designed to affect only certain types of pests, leaving other plants and animals unharmed.

senescence. The aging process. It can refer to lakes in advanced stages of eutrophication.

sensitization. The process in which a substance other than the catalyst is present to facilitate the start of a catalytic reaction.

septic tank. An enclosure that stores and processes wastes where no sewer system exists, as in rural areas or on boats. Bacteria decompose the organic matter into sludge, which is pumped off periodically.

sequestering agent. An agent (usually a chemical compound) that "sequesters" or holds a substance in suspension.

series rinse. A series of tanks that can be individually heated or level controlled.

service water. Raw water that has been treated preparatory to its use in a process or operation; i.e., makeup water.

settleable solids. Materials heavy enough to sink to the bottom of waste water.

settling chamber.   A series of screens placed in the way of flue gases to slow the stream of air, thus helping gravity to pull particles out of the emission into a collection area.

settling ponds.   A large shallow body of water into which industrial waste waters are discharged. Suspended solids settle from the waste waters owing to the larger retention time of water in the pond.

settling tank.   A holding area for waste water, where heavier particles sink to the bottom and can be siphoned off.

sewage.   The organic waste and waste water produced by residential and commercial establishments.

sewage lagoon.   See lagoon.

sewer.   A channel that carries waste water and stormwater runoff from the source to a treatment plant or receiving stream. Sanitary sewers carry household and commercial waste. Storm sewers carry runoff from rain or snow. Combined sewers are used for both purposes.

sewerage.   The entire system of sewage collection, treatment, and disposal. Also applies to all effluent carried by sewers.

shaping.   Using single-point tools fixed to a ram reciprocated in a linear motion past the work. 1. Form shaping—shaping with a tool ground to provide a specific shape. 2. Contour shaping—shaping of an irregular surface, usually with the aid of a tracing mechanism. 3. Internal shaping—shaping of internal forms, such as keyways and guides.

shaving.   1. As a finishing operation, the accurate removal of a thin layer by drawing a cutter in straight-line motion across the work surfaces. 2. Trimming parts like stampings, forgings, and tubes to remove uneven sheared edges or to improve accuracy.

shearing.   The process of severing or cutting of a workpiece by forcing a sharp edge or opposed sharp edges into the workpiece, stressing the material to the point of sheer failure and separation.

shield.   A wall to protect people from exposure to harmful radiation.

shipping.   Transporting.

shot peening.   Dry, abrasive cleaning of metal surfaces by impacting the surfaces with high velocity steel shot.

shredding.   (cutting or stock removal) Material cut, torn or broken up into small parts.

SIC.   Standard Industrial Classification. Defines industries in accordance with the composition and structure of the economy and covers the entire field of economic activity.

significant deterioration.   Pollution from a new source in previously "clean" areas.

silica.   (SiO_2) Dioxide of silicon that occurs inc crystalline form as quartz, cristohalite, tridymite. Used in its pure form for high-grade refractories and high temperature insulators and in impure form (i.e., sand) in silica bricks.

siliconizing.   Diffusing silicon into solid metal, usually steel, at an elevated temperature for the purposes of case hardening, thereby providing a corrosion and wear-resistant surface.

silt.   Fine particles of soil or rock that can be picked up by air or water and deposited as sediment.

silviculture.   Management of forest land for timber. Sometimes contributes to water pollution, as in clear-cutting.

sinking.   Controlling oil spills by using an agent to trap the oil. Both sink to the bottom of the body of water and biodegrade there.

sintering.   The process of forming a mechanical part from a powdered metal by bonding under pressure and heat but below the melting point of the basis metal.

sizing.   1. Secondary forming or squeezing operations, required to square up, set, down, flatten, or otherwise correct surfaces to produce specified dimensions and tolerances. See restriking. 2. Some burnishing, broaching, drawing, and shaving operations are also called sizing. 3. A finishing operation for correcting ovality in tubing. 4. Powder metal. Final pressing of a sintered compact.

skimming.   Using a machine to remove oil or scum from the surface of the water.

slaking.   The process of reacting lime with water to yield a hydrated product.

sludge.   The concentration of solids removed from sewage during waste water treatment.

sludge dewatering. The removal of water from sludge by introducing the water sludge slurry into a centrifuge. The sludge is driven outward with the water remaining near the center. The water is withdrawn, and the dewatered sludge is usually landfilled.

slurry. A watery mixture of insoluble matter that results from some pollution control techniques.

smog. Air pollution associated with oxidants.

smoke. Particles suspended in air after incomplete combustion of materials containing carbon.

snagging. Heavy stock removal of superfluous material from a workpiece by using a portable or swing grinder mounted with a coarse grain abrasive wheel.

SO^x. The chemical symbol or oxides of sulfur.

soft detergents. Cleaning agents that break down in nature.

soil conditioner. An organic material like humus or compost that helps soil absorb water, build a bacterial community, and distribute nutrients and minerals.

solar energy. Power collected from sunlight, used most often for heating purposes but occasionally to generate electricity.

soldering. The process of joining metals by flowing a thin (capillary thickness) layer of nonferrous filler metal into the space between them. Bonding results from the intimate contact produced by the dissolution of a small amount of base metal. The term soldering is used where the temperature range falls below 425°C (800°F).

solids. (Plant waste) Residue material that has been completely dewatered.

solid waste. Useless, unwanted, or discarded material with insufficient liquid to be free-flowing.

solid waste disposal. The final placement of refuse that cannot be salvaged or recycled.

solid waste management. Supervised handling of waste materials from their source through recovery processes to disposal.

solute. A dissolved substance.

solution. Homogeneous mixture of 2 or more components, such as a liquid or a solid in a liquid.

solution treated. (Metallurgical) A process by which it is possible to dissolve micro-constituents by taking certain alloys to an elevated temperature and then keeping them in solution after quenching. (Often a solution treatment is followed by a precipitation or aging treatment to improve the mechanical properties. Most high temperature alloys that are solution treated and aged machine better in the solution treated state just before they are aged.

solvent. A liquid used to dissolve materials. In dilute solutions the component present in large excess is called the solvent, and the dissolved substance is called the solute.

solvent cleaning. Removal of oxides, soils, oils, fats, waxes, greases, etc., by solvents.

solvent degreasing. The removal of oils and grease from a workpiece by means of organic solvents or solvent vapors.

sonic boom. The thunderous noise made when shock waves reach the ground from a jet airplane exceeding the speed of sound.

soot. Carbon dust formed by incomplete combustion.

sorption. The action of soaking up or attracting substances; used in many pollution control processes.

specific conductance. The property of a solution that allows an electric current to flow when a potential difference is applied.

spectrophotometry. A method of analyzing a waste water sample by means of the spectra emitted by its constituents under exposure to light.

spinning. Shaping of seamless hollow cylindrical sheet metal parts by the combined forces of rotation and pressure.

spoil. Dirt or rock that has been removed from its original location, destroying the composition of the soil in the process, as with strip-mining.

spotfacing. Using a rotary, hole-piloted end facing tool to produce a flat surface normal to the axis of rotation of the tool on or slightly below the workpiece.

sprawl. Unplanned development of open land.

spray rinse. A process that utilizes the expulsion of water through a nozzle as a means of rinsing.

sputtering. The process of covering a metallic or nonmetallic workpiece with thin films of metal. The surface to be coated is bombarded with positive ions in a gas discharge tube, which is evacuated to a low pressure.

squeezing. The process of reducing the size of a piece of heated material so that it is smaller but more compressed than it was before.

stabilization. To convert the active organic matter in sludge into inert, harmless material.

stabilization ponds. See lagoon.

stable air. A mass of air that is not moving normally, so that it holds rather than disperses pollutants.

stack. A chimney or smokestack; a vertical pipe that discharges used air.

stack effect. Used air, as in a chimney, that moves upward because it is warmer than the surrounding atmosphere.

stagnation. Lack of motion in a mass of air or water, which tends to hold pollutants.

stainless steels. Steels that have good or excellent corrosion resistance. (One of the common grades contains 18 percent chromium and 8 percent nickel. There are 3 broad classes of stainless steels—ferritic, austenitic, and martensitic. The various classes are produced through the use of several alloying elements in differing quantities.)

staking. A general term covering almost all press operations. It includes blanking, shearing, hot or cold forming, drawing, bending and coining.

stamping compounds. See forming compounds (sheet).

standard of performance. Any restrictions established by the administrator pursuant to Section 306 of the act on quantities, rates, and concentrations of chemical, physical, biological, and other constituents that are or may be discharged from new sources into navigable waters, the waters of the contiguous zone or the ocean.

stannous salt. Tin-based compound used in the acceleration process. Usually stannous chloride.

stationary source. A pollution location that is fixed rather than moving. One point of pollution rather than widespread.

still rinse. See dead rinse.

storm sewer. A system that collects and carries rain and snow runoff to a point where it can soak back into the groundwater or flow into surface waters.

storm water lake. Reservoir for storage of storm water runoff collected from plant site; also, auxiliary source of process water.

stratification. Separating into layers.

stress relieved. The heat treatment used to relieve the internal stresses induced by forming or heat treating operations. (It consists of heating a part uniformly, followed by cooling slow enough so as not to reintroduce stresses. To obtain law stress levels in steels and cast irons, temperature as high at 1,250 degrees F may be required.)

strike. A thin coating of metal (usually less than 0.0001 inch in thickness) to be followed by other coatings.

strip mining. A process that uses machines to scrape soil or rock away from mineral deposits just under the Earth's surface.

stripping. The removal of coatings from metal.

stripcripping. Growing crops in a systematic arrangement of strips or bands that serve as barriers to wind and water erosion.

subcategory or subpart. A segment of a point source for which specific effluent limitations have been established.

submerged tube evaporation. Evaporation of feed material using horizontal steam-heat tubes submerged in solution. Vapors are driven off and condensed while concentrated solution is bled off.

subtractive circuitry. Circuitry produced by the selective etching of a previously deposited copper layer.

substrates. Thin coatings (as of hardened gelatin) that act as a support to facilitate the adhesion of a sensitive emulsion.

sulfur dioxide (SO_2).   A heavy pungent, colorless gas formed primarily by the combustion of fossil fuels. This major air pollutant is unhealthy for plants, animals, and people.

sump.   A depression or tank that catches liquid runoff for drainage or disposal, like a cesspool.

supersonic transport (SST).   A jet airplane that flies above the speed of sound; it may be extremely noisy upon takeoff and landing.

surface tension.   A measure of the force opposing the spread of a thin film of liquid.

surface waters.   Any visible stream or body of water.

surfactant.   A surface active chemical agent, usually made up of phosphates, used in detergents to cause lathering. The phosphates may contribute to water pollution.

surge.   A sudden rise to an excessive value, such as flow, pressure, temperature.

surveillance system.   A series of monitoring devices designed to determine environmental quality.

suspended solids (SS).   Tiny pieces of pollutants floating in sewage that cloud the water and require special treatment to remove.

swaging.   Forming a taper or a reduction on metal products, such as rod and tubing by forging, squeezing, or hammering.

synergism.   A cooperative action of 2 substances that results in a greater effect than both of the substances could have had acting independently.

systematic pesticide.   A chemical that is taken up from the ground or absorbed through the surface and carried through the systems of the organism being protected, making it toxic to pests.

tailings.   Residue of raw materials or waste separated out during the processing of crops or mineral ores.

tank.   A receptacle for holding, transporting, or storing liquids.

tapping.   producing internal threads with a cylindrical cutting tool having 2 or more peripheral cutting elements shaped to cut threads of the desired size and form. By a combination of rotary and axial motion, the leading end of the tap cuts the thread while the tap is supported mainly by the thread it produces.

tempering.   Reheating a quench-hardened or normalized ferrous alloy to a temperature below the transformation range then cooling at any rate desired.

teratogenic.   Substances that are suspected of causing malformations or serious deviations from the normal type, which cannot be inherited, in or on animal embryos or fetuses.

terracing.   Dikes built along the contour of agriculture land to hold runoff and sediment, thus reducing erosion.

tertiary treatment.   Advanced cleaning of waste water that goes beyond the secondary or biological stage. It removes nutrients, such as phosphorus and nitrogen, and most suspended solids.

testing.   The application of thermal, electrical, or mechanical energy to determine the suitability of functionality of a part, assembly, or complete unit.

thermal cutting.   The term "thermal cutting" shall mean the process of cutting, slotting, or piercing a workpiece using an oxyacetylene oxygen lance or an electric arc cutting tool.

thermal infusion.   The process of applying a fused zinc, cadmium, or other metal coating to a ferrous workpiece by imbuing the surface of the workpiece with metal powder or dust in the presence of heat.

thermal pollution.   Discharge of heated water from industrial processes that can affect the life processes of aquatic plants and animals.

thickener.   A device or system wherein the solid contents of slurries or suspensions are increased by gravity settling and mechanical separation fo the phases or by flotation and mechanical separation.

thickening.   (Sludge dewatering) Thickening or concentration is the process of removing water from sludge after the initial separation of the sludge from waste water. The basic objective of thickening is to reduce the volume of liquid sludge to be handled in subsequent sludge disposal processes.

threading.   Producing external threads on a cylindrical surface. 1. Die threading—a process for cutting external threads on cylindrical or tapered surfaces by the use of solid or self-opening dies. 2. Single-point threading—turning threads on a lathe. 3. Thread grinding—see definition under grinding. 4. Thread milling—a method of cutting screw threads with a milling cutter.

threshold dose. The minimum application of a given substance required to produce a measurable effect.

threshold toxicity. Limit upon which a substance becomes toxic or poisonous to a particular organism.

through hole plating. The plating of the inner surfaces of holes in a PC board.

tidal marsh. Low, flat marshlands traversed by interlaced channels and tidal sloughs and subject to tidal inundation; normally, the only vegetation present is salt-tolerant bushes and grasses.

titration. 1. A method of measuring acidity of alkalinity. 2. The determination of a constituent in a known volume of solution by the measured addition of a solution of known strength for completion of the reaction as signaled by observation of an end point.

tolerance. The ability of an organism to cope with changes in its environment. Also the safe level of any chemical applied to crops that will be used as food or feed.

tool steels. Steels used to make cutting tools and dies. Many of these steels have considerable quantities of alloying elements such as chromium, carbon, tungsten, molybdenum, and other elements. They form hard carbides that provide good wearing qualities but at the same time decrease machinability. Tool steels in the trade are classified for the most part by their applications, such as hot work die, cold work die, high speed, shock resisting, mold and special purpose steels.

topography. The physical features of a surface area including relative elevations and the position of natural and manmade features.

total chromium. The sum of chromium in all valences.

total cyanide. The total content of cyanide expressed as the radical CN− or alkali cyanide whether present as simple or complex ions. The sum of both the combined and free cyanide content of a plating solution. In analytical terminology, total cyanide is the sum of cyanide amenable to oxidation by chlorine and that which is not according to standard analytical methods.

total dissolved solids (TDS). The total amount of dissolved solid materials present in an aqueous solution.

total metal. Sum of the metal content in both soluble and insoluble form.

total organic carbon (TOC). TOC is a measure of the amount of carbon in a sample originating from organic matter only. The test is run by burning the sample and measuring the CO_2 produced.

total solids. The sum of dissolved and undissolved constituents in water or waste water, usually stated in milligrams per liter.

total suspended solids (TSS). Solids found in waste water or in the stream, which in most cases can be removed by filtration. The origin of suspended matter may be man-made or of natural sources, such as silt from erosion.

total volatile solids. Volatile residue present in waste water.

toxiccant. A chemical that controls pests by killing rather than repelling them.

toxicity. The degree of danger posed by a substance to animal or plant life.

toxic pollutants. A pollutant or combination of pollutants including disease causing agents, which after discharge and upon exposure, ingestion, inhalation or assimilation into any organism either directly or indirectly cause death, disease, cancer, genetic mutation, physiological malfunctions (including malfunctions in such organisms and their offspring).

toxic substances. A chemical or mixture that may present an unreasonable risk of injury to health or the environment.

treatment facility effluent. Treated process waste water.

trepanning. Cutting with a boring tool so designed as to leave an unmachined core when the operation is completed.

trickling filter. A biological treatment device; waste water is trickled over a bed of stones covered with bacterial growth, the bacteria break down the organic wastes in the sewage and produce cleaner water.

troposphere. The portion of the atmosphere between seven and ten miles from the Earth's surface, where clouds form.

tubidimeter. An instrument for measurement of turbidity in which a standard suspension is usually used for reference.

tumbling. See barrel finishing.

turbidimeter. A device that measures the amount of suspended solids in a liquid.

turbidity. Hazy air caused by the presence of particles and pollutants; a similar cloudy condition in water caused by suspended silt or organic matter.

turning. Generating cylindrical forms by removing metal with a single-point cutting tool moving parallel to the axis of rotation of the work. 1. Single-point turning—using a tool with 1 cutting edge. 2. Face turning—turning a surface perpendicular to the axis of the workpiece. 3. Form turning—using a tool with a special shape. 4. Turning cutoff—severing the workpiece with a special lathe tool. 5. Box tool turning—turning the end of workpiece with 1 or more cutters mounted in a boxlike frame, primarily for finish cuts.

ultrafiltration. A process using semipermeable polymeric membranes to separate molecular or colloidal materials dissolved or suspended in a liquid phase when the liquid is under pressure.

ultrasonic agitation. The agitation of a liquid medium through the use of ultrasonic waves.

ultrasonic cleaning. Immersion cleaning aided by ultrasonic waves that cause microagitation.

ultrasonic machining. Material removal by means of an ultrasonic-vibrating tool usually working in an abrasive slurry in close contact with a workpiece or having diamond or carbide cutting particles on its end.

unit operation. A single, discrete process as part of an overall sequence, e.g., precipitation, settling, and filtration.

urban runoff. Storm water from city streets, usually carrying litter and organic wastes.

vacuum deposition. Condensation of thin metal coatings on the cool surface of work in a vacuum.

vacuum evaporation. A method of coating articles by melting and vaporizing the coating material on an electrically heated conductor in a chamber from which air has been exhausted. The process is only used to produce a decorative effect. Gold, silver, copper and aluminum have been used.

vacuum filtration. A sludge dewatering process in which sludge passes over a drum with a filter medium and a vacuum is applied to the inside of the drum compartments. As the drum rotates, sludge accumulates on the filter surface and the vacuum removes water.

vacuum metalizing. The process of coating a workpiece with metal by flash heating metal vapor in a high-vacuum chamber containing the workpiece. The vapor condenses on all exposed surfaces.

vapor. The gaseous phase of substances that are liquid or solid at atmospheric temperature and pressure, such as steam.

vapor blasting. A method of roughing plastic surfaces in preparation for plating.

vapor degreasing. Removal of soil and grease by a boiling liquid solvent, the vapor being considerably heavier than air. At least 1 constituent of the soil must be soluble in the solvent.

vaporization. The change of a substance from a liquid to a gas.

vapor plating. Deposition of a metal or compound upon a heated surface by reduction or decomposition of a volatile compound at a temperature below the melting points of either the deposit or the basis material.

vapor plumes. Flue gases that are visible because they contain water droplets.

variance. Government permission for a delay or exception in the application of a given law, ordinance, or regulation.

vector. An organism, often an insect, that carries disease.

vinyl chloride. A chemical compound used in producing some plastics. Excessive exposure to the substance may cause cancer.

viscosity. The resistance offered by a real fluid to a shear stress.

volatile. Any substance that evaporates at a low temperature.

volatile substances. Material that is readily vaporizable at a relatively low temperature.

Matter Incorporated by Reference

Matter incorporated by reference is available from the following organizations:

1. ASME: American Society of Mechanical Engineers, United Engineering Center, 345 East 47th Street, New York, NY 10017.
2. American National Standard: American National Standards Institute, Inc., 1430 Broadway, New York, NY 10018.
3. CGA: Compressed Gas Association, Inc., 500 Fifth Avenue, New York, NY 10036.
4. Bureau of Explosives: Association of American Railroads, American Railroads Building, 1920 L Street N.W., Washington, DC 20036.
5. AAR: Association of American Railroads, 59 East Van Buren Street, Chicago, IL 60605.
6. ASTM: American Society for Testing and Materials, 1916 Race Street, Philadelphia, PA 19103.
7. API: American Petroleum Institute, 1801 K Street, N.W., Washington, DC 20006.
8. AISI: American Iron and Steel Institute, 1000 16th Street N.W., Washington, DC 20036.
9. The Chlorine Institute, 342 Madison Avenue, New York, NY 10017.
10. MCA: Manufacturing Chemists' Association, Inc., 1825 Connecticut Avenue, N.W., Washington, DC 20009.
11. NFPA: National Fire Protection Association, 60 Batterymarch Street, Boston, MA 02110.
12. Aluminum Association: The Aluminum Association, 420 Lexington Avenue, New York, NY 10017.
13. NACE: National Association of Corrosion Engineers, 2400 West Loop South, Houston, TX 77027.
14. IME: Institute of Makers of Explosives, 420 Lexington Ave., New York, NY 10017.
15. IAEA: International Atomic Energy Agency, Karnter Ring 11, P.O. Box 590, A-1011, Vienna, Austria (IAEA) publications may be purchased in the United States from: Unipub, Inc., P.O. Box 433, New York, NY 10016.
16. USDOE: United States Department of Energy, Washington, DC 20545. Regulations of the USDOE are available from the Superintendent of Documents, U.S. Government Printing Office, Washington, DC 20402. Other publications by the USDOE may be obtained from the National Technical Information Center, U.S. Department of Commerce, Springfield, VA 22151.
17. Superintendent of Documents, U.S. Government Printing Office, Washington, DC 20402.
18. National Wooden Box Association, P.O. Box 1010, Cumberland, MD 21502.
19. TFI: The Fertilizer Institute, 1015 18th Street, N.W., Washington, DC 20036.
20. AWWA: American Water Works Association, 2 Park Avenue, New York, NY 10016.
21. AWS: American Welding Society, 345 East 47th Street, New York, NY 10016.
22. USDC: U.S. Department of Commerce, National Technical Information Service, 5285 Port Royal Road, Springfield, VA 22151.
23. Inter-governmental Maritime Consultative Organization, 101-104 Piccadilly, London, WIV OAE, England.

volumetric method. A standard method of measuring settleable solids in a aqueo solution.

waste. Unwanted materials left over from manufacturing processes, refuse from pla of human or animal habitation.

waste discharged. The amount (usually expressed as weight) of some residual substa which is suspended or dissolved in the plant effluent.

wastewater. Water carrying dissolved or suspended solids from homes, farms, b nesses and industries.

wastewater constituents. Those materials which are carried by or dissolved in a w stream for disposal.

water pollution. The addition of enough harmful or objectionable material to da water quality.

water quality criteria. The levels of pollutants that affect use of water for drir swimming, raising fish, farming or industrial use.

water quality standard. A management plan that considers (1) what water will b for, (2) setting levels to protect those uses, (3) implementing and enforcing the treatment plans, and (4) protecting existing high-quality waters.

water supply system. The collection, treatment, storage, and distribution of water from source to consumer.

water table. The level of ground water.

zooplankton. Tiny animals that fish feed on.

25. USERDA: United States Energy Research and Development Administration, Washington, DC 20545.
26. USNRC: United States Nuclear Regulatory Commission, Washington, DC 20555.
27. UN: United Nations: United Nations Sales Section, New York, NY 10017.
28. OPPSD: Organic Peroxide Producers' Safety Division, Society of the Plastic Industries, Inc., 355 Lexington Avenue, New York, NY 10017.
29. ISO: International Organization for Standardization, Case Postale 56, CH-1211 Geneva 20, Switzerland. Also available from the American National Standards Institute, Inc., 1430 Broadway, New York, NY 10018.

Information Concerning The Regulations

The Department of Transportation, like all Federal agencies, publishes its new regulations and amendments to its existing regulations in a publication called the Federal Register. The Federal Register is published Monday through Friday by the Office of the Federal Register, an agency of the General Services Administration. It contains not only final rules but also executive orders of the President, proposed rules, legal and informational notices, and other Federal agency documents of public interest. The Federal Register is generally sold on a semi-annual and annual mail subscription basis. Copies of individual issues are also available. Information regarding Federal Register subscriptions and individual issues can be obtained from the Superintendent of Documents, U.S. Government Printing Office, Washington, D.C. 20402 or call (202) 783-3238.

GPO Bookstores Across the Country

ALABAMA
9220-B Parkway East
Birmingham, AL 35206
FTS 229-1056
COM (205) 254-1056

CALIFORNIA
ARCO Plaza, C-Level
505 South Flower Street
Los Angeles, CA 90071
FTS 798-5841
COM (213) 688-5841

CALIFORNIA
Room 1023, Fed. Bldg.
450 Golden Gate Avenue
San Francisco, CA 94102
FTS 556-6657
COM (415) 556-0643

COLORADO
Room 117, Fed. Bldg.
1961 Stout Street
Denver, CO 80294
FTS 327-3964
COM (303) 837-3964

COLORADO
World Savings Building
720 North Main Street
Pueblo, CO 81003
FTS 323-9371
COM (303) 544-3142

FLORIDA
Room 158, Fed. Bldg.
400 West Bay Street
Jacksonville, FL 32202
FTS 946-3801
COM (904) 791-3801

GEORGIA
Room 100, Fed. Bldg.
275 Peachtree St., N.E.
Atlanta, GA 30303
FTS 242-6946
COM (404) 221-6947

ILLINOIS
Room 1365, Fed. Bldg.
219 S. Dearborn Street
Chicago, IL 60604
FTS 353-5133
COM (312) 353-5133

MASSACHUSETTS
Room G-25, Fed. Bldg.
Sudbury Street
Boston, MA 02203
FTS 223-6071
COM (617) 223-6071

MICHIGAN
Suite 160, Fed. Bldg.
477 Michigan Avenue
Detroit, MI 48226
FTS 226-7816
COM (313) 226-7816

MISSOURI
Room 144, Fed. Bldg.
601 East 12th Street
Kansas City, MO 64106
FTS 758-2160
COM (816) 374-2160

NEW YORK
Room 110, Fed. Bldg.
26 Federal Plaza
New York, NY 10278
FTS 264-3825
COM (212) 264-3825

OHIO
First Floor, Fed. Bldg.
1240 East Ninth Street
Cleveland, OH 44199
FTS 942-4922
COM (216) 522-4922

OHIO
Room 207, Fed. Bldg.
200 N. High Street
Columbus, OH 43215
FTS 943-6956
COM (612) 469-6956

PENNSYLVANIA
Room 1214, Fed. Bldg.
660 Arch Street
Philadelphia, PA 19106
FTS 597-0677
COM (215) 597-0677

PENNSYLVANIA
Room 118, Fed. Bldg.
1000 Liberty Avenue
Pittsburgh, PA 15222
FTS 722-2721
COM (412) 644-2721

TEXAS
Room 1050, Fed. Bldg.
1100 Commerce Street
Dallas, TX 75242
FTS 729-0076
COM (214) 767-0076

TEXAS
9319 Gulf Freeway
Houston, TX 77017
FTS 626-7515
COM (713) 229-3515/16

WASHINGTON
Room 194, Fed. Bldg.
915 Second Avenue
Seattle, WA 88174
FTS 339-4270
COM (206) 442-4270

WISCONSIN
Room 190, Fed. Bldg.
517 E. Wisconsin Avenue
Milwaukee, WI 53202
FTS 362-1300
COM (414) 291-1304

WASHINGTON, D.C.
MAIN BOOKSTORE
710 N. Capitol St., N.W.
Washington, D.C. 20401
FTS 275-2091
COM (202) 275-2091

PENTAGON BOOKSTORE
Main Concourse, South End
Room 2E172
Washington, D.C 20310
FTS 557-1821
COM (703) 557-1821

COMMERCE BOOKSTORE
14th and E Streets, N.W.
Room 1604, First Floor
Washington, D.C. 20230
FTS 377-3527
COM (202) 377-3527

STATE BOOKSTORE
Room 2817, North Lobby
21st and C Streets, N.W.
Washington, D.C. 20520
FTS 632-1437
COM (202) 632-1437

HSS BOOKSTORE
Room 1528, HHS N. Bldg.
330 Independence Ave S.W.
Washington, D.C. 20201
FTS 472-74778
COM (202) 472-7478

RETAIL SALES OUTLET-
  LAUREL
8660 Cherry Lane
Laurel, MD 20707
(301) 953-7974

U.S. Environmental Protection Agency Regional Asbestos NESHAPs Contacts

(For information on NESHAPs rule compliance and disposal)

Region 1
Asbestos NESHAPs Contact
Air Management Division
USEPA
JFK Federal Building
Boston, MA 02203
(617) 223-4872

Region 2
Asbestos NESHAPs Contact
Air & Waste Management Division
USEPA
26 Federal Plaza
New York, NY 10007
(212) 264-2611

Region 3
Asbestos NESHAPs Contact
Air Management Division
USEPA
841 Chestnut Street
Philadelphia, PA 19107
(215) 597-6552

Region 4
Asbestos NESHAPs Contact
Air, Pesticide & Toxic Management
USEPA
345 Courtland Street N.E.
Atlanta, GA 30365
(404) 881-3067

Region 5
Asbestos NESHAPs Contact
Air Management Division
USEPA
230 S. Dearborn Street
Chicago, IL 60604
(312) 886-6793

Region 6
Asbestos NESHAPs Contact
Air & Waste Management Division
USEPA
1201 Elm Street
Dallas, TX 75270
(214) 767-9869

Region 7
Asbestos NESHAPs Contact
Air & Waste Management Division
USEPA
726 Minnesota Avenue
Kansas City, KS 66101
(913) 236-2834

Region 8
Asbestos NESHAPs Contact
Air & Waste Management Division
USEPA
1860 Lincoln Street
Denver, CO 80295
(303) 844-3763

Region 9
Asbestos NESHAPs Contact
Air Management Division
USEPA
215 Fremont Street
San Francisco, CA 94105
(415) 974-7648

Region 10
Asbestos NESHAPs Contact
Air & Toxics Management Division
USEPA
1200 Sixth Avenue
Seattle, WA 98101
(206) 442-2724

U.S. Environmental Protection Agency Regional Asbestos Coordinators

(For information on asbestos identification, health effects, abatement options, analytic techniques, monitoring, asbestos in schools, and contract documents)

Region 1
Regional Asbestos Coordinator
USEPA
JFK Federal Building
Boston, MA 02202
(617) 223-0585

Region 2
Regional Asbestos Coordinator
USEPA
Woodbridge Avenue
Edison, NJ 08837
(201) 321-6668

Region 3
Regional Asbestos Coordinator
USEPA
841 Chestnut Street
Philadelphia, PA 19107
(215) 597-9859

Region 4
Regional Asbestos Coordinator
USEPA
345 Courtland Street, N.E.
Atlanta, GA 30365
(404) 881-3864

Region 5
Regional Asbestos Coordinator
USEPA
230 S. Dearborn Street
Chicago, IL 60604
(312) 886-6879

Region 6
Regional Asbestos Coordinator
USEPA
First International Building
1291 Elm Street
Dallas, TX 75270
(214) 767-5314

Region 7
Regional Asbestos Coordinator
USEPA
726 Minnesota Avenue
Kansas City, KS 66101
(913) 236-2838

Region 8
Regional Asbestos Coordinator
USEPA
1860 Lincoln Street
Denver, CO 80295
(303) 837-3926

Region 9
Regional Asbestos Coordinator
USEPA
215 Fremont Street
San Francisco, CA 94105
(415) 454-8588

Region 10
Regional Asbestos Coordinator
USEPA
1200 Sixth Avenue
Seattle, WA 98101
(206) 442-2870

Regional Community Involvement Contacts for Dioxin

Contact		Region
Debra Prybla Office of Public Affairs U.S. EPA Region 1 JFK Federal Building Boston, MA 02203	(617) 223-4906	Connecticut, Maine, Massachusetts, New Hampshire, Rhode Island, Vermont
Richard Cahill Office of Public Affairs U.S. EPA Region 2 26 Federal Plaza New York, NY 10007	(212) 264-2515	New Jersey, New York, Puerto Rico, Virgin Islands
Joe Donovan Office of Public Affairs U.S EPA Region 3 6th and Walnut Sts. Phila., PA 19106	(215) 597-9370	Delaware, Maryland Pennsylvania, Virginia West Virginia, District of Columbia
Hagan Thompson Office of Public Affairs U.S. EPA Region 4 345 Courtland St., NE Atlanta, GA 30308	(404) 881-3004	Alabama, Georgia, Florida, Mississippi, North Carolina, South Carolina, Tennessee, Kentucky
Vanessa Musgrave Office of Public Affairs U.S. EPA Region 5 230 S. Dearborn Chicago, IL 60604	(312) 886-6128	Illinois, Indiana, Ohio, Michigan, Wisconsin, Minnesota
Betty Williamson Office of Public Affairs U.S. EPA Region 6 1201 Elm St. Dallas, TX 75270	(214) 767-9986	Arkansas, Louisiana, Oklahoma, Texas, New Mexico
Steven Wurtz Office of Public Affairs U.S. EPA Region 7 324 E. 11th ST. Kansas City, MO 64106	(816) 374-5894	Iowa, Kansas, Missouri, Nebraska

Nola Cook   (303) 837-5927   Colorado, Utah
Office of Public Affairs   Wyoming, Montana,
U.S. EPA Region 8   North Dakota, South
Suite 900   Dakota
1860 Lincoln St.
Denver, CO 80295

Deanna Wieman   (415) 974-8083   Arizona, California,
Office of External Affairs   Nevada, Hawaii,
U.S. EPA Region 9   American Samoa,
215 Fremont St.   Guam
San Francisco, CA 94105

Bob Jacobson   (206) 442-1203   Alaska, Idaho, Oregon,
Office of Public Affairs   Washington
U.S. EPA Region 10
1200 Sixth Ave.
Seattle, WA 98101

State Solid and Hazardous Waste Agencies

Alabama
Daniel E. Cooper, Director
Land Division
Alabama Dept. of Environmental
  Management
1751 Federal Drive
Montgomery, Alabama 36130
CML (205) 271-7730

Alaska
Stan Hungerford
Air & Solid Waste Management
Dept of Environmental Conservation
Pouch O
Juneau, Alaska 99811
CML (907) 465-2635

American Samoa
Pati Faiai, Executive Secretary
Environmental Quality Commission
American Samoa Government
Pago Pago, American Samoa 96799

Overseas Operator (Commercial Call
  633-4116)

Randy Morris, Deputy Director
Department of Public Works
Pago Pago, American Samoa 96799

Arizona
Ron Miller, Manager
Office of Waste and Water Quality
  Management
Arizona Department of Health Services
2005 North Central Avenue
Phoenix, Arizona 85004
CML (602) 257-2305

Arkansas
Vincent Blubaugh, Chief
Solid & Hazardous Waste Div.
Department of Pollution Control and
  Ecology
P.O. Box 9583
8001 National Drive
Little Rock, Arkansas 72219
CML (501) 562-7444

California
Vacant, Deputy Director
Toxic Substances Control Programs
Department of Health Services
714 P Street, Room 1253
Sacramento, California 95814
CML (916) 322-7202

Michael Campos, Executive Director
State Water Resources Control Board
P.O. Box 100
Sacramento, California 95801
CML (916) 445-1553

Sherman E. Roodzant, Chairman
California Waste Management Board
1020 Ninth Street, Suite 300
Sacramento, California 95814
CML (916) 322-3330

Colorado
Kenneth Waesche, Director
Waste Management Division
Colorado Department of Health
4210 E. 11th Ave.
Denver, Colorado 80220
CML (303) 320-8333

**Commonwealth Of Northern Mariana
  Islands**
George Chan, Administrator
Division of Environmental Quality
Department of Public Health and
  Environmental Services
Commonwealth of the Northern Mariana
  Islands
Saipan, CM 96950

Overseas Operator: 6984
Cable address: GOV. NMI Saipan

Connecticut
Stephen Hitchock, Director
Hazardous Material Management Unit
Department of Environmental Protection

State Office Building
165 Capitol Ave.
Hartford, Connecticut 06106
CML (203) 566-4924

Michael Cawley,
Connecticut Resource Recovery
   Authority
179 Allyn St., Suite 603
Professional Building
Hartford, Connecticut 06103
CML (203) 549-6390

Delaware
William Razor, Supervisor
Solid Waste Management Branch
Department of Natural Resources and
   Environmental Control
89 Kings Highway
P.O. Box 1401
Dover, Delaware 19901
CML (302) 736-4781

District Of Columbia
Angelo Tompros, Chief
Department of Consumer & Regulatory
   Affairs
Pesticides & Hazardous Waste
   Management
Room 112
5010 Overlook Avenue, S.W.
Washington, D.C. 20032
CML (202) 767-8422

Florida
Robert W. McVety, Administrator
Solid & Hazardous Waste Section
Department of Environmental
   Regulation
Twin Towers Office Building
2600 Blair Stone Rd.
Tallahassee, Florida 32301
CML (904) 488-0300

Georgia
John Taylor, Chief
Land Protection Branch
Environmental Protection Division
Department of Natural Resources
270 Washington St. S.W., Room 723
Atlanta, Georgia 30334
CML (404) 656-2833

Guam
James Branch, Administrator
Guam Environmental Protection Agency
P.O. Box 2999
Agana, Guam 96910

Overseas Operator (Commercial Call
   646-8863)

Hawaii
Melvin Koizumi, Deputy Director
Environmental Health Division
Department of Health
P.O. Box 3378
Honolulu, Hawaii 96801

California FTS Operator
8-556-0220
CML (808) 548-4139

Idaho
Steve Provant, Supervisor
Hazardous Materials Bureau
Department of Health and Welfare
   State House
Boise, Idaho 83720
CML (208) 334-2293

Illinois
Robert Kuykendall, Manager
Division of Land Pollution Control
Environmental Protection Agency
2200 Churchill Rd. Room A-104
Springfield, Illinois 62706
CML (217) 782-6760

William Child, Deputy Manager
Division of Land Pollution Control
Environmental Protection Agency
2200 Churchill Rd. Room A-104
Springfield, Illinois 62706
CML (217) 782-6760

Indiana
David Lamm, Director,
Land Pollution Control Division
State Board of Health
1330 West Michigan Street
Indianapolis, Indiana 46206
CML (317) 633-0169

Iowa
Ronald Kolpa
Hazardous Waste Program Coordinator
Dept. of Water, Air & Waste Mgmt.
Henry A. Wallace Building
900 East Grand
Des Moines, Iowa 50319
CML (515) 281-8925

Kansas
Dennis Murphey, Manager
Bureau of Waste Management
Department of Health and Environment
Forbes Field, Building 321
Topeka, Kansas 66620
CML (913) 862-9360

Kentucky
J. Alex Barber, Director
Division of Waste Management
Department of Environmental Protection

Cabinet for Natural Resources and
  Environmental Protection
18 Reilly Rd.
Frankfort, Kentucky 40601
CML (502) 564-6716

Louisiana
Gerald J. Healy, Administrator
Solid Waste Management Division
Department of Environmental Quality
P.O. Box 44307
Baton Rouge, Louisiana 70804
CML (504) 342-1216
Glenn Miller, Administrator
Hazardous Waste Management Division
Department of Environmental Quality
P.O. Box 44307
Baton Rouge, Louisiana 70804
CML (504) 342-1227

Maine
David Boulter, Director
Licensing and Enforcement Division
Bureau of Oil & Hazardous Materials
Department of Environmental Protection

State House — Station 17
Augusta, Maine 04333
CML (207) 289-2651

Maryland
Bernard Bigham
Waste Management Administration
Department of Health & Mental Hygiene
201 W. Preston Street, Room 212
Baltimore, Maryland 21201
CML (301) 225-5649
Alvin Bowles, Chief
Hazardous Waste Division
Waste Management Administration
Department of Health & Mental Hygiene
201 W. Preston Street
Baltimore, M.D. 21201

Ronald Nelson, Director
Waste Management Administration
Office of Environmental Programs
Department of Health & Mental Hygiene
201 West Preston Street - Room 212
Baltimore, Maryland 21201
CML (301) 225-5647

Massachusetts
William Cass, Director
Division of Solid & Hazardous Waste
Department of Environmental Quality
  Engineering
One Winter Street
Boston, Massachusetts 02108
CML (617) 292-5589

Michigan
Delbert Rector, Chief
Hazardous Waste Division
Environmental Protection Bureau
Department of Natural Resources
Box 30028
Lansing, Michigan 48909
CML (517) 373-2730
Allan Howard, Chief
Technical Services Section
Hazardous Waste Division
Department of Natural Resources
Box 30028
Lansing, Michigan 48909
CML (517) 373-8448

Minnesota
Dale L. Wikre, Director
Solid and Hazardous Waste Division
Pollution Control Agency
1935 West County Rd. B-2
Roseville, Minnesota 55113
CML (612) 296-7282

Mississippi
Jack M. McMillan, Director
Division of Solid & Hazardous Waste
  Management
Bureau of Pollution Control
Department of Natural Resources
P.O. Box 10385
Jackson, Mississippi 39209
CML (601) 961-5062

Missouri
Dr. David Bedan, Director
Waste Management Program
Department of Natural Resources
117 East Dunklin Street
P.O. Box 176
Jefferson City, MO 65102
CML (314) 751-3241

Montana
Duane L. Robertson, Chief
Solid Waste Management Bureau
Department of Health and
   Environmental Sciences
Cogswell Bldg.
Helena, Montana 59602
CML (406) 444-2821

Nebraska
Mike Steffensmeier
Section Supervisor
Hazardous Waste Management Section
Department of Environmental Control
State House Station
P.O. Box 94877
Lincoln, Nebraska 68509
CML (402) 471-2186

Nevada
Verne Rosse
Waste Management Program Director
Division of Environmental Protection
Department of Conservation and Natural
   Resources
Capitol Complex
201 South Fall Street
Carson City, Nevada 89710
CML (702) 885-4670

New Hampshire
Dr. Brian Strohm, Assistant Director
Division of Public Health Services
Office of Waste Management
Department of Health and Welfare
Health and Welfare Building
Hazen Drive
Concord, New Hampshire 03301
CML (603) 271-4608

New Jersey
Dr. Marwan Sadat, Director
Division of Waste Management
Department of Environmental Protection
32 E. Hanover Street, CN-027
Trenton, New Jersey 08625
CML (609) 292-1250

New Mexico
Richard Perkins, Acting Chief
Groundwater & Hazardous Waste
   Bureau
Environmental Improvement Division
N.M. Health & Environment Department
P.O. Box 968
Sante Fe, New Mexico 87504-0968
CML (505) 984-0020

Peter Pache, Program Manager
Hazardous Waste Section
Groundwater & Hazardous Waste
   Bureau
Environmental Improvement Division
N.M. Health and Environment
   Department
P.O. Box 968
Santa Fe, New Mexico 87504-0968
CML (505) 984-0020 Ext 340

New York
Norman H. Nosenchuck, Director
Division of Solid & Hazardous Waste
Department of Environmental
   Conservation
50 Wolf Rd., Room 209
Albany, New York 12233
CML (518) 457-6603

North Carolina
William L. Meyer, Head
Solid & Hazardous Waste Management
   Branch
Division of Health Services
Department of Human Resources
P.O. Box 2091
Raleigh, North Carolina 27602
CML (919) 733-2178

North Dakota
Martin Schock, Director
Division of Hazardous Waste
Management and Special Studies
Department of Health
1200 Missouri Ave., 3rd floor
Bismarck, North Dakota 58501
CML (701) 224-2366

Ohio
Steven White, Chief
Divison of Solid & Hazardous Waste
   Management
Ohio EPA
361 East Broad Street
Columbus, Ohio 43215
CML (614) 466-7220

Oklahoma
Dwain Farley, Chief
Waste Management Service
Oklahoma State Dept. of Health
P.O. Box 53551
1000 N.E. 10th St.
Oklahoma City, Ok. 73152
CML (405) 271-5338

Oregon
Mike Downs, Administrator
Hazardous & Solid Waste Division
Department of Environmental Quality
P.O. Box 1760
Portland, Oregon 97207
CML (503) 229-5356

Pennsylvania
Donald A. Lazarchik, Director
Bureau of Solid Waste Management
Department of Environmental Resources

Fulton Building - 8th floor
P.O. Box 2063
Harrisburg, PA 17120
CML (717) 787-9870

Puerto Rico
Santos Rohena, Director
Solid, Toxics, & Hazardous Waste
  Program
Environmental Quality Board
P.O. Box 11488
Santurce, Puerto Rico 00910-1488
CML (809) 725-0439

Rhode Island
John S. Quinn, Jr., Chief
Solid Waste Management Program
Department of Environmental
  Management
204 Cannon Building
75 Davis Street
Providence, Rhode Island 02908
CML (401) 277-2797

South Carolina
Robert E. Malpass, Chief
Bureau of Solid and Haz. Waste Mgtm.
S.C. Dept of Health & Environmental
  Control
2600 Bull Street
Columbia, South Carolina 29201
CML (803) 758-5681

South Dakota
Joel C. Smith, Administrator
Office of Air Quality & Solid Waste
Department of Water & Natural
  Resources
Joe Foss Building
Pierre, South Dakota 57501
CML (605) 773-3329

Tennessee
Tom Tiesler, Director
Division of Solid Waste Management
Bureau of Environmental Services
Tennessee Department of Public Health
150 9th Ave, North
Nashville, Tennessee 37203
CML (615) 741-3424

Texas
Jack Carmichael, Chief
Bureau of Solid Waste Management
Texas Department of Health
1100 West 49th Street, T-602
Austin, Texas 78756-3199
CML (512) 458-7271

Jay Snow, Chief
Solid Waste Section
Texas Department of Water Resources
1700 North Congress
P.O. Box 13087, Capitol Station
Austin, Texas 78711
CML (512) 463-8177

Utah
Dale Parker, Director
Bureau of Solid and Hazardous Waste
  Management
Department of Health
P.O. Box 2500
150 West North Temple
Salt Lake City, Utah 84110
CML (801) 533-4145

Vermont
Richard A. Valentinetti, Director
Air and Solid Waste Programs
Agency of Environmental Conservation
State Office Building
P.O. Box 489
Montpelier, Vermont 05602
CML (802) 828-3395

Virgin Islands
Robert V. Eepoel, Director
Hazardous Waste Program
Division of Natural Resources
Department of Conservation and
   Cultural Affairs
P.O. Box 4340, Charlotte Amalie
St. Thomas, Virgin Islands 00801

D.C. Overseas Operator 472-6620
CML (809) 774-6420

Virginia
William F. Gilley, Director
Division of Solid and Hazardous Waste
   Management
Virginia Department of Health
Monroe Building 11th floor
101 North 14th Street
Richmond, Virginia 23219
CML (804) 225-2667

Washington
Earl Tower, Supervisor
Solid & Hazardous Waste Mgmt. Division

Department of Ecology
Olympia, Washington 98504
CML (206) 459-6316

Linda L. Brothers, Assistant Director
Office of Hazardous Substance & Air
   Quality Programs
Department of Ecology
Olympia, Washington 98504
CML (206) 459-6253

West Virginia
Timothy Laraway, Chief
Division of Water Resources
Department of Natural Resources
1201 Greenbrier Street
Charleston, West Virginia 25311
CML (304) 348-5935

Wisconsin
Paul Didier, Director
Bureau of Solid Waste Management
Department of Natural Resources
P.O. Box 7921
Madison, Wisconsin 53707
CML (608) 266-1327

Wyoming
Charles Porter, Supervisor
Solid Waste Management Program
State of Wyoming
Department of Environmental Quality
Equality State Bank Building
401 West 19th St.
Cheyenne, Wyoming 82002
CML (307) 777-7752

Publications of the U.S. Environmental Protection Administration.

AIR POLLUTION

AUTOMOTIVE IMPORTS—FACT SHEET. October 1982.
CARBON MONOXIDE STUDY BOISE, ID. (Parts 1 & 2) 1978.
CARBON MONOXIDE STUDY SPOKANE, WA. (Executive Summary). February 1981.
DIESEL EMISSION STANDARDS. (Environmental Information Sheet). April 1979.
DO YOU OWN A CAR? April 1980.
GAS MILEAGE GUIDE 1985.
IF YOU LIVE IN A HIGH-ALTITUDE AREA, BUY A HIGH-ALTITUDE CAR. December 1979.
I/M FACT SHEET. June 1984.
MECHANICS . . . A NEW LAW AFFECTS YOU. April 1980.
TRENDS IN THE QUALITY OF THE NATION'S AIR. August 1984.
UNLEADED GASOLINE: THE ONLY WAY. (Revised) April 1984.
UNLEADED GASOLINE: THE ONLY WAY. (Spanish) 1979.
WOOD HEATING AND AIR POLLUTION. June 1984.
WOOD STOVE FEATURES AND OPERATION GUIDELINE FOR CLEANER AIR. September 1983.

GENERAL SUBJECTS

ACCOMPLISHMENTS REPORT 1982.
ASSESSING AND MANAGING RISKS IN THE REAL WORLD. March 25, 1985. (Thomas Speech)
COMMON ENVIRONMENTAL TERMS. (Reprinted) May 1982.
COMPOSTING: A VIABLE METHOD OF RESOURCE RECOVERY. June 1984.
A CONSERVING SOCIETY. (EPA Journal Reprint). March 1980.
CONSUMER'S RESOURCE HANDBOOK, 1984 EDITION.
ENVIRONMENTAL ORGANIZATIONS DIRECTORY. 1983.
ENVIRONMENTAL PROGRESS: A NEW AGENDA. September 1983. (Ruckelshaus Speech)
GET THE MOST FROM YOUR GAS HEATING DOLLAR. August 1979.
GET THE MOST FROM YOUR HEATING OIL DOLLAR. (Revised) November 1984.
HEALTH EFFECTS OF LAND TREATMENT—IS IT REALLY SAFE? March 1980.
MAJOR ACTIONS AND DECISIONS. May 1983–May 1984.
MT. ST. HELENS TECHNICAL INFORMATION BULLETINS (FEMA) 1980.
THE NEXT FOUR YEARS: AN AGENDA FOR ENVIRONMENTAL RESULTS. April 3, 1985. (Thomas Speech)
NORTHWEST ENVIRONMENT. U.S. EPA Region 10. Latest edition.
OUR CHALLENGE. November 1983. (Ruckelshaus Speech)
PRESIDENT'S ENVIRONMENTAL YOUTH AWARDS. 1984.
ROSTER OF MINORITY AND WOMEN ARCHITECTS AND ENGINEERS. (Revised) May 1981.
ROSTER OF MINORITY AND WOMEN CONSTRUCTION CONTRACTORS & SUPPLIERS. (Revised) May 1981.

ROSTER OF MINORITY AND WOMEN ARCHITECTS AND ENGINEERS, CON-
STRUCTION CONTRACTORS AND SUPPLIERS (Addendum to). April 1982.
YOUR GUIDE TO THE ENVIRONMENTAL PROTECTION AGENCY. August 1984.

HAZARDOUS WASTE

EVERYBODY'S PROBLEM: HAZARDOUS WASTE. 1980.
INCINERATION OF CHEMICAL WASTES AT SEA. August 1983.
REQUIREMENTS FOR SMALL QUANTITY HAZARDOUS WASTE GENERATORS:
QUESTIONS AND ANSWERS. March 1985.
SMALL QUANTITY HAZARDOUS WASTE GENERATORS: THE NEW RCRA RE-
QUIREMENTS. March 1985.

PUBLIC LAWS AND RELATED INFORMATION

CLEAN AIR ACT AS AMENDED THROUGH JULY 1981.
CLEAN WATER ACT AS AMENDED THROUGH DECEMBER 1981.
COMPREHENSIVE ENVIRONMENTAL RESPONSE, COMPENSATION, & LIABILITY
ACT OF 1980. (CERCLA) Superfund.
FEDERAL INSECTICIDE, FUNGICIDE, AND RODENTICIDE ACT AS AMENDED.
Revised May 1985. (FIFRA).
THE FEDERAL INSECTICIDE, FUNGICIDE, AND RODENTICIDE ACT AS AMENDED.
Revised May 1985.
FEDERAL PESTICIDE ACT OF 1978: HIGHLIGHTS. August 1979.
HAZARDOUS AND SOLID WASTE AMENDMENTS OF 1984. November 8, 1984.
HIGHLIGHTS OF THE CLEAN AIR ACT AMENDMENTS OF 1977. April 1979.
REGULATIONS FOR IMPLEMENTING THE PROCEDURAL PROVISIONS OF THE
NATIONAL ENVIRONMENTAL POLICY ACT. (NEPA) Council on Environmental
Quality reprint. November 29, 1978.
RESOURCE CONSERVATION AND RECOVERY ACT OF 1976, AMENDED OCTOBER
3, 1984.
SAFE DRINKING WATER ACT AS AMENDED THROUGH DECEMBER 1980.
TOXIC SUBSTANCES CONTROL ACT. (Description) July 1984.
TOXIC SUBSTANCES CONTROL ACT. (TSCA) October 1976.

SPANISH LANGUAGE

ABOUT PESTICIDES. April 1978.
FARMWORKERS PESTICIDE SAFETY. (Spanish and English) April 1980.
KEEP POISON BAITS OUT OF CHILDREN'S REACH. 1980.
PESTICIDE SAFETY FOR FARMWORKERS. (Spanish and English) April 1985.
UNLEADED GASOLINE: THE ONLY WAY. 1979.
YOUR WORLD, YOUR ENVIRONMENT. May 1979.

SUPERFUND

CERCLA: GETTING INTO THE ACT. CONTRACTING AND SUBCONTRACTING
OPPORTUNITIES IN THE SUPERFUND PROGRAM. November 1984.
EPA'S EMERGENCY RESPONSE PROGRAM (2ND EDITION). April 1984.

NATIONAL PRIORITIES LIST. 786 CURRENT AND PROPOSED SITES IN ORDER OF RANKING AND BY STATE. October 1984.
SUPERFUND: WHAT IT IS, HOW IT WORKS. June 1985.
SUPERFUND'S REMEDIAL RESPONSE PROGRAM, 2nd EDITION. October 1983.

TOXIC SUBSTANCES

ASBESTOS (TOXICS INFORMATION SERIES). April 1980.
ASBESTOS WASTE MANAGEMENT GUIDANCE: GENERATION, TRANSPORT, DISPOSAL. May 1985.
DIOXIN FACTS: ANSWERS TO COMMONLY ASKED QUESTIONS. July 1984.
ENVIRONMENTAL FACTS: METHYL ISOCYANATE. (MIC) January 1985.
FARM WORKER'S PESTICIDE SAFETY. (Spanish and English) APril 1980.
FARMERS' RESPONSIBILITIES UNDER THE FEDERAL PESTICIDE LAW. August 1977.
INTEGRATED PEST MANAGEMENT. (Research Summary) September 1980.
KEEP POISON BAITS OUT OF CHILDREN'S REACH. (Spanish and English) February 1980.
LEAKING UNDERGROUND STORAGE TANKS CONTAINING MOTOR FUELS: A CHEMICAL ADVISORY. September 1984.
LEARNING MORE ABOUT DIOXIN. July 1984.
MORE ABOUT LEAKING UNDERGROUND STORAGE TANKS: A BACKGROUND BOOKLET FOR THE CHEMICAL ADVISORY. October 1984.
PCB'S (TOXIC INFORMATION SERIES). June 1980.
PCB'S IN FLUORESCENT LIGHT FIXTURES A FACT SHEET. January 1985.
PESTICIDE SAFETY FOR FARMWORKERS. (Spanish and English) APril 1985.
PESTICIDE WARNING. January 1977.
RECOGNITION AND MANAGEMENT OF PESTICIDE POISONINGS. January 1982.
REGULATING PESTICIDES. (EPA Journal Reprint) June 1984.
REGULATING PESTICIDES. (Pamphlet).
SAFE USE OF SOIL FUMIGANTS. July 1978.
SIX STEPS TO SAFER USE OF PESTICIDES.
STUDY SITE CATEGORIES IN EPA'S NATIONAL DIOXIN STRATEGY (DIOXIN FACTS). July 1984.
SUSPENDED, CANCELLED AND RESTRICTED PESTICIDES. January 1985.
TOXIC CHEMICALS WHAT THEY ARE, HOW THEY AFFECT YOU.

WATER POLLUTION

CATTLE FEEDLOTS IN THE PACIFIC NORTHWEST.
DRINKING WATER IN THE BACKCOUNTRY.
EPA'S ROLE IN DREDGE AND FILL PERMITS.
FOREST CHEMICALS AND WATER QUALITY. 1978.
GROUND-WATER PROTECTION STRATEGY. August 1984.
GROUNDWATER CONTAMINATION FROM LEAKING UNDERGROUND STORAGE TANKS. April 1985.
INFORMATION ON SPCC (SPILL PREVENTION ON CONTROL AND COUNTERMEASURE) PLANS. 40 CFR 112. October 1984.
IS YOUR DRINKING WATER SAFE? March 1985.
OIL SPILLS. (Research Summary) February 1979.
A PRACTICAL TECHNOLOGY: COMPOSTING, A VIABLE METHOD OF RESOURCE RECOVERY. June 1984.
A PRACTICAL TECHNOLOGY: HYDROGRAPH CONTROLLED RELEASE LAGOONS—A PROMISING MODIFICATION. July 1984.

PRIMER FOR WASTEWATER TREATMENT. July 1980.
PROTECTING GROUND WATER: THE HIDDEN RESOURCE. July/August 1984.
PUGET SOUND ALLIANCE.
PUGET SOUND BIBLIOGRAPHY. Region 10 Library. February 1985.
PUGET SOUND NOTES. U.S. EPA Region 10, office of Puget Sound. Latest edition.
STORAGE AND RAFTING IN PUBLIC WATER. 1971.

Bibliography

"Alkaline Zinc Bath Solves Low-CD Problems", *Products Finishing*, Aug., 1980.

Allen, Paul, "Reclaiming Four Plating Solutions", *Products Finishing*, Aug., 1979.

"A Low-Cost Answer to Oil Recycling?", *Factory Management*, January, 1977, pp. 32–33.

Atimion, Leo, "A Program of Conservation, Pollution Abatement", *Plating and Surface Finishing*, March, 1980.

A/V Training Systems, "Hazardous Materials and Wastes", The Operations Council of The American Trucking Assns., ATA, 1616 P Street NW, Washington, DC, 20036.

Barrett, F., "The Electroflotation of Organic Wastes", *Chemistry and Industry*, October 16, 1976, 99. pp. 880–882.

Basily, William, "Industrial Waste Water Treatment Facility, Charleston Plant", General Electric, April 5, 1978.

Belinke, Robert J., "Central Filtration for Coolants", *American Machinists*, December 1976, pp. 86–88.

Bell, John P., "How to Remove Metals from Plating Rinse Waters", *Products Finishing*, Aug., 1979.

Bellis, H.E., and Pearlstein, F., "Electroless Plating of Metals", AES Illustrated Lecture Series, *American Electroplaters Society Inc.*, Winter Park, FL, 1972.

Bolster, Maurice, "How to Maintain Emulsion Coolant Systems", *Modern Machine Shop*, March, 1977, pp. 112–115.

Bowes, H. David, "In-House Solvent Reclamation Eliminates Quality Problems at Low Cost", *Plastics Design & Processing*, May, 1978, pp. 20–32.

Bradner, W., "Asbestos Exposure Assessment in Buildings Inspection Manual", EPA Region 7, Kansas City, MO, October, 1982.

Breslou, Barry R., et al. "Hollow Fiber Ultrafiltration Technology", *Ind. Water Eng.*, Jan./Feb., 1980.

Buaks, S.V., and Dresser, K.J., *Cleaning Alternatives to Solvent Degreasing*, EPA, December 7, 1978.

Carson, Rachel L., *Silent Spring*, Fawcett Crest Books, 1962, 304 pp.

Chin, D.T., and Echert, B., "Destruction of Cyanide Wastes with a Packed-Bed Electrode", *Plating and Surface Finishing*, October, 1976, pp. 38–41.

Chonisby, J., and Kuhn, D., "Practical Oil Reclamation, Purification", *Hydraulics & Pneumatics*, April, 1976, pp. 71–73.

Chua, John P., "Coolant Filtration Systems", *Plant Engineering*, December 23, 1976, pp. 46–51.

"Consumer's Resource Handbook", Published by the U.S. Office of Consumer Affairs, The White House, Washington, DC.

"Dangerous Properties of Industrial Materials", by N. Irving Sax, Van Nostrand Reinhold Co.

DeLatour, Christopher, "Magnetic Separation in Water Pollution Control", *IEEE Transactions on Magnetics*, Volume Mag-9, No. 3, September, 1973, p. 314.

Development Document for Effluent Limitations Guidelines and Standards for the Metal Finishing-Point Source Category by Effluent Guidelines Division, EPA, 440/1-83/091, June, 1983.

Dinius, B., "How to Choose an In-Plant Oil Reclamation System", *Hydraulics and Pneumatics*, July, 1978, pp. 62–64.

Directory of National Environmental Organization, U.S. Environmental Directories, P.O. Box 65156, St. Paul, MN 55165.

"DOD Hazardous Materials Information System Procedures", DOS 6050.5-M, July, 1986.

"Electrotechnology Volume 1, Wastewater Treatment and Separation Methods", Cheremisinoff, Paul N., King, John A., Oullette, Robert P., Ann Arbor Science Publishers, Inc., Ann Arbor, MI, 1978.

"Electrostatic Separation of Solids from Liquids", *Filtration & Separation*, March/April, 1977, pp. 140–144.

Emergency Response Guidebook, 1984, DOT P 5800.3.

"Emerging Technologies for Treatment of Electroplating Wastewaters", for presentation by Stinson, M.K., at AICHE 71st Annual Meeting, Session 69, Miami Beach, FL, November 15, 1978.

Energy/Environment III, October, 1978, EPA-600/9-78-002, Proceedings of the Third National Conference on the EPA-600/0-77-033, 29 pages.

EPA-600/9-77-041, 76 pages, "Alaskan Oil Transportation Issues", October, 1977.

EPA-600/9-77-019, 11 pages, "Oil Shale and the Environment", October, 1977.

EPA-U.S. Directory of Environmental Sources, 4th Edition, National Focal Point of the United Nations Environment Program/International Referral System for Sources of Environmental Information (INFOTERRA), EPA 840-81-011.

Erichson, Paul R., and Throop, William M., "Alkaline Treatment System Reduces Pollution Problems", *Industrial Wastes*, March/April, 1977.

Flynn, B.L. Jr., "Wet Air Oxidation of Waste Streams", *CEP*, April, 1979, pp. 66–69.

Ford, Davis L., and Elton, Richard L., "Removal of Oil and Grease from Industrial Wastewaters", *Chemical Engineering/Deskbook Issue*, October 17, 1977, pp. 49–56.

"Forest Harvesting and Water Quality", EPA 625/5-76-013, available free of charge from Technology Transfer, U.S. EPA, Cincinnati, OH, 45268 (18 page brochure).

"Forest Harvest, Residue Treatment, Reforestation and Protection of Water Quality", EPA 910/90-76-020, April, 1976, through NTIS, PB 253393/AS, ($9.25) Springfield, VA, 22161.

Glover, Harry C., "Are Emulsified Solvents Safer Cleaners?", *Production Engineering*, July, 1978, pp. 41–43.

Graham, A. Kenneth, *Electroplating Engineering Handbook*, 1971, pp. 152–176.

Gransky, Michael, "The Case for Electrodialysis", *Products Finishing*, April, 1980.

Grutsch, James F., and Mallatt, R.C., "Optimizing Granular Media Filtration", *GEP*, April, 1977, pp. 57–66.

Halva, C.J., and Rothschild, B.F., "Plating and Finishing of Printed Wiring/Circuit Boards", AES Illustrated Lecture Series, American Electroplaters Society, Inc., Winter Park, FL, 1976.

"Handbook of Environmental Data on Organic Chemicals", Karel Verschueren, Van Nostrand Reinhold Company, New York, NY, 1977.

Harrison, Albert, "Coil Coater Cuts Effluent Treatment Costs", *Products Finishing*, November, 1980.

Henry, Joseph D. Jr., Lowler, Lee F., and Kuo, C.H. Alex, "A Solid/Liquid Separation Process Based on Cross Flow and Electrofiltration", *AIChE Journal*, Volume 23, No. 6, November, 1977, pp. 851–859.

Hazardous Materials Transportation, Cahners Publishing Company, P.O. Box 716, Back Bay Annex, Boston, MA, 02117.

"Hazardous Waste Training Manual for Supervisors", Harold Swanson, Bureau of Law & Business, Inc., 114 Manhattan Street, Box 1274, Stamford, CT., 06904.

Hochenberry, H.R., and Lieser, J.E., *Practical Application of Membrane Techniques of Waste Oil Treatment*, presented at the 31st Annual Meeting in Philadelphia, PA, May 10–13, 1976, American Society of Lubrication Engineers, Reprint Number 76-AM-28-2.

Hubbell, F.N., "Chemically Deposited Composites—A New Generation of Electroless Coatings", *Plating and Surface Finishing*, American Electroplaters Society, E. Orange, NJ, Vol. 65, Dec., 1978, p. 48.

Humenich, Michael J., and Davis, Barry J., "High Rate Filtration of Refinery Oily Wastewater Emulsions", *Journal WPCF*, August, 1978, pp. 1953–1964.

Hura, LCdr Myron, USN and Mittleman, John, "High Capacity Oil-Water Separator", *Naval Engineers Journal*, December, 1977, pp. 55–62.

"Identification of Components of Energy-Related Wastes and Effluents", January, 1978., EPA-600/7-78-004. 524 pages.

"In Process Pollution Abatement—Upgrading Metal Finishing Facilities to Reduce Pollution", EPA Technology Transfer Seminar Publication, EPA, July, 1973.

"Ion Transfer Method Developed for Metal Plating", *Industrial Finishing*, Hitchcock Publishing Co., Wheaton, OH, April, 1979, p. 95.

Jackson, Lloyd, C., "How to Select a Substrate Cleaning Solvent", *Adhesives Age*, April, 1977, pp. 23–31.

Jackson, Lloyd C., "Removal of Silicone Grease and Oil Contaminants", *Adhesives Age*, April, 1977, pp. 29–32.

Jackson, Lloyd C., "Solvent Cleaning Process Efficiency", *Adhesives Age*, July, 1976, pp. 31–34.

Johnson, Ross E. Jr., *Wastewater Treatment and Oil Reclamation at General Motors, St. Catherines*, pp. 345–357.

Kelley, Ralph, "The Use of Cutting Fluids and Their Effect on Cutting Tools and Grinding Wheels in Solving Production Problems", Cincinnati Milacron/Products Division.

Kim, K.S. and Kuivinen, D.E., "Assessment of Potential Exposure to Friable Insulation Materials Containing Asbestos", NASA Technical Memorandum 81435, April, 1980.

Kitagews, T. and Nishikawa, Y. and Frankenfeld, J.W. and, Liw IMIMW "Wastewater Treatment by Liquid Membrane Process", *Environmental Science Technology*, Volume 11, No. 6, June 1977, pp. 602–605.

Kohn, Philip M., "Photo-Processing Facility Achieves Zero Discharge", *Chem. Eng.*, Dec. 4, 1978.

Kolm, Henry H., "The Large-Scale Manipulation of Small Particles", *IEEE Transactions on Magnetics*, Vol. Mag-11, No. 5, Sept. 1975, pp. 1567–1569.

Kremer, Lawrence N., "Prepaint Final Rinses: Chrome or Chrome-Free?", *Products Finishing*, Nov., 1980.

Ksotura, John D., "Recovery and Treatment of Plating and Anodizing Wastes", *Plating and Surface Finishing*, Aug., 1980.

Lancy, L. E., "Metal Finishing Waste Treatment Aims Accomplished by Process Changes", Chemical Engineering Progress Symposium Series, Vol. 67, 1971, pp. 439–441.

Lancy, L. E., and Steward, F. A., "Disposal of Metal Finishing Sludges—The Segregated Landfill Concept", *Plating and Surface Finishing*, American Electroplaters Society, E. Orange, NJ, Vol. 65, Dec. 1978, p. 14.

Lawes, B. S., and Stevens, W. F., "Treatment of Cyanide and Chromate Rinses", AES Illustrated Lecture Series, American Electroplaters Society, Inc., Winter Park, FL, 1972.

Levadie, B., ed., "Definitions for Asbestos and Other Health-Related Silicates," ASTM, Philadelphia, PA, ASTM Special Technical Publication 834, PCN 04-834000-17, July 1984.

Lewis, Tom A., "How to Electrolytically Recover Metals from Finishing Operations", *Industrial Finishing*, April, 1980.

"Logging Roads and Protection of Water Quality", EPA 910/9-75-007, March 1975, through NTIS, PB 243703/AS, (&9.75) Springfield, VA 22161.

"Logging Roads and Water Quality", EPA 625/5-76-011, available free of charge from Technology Transfer, U.S. EPA, Cincinnati, OH, 45268.

Lowenheim, Frederick A., *Electroplating-Fundamentals of Surface Finishing*, McGraw-Hill, Inc., New York, NY, 1978.

Maloney, J. E., "Low Temperature Cleaning", *Metal Finishing*, June, 1976, pp. 33–35.

Mazzeo, D. A., "Energy Conservation In Plating and Surface Finishing", *Plating and Surface Finishing*, American Electroplaters Society, Inc., Winter Park, FL, July, 1979, pp. 10–12.

McNutt, James E., "Electroplating Waste Control", *Plating and Surface Finishing*, July, 1980.

McNutt, J. E., and Swalheim, D. A., "Recovery and Re-use of Chemicals in Plating Effluents", AES Illustrated Lecture Series, American Electroplaters Society, Inc., Winter Park, FL, 1975.

Metal Cleaning Fundamentals, Materials and Methods, Oakite Products, Inc., F 10646R13-379.

Miranda, Julio G., "Designing Parallel-Plates Separators", *Chemical Engineering*, January 31, 1977.

Mohler, J. B., "The Art and Science of Rinsing", AES Illustrated Lecture Series, American Electroplaters Society, Inc., Winter Park, FL, 1973.

Nakayama, S., et al. "Improved Ozonation in Aqueous Systems", *Ozone: Science and Engineering*, Pergamon Press, 1979.

Natale, A., and H. Levins, "Source Removal and Control", Source Finders and Information Corp., Voorhees, NJ, 1984.

National Wildlife Federation, 1412 Sixteenth St., NW, Washington DC, 20036, Conservation Directory 1985 (A list of organizations, agencies, and officials concerned with National Resource Use and Management.)

Novak, Fred, "Destruction of Cyanide Wastewater by Ozonation", Paper presented at the International Ozone Assn. Conf., Nov., 1979.

Oberteuffer, John A., "High Gradient Magnetic Separation", IEEE Transaction on Magnetics, Volume Mag-9, No. 3, September 1973, pp. 303-306.

Obrzut, John J., "Metal Cleaning Bends with Social Pressures", *Iron Age*, February 17, 1974, pp. 41-44.

Official Air Transport Restricted Articles Tariff No. 6-D, Airline Tariff Publishing Co.

Okamato, S., "Iron Hydroxide as Magnetic Scavengers", Institute of Physical and Chemical Research, Waho-shi, Saitama-hen, 351 Japan.

"Oil Audit and Reuse Manual for the Industrial Plant", Illinois Institute of Natural Resources, Project No. 80.085, Document No. 78/35, November 1978.

Oil Pollution Reports Vol. 5, No 3 (June 1978-September 1978), November 1978. EPA-600/ 7-78-218. (Since July 1974, EPA has published 17 of these quarterly reports. A list of back issues is available from the Oil and Hazardous Materials Spills Branch, U.S. EPA, Edison, NJ, 08817).

"Oil/Water Splitter Snags Emulsified Oil", *Chemical Engineering*, July 18, 1977, p. 77.

"Organic Solvent Cleaning-Background Information for Proposed Standards", US EPA, EPA-450/2-78-045, May 1979.

Ostrow, R., and Kessler, R. B., "A Technical and Economic Comparison of Cyanide and Cyanide-Free Zinc Plating", *Plating*, American Electroplaters Society, Hackensack, NJ, April 1970.

Oulman, Charles S., and Baumann, Robert E., "Polyelectrolyte Coatings for Filter Media", *Industrial Water Engineering*, May 1971, pp. 22-25.

"Packaging Regulations", by Stanley Sacharow, 206 pp, The Center for Professional Advancement, East Brunswick, NJ.

Parker, Konrad, "Renewal of Spent Electroless Nickel Plating Baths", *Plating and Surface Finishing*, March, 1980.

Pearlstein, F. et. al., "Testing and Evaluation of Deposits", AES Illustrated Lecture Series, American Electroplaters Society, Inc., Winter Park, FL, 1974.

"Physiochemical Processes for Water Quality Control", Wiley-Interscience Series, Walter, J. Wever, Jr., John Wiley and Sons, Inc., New York, NY, 1972.

Pietrzak, J., *Unit Operation Discharge Summary for the Mechanical Products Category*, EPA, September 7, 1979.

Pinto, Steven, D., *Ultrafiltration for Dewatering of Waste Emulsified Oils*, Lubrication Challenges in Metalworking and Processing Proceedings, First International Conference, IIT Research Institute, Chicago, IL 60616, USA, June 7-9, 1978.

Piper, S. and M. Grant, NESHAPs "Asbestos Demolition and Renovation Inspection Report", GCA Corporation, Bedford, MA, under EPA Contract No. 68-02-3961, August 1984.

"Plating on Plastics Etchants Regenerated", *Products Finishing*, May, 1979.

"Pollution Control 1978", *Products Finishing*, Gardner Publications, Inc., Cincinnati, OH, August, 1978, pp. 39–41.

Quanstrom, Richard L., "Central Coolant Systems-Closing the Loop on Metalworking Fluids", *Lubrication Engineering*, January 1977, Volume 33, 1, pp. 14–19.

Rajagopal, I., and Rajam, K. S., "A New Addition Agent for Lead Plating", *Metal Finishing*, Metals and Plastics Publication Inc., Hackensack, NJ, December, 1978.

Rasquin, Edgar A. and Lynn, Scott and Hanson, Donald N., "Vacuum Steam Stripping of Volatole, Sparingly Soluble Organic Compounds from Water Streams", *Ind. Eng. Chemical Fundam.*, Vol. 17, No. 3, 1978, pp. 170–174.

"Recovery Pays at Sommer Metalcraft", *Industrial Finishing*, June, 1980.

"Recycling Etchant for Printed Circuits", *Metal Finishing*, Metals and Plastics Publications Inc., Hackensack, NJ, March 1972, pp. 42–43.

Research Highlights 1978, December 1978, EPA-600/9-78-040, 70 pages.

Research Outlook, February 1979, EPA-600/0-79-005. 140 pages.

Rice, Rip G., "Ozone for Industrial Water & Wastewater Treatment", Paper presented at WWEMA Industrial Pollution Control Conf., June, 1980.

Roberts, David A., "Romicon Ultrafiltration for Waste Oil Reclamation", Paper presented to the Water Pollution Control Association of Pennsylvania, June 15, 1977.

Roberts, Vicki, "A Low-Cyanide Zinc for Champion Spark Plugs". *Products Finishing*, Jan. 1979.

Roebuck, A. H., "Safe Chemical Cleaning—The Organic Way", *Chemical Engineering*, July 31, 1978, pp. 107–110.

Rose, Betty A., "Bulk Plater Saves with Evaporative Recovery", *Industrial Finishing*, Jan. 1979.

Rose, Betty A., "Managing Water at Helicopter Plant", *Industrial Finishing*.

Sachs, T. R., "Diversified Finisher Handles Complex Waste Treatment Problem", *Plating and Surface Finishing*, American Electroplaters Society, E. Orange, NJ, Vol. 65, Dec. 1978, p. 36.

Safety Management Planning, Dr. Ted S. Ferry, The Merritt Company, 1661 Ninth Street, P.O. Box 955, Santa Monica, CA 90406.

Seng, W. C., and Kreutzer, G. M., "Resume of Total Operation of Waste Treatment Facility for Animal and Vegetable Oil Refinery", Reprinted from the *Journal of the American Oil Chemists' Society*, Volume 52, No. 1, 1975, pp. 9A–13A.

Shah, B. and Langdon, W., and Wasan, D., "Regeneration of Fibrous Bed Coalescers for Oil-Water Separation", *Environmental Science and Technology*, Volume 11, No. 2, February 1977, pp. 167–170.

Shambaugh, Robert T. and Nelhyh, Peter B., "Removal of Heavy Metals via Ozonation", *Journal WPCF*, Jan. 1978, pp. 113–121.

"Silvicultural Chemicals and Protection of Water Quality", EPA 910/9-77-036, June 1977, through NTIS, PB 271923/AS, (&8.00) Springfield, VA 22161.

"Simple Treatment for Spent Electroless Nickel", *Products Finishing*, Feb., 1981.

"Simple Dragout Recovery Methods", *Products Finishing*, Oct., 1979.

"Slide into Compliance", *Industrial Finishing*, Dec., 1979.

Staebler, C. J., and Simpers, B. F., "Corrosion Resistant Coatings with Low Water Pollution Potential", presented at the EPA/AES First Annual Conference on Advanced Pollution Control for the Metal Finishing Industry, Lake Buena Vista, FL, January 17–19, 1978.

Sundaram, T. R., and Santo, J. E., "Removal of Suspended and Collodial Solids from Waste Streams by the Use of Cross-Flow Microfiltration", *American Society of Mechanical Engineers*, 77-ENAs-51.

"System Strips Solvents, Separates Solids Simultaneously", *Chemical Engineering*, November 22, 1976, pp. 93–94.

Tang, T. L. Don, "Application of Membrane Technology to Power Generation Waters", *Industrial Water Engineering*, Jan./Feb., 1981.

Taller, R. A., and Koleske, J. V., "Energy Conservation in Metal Pretreatment and Coating Operations", *Metal Finishing*, August 1977, pp. 18–19.

Taylor, J. W., "Evaluation of Filter/Separators and Centrifuges for Effects on Properties of Steam Turbine Lubricating Oils", *Journal of Testing and Evaluation*, Volume 5, No. 5, September 1977, pp. 401–405.

Teale, James M., "Fast Payout from In-Plant Recovery of Spent Solvents", *Chemical Engineering*, January 31, 1977, pp. 98–100.

The Condensed Chemical Dictionary, Revised by G. G. Hawley, Van Nostrand Reinhold Co.

"The Electrochemical Removal of Trace Metals for Metal Wastes with Simultaneous Cyanide Destruction", for presentation by H.S.A. Reactors Limited at the First Annual APA/AES Conference on Advanced Pollution Control for the Metal Finishing Industry, Dutch Inn, Lake Buena Vista, FL., Jan. 18, 1978.

The Foundation of the Wall and Ceiling Industry, Washington, DC, "Guide Specifications for the Abatement of Asbestos Release from Spray-or-Trowel-Applied Materials in Buildings and Other Structures", December, 1981.

"Toxic Materials Transport", Business Publishers, Inc., 951 Pershing Drive, Silver Spring, MD, 20910.

"Treating Electroless Plating Effluent", *Products Finishing*, Aug., 1980.

Ukawa, Hiroshi, Koboyashi, Kaseimaza, and Iwata, Minoru, "Analysis of Batch Electrokinetic Filtration", *Journal of Chemical Engineering of Japan*, Volume 9, No. 5, 1976, pp. 396–401.

U.S. Department of Commerce, National Bureau of Standards, "Guidelines for Assessment and Abatement of Asbestos-Containing Materials in Buildings", Center for Building Technology, Washington, DC, NBSIR-83-2688, May, 1983.

U.S. Department of Health, Education and Welfare, "Asbestos Exposure", National Cancer Institute, Bethesda, MD, DHEW Publication No. (NIH) 78-1622, May, 1978.

"Used Oil Recycling in Illinois", Data Book, Illinois Institute of Natural Resources, Project No. 80.085, Document No. 78/34, October, 1978.

U.S. Environmental Protection Agency, "Asbestos-Containing Materials in School Buildings: A Guidance Document", Parts 1 and 2, Office of Toxic Substances, Washington, DC, EPA 450/2-78-014, March, 1979.

U.S. Environmental Protection Agency, "Guidance for Controlling Friable Asbestos-Containing Materials in Buildings", Office of Pesticides and Toxic Substances, Washington, DC, EPA 560/5-83-002, March, 1983.

U.S. Occupational Safety and Health Administration, "Preliminary Regulatory Impact and Regulatory Flexibility Analysis of the Proposed Revisions to the Standard for Regulating Occupational Exposure to Asbestos", PB84-198225, 30 March, 1984.

U.S. Postal Service Publications 1) "Domestic Mail Manual", 121, 123, 124, and 125. 2) Publication 6, "Radioactive Materials". 3) Publication 14, "Plant Quarantines". 4) "International Mail Manual".

Vucich, M. G., "Emulsion Control and Oil Recovery on the Lubricating System of Double-Reduction Mills", *Iron and Steel Engineer*, December 1976, pp. 29–38.

Wahl, James R., Hayes, Thomas C., Kleper, Myles H., and Pinto, Steven D., *Ultrafiltration for Today's Oily Wastewaters: A Survey of Current Ultrafiltration Systems*, presented at the 34th Annual Purdue Industrial Waste Conference, May 8–10, 1979.

"Waste Oil Reclamation", *The Works Managers Guide to Working Fluid Economy*, Alfa-Laval No. 1B40494 E2.

"Waste Oil Recycling—Coming Up a Winner", *Fluid and Lubricant Ideas*, Volume 2, Issue 3, Summer, 1979, p. 8.

"Water and Environmental Technology, Vol. 11.01/Water", ASTM, ISBN 0-8031-0684-X.

Wing, R. E. and others, "Treatment of Complexed Copper Rinsewaters with Insoluble Starch Xanthate", *Plating and Surface Finishing,* Dec., 1978.

"Workbook of Atmospheric Dispersion Estimates" by D. Bruce Turner, EPA, Research Triangle Park, North Carolina.

Yost, Kenneth J., and Scarfi, Anthony, "Factors Affecting Copper Solubility in Electroplating Waste", *Journal WPCF,* Vol. 51, No. 7, July, 1979.

Young, James C., "Removal of Grease and Oil by Biological Treatment Processes", *J1.WPCF,* Vol. 51, No. 8, Aug., 1979.

Zabban, Walter, and Helunick, Robert, "Cyanide Waste Treatment Technology—The Old, the New, and the Practical", *Plating and Surface Finishing,* Aug., 1980.

Index